A BIBLIOGRAPHY
OF QUANTITATIVE ECOLOGY

A BIBLIOGRAPHY OF QUANTITATIVE ECOLOGY

VINCENT SCHULTZ

Department of Zoology
Washington State University
Pullman, Washington

L. L. EBERHARDT, J. M. THOMAS, AND M. I. COCHRAN

Ecosystems Department
Battelle, Pacific Northwest Laboratory
Richland, Washington

Dedication

Douglas G. Chapman, Dean
College of Fisheries
University of Washington
Seattle, Washington

William G. Cochran, Professor
Department of Statistics
Harvard University
Cambridge, Massachusetts

Library of Congress Catalog Card Number: 76-14946
ISBN: 0-87933-254-9

78 77 76 1 2 3 4 5
Manufactured in the United States of America.

Library of Congress Cataloging in Publication Data

Main entry under title:

A bibliography of quantitative eology.

1. Ecology—Indexes. I. Schultz, Vincent.
Z5322.E2B5 [QH541] 574.5'2'018 76-14946
ISBN: 0-87933-254-9

Exclusive distributor: **Halsted Press**
A Division of John Wiley & Sons, Inc.
ISBN: 0-470-15175-7

PREFACE

Scientific and popular interest in ecology have increased tremendously in the last two decades. With that growth has come a slower, but nonetheless very substantial, awareness of the necessity for the use of appropriate quantitative methodology in ecological research as well as in renewable resource management. More recently, there has been a sharp increase in references to mathematics and statistics in the ecological literature, making it difficult for both students and practicing ecologists to keep up with even the readily accessible material.

A few years ago, we began an effort to expand and combine our personal files in the somewhat naive expectation that we might more readily "catch up" with the literature. Our colleague, J. M. Thomas, suggested and instituted a computer retrieval system, and M. I. Cochran managed to keep up with our various entries and changes. When the total exceeded 2000 entries, we began to suppose that the end-product might be useful to others, so a more organized effort was launched. Reading tables of contents and scanning pages, we searched major periodicals from first issues through 1974. In the process of final editing, more recent references were incorporated, but we have not made a systematic effort to cover the 1975 literature. Mr. J. R. Tadlock (Systems Department, Battelle) made the many innovative changes and modifications to the information retrieval system which allowed us to proceed with the expansion and subsequent printing of this bibliography.

We have concentrated on the uses of mathematics and statistics in ecology, but could not, of course, neglect the burgeoning use of computers in the field. We feel renewable resource management is clearly a part of "ecology," but paid less attention to the literature subsumed under headings like "environmental awareness," mostly because of a regrettable lack of quantitative work in such publications. More definitive deficiencies in our coverage are listed in the **Introduction** and they illustrate the wide range of quantitative material that may be relevant in ecology. In general, references employing standard mathematical and statistical procedures are included only if there is also a detailed discussion of such procedures.

Much of the work of preparing this bibliography was supported by the U. S. Atomic Energy Commission (now Energy Research and Development Administration).

Vincent Schultz
L. L. Eberhardt

CONTENTS

A BIBLIOGRAPHY OF QUANTITATIVE ECOLOGY

INTRODUCTION

This bibliography is arranged in subject-matter categories. In our experience, indexes by subject-matter are not very useful because of the labor involved in looking up more than twenty or thirty entries. Permuted titles are useful as guides to the current literature but require too much space for a large bibliography since they require multiple entries. In our opinion, there are two effective ways to use a bibliography. One is simply to scan titles. This is facilitated by a classification into categories that can each be examined in a moderate length of time. The other is by use of an author index that permits the user to locate titles by authors known to be working in a particular area.

The disadvantage of an arrangement by categories is that such a classification depends a great deal on the particular interest of the compilers and the difficulties of assigning papers to a single category. A discussion of the categories used in this bibliography follows. It should be helpful in trying to narrow down the field of search. Certain obvious limitations exist in the list. One is in the area of genetics and evolutionary theory, where we have not attempted to cover the large literature. There are similar limitations in our coverage of epidemiology, animal behavior, hydrology, and oceanography. References in these fields mostly appear in the bibliography in connection with other topics.

Various key words were used in preparing the bibliography and have been listed with the individual entries. Hopefully these words and phrases will be useful in deciding whether or not to look up a particular item, but we have not attempted a thorough listing of key words. Those we did use were intended to help in classification by supplementing the titles and our knowledge of the contents of a particular paper.

The bibliography was produced by use of a computer program. Since the particular program used was not suitable for a bibliography of this size and various modifications had to be introduced, there is not much point in describing the program or procedures used to modify it. Certain drawbacks in typography will be obvious to the reader, such as the lack of a capability for distinguishing species nomenclature, capitalization, and others. Some difficulties with output devices (line printers) resulted in the appearance of peculiar symbols in early drafts. We believe errors of this kind have been reduced to the point of trivial importance, but unfortunately once an error of any kind is introduced into a computer system, it tends to beget others. We have made repeated attempts to eliminate such gremlins, but suspect we have not caught them all.

The categories in the bibliography are arranged alphabetically according to certain key words or phrases. We considered an arrangement of the individual categories under broader classifications, but found that there are too many interconnections between categories to make such an arrangement worthwhile. Once the user becomes familiar with the main categories and has done a little browsing through the bibliography, the table of contents can be used to find sections relevant to a particular purpose. First-time users may be aided by a brief discussion of the contents of each category given below.

The final section includes books and book reviews, some other bibliographies, and a few references concerned with the educational aspects of quantitative ecology.

CONTENTS OF CATEGORIES

Age

This brief category includes those entries having specific reference to age, age structure, and age ratios. Many of these items are also relevant to the section on population dynamics.

Compartment models

This category includes references concerning the kinetics of various trace materials in biological systems and some related topics.

Competition, density-dependence, and regulation

This broad grouping also includes predator-prey and host-parasite models, as well as various other kinds of interactions. It seemed to us that the interconnections between the several topics included here are such that grouping is preferable to splitting into the several logical subdivisions. An important class of references not included is that of the stock-recruitment models of the fisheries biologist. Since these refer specifically to exploited populations, we have included them in the section on exploitation, management, and control.

Computers and data-processing

The references in this section are concerned primarily with direct applications of computers and include most of the material on computer programs. References dealing with simulation appear under the next heading since this is a major and convenient subdivision of computer uses in ecology.

Computer simulation

Simulation models are listed separately here. Some references dealing specifically with computer programs are included, but the section on computers should also be consulted for programs and further models. An occasional reference listed here does not deal specifically with simulation on a computer, but nonetheless seemed best included.

Cycles

This is a small section containing specific references to cyclic phenomena.

Diversity

The references in this category mainly pertain to the topics of species diversity and the relative abundance of species. A useful companion section is that on frequency distributions.

Energetics and productivity

Topics having to do with energetics and energy flow are difficult to separate effectively from references to production and productivity. Metabolism and feeding-rate references appear here, but may also appear under several other categories (such as growth), depending on our judgment as to the main content of the paper. Much related work on productivity in forestry and agriculture appears in the next section.

Exploitation, management, and control

This is a rather large grouping that we felt would be examined more usefully as a whole rather than in subdivision. Our reasoning is that the three topics have considerable structural similarity, so that the fisheries specialist or forester might well benefit by at least encountering titles in a field not normally represented in his usual journals. As noted above, the section on energetics and productivity includes some references relevant here, as does the one on competition, density-dependence, and regulation.

Frequency distributions

This section primarily contains references dealing specifically with frequency or probability distributions (binomial, multinomial, Poisson, hypergeometric, logarithmic, geometric, and so forth). There are, of course, many references to specific distributions in the literature, and we have attempted to include such major uses in the key words following citations throughout the bibliography.

Growth of individuals

This category contains references dealing with growth rates, growth curves, and so on that pertain to individual organisms. The following section contains references predominantly concerned with similar analyses directed at **populations** of organisms or cells. The usual growth models do not, however, distinguish between individuals and populations (except, strictly speaking, the distinction between continuous and discrete phenomena), so the two sections should be used together for many purposes. Also, we suspect our discrimination between the two categories will occasionally be in error, since the subdivision was done after an initial classification when it was no longer practicable to go back and check each reference.

Growth of populations

As noted above, this and the preceding section often may need to be used together. However, some investigators will undoubtedly want to refer mainly to one section or the other depending on their primary interest. References given here also have close connections with those in the sections on population models and population dynamics, since most of the growth equations can be looked at alternatively as population models.

Models (mathematical)

This section is somewhat larger than most of the others. Its size reflects the difficulty of usefully and simply classifying mathematical models, but is also a consequence of our not having attempted full coverage of topics such as genetics, epidemiology, evolution, behavior, pollution, and so on. Also, various kinds of mathematical and statistical techniques that seemed useful but not particularly adapted to categorization tended to land here by virtue of the wide scope of the phrase "mathematical model." Two other modeling sections follow immediately after this one, and a third is, of course, computer simulation, which we thought best bracketed with computers and data-processing earlier in the bibliography.

Models (population)

This section mainly concerns references directly concerned with population models. There is, however, an unavoidable overlap with other categories, including competition, density-dependence, and regulation; population dynamics, and population growth.

Models (stochastic)

We have attempted to group those references dealing primarily with stochastic or probabilistic models in this section. Some overlap is again unavoidable; for example, many papers contain both deterministic and stochastic models or elements thereof.

Operations research and programming

This section includes those references explicitly dealing with operations research or one of the forms of programming (linear, dynamic, adaptive, and so forth). Some references to systems analysis also appear here, while others were judged to be included more usefully under other headings.

Ordination

The rather special uses of ordination, factor analysis, and principal component analysis led us to segregate these topics in one section. Most users should examine the section on patterns in connection with the present section.

Patterns

This section provides a more general array of papers dealing with patterns in space, time, and among species. It cannot be separated effectively from the preceding one (ordination). References on home range and territory are included here since they really deal with patterns.

Plotless sampling

References in this category pertain to the several methods used for estimating density of individuals that do not involve plots or quadrats—the "distance" or "nearest-neighbor" methods, the Bitterlich (angle-gauge) method, point methods, and line transects. Many of these references should be of special interest in relation to the previous section on patterns.

Plots and quadrats

This small section is included to segregate references dealing specifically with the use of plots or quadrats. Additional important references appear in the category on frequency distributions.

Population dynamics

This is one of a number of sections that deal specifically with populations. These papers are concerned with the fluctuations of numbers in a population as a consequence of reproductive and survival rates. Further relevant references appear in the sections on population growth, population models under the headings of reproduction and survival, and population estimation.

Population estimation (capture-recapture methods)

This category contains the references that deal with the uses of tagging and marking in population estimation. It includes the specific methods, such as those due to Petersen (Lincoln Index), Schnabel, Seber-Jolly, Fisher-Ford, Manly-Parr, and others, as well as references on tagging, marking and tag loss, tests of assumptions and so on. The following category is a companion one, mainly covering methods not based on recovery (or resighting) of marked individuals. The section on computers should be consulted for computer programs for these methods. Tagging and marking methods are also used widely in efforts to measure survival and thus may appear in papers in that section.

Population estimation (other methods)

References in this category are mostly concerned with population estimation without the use of tags or marks, but some of the papers also include that aspect. For example, the various removal methods may be used with or without tags.

Reproduction

This category contains references dealing fairly specifically with reproduction. It should be considered in conjunction with the sections on population dynamics and population models.

Sampling

This section contains references concerned specifically with sampling techniques. The sections on plotless sampling and plots and quadrats should also be consulted for related references, as should the frequency distribution category.

Survival and mortality

This section deals specifically with references on survival and mortality. These topics will, of course, also appear in several other sections above.

Taxonomy

We have attempted to group those references pertaining mainly to taxonomy and systematics in this section. Reference should also be made to the sections on ordination and patterns.

Books, bibliographies, and educational references

Books are not readily classed in the categories given above. Occasionally a chapter or section of a book is referenced, but by and large, those books and book reviews that are appropriate for inclusion appear here. There are also a few bibliographies that seem to be useful—some old, some appearing in the back of a book, and so forth—so these are appended here. Many readers may be interested in papers having to do with education and training; these also appear in this final section.

AGE

ALLEN, K. R. 1966. DETERMINATION OF AGE DISTRIBUTION FROM AGE - LENGTH KEYS AND LENGTH DISTRIBUTIONS, IBM 7090, 7094 FORTRAN IV. TRANS. AMER. FISH. SOC. 95(2), 230-231.

ALLEN, R. L. AND P. BASASIBWAKI. 1974. PROPERTIES OF AGE STRUCTURE MODELS FOR FISH POPULATIONS. J. FISH. RES. BOARD CAN. 31(6), 1119-1125. [SIMULATION; STOCK-RECRUITMENT; STABILITY; RICKER'S MODEL]

ANDERSEN, F. S. 1960. COMPETITION IN POPULATIONS CONSISTING OF ONE AGE GROUP. BIOMETRICS 16(1), 19-27.

BARTLETT, M. S. 1970. AGE DISTRIBUTIONS. BIOMETRICS 26(3), 377-385.

CAUGHLEY, G. 1974. INTERPRETATION OF AGE RATIOS. J. WILDL. MANAGE. 38(3), 557-562. [COMPUTER; SIMULATION; MODEL]

CAUGHLEY, G. AND J. CAUGHLEY. 1974. ESTIMATING MEDIAN DATE OF BIRTH. J. WILDL. MANAGE. 38(3), 552-556. [PROBIT ANALYSIS]

COUTINHO, A. B. AND F. A. COUTINHO. 1968. THE CONSERVATION OF THE BIOMASS DENSITY AND THE AGE STRUCTURE OF THE POPULATIONS. BULL. MATH. BIOPHYS. 30, 553-563.

DAPSON, R. W. 1971. QUANTITATIVE COMPARISON OF POPULATIONS WITH DIFFERENT AGE STRUCTURES. ANN. ZOOL. FENN. 8(1), 75-79. [PROBABILITY; CONFIDENCE INTERVAL]

DOI, T. 1949. AN INVESTIGATION ON THE AGE COMPOSITION OF FISH STOCK BY THE ANALYSIS OF REGRESSION. BULL. JAPANESE SOC. SCI. FISH. 15(7), 306-310. [SIMPLE LINEAR REGRESSION]

EBERHARDT, L. AND R. I. BLOUCH. 1955. ANALYSIS OF PHEASANT AGE RATIOS. TRANS. N. AMER. WILDL. CONF. 20, 357-367.

EMLEN, J. M. 1970. AGE SPECIFICITY AND ECOLOGICAL THEORY. ECOLOGY 51(4), 588-601.

FREDRICKSON, A. G. 1971. A MATHEMATICAL THEORY OF AGE STRUCTURE IN SEXUAL POPULATIONS: RANDOM MATING AND MONOGAMOUS MARRIAGE MODELS. MATH. BIOSCI. 10, 117-143.

GANI, J. 1973. ON THE AGE STRUCTURE OF SOME STOCHASTIC PROCESSES. BULL. INTERN. STAT. INST. 45(1), 434-436. [MULTIPLE BRANCHING PROCESS; LESLIE MATRIX]

GOLUBITSKY, M., E. B. KEELER AND M. ROTHSCHILD. 1975. CONVERGENCE OF THE AGE STRUCTURE: APPLICATIONS OF THE PROJECTIVE METRIC. THEOR. POPUL. BIOL. 7(1), 84-93.

HAYASHI, Y. 1970. STUDIES ON THE MATURITY CONDITION OF THE COMMON SQUID-I. A METHOD OF EXPRESSING MATURITY CONDITION BY NUMERICAL VALUES. BULL. JAPANESE SOC. SCI. FISH. 36(10), 995-999.

HETT, J. M. 1971. A DYNAMIC ANALYSIS OF AGE IN SUGAR MAPLE SEEDLINGS. ECOLOGY 52(6), 1071-1074. [POWER FUNCTION MODEL]

KELKER, G. H. 1950. SEX AND AGE CLASS RATIOS AMONG VERTEBRATE POPULATIONS. PROC. UTAH ACAD. SCI., ARTS LETT. 27, 12-21.

KETCHEN, K. S. 1950. STRATIFIED SUBSAMPLING FOR DETERMINING AGE DISTRIBUTIONS. TRANS. AMER. FISH. SOC. 79, 205-212. [CHI-SQUARE; POLYGON; PROBABILITY]

KUTKUHN, J. H. 1963. ESTIMATING ABSOLUTE AGE COMPOSITION OF CALIFORNIA SALMON LANDINGS. CALIF. DEPT. FISH GAME FISH BULL. NO. 120, 42 PP. [TWO-PHASE STRATIFIED SAMPLING; VARIANCE; SAMPLE SIZE]

LOTKA, A. J. 1922. THE STABILITY OF THE NORMAL AGE DISTRIBUTION. PROC. NATL. ACAD. SCI. U. S. A. 8(11), 339-345.

LOTKA, A. J. 1937. POPULATION ANALYSIS: A THEOREM REGARDING THE STABLE AGE DISTRIBUTION. J. WASH. ACAD. SCI. 27(7), 299-303.

MACARTHUR, R. H. 1958. A NOTE ON STATIONARY AGE DISTRIBUTIONS IN SINGLE-SPECIES POPULATIONS AND STATIONARY SPECIES POPULATIONS IN A COMMUNITY. ECOLOGY 39(1), 146-147.

MANLY, B. F. J. 1974. A COMPARISON OF METHODS FOR THE ANALYSIS OF INSECT STAGE-FREQUENCY DATA. OECOLOGIA 17(4), 335-348. [COMPUTER; SIMULATION]

MOSER, J. W., JR. 1972. DYNAMICS OF AN UNEVEN-AGED FOREST STAND. FOREST SCI. 18(3), 184-191. [MODEL]

MOSER, J. W., JR. AND O. F. HALL. 1969. DERIVING GROWTH AND YIELD FUNCTIONS FOR UNEVEN-AGED FOREST STANDS. FOREST SCI. 15(2), 183-188. [MODEL]

NAKAI, Z. AND S. HAYASHI. 1962. ON THE AGE COMPOSITION OF THE JAPANESE SARDINE CATCH, 1949 THROUGH 1951 - WITH AN ATTEMPT OF POPULATION ANALYSIS-. BULL. TOKAI REG. FISH. RES. LAB. 9, 61-84. [CONDITION COEFFICIENT EQUATIONS; SURVIVAL RATE; MORTALITY RATE]

NAMKOONG, G. AND J. H. ROBERDS. 1974. EXTINCTION PROBABILITIES AND THE CHANGING AGE STRUCTURE OF REDWOOD FORESTS. AMER. NAT. 108(961), 355-368. [MODEL; LESLIE MATRIX; STOCHASTIC PROCESS]

NOONEY, G. C. 1968. AGE DISTRIBUTIONS IN STOCHASTICALLY DIVIDING POPULATIONS. J. THEOR. BIOL. 20(3), 314-320. [PROBABILITY GENERATING FUNCTION]

PEARL, R. AND T. J. LEBLANC. 1922. A FURTHER NOTE ON THE AGE INDEX OF A POPULATION. PROC. NATL. ACAD. SCI. U. S. A. 8(10), 300-303.

PETRIDES, G. A. 1949. VIEWPOINTS ON THE ANALYSIS OF OPEN SEASON SEX AND AGE RATIOS. TRANS. N. AMER. WILDL. NAT. RESOUR. CONF. 14, 391-410.

RAMAKRISHNAN, A. AND S. K. SRINIVASAN. 1958. ON AGE DISTRIBUTION IN POPULATION GROWTH. BULL. MATH. BIOPHYS. 20, 289-303.

RUESINK, W. G. 1975. ESTIMATING TIME-VARYING SURVIVAL OF ARTHROPOD LIFE STAGES FROM POPULATION DENSITY. ECOLOGY 56(1), 244-247. [COMPUTER; MODEL]

SCHAFFER, W. M. 1974. SELECTION FOR OPTIMAL LIFE HISTORIES: THE EFFECTS OF AGE STRUCTURE. ECOLOGY 55(2), 291-303. [MODEL; STABILITY ANALYSIS; MATRIX]

SELLECK, D. M. AND C. M. HART. 1957. CALCULATING THE PERCENTAGE OF KILL FROM SEX AND AGE RATIOS. CALIF. FISH GAME 43(4), 309-316.

SHENTYAKOVA, L. F. 1969. THE VERIFICATION BY MATHEMATICAL TESTS OF THE HYPOTHESIS THAT THE RELATIONSHIP BETWEEN BODY GROWTH AND THE SCALES OF FISHES IS CONSTANT WITHIN THE SPECIES. PROBLEMS OF ICHTHYOLOGY 9(3), 338-354. [REGRESSION; CHI-SQUARE; VARIANCE]

SINKO, J. W. AND W. STREIFER. 1967. A NEW MODEL FOR AGE-SIZE STRUCTURE OF A POPULATION. ECOLOGY 48(6), 910-918.

SINKO, J. W. AND W. STREIFER. 1969. APPLYING MODELS INCORPORATING AGE-SIZE STRUCTURE OF A POPULATION TO DAPHNIA. ECOLOGY 50(4), 608-615.

WAUGH, W. A. O'N. 1955. AN AGE-DEPENDENT BIRTH AND DEATH PROCESS. BIOMETRIKA 42(3/4), 291-306.

WAUGH, W. A. O'N. 1968. AGE-DEPENDENT BRANCHING PROCESSES UNDER A CONDITION OF ULTIMATE EXTINCTION. BIOMETRIKA 55(2), 291-296. [MODEL; MARKOV PROCESS]

WEISS, G. H. 1968. EQUATIONS FOR THE AGE STRUCTURE OF GROWING POPULATIONS. BULL. MATH. BIOPHYS. 30, 427-435. [CELL POPULATIONS; MARKOVIAN THEORY; KENDALL'S MODEL]

ZAIKA, V. E. 1968.
AGE-STRUCTURE DEPENDENCE OF THE
"SPECIFIC PRODUCTION" IN
ZOOPLANKTON POPULATIONS. MARINE
BIOL. 1(4), 311-315. [PRODUCTION;
VON BERTALANFFY GROWTH THEORY]

COMPARTMENT MODELS

AOYAMA, I., S. YOSHIKAWA, Y. INOUE AND S. IWAO. 1970. DYNAMIC CONCENTRATION PROCESS OF RADIOACTIVE SUBSTANCES IN AQUATIC ORGANISMS. TENTATIVE APPROACH FROM STOCHASTIC THEORY. HOKEN BUTSURI 5, 135-140.

ARMSTRONG, N. E. AND E. F. GLOYNA. 1969. MATHEMATICAL MODELS FOR THE DISPERSION OF RADIONUCLIDES IN AQUATIC SYSTEMS. PP. 329-335. IN, SYMPOSIUM ON RADIOECOLOGY. (D. J. NELSON AND F. C. EVANS, EDS.). U. S. AEC REPT. CONF-670503.

AYZATULLIN, T. A. AND K. M. KHAYLOV. 1970. KINETICS OF THE ENZYMATIC HYDROLYSIS IN THE PRESENCE OF BACTERIA OF MACROMOLECULES DISSOLVED IN SEA WATER. HYDROBIOL. J. 6(6), 37-43. (ENGLISH TRANSL.)

BERNARD, S. R., L. R. SHENTON AND V. R. R. UPPULURI. 1967. STOCHASTIC MODELS FOR THE DISTRIBUTION OF RADIOACTIVE MATERIALS IN A CONNECTED SYSTEM OF COMPARTMENTS. PP. 481-510. IN, BIOLOGY AND PROBLEMS OF HEALTH. PROC. FIFTH BERKELEY SYMP. MATH. STAT. PROBAB. VOL. 4. UNIVERSITY OF CALIFORNIA PRESS, BERKELEY, CALIFORNIA.

BLEDSOE, L. J. 1968. COMPARTMENT MODELS AND THEIR USE IN THE SIMULATION OF SECONDARY SUCCESSION. M. S. THESIS, COLORADO STATE UNIVERSITY, FT. COLLINS, COLORADO. 112 PP.

BLEDSOE, L. J. AND G. M. VAN DYNE. 1971. A COMPARTMENT MODEL SIMULATION OF SECONDARY SUCCESSION. PP. 479-511. IN, SYSTEMS ANALYSIS AND SIMULATION IN ECOLOGY, VOLUME I. (B. C. PATTEN, ED.), ACADEMIC PRESS, NEW YORK, NEW YORK.

BLOOM, S. G. AND G. E. RAINES. 1971. MATHEMATICAL MODELS PREDICTING THE TRANSPORT OF RADIONUCLIDES IN A MARINE ENVIRONMENT. BIOSCIENCE 21(12), 691-696. [COMPUTER]

BLOOM, S. G., A. A. LEVIN, W. E. MARTIN AND G. E. RAINES. 1970. MATHEMATICAL METHODS FOR EVALUATING THE TRANSPORT AND ACCUMULATION OF RADIONUCLIDES. AEC REPT. IOCS MEMORANDUM BMI-35. BATTELLE MEMORIAL INSTITUTE, COLUMBUS, OHIO. 40 PP.

CASWELL, H., H. E. KOENIG, J. A. RESH AND Q. E. ROSS. 1972. AN INTRODUCTION TO SYSTEMS SCIENCE FOR ECOLOGISTS. PP. 3-78. IN, SYSTEMS ANALYSIS AND SIMULATION IN ECOLOGY, VOLUME II. (B. C. PATTEN, ED.), ACADEMIC PRESS, NEW YORK, NEW YORK. [MATRIX; MODEL]

CONOVER, R. J. AND V. FRANCIS. 1973. THE USE OF RADIOACTIVE ISOTOPES TO MEASURE THE TRANSFER OF MATERIALS IN AQUATIC FOOD CHAINS. MARINE BIOL. 18(4), 272-283. [MODEL; COMPUTER; SIMULATION; MATRIX]

CROSSLEY, D. A., JR. AND C. S. GIST. 1973. USE OF RADIOISOTOPES IN MODELING SOIL MICROCOMMUNITIES. PP. 258-278. IN, PROC. FIRST SOIL MICROCOMMUNITIES CONFERENCE, SYRACUSE, NEW YORK, OCTOBER 18-20, 1971. (D. L. DINDAL, ED.). U. S. AEC REPT. CONF-711076. [MODEL]

DAYKIN, P. N. 1965. APPLICATION OF MASS TRANSFER THEORY TO THE PROBLEM OF RESPIRATION OF FISH EGGS. J. FISH. RES. BOARD CAN. 22(1), 159-171. [EQUATIONS]

DENMEAD, O. T. 1970. TRANSFER PROCESSES BETWEEN VEGETATION AND AIR: MEASUREMENTS, INTERPRETATION AND MODELLING. PP. 149-164. IN, PREDICTION AND MEASUREMENT OF PHOTOSYNTHETIC PRODUCTIVITY. (I. SETLIK, ED.). CENTRE FOR AGRICULTURAL PUBLISHING AND DOCUMENTATION, WAGENINGEN, NETHERLANDS. [MODEL]

EBERHARDT, L. L. 1969. SIMILARITY, ALLOMETRY AND FOOD CHAINS. J. THEOR. BIOL. 24(1), 43-55.

EBERHARDT, L. L. 1970. SOME BIOMETRIC ASPECTS OF POLLUTION PROBLEMS. PP. 178-189. IN, PROCEEDINGS OF THE THE SYMPOSIUM ON THE DEVELOPMENT AND IMPLEMENTATION OF COURSES AND CURRICULA IN NATURAL-RESOURCES BIOMETRY. (W. E. FRAYER, ED.). COLORADO STATE UNIVERSITY, FT. COLLINS, COLORADO. [SYSTEMS ANALYSIS]

EBERHARDT, L. L. 1973. MODELING RADIONUCLIDES AND PESTICIDES IN FOOD CHAINS. PP. 894-897. IN, RADIONUCLIDES IN ECOSYSTEMS. (D. J. NELSON, ED.). U. S. AEC REPT. CONF-710501-P2. [MODEL; COMPUTER; SIMULATION; SAMPLING]

EBERHARDT, L. L. AND R. E. NAKATANI. 1969. MODELING THE BEHAVIOR OF RADIONUCLIDES IN SOME NATURAL SYSTEMS. PP. 740-750. IN, SYMPOSIUM ON RADIOECOLOGY. (D. J. NELSON AND F. C. EVANS, EDS.), U. S. AEC REPT. CONF-670503. [GEOMETRIC MEAN]

EBERHARDT, L. L. AND R. O. GILBERT 1973. GAMMA AND LOGNORMAL DISTRIBUTIONS AS MODELS IN STUDYING FOOD-CHAIN KINETICS. AEC REPT. BNWL-1747. BATTELLE-NORTHWEST LAB., RICHLAND, WASHINGTON, III, 95 PP.

EMERSON, S., W. BROECKER AND D. W. SCHINDLER. 1973. GAS-EXCHANGE RATES IN A SMALL LAKE AS DETERMINED BY THE RADON METHOD. J. FISH. RES. BOARD CAN. 30(10), 1475-1484. [MODEL]

ENDELMAN, F. J., M. L. NORTHUP, D. R. KENNEY, J. R. BOYLE AND R. R. HUGHES. 1972. A SYSTEMS APPROACH TO AN ANALYSIS OF THE TERRESTRIAL NITROGEN CYCLE. J. ENVIRON. SYSTEMS 2(1), 3-19. [MODEL]

ERIKSSON, E. 1971. COMPARTMENT MODELS AND RESERVOIR THEORY. PP. 67-84. IN, ANNUAL REVIEW OF ECOLOGY AND SYSTEMATICS, VOLUME 3. (R. F. JOHNSTON, P. W. FRANK AND C. D. MICHENER, EDS.). ANNUAL REVIEWS, INC., PALO ALTO, CALIFORNIA.

FUNDERLIE, R. E. AND M. T. HEATH. 1971. LINEAR COMPARTMENTAL ANALYSIS OF ECOSYSTEMS. AEC REPT. ORNL-IBP-71-4. OAK RIDGE NATL. LAB., OAK RIDGE, TENNESSEE. V, 50 PP.

GALLOPIN, G. C. 1972. STRUCTURAL PROPERTIES OF FOOD WEBS. PP. 241-282. IN, SYSTEMS ANALYSIS AND SIMULATION IN ECOLOGY, VOLUME II. (B. C. PATTEN, ED.) ACADEMIC PRESS, NEW YORK, NEW YORK. [SET THEORY; GRAPHIC; MATRIX]

GIST, C. S. AND F. W. WHICKER. 1971. RADIOIODINE UPTAKE AND RETENTION BY THE MULE DEER THYROID. J. WILDL. MANAGE. 35(3), 461-468. [MODEL]

GIST, C. S., K. M. DUKE AND B. C. PATTEN. 1973. A SYSTEMS ANALYSIS OF A MODEL FOR RADIOIODINE MOVEMENT IN MULE DEER. PP. 909-914. IN, RADIONUCLIDES IN ECOSYSTEMS. (D. J. NELSON, ED.). U. S. AEC REPT. CONF-710501-P2. [COMPUTER; SIMULATION]

GOLDSTEIN, R. A. AND J. W. ELWOOD. 1971. TWO COMPARTMENT, THREE-PARAMETER MODEL FOR THE ABSORPTION AND RETENTION OF INGESTED ELEMENTS BY ANIMALS. ECOLOGY 52(5), 935-939.

GRENNEY, W. J. AND D. A. BELLA. 1972. FIELD STUDY AND MATHEMATICAL MODEL OF THE SLACK-WATER BUILDUP OF A POLLUTANT IN A TIDAL RIVER. LIMNOL. OCEANOGR. 17(2), 229-236. [COMPUTER; SIMULATION]

GUS'KOVA, V. N., O. N. PROKOF'YEV, A. A. ZASEDATELEV, B. N. IL'IN AND A. I. TIKHONOVA. 1971. DYNAMICS OF CONCENTRATION OF STRONTIUM-89 IN WATER AND CARP TISSUES AFTER A SINGLE CONTAMINATION OF A BODY OF WATER. HYDROBIOL. J. 7(1), 53-56. (ENGLISH TRANSL.). [EQUATIONS]

HALFON, E. 1974. SYSTEMS IDENTIFICATION: A THEORETICAL METHOD APPLIED TO TRACER KINETICS IN AQUATIC MICROCOSMS. PP. 262-265. IN, PROC. FIRST INTERN. CONGR. ECOL. CENTRE FOR AGRICULTURAL PUBLISHING AND DOCUMENTATION, WAGENINGEN, NETHERLANDS. [MODEL]

HANSON, W. C. AND L. L. EBERHARDT. 1969. EFFECTIVE HALF-TIMES OF RADIONUCLIDES IN ALASKAN LICHEN AND ESKIMOS. PP. 627-634. IN, SYMPOSIUM ON RADIOECOLOGY. (D. J. NELSON AND F. C. EVANS, EDS.). U. S. AEC REPT. CONF-670503. [MODEL; POISSON DISTRIBUTION]

HAUSSMANN, U. G. 1971. ABSTRACT FOOD WEBS IN ECOLOGY. MATH. BIOSCI. 11, 291-316. [SET; DENSITY FUNCTION; VOLTERRA'S EQUATIONS; MODEL; COMPETITION]

HAVSTEEN, B. 1974. NADP+/ISOCITRATE DEHYDROGENASE FROM IDUS IDUS (PISCES). III. DISCUSSION OF TEMPERATURE DEPENDENCE OF KINETIC PARAMETERS. MARINE BIOL. 25(1), 77-83. [EQUATIONS]

HAYES, F. R. AND M. A. MACAULAY. 1959. LAKE WATER AND SEDIMENT V. OXYGEN CONSUMED IN WATER OVER SEDIMENT CORES. LIMNOL. OCEANOGR. 4(3), 291-298. [EQUATIONS]

HAYES, F. R., J. A. MCCARDTER, M. L. CAMERON AND D. A. LIVINGSTONE. 1952. ON THE KINETICS OF PHOSPHORUS EXCHANGE IN LAKES. J. ECOL. 40(1), 202-216. [EQUATIONS]

HOFFMAN, G. R. 1969. STEADY STATE EXPRESSIONS FOR MATERIAL TRANSFERS IN TWO-COMPARTMENT MODEL ECOSYSTEMS. PROC. SOUTH DAKOTA ACAD. SCI. 48, 110-118. [SYSTEMS ANALYSIS]

ISAACS, J. D. 1973. POTENTIAL TROPHIC BIOMASSES AND TRACE-SUBSTANCE CONCENTRATIONS IN UNSTRUCTURED MARINE FOOD WEBS. MARINE BIOL. 22(2), 97-104. [MODEL]

JACOBI, W. 1971. TRANSFER OF FISSION PRODUCTS FROM ATMOSPHERIC FALLOUT INTO RIVER WATER. PP. 1153-1163. IN, PROC. INTERN. SYMP. RADIOECOLOGY APPLIED TO THE PROTECTION OF MAN AND HIS ENVIRONMENT. COMM. OF THE EUROPEAN COMMUNITIES, REPT. EUR-4800 D-F-I-E. [MODEL]

JORDAN, C. F., J. R. KLINE AND D. S. SASSCER. 1972. RELATIVE STABILITY OF MINERAL CYCLES IN FOREST ECOSYSTEMS. AMER. NAT. 106(948), 237-253. [COMPUTER; MODEL]

JOST, J. L., J. F. DRAKE, H. M. TSUCHIYA AND A. G. FREDRICKSON. 1973. MICROBIAL FOOD CHAINS AND FOOD WEBS. J. THEOR. BIOL. 41(3), 461-484. [MODEL; MONOD'S MODEL; PREDATOR-PREY MODEL; DIFFERENTIAL EQUATION]

KAYE, S. V. AND S. J. BALL. 1969. SYSTEMS ANALYSIS OF A COUPLED COMPARTMENT MODEL FOR RADIONUCLIDE TRANSFER IN A TROPICAL ENVIRONMENT. PP. 731-739. IN, SYMPOSIUM ON RADIOECOLOGY. (D. J. NELSON AND F. C. EVANS, EDS.). U. S. AEC REPT. CONF-670503.

KETCHUM, B. H. 1951. THE EXCHANGES OF FRESH AND SALT WATERS IN TIDAL ESTUARIES. J. MARINE RES. 10(1), 18-38.

KLINE, J. R., M. L. STEWART AND C. F. JORDAN. 1972. ESTIMATION OF BIOMASS AND TRANSPIRATION IN CONIFEROUS FORESTS USING TRITIATED WATER. PP. 159-166. IN, PROCEEDINGS-RESEARCH ON CONIFEROUS FOREST ECOSYSTEMS-A SYMPOSIUM. (J. F. FRANKLIN, L. J. DEMPSTER AND R. H. WARING, EDS.). U. S. FOR. SERV. PACIFIC NORTHWEST FOREST EXPT. STN., PORTLAND, OREGON.

KULIKOV, N. V., V. S. BEZEL AND L. N. OZHEGOV. 1970. ACCUMULATION OF RADIOISOTOPES BY DEVELOPING ROE OF TENCH (TINCA TINCA L.) AND PERCH (PERCA FLUVIATILIS L.). EKOLOGIYA 1(5), 73-77. (ENGLISH TRANSL. PP. 425-428). [MODEL]

LERMAN, A. 1973. TRANSPORT OF RADIONUCLIDES IN SEDIMENTS. PP. 936-944. IN, RADIONUCLIDES IN ECOSYSTEMS. (D. J. NELSON, ED.). U. S. AEC REPT. CONF-710501-P2. [MODEL]

LERMAN, A. AND B. F. JONES. 1973. TRANSIENT AND STEADY-STATE SALT TRANSPORT BETWEEN SEDIMENTS AND BRINE IN CLOSED LAKES. LIMNOL. OCEANOGR. 18(1), 72-85. [MODEL]

MARTIN, W. E. AND G. E. RAINES. 1969. ECOLOGICAL TRANSFER MECHANISMS-TERRESTRIAL. PP. 401-433. IN, PROCEEDINGS FOR THE SYMPOSIUM ON PUBLIC HEALTH ASPECTS OF PEACEFUL USES OF NUCLEAR EXPLOSIVES. U. S. PUBLIC HEALTH SERVICE REPORT SWRHL-82.

MATIS, J. H. AND M. W. CARTER. 1972. MULTI-COMPARTMENTAL ANALYSIS IN STEADY STATE AS A STOCHASTIC PROCESS. ACTA BIOTHEOR. 21(1/2), 2-23. [MODEL]

MAY, P. F., A. R. TILL AND M. J. CUMMING. 1972. SYSTEMS ANALYSIS OF 35-SULFUR KINETICS IN PASTURES GRAZED BY SHEEP. J. APPL. ECOL. 9(1), 25-49.

MAY, R. M. 1973. MASS AND ENERGY FLOW IN CLOSED ECOSYSTEMS: A COMMENT. J. THEOR. BIOL. 39(1), 155-163.

MCCOLL, J. G. 1973. A MODEL OF ION TRANSPORT DURING MOISTURE FLOW FROM A DOUGLAS-FIR FOREST FLOOR. ECOLOGY 54(1), 181-187.

MULHOLLAND, R. J. AND M. S. KEENER. 1974. ANALYSIS OF LINEAR COMPARTMENT MODELS FOR ECOSYSTEMS. J. THEOR. BIOL. 44(1), 105-116.

NEAL, E. C., B. C. PATTEN AND C. E. DE POE. 1967. PERIPHYTON GROWTH ON ARTIFICIAL SUBSTRATES IN A RADIOACTIVELY CONTAMINATED LAKE. ECOLOGY 48(6), 918-924. [MODEL]

O'NEILL, R. V. 1971. EXAMPLES OF ECOLOGICAL TRANSFER MATRICES. AEC REPT. ORNL-IBP-71-3. OAK RIDGE NATL. LAB., OAK RIDGE, TENNESSEE. 26 PP.

O'NEILL, R. V. 1971. A STOCHASTIC MODEL OF ENERGY FLOW IN PREDATOR COMPARTMENTS OF AN ECOSYSTEM. PP. 107-121. IN, STATISTICAL ECOLOGY, VOLUME 3, MANY SPECIES POPULATIONS, ECOSYSTEMS, AND SYSTEMS ANALYSIS. (G. P. PATIL, E. C. PIELOU AND W. E. WATERS, EDS.). PENNSYLVANIA STATE UNIVERSITY PRESS, UNIVERSITY PARK, PENNSYLVANIA.

O'NEILL, R. V. 1971. SYSTEMS APPROACHES TO THE STUDY OF FOREST FLOOR ARTHROPODS. PP. 441-477. IN, SYSTEMS ANALYSIS AND SIMULATION IN ECOLOGY, VOLUME I. (B. C. PATTEN, ED.). ACADEMIC PRESS, NEW YORK, NEW YORK.

O'NEILL, R. V. 1973. ERROR ANALYSIS OF ECOLOGICAL MODELS. PP. 898-908. IN, RADIONUCLIDES IN ECOSYSTEMS. (D. J. NELSON, ED.). U. S. AEC REPT. CONF-710501-P2. [SYSTEMS ANALYSIS; VARIANCE; PROBABILITY; VECTOR; MONTE CARLO; ALGORITHM; MATRIX; COMPUTER]

O'NEILL, R. V. AND C. E. STYRON. 1970. APPLICATIONS OF COMPARTMENT MODELING TECHNIQUES TO COLLEMBOLA POPULATION STUDIES. AMER. MIDL. NAT. 83(2), 489-495. [SIMULATION; MODEL]

O'NEILL, R. V. AND O. W. BURKE. 1971. A SIMPLE SYSTEMS MODEL FOR DDT AND DDE MOVEMENT IN THE HUMAN FOOD-CHAIN. AEC REPT. ORNL-IBP-71-9. OAK RIDGE NATL. LAB., OAK RIDGE, TENNESSEE. IV, 18 PP.

ODUM, H. T. 1960. ECOLOGICAL POTENTIAL AND ANALOGUE CIRCUITS FOR THE ECOSYSTEM. AMER. SCI. 48(1), 1-8.

OKUBO, A. 1964. EQUATIONS DESCRIBING THE DIFFUSION OF AN INTRODUCED POLLUTANT IN A ONE-DEMENSIONAL ESTUARY. PP. 216-226. IN, STUDIES ON OCEANOGRAPHY. (K. YOSHIDA, ED.). UNIVERSITY OF WASHINGTON PRESS, SEATTLE, WASHINGTON.

OLSON, J. S. 1965. EQUATIONS FOR CESIUM TRANSFER IN A LIRIODENDRON FOREST. HEALTH PHYSICS 11(12), 1385-1392. [FORTRAN]

OLSON, J. S. AND V. R. R. UPPULURI. 1967. ECOSYSTEM MAINTENANCE AND TRANSFORMATION MODELS AS MARKOV PROCESSES WITH ABSORBING BARRIERS. PP. PS3.1-PS3.5. IN, SOME MATHEMATICAL MODELS IN BIOLOGY. (REVISED EDITION). (R. M. THRALL, J. A. MORTIMER, K. R. REBMAN AND R. F. BAUM, EDS.). UNIVERSITY OF MICHIGAN, ANN ARBOR, MICHIGAN.

PARKER, R. A. 1974. EMPIRICAL FUNCTIONS RELATING METABOLIC PROCESSES IN AQUATIC SYSTEMS TO ENVIRONMENTAL VARIABLES. J. FISH. RES. BOARD CAN. 31(9), 1550-1552. [MODEL]

PATTEN, B. C. AND M. WITKAMP. 1967. SYSTEMS ANALYSIS OF 134-CESIUM KINETICS IN TERRESTRIAL MICROCOSMS. ECOLOGY 48(5), 813-824.

PISKUNOV, L. I. 1970. STATISTICAL STUDY OF THE ACCUMULATION OF RADIOISOTOPES IN AQUATIC ORGANISMS. HYDROBIOL. J. 6(5), 58-60. (ENGLISH TRANSL.)

PISKUNOV, L. I. 1971. QUANTITATIVE ASPECTS OF THE ACCUMULATION OF RADIOISOTOPES BY FRESHWATER ORGANISMS. HYDROBIOL. J. 7(1), 49-52. (ENGLISH TRANSL.)

PROKHOROV, V. M. AND N. G. SAFRONOVA. 1973. KINETICS OF THE SELF-PURIFICATION OF A BODY OF WATER CONTAINING STRONTIUM-90 AS A RESULT OF ABSORPTION OF THE RADIONUCLIDE BY THE BOTTOM DEPOSITS. EKOLOGIYA 4(2), 12-18. (ENGLISH TRANSL. PP. 101-105). [MODEL]

RAINES, G. E., S. G. BLOOM AND A. A. LEVIN. 1969. ECOLOGICAL MODELS APPLIED TO RADIONUCLIDE TRANSFER IN TROPICAL ECOSYSTEMS. BIOSCIENCE 19(12), 1086-1091.

REICHLE, D. E. 1967. RADIOISOTOPE TURNOVER AND ENERGY FLOW IN TERRESTRIAL ISOPOD POPULATIONS. ECOLOGY 48(3), 351-366. [MODEL]

REICHLE, D. E. 1969. MEASUREMENT OF ELEMENTAL ASSIMILATION BY ANIMALS FROM RADIOISOTOPE RETENTION PATTERNS. ECOLOGY 50(6), 1102-1104. [EQUATIONS]

RODIONOVA, L. F. AND S. YA. SUKAL'SKAYA. 1969. THE ACCUMULATION OF BA-140 AND LA-140 BY PLANKTONIC ORGANISMS FROM FRESH WATER. HYDROBIOL. J. 5(6), 70-74. (ENGLISH TRANSL.). [EQUATIONS; COEFFICIENT OF ACCUMULATION]

RUZIC, I. 1972. TWO-COMPARTMENT MODEL OF RADIONUCLIDE ACCUMULATION INTO MARINE ORGANISMS. I. ACCUMULATION FROM A MEDIUM OF CONSTANT ACTIVITY. MARINE BIOL. 15(2), 105-112.

SARMA, T. P., T. M. KRISHNAMOORTHY AND V. N. SASTRY. 1971. AN APPROACH TO THE CALCULATION OF THE ALLOWABLE SPECIFIC ACTIVITIES IN MARINE FISHES. HEALTH PHYSICS 20(1), 23-30. [VON BERTALANFFY'S EQUATION]

SASSCER, D. S., C. F. JORDAN AND J. R. KLINE. 1973. MATHEMATICAL MODEL OF TRITIATED AND STABLE WATER MOVEMENT IN AN OLD-FIELD ECOSYSTEM. PP. 915-923. IN, RADIONUCLIDES IN ECOSYSTEMS. (D. J. NELSON, ED.). U. S. AEC REPT. CONF.-710501-P2. [SIMULATION; COMPUTER; MATRIX]

SINGH, J. S. 1973. A COMPARTMENT MODEL OF HERBAGE DYNAMICS FOR INDIAN TROPICAL GRASSLANDS. OIKOS 24(3), 367-372. [LINEAR MODEL; COMPUTER]

SMITH, D. F. 1974. QUANTITATIVE ANALYSIS OF THE FUNCTIONAL RELATIONSHIPS EXISTING BETWEEN ECOSYSTEM COMPONENTS I. ANALYSIS OF THE LINEAR INTERCOMPONENT MASS TRANSFERS. OECOLOGIA 16(2), 97-106. [SIMULATION; TRACER MODEL; MATRIX]

SMITH, D. F. 1974. QUANTITATIVE ANALYSIS OF THE FUNCTIONAL RELATIONSHIPS EXISTING BETWEEN ECOSYSTEM COMPONENTS II. ANALYSIS OF NON-LINEAR RELATIONSHIPS. OECOLOGIA 16(2), 107-117. [SIMULATION; TRACER MODEL]

TURNER, F. B. 1965. UPTAKE OF FALLOUT RADIONUCLIDES BY MAMMALS AND A STOCHASTIC SIMULATION OF THE PROCESS. PP. 800-820. IN, RADIOACTIVE FALLOUT FROM NUCLEAR WEAPONS TESTS. (A. W. KLEMENT, JR., ED.). U. S. AEC SYMPOSIUM SERIES NO. 5. [COMPUTER]

VROCHINSKIY, K. K. 1972. SYMPOSIUM ON THE USE OF MATHEMATICAL METHODS TO ASSESS AND FORECAST THE ACTUAL DANGER OF PESTICIDE ACCUMULATION IN THE ENVIRONMENT AND BODY. (DECEMBER 14-15, 1971, KIEV). HYDROBIOL. J. 8(4), 117-119. (ENGLISH TRANSL.)

WAIDE, J. B., J. E. KREBS, S. P. CLARKSON AND E. M. SETZLER. 1974. A LINEAR SYSTEMS ANALYSIS OF THE CALCIUM CYCLE IN A FORESTED WATERSHED ECOSYSTEM. PP. 261-345. IN, PROGRESS IN THEORETICAL BIOLOGY, VOL. 3. (R. ROSEN AND F. M. SNELL, EDS.). ACADEMIC PRESS, NEW YORK, NEW YORK. [MODEL; SIMULATION; SYSTEMS ANALYSIS; MATRIX; COMPUTER]

WHITTAKER, R. H. 1961. EXPERIMENTS WITH RADIOPHOSPHORUS TRACER IN AQUARIUM MICROCOSMS. ECOL. MONOGR. 31(2), 157-188. [DIFFERENTIAL EQUATION]

WILLIAMS, R. B. 1972. STEADY-STATE EQUILIBRIUMS IN SIMPLE NONLINEAR FOOD WEBS. PP. 213-240. IN, SYSTEMS ANALYSIS AND SIMULATION IN ECOLOGY, VOLUME II. (B. C. PATTEN, ED.) ACADEMIC PRESS, NEW YORK, NEW YORK. [MODEL; COMPUTER; SYSTEMS]

WILLIAMS, R. B. AND M. B. MURDOCH. 1972. COMPARTMENTAL ANALYSIS OF THE PRODUCTION OF JUNCUS ROEMERIANUS IN A NORTH CAROLINA SALT MARSH. CHESAPEAKE SCIENCE 13(2), 69-79. [COMPUTER]

WNEK, W. J. AND E. G. FOCHTMAN. 1972. MATHEMATICAL MODEL FOR FATE OF POLLUTANTS IN NEAR-SHORE WATERS. ENVIRON. SCI. TECHNOL. 6(4), 331-337.

COMPETITION, DENSITY-DEPENDENCE AND REGULATION

ALEKSEEV, V. V, AND V. A. SVETLOSANOV. 1974. ESTIMATION OF THE LIFE SPAN OF THE PREDATOR-VICTIM SYSTEM UNDER RANDOM MIGRATION OF VICTIMS. EKOLOGIYA 5(1), 91-95. (ENGLISH TRANSL. PP. 74-77). [PREDATOR-PREY MODEL; VOLTERRA'S EQUATIONS]

ALLAN, J. D. 1973. COMPETITION AND THE RELATIVE ABUNDANCES OF TWO CLADOCERANS. ECOLOGY 54(3), 484-498. [MODEL]

ANDERSEN, F. S. 1957. THE EFFECT OF DENSITY ON THE BIRTH AND DEATH RATE. P. 27. IN, ANNUAL REPORT 1954-1955, STATENS SKADEDYRLABORATORIUM, SPRINGFORBI, DENMARK. [MODEL; LOGISTIC OF VERHULST AND PEARL; KOSTITZIN'S EQUATIONS]

ANDERSEN, F. S. 1961. EFFECT OF DENSITY ON ANIMAL SEX RATIO. OIKOS 12(1), 1-16.

ANDREWARTHA, H. G. 1959. DENSITY-DEPENDENT FACTORS IN ECOLOGY. NATURE 183(4655), 200. [LOTKA-VOLTERRA MODEL; NICHOLSON'S MODEL]

ANDREWARTHA, H. G. 1963. DENSITY-DEPENDENCE IN THE AUSTRALIAN THRIPS. ECOLOGY 44(1), 218-220. [CORRELATION COEFFICIENT]

ARNOLD, S. J. 1972. SPECIES DENSITIES OF PREDATORS AND THEIR PREY. AMER. NAT. 106(948), 220-236. [PATH COEFFICIENT; CORRELATION COEFFICIENT]

AYALA, F. J., M. E. GILPIN AND J. G. EHRENFELD. 1973. COMPETITION BETWEEN SPECIES: THEORETICAL MODELS AND EXPERIMENTAL TESTS. THEOR. POPUL. BIOL. 4(3), 331-356. [LOTKA-VOLTERRA MODEL; DROSOPHILA; COMPUTER]

BAILEY, V. A., A. J. NICHOLSON AND E. J. WILLIAMS. 1962. INTERACTION BETWEEN HOSTS AND PARASITES WHEN SOME HOST INDIVIDUALS ARE MORE DIFFICULT TO FIND THAN OTHERS. J. THEOR. BIOL. 3(1), 1-18. [MODEL]

BARNETT, V. D. 1962. THE MONTE CARLO SOLUTION OF A COMPETING SPECIES PROBLEM. BIOMETRICS 18(1), 76-103. [BIRTH RATE; DEATH RATE; LESLIE MATRIX; GOWER; BARTLETT]

BARTLETT, M. S. 1957. ON THEORETICAL MODELS FOR COMPETITIVE AND PREDATORY BIOLOGICAL SYSTEMS. BIOMETRIKA 44(1/2), 27-42.

BECKER, N. G. 1970. INTERACTIONS BETWEEN SPECIES: SOME COMPARISONS BETWEEN DETERMINISTIC AND STOCHASTIC MODELS. CENTER FOR ENVIRONMENTAL QUALITY MANAGEMENT, CORNELL UNIVERSITY, ITHACA, NEW YORK, REPORT NO. 1021. 30 PP. [LOTKA-VOLTERRA EQUATION; MONTE CARLO STUDIES; PREDATOR-PREY MODEL; PROBABILITY GENERATING FUNCTION; SIMULATION]

BEDDINGTON, J. R. 1975. MUTUAL INTERFERENCE BETWEEN PARASITES OR PREDATORS AND ITS EFFECT ON SEARCHING EFFICIENCY. J. ANIM. ECOL. 44(1), 331-340. [MODEL; PREDATOR-PREY MODEL]

BELLA, I. E. 1971. A NEW COMPETITION MODEL FOR INDIVIDUAL TREES. FOREST SCI. 17(3), 364-372. [COMPUTER]

BENSON, J. F. 1973. SOME PROBLEMS OF TESTING FOR DENSITY-DEPENDENCE IN ANIMAL POPULATIONS. OECOLOGIA 13, 183-190 [REGRESSION; ERRORS]

BIGGER, M. 1973. AN INVESTIGATION BY FOURIER ANALYSIS INTO THE INTERACTION BETWEEN COFFEE LEAF-MINERS AND THEIR LARVAL PARASITES. J. ANIM. ECOL. 42(2), 417-434. [MODEL; LOTKA-VOLTERRA EQUATION; LESLIE AND GOWER EQUATIONS]

BILLARD, L. 1974. COMPETITION BETWEEN SPECIES. STOCHASTIC PROCESSES AND THEIR APPLICATIONS 2(4), 391-398. [STOCHASTIC MODEL; MOMENTS]

BRADLEY, D. J. 1974. STABILITY IN HOST-PARASITE SYSTEMS. PP. 71-87. IN, ECOLOGICAL STABILITY. (M. B. USHER AND M. H. WILLIAMSON, EDS.). HALSTED PRESS, NEW YORK, NEW YORK. [DISCUSSION; CONSIDERS MATHEMATICAL MODEL]

BRAUER, F. 1974. ON THE POPULATION OF COMPETING SPECIES. MATH. BIOSCI. 19, 299-306. [NON-LINEAR GROWTH RATE MODEL]

BRIAN, M. V. 1956. EXPLOITATION AND INTERFERENCE IN INTERSPECIES COMPETITION. J. ANIM. ECOL. 25(2), 339-347. [WINSOR MODEL; LOTKA-VOLTERRA MODEL]

BROCKELMAN, W. Y. AND R. M. FAGEN. 1972. ON MODELING DENSITY-INDEPENDENT POPULATION CHANGE. ECOLOGY 53(5), 944-948. [MORRIS'S TEST; REGRESSION]

BROCKSEN, R. W., G. E. DAVIS AND C. E. WARREN. 1970. ANALYSIS OF TROPHIC PROCESSES ON THE BASIS OF DENSITY-DEPENDENT FUNCTIONS. PP. 468-498. IN, MARINE FOOD CHAINS, (J. H. STEELE, ED.). UNIVERSITY OF CALIFORNIA PRESS, BERKELEY, CALIFORNIA. [PRODUCTION MODEL]

BUFFINGTON, J. D. 1971. PREDATION, COMPETITION, AND PLEISTOCENE MEGAFAUNA EXTINCTION. BIOSCIENCE 21(4), 167-170. [MODEL]

BULGAKOVA, T. I. 1968. CONCERNING MODELS OF THE COMPETITION OF SPECIES. PROBLEMY KIBERNETIKI 20, 263-279. (ENGLISH TRANSL., 1969. JPRS 48396, PP. 50-58.).

BULGAKOVA, T. I. 1968. CONCERNING THE STABILITY OF THE SIMPLEST MODEL OF BIOGEOCENOSIS. PROBLEMY KIBERNETIKI 20, 271-276. (ENGLISH TRANSL., 1969. JPRS 48396, PP. 59-65.).

BULMER, M. G. 1974. DENSITY-DEPENDENT SELECTION AND CHARACTER DISPLACEMENT. AMER. NAT. 108(959), 45-58. [MODEL; COMPETITION; PROBABILITY; NORMAL DISTRIBUTION]

CALHOUN, J. B. 1956. A COMPARATIVE STUDY OF THE SOCIAL BEHAVIOR OF TWO INBRED STRAINS OF HOUSE MICE. ECOL. MONOGR. 26(1), 81-103. [PROBABILITY]

CANALE, E. P. 1970. AN ANALYSIS OF MODELS DESCRIBING PREDATOR-PREY INTERACTION. BIOTECHÑOL. BIOENG. 12(3), 353-378. [LOTKA-VOLTERRA MODEL; PHASE PLANE ANALYSIS]

CASE, T. J. AND M. E. GILPIN. 1974. INTERFERENCE COMPETITION AND NICHE THEORY. PROC. NATL. ACAD. SCI. U.S.A. 71(8), 3073-3077. [LINEAR MODEL; MATRIX]

CHARLESWORTH, B. 1971. SELECTION IN DENSITY-REGULATED POPULATIONS. ECOLOGY 52(3), 469-474.

CHESSEL, D. 1971. MODELS FOR HOST-PARASITE RELATIONSHIP. ACTA BIOTHEOR. 20(1/2), 2-17.

CHEWNING, W. C. 1975. MIGRATORY EFFECTS IN PREDATOR-PREY MODELS. MATH. BIOSCI. 23(3/4), 253-262. [MATRIX; EIGENVECTORS]

CHIANG, C. L. 1954. COMPETITION AND OTHER INTERACTIONS BETWEEN SPECIES. PP. 197-215. IN, STATISTICS AND MATHEMATICS IN BIOLOGY. (O. KEMPTHORNE, T. A. BANCROFT, J. W. GOWEN AND J. L. LUSH, EDS.). IOWA STATE COLLEGE PRESS, AMES, IOWA. [MODEL; LOTKA-VOLTERRA DETERMINISTIC MODEL; VERHULST-PEARL-REED LOGISTIC CURVE; STOCHASTIC MODEL]

CLIFFORD, P. AND A. SUDBURY. 1973. A MODEL FOR SPATIAL CONFLICT. BIOMETRIKA 60(3), 581-588. [COMPETITION; MODEL; INVASION; STOCHASTIC MODEL]

CODY, M. L. 1968. ON THE METHODS OF RESOURCE DIVISION IN GRASSLAND BIRD COMMUNITIES. AMER. NAT. 102(924), 107-147. [MODEL; DISCRIMINANT FUNCTION]

COLE, L. C. 1960. COMPETITIVE EXCLUSION. SCIENCE 132(3423), 348-349. [POISSON DISTRIBUTION; LOGISTIC; VOLTERRA-LOTKA MODEL]

COLWELL, R. K. AND D. K. FUTUYAMA. 1971. ON THE MEASUREMENT OF NICHE BREATH AND OVERLAP. ECOLOGY 52(4), 567-576. [MATRIX; SHANNON-WIENER MEASURE OF INFORMATION; SIMPSON'S INDEX OF DIVERSITY; PROBABILITY]

COUTLEE, E. L. AND R. I. JENNRICH. 1968. THE RELEVANCE OF LOGARITHMIC MODELS FOR POPULATION INTERACTION. AMER. NAT. 102(926), 307-321.

CRAMER, N. F. AND R. M. MAY. 1971. INTERSPECIFIC COMPETITION, PREDATION AND SPECIES DIVERSITY: A COMMENT. J. THEOR. BIOL. 34(2), 289-293. [MODEL; PREDATOR-PREY MODEL]

CROFTON, H. D. 1971. A QUANTITATIVE APPROACH TO PARASITISM. PARASITOLOGY 62(2), 179-193. [NEGATIVE BINOMIAL DISTRIBUTION; TRUNCATED]

CULVER, D. C. 1970. ANALYSIS OF SIMPLE CAVE COMMUNITIES, NICHE SEPARATION AND SPECIES PACKING. ECOLOGY 51(6), 949-958. [LEVINS' COMPETITION MODEL]

CULVER, D. C. 1973. COMPETITION IN SPATIALLY HETEROGENEOUS SYSTEMS: AN ANALYSIS OF SIMPLE CAVE COMMUNITIES. ECOLOGY 54(1), 102-110. [COMPETITION MODELS]

CULVER, D. C. 1974. SPECIES PACKING IN CARIBBEAN AND NORTH TEMPERATE ANT COMMUNITIES. ECOLOGY 55(5), 974-988. [MATRIX; COMPETITION COEFFICIENT]

CUSHING, D. H. AND J. G. K. HARRIS. 1973. STOCK AND RECRUITMENT AND THE PROBLEM OF DENSITY DEPENDENCE. RAPP. PROCES-VERB. REUNIONS J. CONS. PERM. INT. EXPLOR. MER 164, 142-155. [MODEL]

DABROWSKA-PROT, E., J. LUCZAK AND K. TARWID. 1968. PREY AND PREDATOR DENSITY AND THEIR REACTIONS IN THE PROCESS OF MOSQUITO REDUCTION BY SPIDERS IN FIELD EXPERIMENTS. EKOLOGIYA POLSKA-SERIA A 16(40), 773-819. [INDEX OF MORTALITY; MODEL]

DE WIT, C. T. 1971. ON THE MODELLING OF COMPETITIVE PHENOMENA. PP. 269-280. IN, DYNAMICS OF POPULATIONS. (P. J. DEN BOER AND G. R. GRADWELL, EDS.). CENTRE FOR AGRICULTURAL PUBLISHING AND DOCUMENTATION, WAGENINGEN, NETHERLANDS.

DEAKIN, M. A. B. 1971. RESTRICTIONS ON THE APPLICABILITY OF VOLTERRA'S ECOLOGICAL EQUATIONS. BULL. MATH. BIOPHYS. 33, 571-577. [PREDATOR-PREY MODEL]

DEMETRIUS, L. 1969. ON COMMUNITY STABILITY. MATH. BIOSCI. 5, 321-325. [MODEL; MARKOV CHAIN; VECTOR TIME LAG; PREDATOR-PREY MODEL]

DIAMOND, P. 1974. THE STABILITY OF THE INTERACTION BETWEEN ENTOMOPHAGUS PARASITES AND THEIR HOST. MATH. BIOSCI. 19, 121-129. [NICHOLSON'S MODEL; HASSELL-VARLEY MODEL]

EBERHARDT, L. L. 1970. CORRELATION, REGRESSION, AND DENSITY DEPENDENCE. ECOLOGY 51(2), 306-310.

EDWARDS, R. L. 1961. THE AREA OF DISCOVERY OF TWO INSECT PARASITES, NASONIA VITRIPENNIS (WALKER) AND TRICHOGRAMMA EVANESCENS WESTWOOD, IN AN ARTIFICIAL ENVIRONMENT. CAN. ENTOMOL. 93(6), 475-481. [HOST-PARASITE MODEL; AREA OF DISCOVERY; NICHOLSON'S MODEL]

ENGSTROM-HEG, V. L. 1970. PREDATION, COMPETITION AND ENVIRONMENTAL VARIABLES: SOME MATHEMATICAL MODELS. J. THEOR. BIOL. 27(2), 175-195. [SIMULATION; COMPUTER]

ESTABROOK, G. F. AND D. C. JESPERSEN. 1974. STRATEGY FOR A PREDATOR ENCOUNTERING A MODEL-MIMIC SYSTEM. AMER. NAT. 108(962), 443-457. [MODEL; PROBABILITY; MATRIX; PREDATOR-PREY MODEL]

FELDMAN, M. W. AND J. ROUGHGARDEN. 1975. A POPULATION'S STATIONARY DISTRIBUTION IN A STOCHASTIC ENVIRONMENT WITH REMARKS ON THE THEORY OF SPECIES PACKING. THEOR. POPUL. BIOL. 7(2), 197-207. [LOGISTIC EQUATION; MODEL; LOTKA-VOLTERRA EQUATION; MAY'S ANALYSIS]

FELLER, W. 1941. DIE GRUNDLAGEN DER VOLTERRASCHEN THEORIE DES KAMPFES UMS DASEIN IN WAHRSCHEINLICHKEITSTHEORETISCHER BEHANDLUNG. ACTA BIOTHEOR. 5(1), 11-40. [VOLTERRA; MARKOV CHAIN; PEARL; KOLOMOGOROFF; MODEL]

FRAME, J. S. 1974. EXPLICIT SOLUTIONS IN TWO SPECIES VOLTERRA SYSTEMS. J. THEOR. BIOL. 43(1), 73-81. [PREDATOR-PREY MODEL; BESSEL FUNCTION]

FRANK, P. 1968. LIFE HISTORIES AND COMMUNITY STABILITY. ECOLOGY 49(2), 355-357. [REPRODUCTIVE VALUE FUNCTION OF FISHER]

FREDRICKSON, A. G., J. L. JOST, J. M. TSUCHIYA AND P-H. HSU. 1973. PREDATOR-PREY INTERACTIONS BETWEEN MALTHUSIAN POPULATIONS. J. THEOR. BIOL. 38(3), 487-526. [MODEL; LOTKA-VOLTERRA MODEL; STABILITY MATRIX]

FUJII, K. 1965. A STATISTICAL MODEL OF THE COMPETITION CURVE. RES. POPUL. ECOL.-KYOTO 7(2), 118-125.

FUJITA, H. 1953. FACTORS AFFECTING THE TYPE OF POPULATION DENSITY EFFECT UPON AVERAGE RATE OF OVIPOSITION. RES. POPUL. ECOL.-KYOTO 2, 1-7. [MODEL]

FUJITA, H. 1954. AN INTERPRETATION OF THE CHANGES IN TYPE OF THE POPULATION DENSITY EFFECT UPON THE OVIPOSITION RATE. ECOLOGY 35(2), 253-257. [MODEL]

FUJITA, H. AND S. UTIDA. 1952. THE EFFECT OF POPULATION DENSITY ON THE GROWTH OF AN ANIMAL POPULATION. RES. POPUL. ECOL.-KYOTO 1, 1-14. [LOGISTIC EQUATION]

GADGIL, M. AND O. T. SOLBRIG. 1972. THE CONCEPT OF R- AND K-SELECTION: EVIDENCE FROM WILD FLOWERS AND SOME THEORETICAL CONSIDERATIONS. AMER. NAT. 106(947), 14-31. [BIRTH RATE; DEATH RATE]

GALLOPIN, G. C. 1971. A GENERALIZED MODEL OF A RESOURCE-POPULATION SYSTEM. II. STABILITY ANALYSIS . OECOLOGIA 7(4), 414-432.

GAUSE, G. F., N. P. SMARAGDOVA AND A. A. WITT. 1936. FURTHER STUDIES OF INTERACTION BETWEEN PREDATORS AND PREY. J. ANIM. ECOL. 5(1), 1-18. [MODEL]

GILBERT, N. AND A. P. GUTIERREZ. 1973. A PLANT-APHID-PARASITE RELATIONSHIP. J. ANIM. ECOL. 42(2), 323-340. [MODEL; COMPUTER]

GILL, D. E. 1974. INTRINSIC RATE OF INCREASE, SATURATION DENSITY, AND COMPETITIVE ABILITY. II. THE EVOLUTION OF COMPETITIVE ABILITY. AMER. NAT. 108(959), 103-116. [LOTKA-VOLTERRA EQUATION; R-SELECTION; K-SELECTION; MALTHUSIAN PARAMETER; LOTKA'S EQUATION; ALPHA-SELECTION]

GILPIN, M. E. 1973. DO HARES EAT LYNX? AMER. NAT. 107(957), 727-730. [PREDATOR-PREY MODEL; MODEL; COMPUTER; LOTKA-VOLTERRA; MODEL]

GILPIN, M. E. 1974. HABITAT SELECTION AND A LIAPUNOV FUNCTION FOR COMPETITION COMMUNITIES. PP. 62-65. IN, MATHEMATICAL PROBLEMS IN BIOLOGY. LECTURE NOTES IN BIOMATHEMATICS 2. (P. VAN DEN DRIESSCHE, ED.), SPRINGER-VERLAG NEW YORK INC., NEW YORK, NEW YORK. [GAME THEORY; LOTKA-VOLTERRA EQUATION]

GILPIN, M. E. 1974. A LIAPUNOV FUNCTION FOR COMPETITION COMMUNITIES. J. THEOR. BIOL. 44(1), 35-48. [ALGORITHM; LOTKA-VOLTERRA MODEL]

GILPIN, M. E. 1974. A MODEL OF THE PREDATOR-PREY RELATIONSHIP. THEOR. POPUL. BIOL. 5(3), 333-344. [NON-LINEAR MODEL; COMPUTER]

GILPIN, M. E. 1975. LIMIT CYCLES IN COMPETITION COMMUNITIES. AMER. NAT. 109(965), 51-60. [LOTKA-VOLTERRA MODEL; MATRIX; PREDATOR-PREY MODEL]

GILPIN, M. E. AND F. J. AYALA. 1973. GLOBAL MODELS OF GROWTH AND COMPETITION. PROC. NATL. ACAD. SCI. U.S.A. 70(12; PART I), 3590-3593. [NICHE THEORY; LOTKA-VOLTERRA MODEL; CURVILINEAR REGRESSION; DROSOPHILA; MODEL; CURVILINEAR REGRESSION; DROSOPHILA]

GILPIN, M. E. AND K. E. JUSTICE. 1973. A NOTE ON NONLINEAR COMPETITION MODELS. MATH. BIOSCI. 17(1/2), 57-63. [LOTKA-VOLTERRA MODEL; COMPETITION MODELS]

GILPIN, M. E. AND M. L. ROSENZWEIG. 1972. ENRICHED PREDATOR-PREY SYSTEMS: THEORETICAL STABILITY. SCIENCE 177(4052), 902-904. [DETERMINISTIC MODEL; LIMIT CYCLE; COMPUTER; MODEL; SIMULATION]

GINZBERG, L. R., Y. I. GOL'DMAN AND A. I. RAILKIN. 1971. MATHEMATICAL MODEL OF INTERACTION BETWEEN TWO POPULATIONS: I. PREDATOR-PREY. ZH. OBSCH. BIOL. 32(6), 724-730.

GLEN, D. M. 1975. SEARCHING BEHAVIOUR AND PREY-DENSITY REQUIREMENTS OF BLEPHARIDOPTERUS ANGULATUS (FALL.) (HETEROPTERA: MIRIDAE) AS A PREDATOR OF THE LIME APHID, EUCALLIPTERUS TILIAE (L.), AND LEAFHOPPER, ALNETOIDEA ALNETI (DAHLBOM). J. ANIM. ECOL. 44(1), 115-134. [PREDATOR-PREY MODEL; MODEL]

GOEL, N. S., S. C. MAITRA AND E. W. MONTROLL. 1971. ON THE VOLTERRA AND OTHER NONLINEAR MODELS OF INTERACTING POPULATIONS. REVIEWS OF MODERN PHYSICS 43(2), 231-276. (REISSUED ACADEMIC PRESS, 1971.).

GOH, B. S., G. LEITMANN AND T. L. VINCENT. 1974. OPTIMAL CONTROL OF A PREY-PREDATOR SYSTEM. MATH. BIOSCI. 19, 263-286. [OPTIMIZATION; DIFFERENTIAL EQUATION]

GOLDSTEIN, R. A. 1972. MATHEMATICAL PROPERTIES OF SEVERAL POPULATION DYNAMICS MODELS. (REVIEW: N. S. GOEL, S. C. MAITRA AND E. W. MONTROLL). ECOLOGY 53(4), 762-763.

GOMATAM, J. 1974. A NEW MODEL FOR INTERACTING POPULATIONS-II: PRINCIPLE OF COMPETITIVE EXCLUSION. BULL. MATH. BIOL. 36(4), 355-364. [GAUSE-VOLTERRA'S MODEL; VOLTERRA'S EQUATIONS; VERHULST EQUATIONS; PREDATOR-PREY MODEL; GOMPERTZ EQUATION]

GOMATAM, J. 1974. A NEW MODEL FOR INTERACTING POPULATIONS-I: TWO-SPECIES SYSTEMS. BULL. MATH. BIOL. 36(4), 347-353. [GOMPERTZ EQUATION; VOLTERRA'S EQUATIONS; VERHULST EQUATIONS; PREDATOR-PREY MODEL]

GRANT, P. R. 1971. INTERACTIVE BEHAVIOUR OF PUFFINS (FRATERCULA ARCTICA L.) AND SKUAS (STEROCORARIUS PARASITICUS L.) BEHAVIOUR 40, 263-281. [HEIGHT EQUATION]

GRASMAN, J. AND E. VELING. 1973. AN ASYMPTOTIC FORMULA FOR THE PERIOD OF A VOLTERRA-LOTKA SYSTEM. MATH. BIOSCI. 18, 185-189.

GREEN, R. H. 1974. MULTIVARIATE NICHE ANALYSIS WITH TEMPORALLY VARYING ENVIRONMENTAL FACTORS. ECOLOGY 55(1), 73-83. [DISCRIMINANT-COVARIANCE ANALYSIS]

GRENNEY, W. J., D. A. BELLA AND H. C. CURL, JR. 1973. A THEORETICAL APPROACH TO INTERSPECIFIC COMPETITION IN PHYTOPLANKTON COMMUNITIES. AMER. NAT. 107(955), 405-425. [MODEL; DIGITAL COMPUTER; PROGRAM; SIMULATION]

GRIFFITHS, K. J. AND C. S. HOLLING. 1969. A COMPETITION SUBMODEL FOR PARASITES AND PREDATORS. CAN. ENTOMOL. 101(8), 785-818. [CONTAGION; COMPUTER; NEGATIVE BINOMIAL DISTRIBUTION; POISSON DISTRIBUTION; ANALYSIS OF VARIANCE]

HAIGH, J. AND J. MAYNARD SMITH. 1972. CAN THERE BE MORE PREDATORS THAN PREY? THEOR. POPUL. BIOL. 3(3), 290-299. [PREDATOR-PREY MODEL; VOLTERRA ASSUMPTIONS; MACARTHUR-LEVINS; RESCIGNO-RICHARDSON AND LEVINS VERSIONS]

HAIRSTON, N. G. 1967. STUDIES ON THE LIMITATION OF A NATURAL POPULATION OF PARAMECIUM AURELIA. ECOLOGY 48(6), 904-910. [LOGISTIC EQUATION; INTRINSIC RATE OF INCREASE]

HALDANE, J. B. S. 1953. ANIMAL POPULATIONS AND THEIR REGULATION. NEW BIOL. 15, 9-24. [MODEL]

HAMILTON, W. J., III. AND K. E. F. WATT. 1970. REFUGING. PP. 263-286. IN, ANNUAL REVIEW OF ECOLOGY AND SYSTEMATICS, VOLUME 1. (R. F. JOHNSTON , P. W. FRANK AND C. D. MICHENER, EDS.). ANNUAL REVIEWS, INC., PALO ALTO, CALIFORNIA.

HARDMAN, J. M. AND A. L. TURNBULL. 1974. THE INTERACTION OF SPATIAL HETEROGENEITY, PREDATOR COMPETITION AND THE FUNCTIONAL RESPONSE TO PREY DENSITY IN A LABORATORY SYSTEM OF WOLF SPIDERS (ARANEAE: LYCOSIDAE) AND FRUIT FLIES (DIPTERA: DROSOPHILIDAE). J. ANIM. ECOL. 43(1), 155-171. [COMPUTER; SIMULATION; MODEL]

HARPER, J. L. 1967. A DARWINIAN APPROACH TO PLANT ECOLOGY. J. ANIM. ECOL. 36(3), 495-518. [INTRINSIC RATE OF INCREASE; COMPETITION; MODIFIED LOGISTIC]

HARRIS, J. R. W. 1974. THE KINETICS OF POLYPHAGY. PP. 123-139. IN, ECOLOGICAL STABILITY. (M. B. USHER AND M. H. WILLIAMSON, EDS.). HALSTED PRESS, NEW YORK, NEW YORK. [MODEL; PREDATOR-PREY MODEL]

HARTE, J. AND D. LEVY. 1974. ON THE VULNERABILITY OF ECOSYSTEMS DISTURBED BY MAN. LAWRENCE BERKELEY LAB., UNIVERSITY OF CALIFORNIA, BERKELEY, CALIFORNIA. REPORT LBL-3214. 30 PP. [LIAPUNOV STABILITY THEORY; MODEL; MATRIX; LOTKA-VOLTERRA MODEL; PREDATOR-PREY MODEL; DETERMINANT]

HASSELL, M. P. 1975. DENSITY-DEPENDENCE IN SINGLE-SPECIES POPULATIONS. J. ANIM. ECOL. 44(1), 283-295. [LOGISTIC; STABILITY ANALYSIS; MODEL; COMPETITION]

HASSELL, M. P. AND C. B. HUFFAKER. 1969. THE APPRAISAL OF DELAYED AND DIRECT DENSITY-DEPENDENCE. CAN. ENTOMOL. 101(4), 353-361. [NICHOLSON'S MODEL; MORRIS' METHOD; NICHOLSON-BAILEY MODEL]

HASSELL, M. P. AND R. M. MAY. 1973. STABILITY IN INSECT HOST-PARASITE MODELS. J. ANIM. ECOL. 42(3), 693-726. [RANDOM AND NON-RANDOM SEARCH; PROBABILITY; COMPUTER]

HASSELL, M. P. AND R. M. MAY. 1974. AGGREGATION OF PREDATORS AND INSECT PARASITES AND ITS EFFECT ON STABILITY. J. ANIM. ECOL. 43(2), 567-594. [STABILITY ANALYSIS; PREDATOR-PREY MODEL; NICHOLSON-BAILEY MODEL; RANDOM WALK; MURDOCH-OATEN MODEL; PROBABILITY]

HAUSSMANN, U. G. 1973. ON THE PRINCIPLE OF COMPETITIVE EXCLUSION. THEOR. POPUL. BIOL. 4(1), 31-41. [MODEL]

HAUSSMANN, U. G. 1974. COEXISTENCE OF SPECIES IN A DISCRETE SYSTEM. PP. 73-82. IN, MATHEMATICAL PROBLEMS IN BIOLOGY. LECTURE NOTES IN BIOMATHEMATICS 2. (P. VAN DEN DRIESSCHE, ED.). SPRINGER-VERLAG NEW YORK INC., NEW YORK, NEW YORK. [COMPETITIVE EXCLUSION; MODEL; DIFFERENCE EQUATION]

HOLLING, C. S. 1959. SOME CHARACTERISTICS OF SIMPLE TYPES OF PREDATION AND PARASITISM. CAN. ENTOMOL. 91(7), 385-398. [DISC EQUATION; PREDATOR-PREY MODEL; MODEL]

HOLLING, C. S. 1961. PRINCIPLES OF INSECT PREDATION. ANNU. REV. ENTOMOL. 6, 163-182. [MODEL]

HOLLING, C. S. 1964. THE ANALYSIS OF COMPLEX POPULATION PROCESSES. CAN. ENTOMOL. 96(1/2), 335-347.

HOLLING, C. S. 1965. THE FUNCTIONAL RESPONSE OF PREDATORS TO PREY DENSITY AND ITS ROLE IN MIMICRY AND POPULATION REGULATION. MEM. ENTOMOL. SOC. CAN. 45. 60 PP. [MULLERIAN MIMICRY; MODEL; SIMULATION; BATESIAN MIMICRY]

HOLLING, C. S. 1966. THE FUNCTIONAL RESPONSE OF INVERTEBRATE PREDATORS TO PREY DENSITY. MEM. ENTOMOL. SOC. CAN. 48. 86 PP. [COMPUTER; MODEL]

HOLLING, C. S. 1973. DESCRIPTION OF THE PREDATION MODEL: PREDATOR-PREY FUNCTIONAL RESPONSE. INTERN. INST. FOR APPLIED SYSTEMS ANALYSIS, RESEARCH MEMORANDUM RM-73-1. 41 PP.

HOLLING, C. S. 1973. RESILIENCE AND STABILITY OF ECOLOGICAL SYSTEMS. PP. 1-23. IN, ANNUAL REVIEW OF ECOLOGY AND SYSTEMATICS, VOLUME 4. (R. F. JOHNSTON, P. W. FRANK AND C. D. MICHENER, EDS.). ANNUAL REVIEWS, INC., PALO ALTO, CALIFORNIA. [MODEL; TIME LAG; PROCESS ANALYSIS; SIMULATION]

HORN, H. S. 1968. REGULATION OF ANIMAL NUMBERS: A MODEL COUNTER-EXAMPLE. ECOLOGY 49(4), 776-778.

HORN, H. S. AND R. H. MACARTHUR. 1972. COMPETITION AMONG FUGITIVE SPECIES IN A HARLEQUIN ENIVRONMENT. ECOLOGY 53(4), 749-752. [LOTKA-VOLTERRA EQUATION; COMPETITION]

HUTCHINSON, G. E. 1947. A NOTE ON THE THEORY OF COMPETITION BETWEEN TWO SOCIAL SPECIES. ECOLOGY 28(3), 319-321.

HUTCHINSON, G. E. 1948. CIRCULAR CASUAL SYSTEMS IN ECOLOGY. ANN. N. Y. ACAD. SCI. 50(ART. 4), 221-246. [VOLTERRA; GAUSE MODEL; LOGISTIC; PEARL-VERHULST]

HUTCHINSON, G. E. 1957. CONCLUDING REMARKS. COLD SPRING HARBOR SYMPOSIA ON QUANTITATIVE BIOLOGY 22, 415-427. [VOLTERRA-GAUSE MODEL; NICHE; HYPERVOLUME; SET; COMPUTER]

HUTCHINSON, G. E. 1961. THE PARADOX OF THE PLANKTON. AMER. NAT. 95(882), 137-145. [COMPETITIVE EXCLUSION; MODEL; MACARTHUR'S DIVERSITY MODEL]

HUTCHINSON, G. E. AND R. H. MACARTHUR. 1959. A THEORETICAL ECOLOGICAL MODEL OF SIZE DISTRIBUTIONS AMONG SPECIES OF ANIMALS. AMER. NAT. 93(869), 117-125.

ITO, Y. 1971. SOME NOTES ON THE COMPETITIVE EXCLUSION PRINCIPLE. RES. POPUL. ECOL.-KYOTO 13(1), 46-54. [VOLTERRA; GAUSE MODEL]

ITO, Y. 1972. ON THE METHODS FOR DETERMINING DENSITY-DEPENDENCE BY MEANS OF REGRESSION. OECOLOGIA 10(4), 347-372.

JEFFRIES, C. 1974. PROBABILISTIC LIMIT CYCLES. PP. 123-131. IN, MATHEMATICAL PROBLEMS IN BIOLOGY. LECTURE NOTES IN BIOMATHEMATICS 2. (P. VAN DEN DRIESSCHE, ED.). SPRINGER-VERLAG NEW YORK INC., NEW YORK, NEW YORK. [MODEL; PREDATOR-PREY SYSTEM]

JENSEN, A. L. 1974. PREDATOR-PREY AND COMPETITION MODELS WITH STATE VARIABLES: BIOMASS, NUMBER OF INDIVIDUALS, AND AVERAGE INDIVIDUAL WEIGHT. J. FISH. RES. BOARD CAN. 30(10), 1669-1674. [KOSTITZIN'S EQUATIONS; LOTKA-VOLTERRA EQUATION]

JONES, R. 1973. DENSITY DEPENDENT REGULATION OF THE NUMBERS OF COD AND HADDOCK. RAPP. PROCES-VERB. REUNIONS J. CONS. PERM. INT. EXPLOR. MER 164, 156-173. [MODEL]

JORGENSEN, C. D. 1968. HOME RANGE AS A MEASURE OF PROBABLE INTERACTIONS AMONG POPULATIONS OF SMALL MAMMALS. J. MAMMAL. 49(1), 104-112.

KELKER, G. H. 1939. A MATHEMATICAL STUDY OF A PREDATORY-PREY RELATIONSHIP. PROC. UTAH ACAD. SCI., ARTS LETT. 16, 77-81.

KELKER, G. H. 1940. FURTHER MATHEMATICAL STUDIES IN PREY-PREDATOR RELATIONSHIPS. PROC. UTAH ACAD. SCI., ARTS LETT. 17, 59-64.

KILMER, W. L. 1972. ON SOME REALISTIC CONSTRAINTS IN PREY-PREDATOR MATHEMATICS. J. THEOR. BIOL. 36(1), 9-22. [MODEL; MATRIX; COMPUTER; SIMULATION; LIMIT CYCLE]

KOCH, A. L. 1974. COEXISTENCE RESULTING FROM AN ALTERNATION OF DENSITY DEPENDENT AND DENSITY INDEPENDENT GROWTH. J. THEOR. BIOL. 44(2), 373-386. [VOLTERRA-LOTKA AND OTHER MODELS]

KOCH, A. L. 1974. COMPETITIVE COEXISTENCE OF TWO PREDATORS UTILIZING THE SAME PREY UNDER CONSTANT ENVIRONMENTAL CONDITIONS. J. THEOR. BIOL. 44(2), 387-395. [MODEL; SIMULATION; COMPUTER]

KOLMAN, W. A. 1960. THE MECHANISM OF NATURAL SELECTION FOR THE SEX RATIO. AMER. NAT. 94(878), 373-377. [MODEL]

KOLMOGOROFF, A. 1936. SULLA TEORIA DI VOLTERRA DELLA LOTTA PER L'ESISTENZA. GIORNALE DELL'ISTITUTO ITALIANO DEGLI ATTUARI 7, 74-80. [PREDATOR-PREY MODEL]

KRAPIVIN, V. F. 1972. INVESTIGATION OF GENERALIZED MATHEMATICAL PREDATOR-PREY MODEL. EKOLOGIYA 3(3), 28-37. (ENGLISH TRANSL. PP. 215-221). [IVLEV COMPETITION THEORY; POISSON DISTRIBUTION; PROBABILITY]

KUNO, E. 1973. STATISTICAL CHARACTERISTICS OF THE DENSITY-INDEPENDENT POPULATION FLUCTUATION AND THE EVALUATION OF DENSITY-DEPENDENCE AND REGULATION IN ANIMAL POPULATIONS. RES. POPUL. ECOL.-KYOTO 15(1), 99-120. [BIAS; KEY FACTOR ANALYSIS; LIFE TABLE; REGRESSION]

LARKIN, P. A. 1963. INTERSPECIFIC COMPETITION AND EXPLOITATION. J. FISH. RES. BOARD CAN. 20(3), 647-678.

LARKIN, P. A., R. F. RALEIGH AND N. J. WILIMOVSKY. 1964. SOME ALTERNATIVE PREMISES FOR CONSTRUCTING THEORETICAL REPRODUCTION CURVES. J. FISH. RES. BOARD CAN. 21(3), 477-484. [RICKER'S CURVE; BEVERTON-HOLT EQUATION]

LAWTON, J. H., J. R. BEDDINGTON AND R. BONSER. 1974. SWITCHING IN INVERTEBRATE PREDATORS. PP. 141-158. IN, ECOLOGICAL STABILITY. (M. B. USHER AND M. H. WILLIAMSON, EDS.). HALSTED PRESS, NEW YORK, NEW YORK. [MODEL]

LEIGH, E. G., JR. 1968. THE ECOLOGICAL ROLE OF VOLTERRA'S EQUATIONS. PP. 1-61. IN, SOME MATHEMATICAL PROBLEMS IN BIOLOGY. (M. GERSTENHABER, ED.). AMER. MATH. SOC., PROVIDENCE, RHODE ISLAND.

LEON, J. A. 1974. SELECTION IN CONTEXTS OF INTERSPECIFIC COMPETITION. AMER. NAT. 108(964), 739-757. [GAUSE AND WITT EQUATIONS; ROUGHGARDEN'S ADAPTIVE VALUES; MODEL; GENETICS; CARRYING CAPACITY; INTRINSIC RATE OF INCREASE]

LEVIN, B. R. 1969. A MODEL FOR SELECTION IN SYSTEMS OF SPECIES COMPETITION. PP. 237-275. IN, CONCEPTS AND MODELS OF BIOMATHEMATICS: SIMULATION TECHNIQUES AND METHODS, VOLUME 1. (F. HEINMETS, ED.). MARCEL DEKKER, NEW YORK, NEW YORK. [STOCHASTIC; DETERMINISTIC MODEL; VERHULST-PEARL EQUATION; GAUSE MODEL; COMPUTER]

LEVIN, D. A. AND W. W. ANDERSON. 1970. COMPETITION FOR POLLINATORS BETWEEN SIMULTANEOUSLY FLOWERING SPECIES. AMER. NAT. 104(939), 455-467. [MODEL]

LEVIN, S. A. 1970. COMMUNITY EQUILIBRIA AND STABILITY AND AN EXTENSION OF THE COMPARATIVE EXCLUSION PRINCIPLE. AMER. NAT. 104(939), 413-423. [MODEL; LOTKA'S EQUATION; PREDATOR-PREY MODEL]

LEVIN, S. A. 1974. DISPERSION AND POPULATION INTERACTIONS. AMER. NAT. 108(960), 207-228. [COMPETITION; MODEL; STABILITY ANALYSIS; PREDATOR-PREY MODEL]

LEVINE, R. L., J. E. HUNTER AND P. L. BORCHELT. 1974. DUSTBATHING AS A REGULATORY MECHANISM. BULL. MATH. BIOL. 36(5/6), 545-553. [BIRDS; MODEL]

LEVINS, R. 1965. GENETIC CONSEQUENCES OF NATURAL SELECTION. PP. 371-387. IN, THEORETICAL AND MATHEMATICAL BIOLOGY. (H. J. MOROWITZ AND T. H. WATERMAN, EDS.). BLAISDELL PUBLISHING CO., NEW YORK, NEW YORK. [MODEL; NICHE; PROBABILITY; MATRIX]

LEVINS, R. 1969. SOME DEMOGRAPHIC AND GENETIC CONSEQUENCES OF ENVIRONMENTAL HETEROGENEITY FOR BIOLOGICAL CONTROL. BULL. ENTOMOL. SOC. AMER. 15(3), 237-240. [MODEL; PREDATOR-PREY MODEL; PROBABILITY DISTRIBUTION]

LEVINS, R. AND D. CULVER. 1971. REGIONAL COEXISTENCE OF SPECIES AND COMPETITION BETWEEN RARE SPECIES. PROC. NATL. ACAD. SCI. U.S.A. 68(6), 1246-1248. [MODEL]

LLOYD, M. 1968. SELF REGULATION OF ADULT NUMBERS BY CANNIBALISM IN TWO LABORATORY STRAINS OF FLOUR BEETLES (TRIBOLIUM CASTANEUM). ECOLOGY 49(2), 245-259. [MODEL]

LOMNICKI, A. 1971. ANIMAL POPULATION REGULATION BY THE GENETIC FEEDBACK MECHANISM: A CRITIQUE OF THE THEORETICAL MODEL. AMER. NAT. 105(945), 413-421. [PIMENTEL'S MODEL; PREDATOR-PREY MODEL]

LOMNICKI, A. 1974. EVOLUTION OF THE HERBIVORE-PLANT, PREDATOR-PREY, AND PARASITE-HOST SYSTEMS: A THEORETICAL MODEL. AMER. NAT. 108(960), 167-180. [PIMENTEL GENETIC FEEDBACK MODEL]

MACARTHUR, R. 1955. FLUCTUATIONS OF ANIMAL POPULATIONS AND A MEASURE OF COMMUNITY STABILITY. ECOLOGY 36(3), 533-536. [MARKOV CHAIN; ENTROPY]

MACARTHUR, R. 1970. SPECIES PACKING AND COMPETITIVE EQUILIBRIUM FOR MANY SPECIES. THEOR. POPUL. BIOL. 1(1), 1-11.

MACARTHUR, R. AND R. LEVINS. 1964. COMPETITION, HABITAT SELECTION, AND CHARACTER DISPLACEMENT IN A PATCHY ENVIRONMENT. PROC. NATL. ACAD. SCI. U.S.A. 51(6), 1207-1210. [MODEL]

MACARTHUR, R. AND R. LEVINS. 1967. THE LIMITING SIMILARITY, CONVERGENCE, AND DIVERGENCE OF COEXISTING SPECIES. AMER. NAT. 101(921), 377-385. [VOLTERRA'S EQUATIONS; COMPETITION; SPECIES PACKING]

MACARTHUR, R. H. 1958. POPULATION ECOLOGY OF SOME WARBLERS OF NORTHEASTERN CONIFEROUS FORESTS. ECOLOGY 39(4), 599-619. [DENSITY DEPENDENCE; RUN TEST]

MACARTHUR, R. H. 1960. ON THE RELATION BETWEEN REPRODUCTIVE VALUE AND OPTIMAL PREDATION. PROC. NATL. ACAD. SCI. U.S.A. 46(1), 143-145.

MACARTHUR, R. H. 1965. ECOLOGICAL CONSEQUENCES OF NATURAL SELECTION. PP. 388-397. IN, THEORETICAL AND MATHEMATICAL BIOLOGY. (H. J. MOROWITZ AND T. H. WATERMAN, EDS.). BLAISDELL PUBLISHING CO., NEW YORK, NEW YORK. [PREDATOR-PREY MODEL; BIRTH RATE; DEATH RATE; MODEL]

MACARTHUR, R. H. 1968. THE THEORY OF THE NICHE. PP. 159-176. IN, POPULATION BIOLOGY AND EVOLUTION. (R. C. LEWONTIN, ED.). SYRACUSE UNIVERSITY PRESS, SYRACUSE, NEW YORK. [VOLTERRA'S EQUATIONS; MATRIX]

MACARTHUR, R. H. 1969. SPECIES PACKING AND WHAT INTERSPECIES COMPETITION MINIMIZES. PROC. NATL. ACAD. SCI. U.S.A. 64(4), 1369-1371.

MACARTHUR, R. H. AND E. S. PIANKA. 1966. ON OPTIMAL USE OF A PATCHY ENVIRONMENT. AMER. NAT. 100(916), 603-609. [COMPETITION; GRAPHICAL METHOD]

MACDONALD, P. D. M. AND L. CHENG. 1970. A METHOD OF TESTING FOR SYNCHRONIZATION BETWEEN HOST AND PARASITE POPULATIONS. J. ANIM. ECOL. 39(2), 321-331.

MACHIN, D. 1970. REGRESSION WITH CORRELATED RESIDUALS: AN EXAMPLE FROM COMPETITION EXPERIMENTS. BIOMETRICS 26(4), 835-840. [INSECT]

MAELZER, D. A. 1970. THE REGRESSION OF LOG (N(N-1)) ON LOG (N(N)) AS A TEST OF DENSITY DEPENDENCE: AN EXERCISE WITH COMPUTER-CONSTRUCTED DENSITY-INDEPENDENT POPULATIONS. ECOLOGY 51(5), 810-822. [COMPUTER]

MAGUIRE, B., JR. 1967. A PARTIAL ANALYSIS OF THE NICHE. AMER. NAT. 101(922), 515-523. [MODEL; HYPERVOLUME VERSATILITY INDEX]

MAGUIRE, B., JR. 1973. NICHE RESPONSE STRUCTURE AND THE ANALYTICAL POTENTIALS OF ITS RELATIONSHIP TO THE HABITAT. AMER. NAT. 107(954), 213-246. [HYPERVOLUME; MODEL; COMPETITION; SYSTEMS; TIME LAG]

MANLY, B. F. J. 1972. ESTIMATING SELECTIVE VALUES FROM FIELD DATA. BIOMETRICS 28(4), 1115-1125. [INSECT]

MANLY, B. F. J. 1972. TABLES FOR THE ANALYSIS OF SELECTIVE PREDATION EXPERIMENTS. RES. POPUL. ECOL.-KYOTO 14(1), 74-81.

MANLY, B. F. J. 1973. A LINEAR MODEL FOR FREQUENCY-DEPENDENT SELECTION BY PREDATORS. RES. POPUL. ECOL.-KYOTO 14(2), 137-150.

MANLY, B. F. J., P. MILLER AND L. M. COOK. 1972. ANALYSIS OF A SELECTIVE PREDATION EXPERIMENT. AMER. NAT. 106(952), 719-736. [PREDATOR-PREY MODEL; PROBABILITY]

MANN, S. H. 1970. A MATHEMATICAL THEORY FOR THE HARVEST OF NATURAL ANIMAL POPULATIONS WHEN BIRTH RATES ARE DEPENDENT ON TOTAL POPULATION SIZE. MATH. BIOSCI. 7, 97-110. [MARKOVIAN DISTRIBUTION PROCESS]

MARTEN, G. G. 1973. AN OPTIMIZATION EQUATION FOR PREDATION. ECOLOGY 54(1), 92-101. [COMPUTER; MATRIX; VECTOR]

MAY, R. M. 1971. STABILITY IN MODEL ECOSYSTEMS. PROC. ECOL. SOC. AUSTRALIA 6, 17-56.

MAY, R. M. 1971. STABILITY IN MULTISPECIES COMMUNITY MODELS. MATH. BIOSCI. 12, 59-79. [VOLTERRA-LOTKA MODEL; PREDATOR-PREY MODEL; MATRIX; INFORMATION THEORY]

MAY, R. M. 1972. LIMIT CYCLES IN PREDATOR-PREY COMMUNITIES. SCIENCE 177(4052), 900-902. [MODEL; LIMIT CYCLE; LOTKA-VOLTERRA]

MAY, R. M. 1972. WILL A LARGE COMPLEX SYSTEM BE STABLE? NATURE 238(5364), 413-414. [MATRIX]

MAY, R. M. 1973. ON RELATIONSHIPS AMONG VARIOUS TYPES OF POPULATION MODELS. AMER. NAT. 107(953), 46-57. [LOTKA-VOLTERRA; MATRIX; NICHOLSON-BAILEY EQUATION; POISSON DISTRIBUTION; CANONICAL FORM;PREDATOR-PREY MODEL]

MAY, R. M. 1973. QUALITATIVE STABILITY IN MODEL ECOSYSTEMS. ECOLOGY 54(3), 638-641. [MATRIX]

MAY, R. M. 1973. STABILITY IN RANDOMLY FLUCTUATING VERSUS DETERMINISTIC ENVIRONMENTS. AMER. NAT. 107(957), 621-650. [MODEL; STOCHASTIC MODEL; MATRIX; LOGISTIC; PROBABILITY; PREDATOR-PREY MODEL; COMPUTER; LIMIT CYCLE]

MAY, R. M. 1974. ON THE THEORY OF NICHE OVERLAP. THEOR. POPUL. BIOL. 5(3), 297-332. [DETERMINISTIC AND STOCHASTIC STABILITY; MODEL; MATRIX; SPECIES PACKING]

MAY, R. M. AND R. H. MACARTHUR. 1972. NICHE OVERLAP AS A FUNCTION OF ENVIRONMENTAL VARIABILITY. PROC. NATL. ACAD. SCI. U.S.A. 69(5), 1109-1113. [MODEL; MATRIX; LOTKA-VOLTERRA EQUATION; STOCHASTIC; COMPETITION]

MAY, R. M., G. R. CONWAY, M. P. HASSELL AND T. R. E. SOUTHWOOD. 1974. TIME DELAYS, DENSITY-DEPENDENCE AND SINGLE-SPECIES OSCILLATIONS. J. ANIM. ECOL. 43(3), 747-770. [SINGLE AGE-CLASS MODELS; MULTIPLE AGE-CLASS MODELS; MATRIX; DIFFERENCE AND DIFFERENTIAL EQUATIONS]

MAYNARD SMITH, J. 1972. THE STABILITY OF PREDATOR-PREY SYSTEMS. (ABSTRACT). BIOMETRICS 28(4), 1158.

MCGILCHRIST, C. 1967. ANALYSIS OF PLANT COMPETITION EXPERIMENTS FOR DIFFERENT RATIOS OF SPECIES. BIOMETRIKA 54(3/4), 471-477. [MODEL; ANALYSIS OF VARIANCE; VARIANCE MAXIMUM LIKELIHOOD]

MEAD, R. 1967. A MATHEMATICAL MODEL FOR THE ESTIMATION OF INTER-PLANT COMPETITION. BIOMETRICS 23(2), 189-205.

MEAD, R. 1968. MEASUREMENT OF COMPETITION BETWEEN INDIVIDUAL PLANTS IN A POPULATION. J. ECOL. 56(1), 35-45. [MODEL; ANALYSIS OF VARIANCE]

MEAD, R. 1971. MODELS FOR INTERPLANT COMPETITION IN IRREGULARLY DISTRIBUTED POPULATIONS. PP. 13-30. IN, STATISTICAL ECOLOGY, VOLUME 2. SAMPLING AND MODELING BIOLOGICAL POPULATIONS AND POPULATION DYNAMICS. (G. P. PATIL, E. C. PIELOU AND W. E. WATERS, EDS.). PENNSYLVANIA STATE UNIVERSITY PRESS, UNIVERSITY PARK, PENNSYLVANIA.

MILLER, C. A. 1959. THE INTERACTION OF THE SPRUCE BUDWORM, CHORISTONEURA FUMIFERANA (CLEM.), AND THE PARASITE APANTELES FUMIFERANAE VIER. CAN. ENTOMOL. 91(8), 457-477. [HOST-PARASITE MODEL]

MILLER, R. S. 1964. INTERSPECIES COMPETITION IN LABORATORY POPULATIONS OF DROSOPHILA MELANOGASTER AND DROSOPHILA SIMULANS. AMER. NAT. 98(901), 221-238. [MODEL; NICHE; GAUSE MODEL; HUTCHINSON'S MODEL]

MILLER, R. S. 1964. LARVAE COMPETITION IN DROSOPHILA MELANOGASTER AND D. SIMULANS. ECOLOGY 45(1), 132-148. [GAUSE COMPETITION MODEL; NICHE]

MILNE, A. 1957. THE NATURAL CONTROL OF INSECT POPULATIONS. CAN. ENTOMOL. 89(5), 193-213. [HOST-PARASITE MODEL; VARLEY'S MODEL]

MILNE, A., W. B. YAPP AND G. C. VARLEY. 1959. A CONTROVERSIAL EQUATION IN POPULATION ECOLOGY. NATURE 184(4698), 1582-1583.

MONTROLL, E. W. 1972. SOME STATISTICAL ASPECTS OF THE THEORY OF INTERACTING SPECIES. PP. 99-143. IN, SOME MATHEMATICAL QUESTIONS IN BIOLOGY. III. (J. D. COWAN, ED.). AMER. MATH. SOC., PROVIDENCE, RHODE ISLAND. [LOTKA-VOLTERRA MODEL; MODEL; GOMPERTZ]

MOORE, W. S. AND F. E. MCKAY. 1971. COEXISTENCE IN UNISEXUAL-BISEXUAL SPECIES COMPLEXES OF POECILIOPSIS (PISCES: POELILIIDAE). ECOLOGY 52(5), 791-799. [MODEL; COMPUTER; SIMULATION]

MORAN, P. A. P. 1954. THE LOGIC OF THE MATHEMATICAL THEORY OF ANIMAL POPULATIONS. J. WILDL. MANAGE. 18(1), 60-66.

MUKERJI, M. K. AND E. J. LEROUX. 1969. THE EFFECT OF PREDATOR AGE ON THE FUNCTIONAL RESPONSE OF PODISUS MACULIVENTRIS TO THE PREY SIZE OF GALLERIA MELLONELLA. CAN. ENTOMOL. 101(3), 314-327. [DISC EQUATION]

MURAI, M. 1975. THEORETICAL STUDIES ON THE ROLE OF FOOD EXPLOITATION FOR THE COMPETITION OF SCRAMBLE TYPE. RES. POPUL. ECOL.-KYOTO 16(2), 289-308. [MODEL]

MURAI, M. AND K. FUJII. 1970. EXAMINATION OF THE INFLUENCES OF DENSITY PRESSURE ON THE PATTERN OF ADULT EMERGENCE WITH REFERENCE TO THE AZUKI BEAN WEEVIL, CALLOSOBRUCHUS CHINENSIS. RES. POPUL. ECOL.-KYOTO 12(2), 219-232.

MURDIE, G. AND M. P. HASSELL. 1973. FOOD DISTRIBUTION, SEARCHING SUCCESS AND PREDATOR-PREY MODELS. PP. 87-101. IN, THE MATHEMATICAL THEORY OF THE DYNAMICS OF BIOLOGICAL POPULATIONS. (M. S. BARTLETT AND R. W. HIORNS, EDS.). ACADEMIC PRESS, NEW YORK, NEW YORK. [RANDOM ATTACK; NICHOLSON-BAILEY MODEL; SIMULATION; VARIANCE; DETERMINISTIC MODEL; STOCHASTIC; GRAPHICAL CALCOMP PLOTTER; COMPUTER]

MURDOCH, W. W. 1969. SWITCHING IN GENERAL PREDATORS. EXPERIMENTS ON PREDATOR SPECIFICITY AND STABILITY OF PREY POPULATIONS. ECOL. MONOGR. 39(4), 335-354. [MODEL]

MURDOCH, W. W. 1970. POPULATION REGULATION AND POPULATION INERTIA. ECOLOGY 51(3), 497-502. [MODEL]

MURDOCH, W. W. 1971. THE DEVELOPMENTAL RESPONSE OF PREDATORS TO CHANGES IN PREY DENSITY. ECOLOGY 52(1), 132-137. [PREDATOR-PREY MODEL; MODEL]

MURDOCH, W. W. 1973. THE FUNCTIONAL RESPONSE OF PREDATORS. J. APPL. ECOL. 10(1), 335-342. [SIGMOID CURVE; HOLLING MODEL; SOLOMON'S MODEL]

MURDOCH, W. W. AND J. R. MARKS. 1973. PREDATION BY COCCINELLID BEETLES: EXPERIMENTS ON SWITCHING. ECOLOGY 54(1), 160-167. [PREDATOR-PREY MODEL; MODEL]

NAKAMURA, K. 1974. A MODEL OF THE FUNCTIONAL RESPONSE OF A PREDATOR TO PREY DENSITY INVOLVING THE HUNGER EFFECT. OECOLOGIA 16(4), 265-278. [IVLEV'S EQUATION; HOLLING'S DISC EQUATION]

NEILL, W. E. 1974. THE COMMUNITY MATRIX AND INTERDEPENDENCE OF THE COMPETITION COEFFICIENTS. AMER. NAT. 108(962), 399-408. [LOTKA-VOLTERRA EQUATION; MODEL; MACARTHUR-LEVINS MODEL]

NICHOLSON, A. J. 1933. THE BALANCE OF ANIMAL POPULATIONS. J. ANIM. ECOL. 2(1), 132-178. [COMPETITION; OSCILLATION]

NICHOLSON, A. J. AND V. A. BAILEY. 1935. THE BALANCE OF ANIMAL POPULATIONS- PART I. PROC. ZOOL. SOC. (LONDON) 1935(3), 551-598. [MODEL; COMPETITION; PROBABILITY]

NICOLIS, G. AND I. PRIGOGINE. 1971. FLUCTUATIONS IN NONEQUILIBRIUM SYSTEMS. PROC. NATL. ACAD. SCI. U.S.A. 68(9), 2102-2107. [NON-LINEAR SYSTEMS; BIRTH AND DEATH MODEL; VOLTERRA-LOTKA MODEL]

ODUM, H. T. AND W. C. ALLEE. 1954. A NOTE ON THE STABLE POINT OF POPULATIONS SHOWING BOTH INTRASPECIFIC COOPERATION AND DISOPERATION. ECOLOGY 35(1), 95-97. [MODEL]

ORIANS, G. H. AND H. S. HORN. 1969. OVERLAP IN FOODS AND FORAGING OF FOUR SPECIES OF BLACKBIRDS IN THE POTHOLES OF CENTRAL WASHINGTON. ECOLOGY 50(5), 930. [PREDATOR-PREY MODEL; PROBABILITY; OVERLAP INDICES]

OSTER, G. AND Y. TAKAHASHI. 1974. MODELS FOR AGE-SPECIFIC INTERACTIONS IN A PERIODIC ENVIRONMENT. ECOL. MONOGR. 44(4), 483-501. [HOST-PARASITE MODEL; BALANCE EQUATIONS; HERBIVORE-RESOURCE MODEL]

PAPENTIN, F. 1973. A DARWINIAN EVOLUTIONARY SYSTEM. III. EXPERIMENTS ON THE EVOLUTION OF FEEDING PATTERNS. J. THEOR. BIOL. 39(2), 431-445.

PARK, T., P. H. LESLIE AND D. B. MERTZ. 1964. GENETIC STRAINS AND COMPETITION IN POPULATIONS OF TRIBOLIUM. PHYSIOL. ZOOL. 37(2), 97-162. [MODEL; LOGISTIC; INTRINSIC RATE OF INCREASE; SERIAL CORRELATION; MONTE CARLO]

PATTEN, B. C. 1961. COMPETITIVE EXCLUSION. SCIENCE 134(3490), 1599-1601.

PEARCE, C. 1970. A NEW DETERMINISTIC MODEL FOR THE INTERACTION BETWEEN PREDATOR AND PREY. BIOMETRICS 26(3), 387-392.

PEARSON, E. S. 1927. THE APPLICATION OF THE THEORY OF DIFFERENTIAL EQUATIONS TO THE SOLUTION OF PROBLEMS CONNECTED WITH THE INTERDEPENDENCE OF SPECIES. BIOMETRIKA 19(1/2) 216-222. [VOLTERRA'S EQUATIONS; LOTKA'S EQUATION; PREDATOR-PREY MODEL; LOGISTIC CURVE]

PHILIP, J. R. 1957. SOCIALITY AND SPARSE POPULATIONS. ECOLOGY 38(1), 107-111.

PHILLIPS, O. M. 1973. THE EQUILIBRIUM AND STABILITY OF SIMPLE MARINE BIOLOGICAL SYSTEMS. I. PRIMARY NUTRIENT CONSUMERS. AMER. NAT. 107(953), 73-93. [VOLTERRA; COMPETITION]

PIANKA, E. R. 1972. R AND K SELECTION OR B AND D SELECTION. AMER. NAT. 106(951), 581-588. [VERHULST-PEARL LOGISTIC EQUATION; BIRTH RATE; DEATH RATE; CARRYING CAPACITY; MODEL]

PIANKA, E. R. 1974. NICHE OVERLAP AND DIFFUSE COMPETITION. PROC. NATL. ACAD. SCI. U.S.A. 71(5), 2141-2145. [EQUATIONS]

PIELOU, D. P. AND E. C. PIELOU. 1967. THE DETECTION OF DIFFERENT DEGREES OF COEXISTENCE. J. THEOR. BIOL. 16(3), 427-437. [TEST TO DISTINGUISH AMONG TYPES OF COEXISTENCE]

PIELOU, E. C. 1972. MEASUREMENT OF STRUCTURE IN ANIMAL COMMUNITIES. PP. 113-133. IN. ECOSYSTEM STRUCTURE AND FUNCTION. (J. A. WIENS, ED.), OREGON STATE UNIVERSITY PRESS, CORVALLIS, OREGON.

PIELOU, E. C. 1972. NICHE WIDTH AND NICHE OVERLAP: A METHOD FOR MEASURING THEM. ECOLOGY 53(4), 687-692. [BRILLOUIN'S INFORMATION MEASURE]

PIELOU, E. C. 1973. GEOGRAPHIC VARIATION IN HOST-PARASITE SPECIFICITY. PP. 103-123. IN. THE MATHEMATICAL THEORY OF THE DYNAMICS OF BIOLOGICAL POPULATIONS. (M. S. BARTLETT AND R. W. HIORNS, ED.S), ACADEMIC PRESS, NEW YORK, NEW YORK. [PROBABILITY; CHI-SQUARE; MEASURE OF ASSOCIATION; VECTOR; MATRIX]

PIELOU, E. C. 1974. BIOGEOGRAPHIC RANGE COMPARISONS AND EVIDENCE OF GEOGRAPHIC VARIATION IN HOST-PARASITE RELATIONS. ECOLOGY 55(6), 1359-1367. [TEST; PROBABILITY]

PIELOU, E. C. 1974. COMPETITION ON AN ENVIRONMENTAL GRADIENT. PP. 184-204. IN. MATHEMATICAL PROBLEMS IN BIOLOGY. LECTURE NOTES IN BIOMATHEMATICS 2. (P. VAN DEN DRIESSCHE, ED.), SPRINGER-VERLAG NEW YORK INC., NEW YORK, NEW YORK. [MODEL; LESLIE MATRIX; STABILITY]

PIMBLEY, G. H., JR. 1974. PERIODIC SOLUTIONS OF PREDATOR-PREY EQUATIONS SIMULATING AN IMMUNE RESPONSE.I. MATH. BIOSCI. 20, 27-51. [BELL'S MODEL]

PIMBLEY, G. H., JR. 1974. PERIODIC SOLUTIONS OF PREDATOR-PREY EQUATIONS SIMULATING AN IMMUNE RESPONSE, II. MATH. BIOSCI. 21, 251-277.

PIMENTEL, D. 1961. ANIMAL POPULATION REGULATION BY THE GENETIC FEED-BACK MECHANISM. AMER. NAT. 95(881), 65-79. [MODEL]

PODOLER, H. AND D. ROGERS. 1975. A NEW METHOD FOR THE IDENTIFICATION OF KEY FACTORS FROM LIFE-TABLE DATA. J. ANIM. ECOL. 44(1), 85-114. [MODEL; CORRELATION COEFFICIENT; GRAPHICAL METHOD; REGRESSION]

PONSSARD, J.-P. 1974. THE VALUE OF INFORMATION IN STRICTLY COMPETITIVE SITUATIONS. INTERN. INST. FOR APPLIED SYSTEMS ANALYSIS, RESEARCH MEMORANDUM RM-74-11, 21 PP. [MODEL; INFORMATION THEORY]

POWERS, J. E. 1974. COMPETITION FOR FOOD: AN EVALUATION OF IVLEV'S MODEL. TRANS. AMER. FISH. SOC. 103(4), 772-776. [COMPUTER; REGRESSION]

RACTLIFFE, L. H., H. M. TAYLOR, J. H. WHITLOCK AND W. R. LYNN. 1969. SYSTEMS ANALYSIS OF A HOST-PARASITE INTERACTION. PARASITOLOGY 59(3), 649-661. [MODEL; SIMULATION]

RAPPORT, D. J. 1971. AN OPTIMIZATION MODEL OF FOOD SELECTION. AMER. NAT. 105(946), 575-587. [PREDATOR-PREY MODEL]

RESCIGNO, A. 1968. THE STRUGGLE FOR LIFE: II. THREE COMPETITORS. BULL. MATH. BIOPHYS. 30, 291-298. [MODEL; DETERMINANT]

RESCIGNO, A. AND I. W. RICHARDSON. 1965. ON THE COMPETITIVE EXCLUSION PRINCIPLE. BULL. MATH. BIOPHYS. 27(SPECIAL ISSUE), 85-89. [VOLTERRA'S MODEL; COMPETITION]

RESCIGNO, A. AND I. W. RICHARDSON. 1967. THE STRUGGLE FOR LIFE; I. TWO SPECIES. BULL. MATH. BIOPHYS. 29, 377-388. [VOLTERRA; PREDATOR-PREY MODEL; MODEL]

RESCIGNO, A. AND K. G. JONES. 1972. THE STRUGGLE FOR LIFE: III. A PREDATOR-PREY CHAIN. BULL. MATH. BIOPHYS. 34, 521-532. [MODEL]

RICH, E. R. 1956. EGG CANNIBALISM AND FECUNDITY IN TRIBOLIUM. ECOLOGY 37(1), 109-120. [MODEL]

RIEBESELL, J. F. 1974. PARADOX OF ENRICHMENT IN COMPETITIVE SYSTEMS. ECOLOGY 55(1), 183-187. [GAUSE'S EQUATIONS; CARRYING CAPACITY; MODEL; MATRIX]

RILEY, G. A. 1953. THEORY OF GROWTH AND COMPETITION IN NATURAL POPULATIONS. J. FISH. RES. BOARD CAN. 10(5), 211-223. [PEARL AND REED LOGISTIC; COMPETITION; VOLTERRA'S EQUATIONS]

ROGERS, D. AND S. HUBBARD. 1974. HOW THE BEHAVIOUR OF PARASITES AND PREDATORS PROMOTES POPULATION STABILITY. PP. 99-119. IN, ECOLOGICAL STABILITY. (M. B. USHER AND M. H. WILLIAMSON, ED.S). HALSTED PRESS, NEW YORK, NEW YORK. [MODEL; NICHOLSON'S MODEL]

ROGERS, D. J. AND M. P. HASSELL. 1974. GENERAL MODELS FOR INSECT PARASITE AND PREDATOR SEARCHING BEHAVIOUR: INTERFENCE. J. ANIM. ECOL. 43(1), 239-253. [COMPUTER; SIMULATION]

ROSENZWEIG, M. L. 1969. WHY THE PREY CURVE HAS A HUMP. AMER. NAT. 103(929), 81-87.

ROSENZWEIG, M. L. 1971. PARADOX OF ENRICHMENT: DESTABILIZATION OF EXPLOITATION ECOSYSTEMS IN ECOLOGICAL TIME. SCIENCE 171(3969), 385-387. [MODEL; LOGISTIC; LOTKA-VOLTERRA; DETERMINISTIC MODEL]

ROSENZWEIG, M. L. AND P. W. STERNER. 1970. POPULATION ECOLOGY OF DESERT RODENT COMMUNITIES: BODY SIZE AND SEED-HUSKING AS BASES FOR HETEROMYID COEXISTENCE. ECOLOGY 51(2), 217-224. [ENERGY EQUATION; COMPUTER]

ROSENZWEIG, M. L. AND R. H. MACARTHUR. 1963. GRAPHICAL REPRESENTATION AND STABILITY CONDITIONS OF PREDATOR-PREY INTERACTIONS. AMER. NAT. 97(895), 209-223.

ROSS, G. G. 1973. A MODEL FOR THE COMPETITIVE GROWTH OF TWO DIATOMS. J. THEOR. BIOL. 42(2), 307-331. [DIFFERENTIAL EQUATION]

ROSS, G. G. 1973. A POPULATION MODEL FOR LIMITED FOOD COMPETITION. J. THEOR. BIOL. 42(2), 333-347. [DETERMINISTIC DIFFERENTIAL EQUATION SYSTEM]

ROTHSTEIN, S. I. 1973. THE NICHE-VARIATION MODEL-IS IT VALID? AMER. NAT. 107(957), 598-620. [COEFFICIENT OF VARIATION; CHI-SQUARE; T-TEST]

ROUGHGARDEN, J. 1971. DENSITY-DEPENDENT NATURAL SELECTION. ECOLOGY 52(3), 453-468.

ROUGHGARDEN, J. 1972. EVOLUTION OF NICHE WIDTH. AMER. NAT. 106(952), 683-718. [MODEL; LIZARDS; COMMUNITY MATRIX]

ROUGHGARDEN, J. 1974. THE FUNDAMENTAL AND REALIZED NIC[illegible] OF A SOLITARY POPULATION. AMER. NAT. 108(960), 232-235. [COMPETITION FUNCTION; VARIANCE; CARRYING CAPACITY; RESOURCE FUNCTION; COMPETITION COEFFICIENT]

ROUGHGARDEN, J. 1974. NICHE WIDTH: BIOGEOGRAPHIC PATTERNS AMONG ANOLIS LIZARD POPULATIONS. AMER. NAT. 108(962), 429-442. [MEASURE; UTILIZATION FUNCTION; COMPETITION COEFFICIENT; REGRESSION MODEL]

ROUGHGARDEN, J. 1974. SPECIES PACKING AND THE COMPETITION FUNCTION WITH ILLUSTRATIONS FROM CORAL REEF FISH. THEOR. POPUL. BIOL. 5(2), 163-186. [LOTKA-VOLTERRA MODEL; MACARTHUR-LEVINS; COMPETITION COEFFICIENT; COMPUTER]

ROUGHGARDEN, J. 1975. POPULATION DYNAMICS IN A STOCHASTIC ENVIRONMENT: SPECTRAL THEORY FOR THE LINEARIZED N-SPECIES LOTKA-VOLTERRA COMPETITION EQUATIONS. THEOR. POPUL. BIOL. 7(1), 1-12. [CARRYING CAPACITY]

ROYAMA, T. 1971. A COMPARATIVE STUDY OF MODELS FOR PREDATION AND PARASITISM. RES. POPUL. ECOL.-KYOTO, SUPPL. NO. 1, 91 PP. (ERRATA: 15(1), 121.)

ROYAMA, T. 1971. EVOLUTIONARY SIGNIFICANCE OF PREDATORS' RESPONSE TO LOCAL DIFFERENCES IN PREY DENSITY: A THEORETICAL STUDY. PP. 344-355. IN, DYNAMICS OF POPULATIONS. (P. J. DEN BOER AND G. R. GRADWELL, EDS.), CENTRE FOR AGRICULTURAL PUBLISHING AND DOCUMENTATION, WAGENINGEN, NETHERLANDS.

ROYAMA, T. 1973. CORRIGENDUM AND ERRATA FOR T. ROYAMA, "A COMPARATIVE STUDY OF MODELS FOR PREDATION AND PARASITISM", RES. POPUL. ECOL.-KYOTO 1971, SUPPLEMENT 1. RES. POPUL. ECOL.-KYOTO 15(1), 121.

RUBIN, E. 1960. THE QUANTITATIVE DATA AND METHODS OF THE REV. T. R. MALTHUS. AMER. STATISTICIAN 14(1), 28-31.

SAILA, S. B. AND J. D. PARRISH. 1972. EXPLOITATION EFFECTS UPON INTERSPECIFIC RELATIONSHIPS IN MARINE ECOSYSTEMS. U. S. NATL. MAR. FISH. SERV. FISH. BULL. 70(2), 383-393. [PREDATOR-PREY MODEL; NETWORK ANALYSIS; GRAPH THEORY; MODEL; MATRIX; COMPETITION]

SALE, P. F. 1974. OVERLAP IN RESOURCE USE, AND INTERSPECIFIC COMPETITION. OECOLOGIA 17(3), 245-256. [EQUATIONS; NICHE OVERLAP]

SALT, G. W. 1966. AN EXAMINATION OF LOGARITHMIC REGRESSION AS A MEASURE OF POPULATION DENSITY RESPONSE. ECOLOGY 47(6), 1035-1039.

SALT, G. W. 1967. PREDATION IN AN EXPERIMENTAL PROTOZOAN POPULATION (WOODRUFFIA - PARAMECIUM) ECOL. MONOGR. 37(2), 113-144. [SEARCHING MODEL; HOLLING MODEL; STANLEY MODEL; SKELLAM'S MODEL]

SAMUELSON, P. A. 1971. GENERALIZED PREDATOR-PREY OSCILLATIONS IN ECOLOGICAL AND ECONOMIC EQUILIBRIUM. PROC. NATL. ACAD. SCI. U.S.A. 68(5), 980-983. [HAMILTONIAN FORM; TWO-SPECIES AND MULTIPLE-SPECIES MODELS; LIMIT CYCLE]

SAMUELSON, P. A. 1974. A BIOLOGICAL LEAST-ACTION PRINCIPLE FOR THE ECOLOGICAL MODEL OF VOLTERRA-LOTKA. PROC. NATL. ACAD. SCI. U.S.A. 71(8), 3041-3044. [PREDATOR-PREY MODEL; HAMILTON'S FUNCTION AND FORM; CANONICAL FORM]

SCHOENER, T. W. 1969. MODELS OF OPTIMAL SIZE FOR SOLITARY PREDATORS. AMER. NAT. 103(931), 277-313.

SCHOENER, T. W. 1969. OPTIMAL SIZE AND SPECIALIZATION IN CONSTANT AND FLUCTUATING ENVIRONMENTS: AN ENERGY-TIME APPROACH. PP. 103-114. IN, DIVERSITY AND STABILITY IN ECOLOGICAL SYSTEMS. BROOKHAVEN SYMPOSIUM IN BIOLOGY NO. 22. [MODEL]

SCHOENER, T. W. 1971. THEORY OF FEEDING STRATEGIES. PP. 369-404. IN, ANNUAL REVIEW OF ECOLOGY AND SYSTEMATICS, VOLUME 2. (R. F. JOHNSTON, P. W. FRANK AND C. D. MICHENER, EDS.). ANNUAL REVIEWS INC., PALO ALTO, CALIFORNIA.

SCHOENER, T. W. 1973. POPULATION GROWTH REGULATED BY INTRASPECIFIC COMPETITION FOR ENERGY OR TIME: SOME SIMPLE REPRESENTATIONS. THEOR. POPUL. BIOL. 4(1), 56-84. [MODEL; LOGISTIC; PREDATOR-PREY MODEL]

SCHOENER, T. W. 1974. COMPETITION AND THE FORM OF HABITAT SHIFT. THEOR. POPUL. BIOL. 6(3), 265-307. [MODEL; LOTKA-VOLTERRA MODEL; MACARTHUR-LEVINS; REGRESSION METHOD]

SCHOENER, T. W. 1974. THE COMPRESSION HYPOTHESIS AND TEMPORAL RESOURCE PARTITIONING. PROC. NATL. ACAD. SCI. U.S.A. 71(10), 4169-4172. [CONTINGENCY MODELS OF FEEDING; COMPETITION]

SCHOENER, T. W. 1974. SOME METHODS FOR CALCULATING COMPETITION COEFFICIENTS FROM RESOURCE-UTILIZATION SPECTRA. AMER. NAT. 108(961), 332-340. [VOLTERRA'S EQUATIONS; GAUSE'S EQUATIONS; LOTKA'S EQUATION; MACARTHUR-LEVINS MEASURE; PREDATOR-PREY MODEL; MACARTHUR'S MODEL]

SCUDO, F. M. 1971. VITO VOLTERRA AND THEORETICAL ECOLOGY. THEOR. POPUL. BIOL. 2(1), 1-23.

SHUGART, H. H., JR. AND B. C. PATTEN 1972. NICHE QUANTIFICATION AND THE CONCEPT OF NICHE PATTERN. PP. 283-327. IN, SYSTEMS ANALYSIS AND SIMULATION IN ECOLOGY, VOLUME II. (B. C. PATTEN, ED.) ACADEMIC PRESS, NEW YORK, NEW YORK. [DISCRIMINANT FUNCTION; MODEL; EUCLIDEAN SPACE; MATRIX; HUTCHINSON'S MODEL; INFORMATION THEORY]

SIMBERLOFF, D. S. 1969. EXPERIMENTAL ZOOGEOGRAPHY OF ISLANDS: A MODEL FOR INSULAR COLONIZATION. ECOLOGY 50(2), 296-314. [STOCHASTIC MODEL; VARIANCE; PROBABILITY; MACARTHUR-WILSON EQUILIBRIUM MODEL; BOSSERT-HOLLAND MODEL; SIMULATION; COMPUTER]

SIMBERLOFF, D. S. AND E. O. WILSON. 1969. EXPERIMENTAL ZOOGEOGRAPHY OF ISLANDS: THE COLONIZATION OF EMPTY ISLANDS. ECOLOGY 50(2), 278-296. [MACARTHUR-WILSON EQUILIBRIUM MODEL; STOCHASTIC MODEL; BOSSERT-HOLLAND MODEL]

SINCLAIR, A. R. E. 1973. REGULATION AND POPULATION MODELS FOR A TROPICAL RUMINANT. EAST AFR. WILDL. J. 11(3/4), 307-316. [LIFE TABLE; LOG TRANSFORMATION; REGRESSION; KEY FACTOR OF VARLEY AND GRADWELL]

SLATKIN, M. 1974. COMPETITION AND REGIONAL COEXISTENCE. ECOLOGY 55(1), 128-134. [MODEL; PREDATOR-PREY MODEL; COHEN'S MODEL; LEVINS AND CULVER MODEL]

SLOBODKIN, L. B. 1953. ON SOCIAL SINGLE SPECIES POPULATIONS. ECOLOGY 34(2), 430-434.

SLOBODKIN, L. B. 1974. PRUDENT PREDATION DOES NOT REQUIRE GROUP SELECTION. AMER. NAT. 108(963), 665-678. [PREDATOR-PREY MODEL; MODEL; REPRODUCTIVE VALUE; INTRINSIC RATE OF INCREASE]

SMITH, F. E. 1963. DENSITY-DEPENDENCE. ECOLOGY 44(1), 220. [CORRELATION COEFFICIENT]

SMITH, R. H. AND R. MEAD. 1974. AGE STRUCTURE AND STABILITY IN MODELS OF PREY-PREDATOR SYSTEMS. THEOR. POPUL. BIOL. 6(3), 308-322. [SIMULATION; STOCHASTIC AND DETERMINISTIC MODELS; COMPUTER]

SMOUSE, P. E. 1971. THE EVOLUTIONARY ADVANTAGES OF SEXUAL DIMORPHISM. THEOR. POPUL. BIOL. 2(4), 469-481. [MODEL; VOLTERRA'S MODEL]

SMOUSE, P. E. 1972. THE CANONICAL ANALYSIS OF MULTIPLE SPECIES HYBRIDIZATION. BIOMETRICS 28(2), 361-371.

SOLOMON, M. E. 1968. LOGARITHMIC REGRESSION AS A MEASURE OF POPULATION DENSITY RESPONSE: COMMENT ON A REPORT BY G. W. SALT. ECOLOGY 49(2), 357-358.

SOUTHWOOD, T. R. E., R. M. MAY, M. P. HASSELL AND G. R. CONWAY. 1974. ECOLOGICAL STRATEGIES AND POPULATION PARAMETERS. AMER. NAT. 108(964), 791-804. [R-STRATEGY; K-STRATEGY; LOGISTIC EQUATION; DIFFERENCE EQUATION; HABITAT STABILITY]

ST. AMANT, J. L. S. 1970. THE DETECTION OF REGULATION IN ANIMAL POPULATIONS. ECOLOGY 51(5), 823-828. [LOG TRANSFORMATION; REGRESSION; EXPECTED VALUE]

ST. AMANT, J. L. S. 1970. NOTES ON THE MATHEMATICS OF PREDATOR - PREY INTERACTIONS. M. A. THESIS. UNIVERSITY OF CALIFORNIA, SANTA BARBARA, CALIFORNIA.

STERN, K. AND L. ROCHE. 1974. I. THE ECOLOGICAL NICHE. 1. FORMAL CONCEPT OF THE NICHE. PP. 3-10. IN. GENETICS OF FOREST ECOSYSTEMS. SPRINGER-VERLAG NEW YORK INC., NEW YORK, NEW YORK. [HYPERSPACE OF HUTCHINSON; VOLTERRA-GAUSE COMPETITION MODEL; SET THEORY; MACARTHUR'S MODEL; LEVINS' MEASURE]

STEWART, F. M. 1971. EVOLUTION OF DIMORPHISM IN A PREDATOR-PREY MODEL. THEOR. POPUL. BIOL. 2(4), 493-506. [COMPUTER]

STEWART, F. M. AND B. R. LEVIN. 1973. PARTITIONING OF RESOURCES AND THE OUTCOME OF INTERSPECIFIC COMPETITION: A MODEL AND SOME GENERAL CONSIDERATIONS. AMER. NAT. 107(954), 171-198.

STINNER, R. E. 1972. PRIMARY CONSUMERS OF TREES-A MODEL FOR INSECT POPULATION INTERACTIONS. PP. 185-188. IN. MODELING THE GROWTH OF TREES. (C. E. MURPHY, JR., J. D. HESKETH AND B. R. STRAIN, EDS.). AEC REPT. EDFB-IBP-72-11. OAK RIDGE NATL. LAB., OAK RIDGE, TENNESSEE. [POISSON DISTRIBUTION]

STOY, R. H. 1932. APPENDIX: SUPERPARASITISM BY COLLYRIA CALCITRATOR, GRAV. BULL. ENTOMOL. RES. 23(2), 215-216. [HOST-PARASITE MODEL]

STRAW, R. M. 1972. A MARKOV MODEL FOR POLLINATOR CONSTANCY AND COMPETITION. AMER. NAT. 106(951), 597-620. [POPULATION MODEL; MATRIX; PROBABILITY]

STREBEL, D. E. AND N. S. GOEL 1973. ON THE ISOCLINE METHODS FOR ANALYZING PREY-PREDATOR INTERACTIONS. J. THEOR. BIOL. 39(1), 211-239. [VOLTERRA]

STRICKFADEN, W. B. AND B. A. LAWRENCE. 1975. SOLVABLE LIMIT CYCLE IN A VOLTERRA-TYPE MODEL OF INTERACTING POPULATIONS. MATH. BIOSCI. 23(3/4), 273-279. [VAN DER POL EQUATION]

STROBECK, C. 1973. N SPECIES COMPETITION. ECOLOGY 54(3), 650-654. [LOTKA-VOLTERRA EQUATION; STABILITY THEORY; DIFFERENTIAL EQUATION]

SUTHERST, R. W., K. B. W. UTECH, M. J. DALLWITZ AND J. D. KERR. 1973. INTRA-SPECIFIC COMPETITION OF BOOPHILUS MICROPLUS (CANESTRINI) ON CATTLE. J. APPL. ECOL. 10(3), 855-862. [MODEL]

TAYLOR, L. R., R. A. FRENCH AND E. D. M. MACAULAY. 1973. LOW-ALTITUDE MIGRATION AND DIURNAL FLIGHT PERIODICITY; THE IMPORTANCE OF PLUSIA GAMMA L. (LEPIDOPTERA: PLUSIIDAE). J. ANIM. ECOL. 42(3), 751-760. [COMPUTER GRAPHICS]

TAYLOR, N. W. 1965. A THEORETICAL STUDY OF POPULATION REGULATION IN TRIBOLIUM CONFUSUM. ECOLOGY 46(3), 334-340. [DETERMINISTIC MODEL]

TAYLOR, N. W. 1968. A MATHEMATICAL MODEL FOR TWO TRIBOLIUM POPULATIONS IN COMPETITION. ECOLOGY 49(5), 843-848. [COMPUTER; PROBABILITY]

TAYLOR, R. J. 1974. ROLE OF LEARNING IN INSECT PARASITISM. ECOL. MONOGR. 44(1), 89-104. [PREDATOR-PREY MODEL; MODEL]

TEN, V. S. 1970. ON POLYPHAGISM. MARINE BIOL. 5(3), 169-171. [VOLTERRA'S MODEL; MODEL OF POLYPHAGISM; PREDATOR-PREY MODEL]

THOMAS, V. J. 1970. A MATHEMATICAL APPROACH TO FITTING PARAMETERS IN A COMPETITION MODEL. J. APPL. ECOL. 7(3), 487-496.

THOMPSON, W. A., JR. AND G. H. WEISS. 1963. TRANSIENT BEHAVIOR OF POPULATION DENSITY WITH COMPETITION FOR RESOURCES. BULL. MATH. BIOPHYS. 25, 203-211. [SKELLAM AND KIMURA EQUATIONS]

THOMPSON, W. R. 1939. BIOLOGICAL CONTROL AND THE THEORIES OF THE INTERACTIONS OF POPULATIONS. PARASITOLOGY 31(3), 299-388.

THOMPSON, W. R. 1960. NOTES ON THE VOLTERRA EQUATIONS. CAN. ENTOMOL. 92(8), 582-594.

TRUBATCH, S. L. AND A. FRANCO. 1974. CANONICAL PROCEDURES FOR POPULATION DYNAMICS. J. THEOR. BIOL. 48(2), 299-324. [LAGRANGIANS; HAMILTONIAN FORM; INTERACTING SPECIES; MODEL; VOLTERRA-LOTKA EQUATIONS; GOMPERTZ EQUATION; VERHULST EQUATIONS; LESLIE AND GOWER EQUATIONS]

TURNER, J. E. AND D. J. RAPPORT. 1974. AN ECONOMIC MODEL OF POPULATION GROWTH AND COMPETITION IN NATURAL COMMUNITIES. PP. 236-240. IN, MATHEMATICAL PROBLEMS IN BIOLOGY. LECTURE NOTES IN BIOMATHEMATICS 2. (P. VAN DEN DRIESSCHE, ED.), SPRINGER-VERLAG NEW YORK INC., NEW YORK. [RESOURCE COMPETITION AND CONSUMPTION]

UTIDA, S. 1950. ON THE EQUILIBRIUM STATE OF THE INTERACTING POPULATION OF AN INSECT AND ITS PARASITE. ECOLOGY 31(2), 165-175. [LEXIS' RATIO; PROBABLE AFTEREFFECT; KAMEDA'S EQUATION]

UTIDA, S. 1953. POPULATION FLUCTUATION IN THE SYSTEM OF HOST-PARASITE INTERACTION. RES. POPUL. ECOL.-KYOTO 2, 22-46. [MODEL]

UTIDA, S. 1955. FLUCTUATIONS IN THE INTERACTING POPULATIONS OF HOST AND PARASITE IN RELATION TO THE BIOTIC POTENTIAL OF THE HOST. ECOLOGY 36(2), 202-206. [LOTKA-VOLTERRA MODEL; LOGISTIC GROWTH]

UTIDA, S. 1955. POPULATION FLUCTUATION IN THE SYSTEM OF HOST-PARASITE INTERACTION. MEM. COLL. AGRIC. KYOTO UNIV. NO. 71 (ENTOMOL. SERIES NO. 11), 34 PP. [MODEL]

VALIELA, I. 1971. FOOD SPECIFICITY AND COMMUNITY SUCCESSION: PRELIMINARY ORNTHOLOGICAL EVIDENCE FOR A GENERAL FRAMEWORK. GENERAL SYSTEMS 16, 77-84. [MARGALEFF'S CONCEPT OF STABILITY; MCARTHUR'S CONCEPT OF STABILITY; MODEL; PROBABILITY]

VANDERMEER, J. H. 1969. THE COMPETITIVE STRUCTURE OF COMMUNITIES: AN EXPERIMENTAL APPROACH WITH PROTOZOA. ECOLOGY 50(3), 362-371. [LOTKA-VOLTERRA EQUATION; PREDATOR-PREY MODEL; GAUSE'S EQUATIONS; LOGISTIC EQUATION]

VANDERMEER, J. H. 1970. THE COMMUNITY MATRIX AND THE NUMBER OF SPECIES IN A COMMUNITY. AMER. NAT. 104(935), 73-83. [LOTKA-VOLTERRA EQUATION; COMPETITION; DETERMINANT]

VANDERMEER, J. H. 1972. NICHE THEORY. PP. 107-132. IN, ANNUAL REVIEW OF ECOLOGY AND SYSTEMATICS, VOLUME 3. (R. F. JOHNSTON, P. W. FRANK AND C. D. MICHENER, EDS.), ANNUAL REVIEWS INC., PALO ALTO, CALIFORNIA.

VANDERMEER, J. H. 1973. GENERALIZED MODELS OF TWO SPECIES INTERACTIONS: A GRAPHICAL ANALYSIS. ECOLOGY 54(4), 809-818. [MATRIX]

VANDERMEER, J. H. 1973. ON THE REGIONAL STABILIZATION OF LOCALLY UNSTABLE PREDATOR-PREY RELATIONSHIPS. J. THEOR. BIOL. 41(1), 161-170. [MODEL]

VARLEY, G. C. 1970. THE NEED FOR LIFE TABLES FOR PARASITES AND PREDATORS. PP. 59-68. IN, CONCEPTS OF PEST MANAGEMENT. (R. L. RABB AND F. E. GUTHRIE, EDS.), NORTH CAROLINA STATE UNIVERSITY, RALEIGH, NORTH CAROLINA. [NICHOLSON-BAILEY MODEL; MODEL]

VINE, I. 1971. RISK OF VISUAL DETECTION AND PURSUIT BY A PREDATOR AND THE SELECTIVE ADVANTAGE OF FLOCKING BEHAVIOUR. J. THEOR. BIOL. 30(2), 405-422. [MODEL]

VINE, I. 1973. DETECTION OF PREY FLOCKS BY PREDATORS. J. THEOR. BIOL. 40(2), 207-210. [MODEL]

VINE, I. 1974. REPLY TO G. F. ESTABROOK AND H. H. ROBINSON. J. THEOR. BIOL. 47(1), 249-250. [VINE'S PREDATOR-PREY VISUAL DETECTION MODEL]

VOLTERRA, V. 1926. VARIAZIONI E FLUTTUAZIONI DEL NUMERO D'INDIVIDUI IN SPECIE ANIMALI CONVIVENTI. MEM. R. ACCAD. NAZ. DEI. LINCEI. SER VI, 2, 31-113.

VOLTERRA, V. 1927. VARIAZIONI E FLUTTUAZIONI DEL NUMERO D'INDIVIDUI IN SPECIE ANIMALI CONVIVENTI. MEMORIE DEL R. COMITATO TALASSOGRAFICO ITALIANO 131, 1-142. (ENGLISH TRANSL. BY M. G. WELLS 1928, J. CONS. PERM. INT. EXPLOR. MER 3, 3-51.; 1931. PP. 409-448. IN, ANIMAL ECOLOGY. (R. N. CHAPMAN), MCGRAW-HILL BOOK CO., NEW YORK, NEW YORK). [PREDATOR-PREY MODEL; COMPETITION]

WALTER, C. 1974. THE GLOBAL ASYMPTOTIC STABILITY OF PREY-PREDATOR SYSTEMS WITH SECOND-ORDER DISSIPATION. BULL. MATH. BIOL. 36(2), 215-217. [MODEL; LOTKA-VOLTERRA MODEL; SIMULATION]

WANGERSKY, P. J. 1965. A MODEL OF COMPETITION BETWEEN GENETICALLY VARIABLE POPULATIONS. AEC REPT. NYO-3216-1, FINAL REPT. CONTRACT AT(30-1) - 3216. 8 PP.

WARE, D. M. 1973. RISK OF EPIBENTHIC PREY TO PREDATION BY RAINBOW TROUT (SALMO GAIRDNERI). J. FISH. RES. BOARD CAN. 30(6), 787-797. [PREDATOR-PREY MODEL; MODEL]

WATSON, A. 1971. KEY FACTOR ANALYSIS, DENSITY DEPENDENCE AND POPULATION LIMITATION IN RED GROUSE. PP. 548-559. IN, DYNAMICS OF POPULATIONS. (P. J. DEN BOER AND G. R. GRADWELL, EDS.), CENTRE FOR AGRICULTURAL PUBLISHING AND DOCUMENTATION, WAGENINGEN, NETHERLANDS.

WATT, K. E. F. 1959. A MATHEMATICAL MODEL FOR THE EFFECT OF DENSITIES OF ATTACKED AND ATTACKING SPECIES ON THE NUMBER ATTACKED. CAN. ENTOMOL. 91(3), 129-144.

WATT, K. E. F. 1964. COMMENTS ON FLUCTUATIONS OF ANIMAL POPULATIONS AND MEASURES OF COMMUNITY STABILITY. CAN. ENTOMOL. 96(11), 1434-1442. [STERLING'S FORMULA; INFORMATION CONTENT; REGRESSION; COMPUTER]

WATT, K. E. F. 1964. DENSITY DEPENDENCE IN POPULATION FLUCTUATIONS. CAN. ENTOMOL. 96(8), 1147-1148. [LOG TRANSFORMATION; CURVILINEAR REGRESSION; MULTIPLE REGRESSION]

WEISS, G. H. 1963. COMPARISON OF A DETERMINISTIC AND A STOCHASTIC MODEL FOR INTERACTION BETWEEN ANTAGONISTIC SPECIES. BIOMETRICS 19(4), 595-602.

WERNER, E. E. AND D. J. HALL. 1974. OPTIMAL FORAGING AND THE SIZE SELECTION OF PREY BY THE BLUEGILL SUNFISH (LEPOMIS MACROCHIRUS). ECOLOGY 55(5), 1042-1052. [MODEL]

WILBERT, H. 1970. FEIND-BEUTE-SYSTEME IN KYBERNETISCHER SICHT. OECOLOGIA 5, 347-373. [CYBERNETICS; PREDATOR-PREY MODEL; MODEL]

WILBERT, H. 1971. FEEDBACK CONTROL BY COMPETITION. PP. 174-186. IN, DYNAMICS OF POPULATIONS. (P. J. DEN BOER AND G. R. GRADWELL, EDS.). CENTRE FOR AGRICULTURAL PUBLISHING AND DOCUMENTATION, WAGENINGEN, NETHERLANDS.

WILBUR, H. M. 1972. COMPETITION, PREDATION, AND THE STRUCTURE OF THE AMBYSTOMA-RANA SYLVATICA COMMUNITY. ECOLOGY 53(1), 3-21. [LOTKA-VOLTERRA MODEL; COMPETITION; EUCLIDEAN DISTANCE; LOGISTIC EQUATION]

WILBUR, H. M., D. W. TINKLE AND J. P. COLLINS. 1974. ENVIRONMENTAL CERTAINTY, TROPHIC LEVEL, AND RESOURCE AVAILABILITY IN LIFE HISTORY EVOLUTION. AMER. NAT. 108(964), 805-817. [R-SELECTION; K-SELECTION; CARRYING CAPACITY; INTRINSIC RATE OF INCREASE]

WILLIAMSON, G. B. AND C. E. NELSON. 1972. FITNESS SET ANALYSIS OF MIMETIC ADAPTIVE STRATEGIES. AMER. NAT. 106(950), 525-537. [MODEL]

WILLIAMSON, M. H. 1957. AN ELEMENTARY THEORY OF INTERSPECIFIC COMPETITION. NATURE 180(4583), 422-425.

WUENSCHER, J. E. 1969. NICHE SPECIFICATION AND COMPETITION MODELING. J. THEOR. BIOL. 25(3), 436-443. [HYPERVOLUME; VECTOR SPACE]

YEATON, R. I. AND M. L. CODY. 1974. COMPETITIVE RELEASE IN ISLAND SONG SPARROW POPULATIONS. THEOR. POPUL. BIOL. 5(1), 42-58. [MACARTHUR; LEVINS; MATRIX; COMPETITION]

YORKE, J. A. AND W. N. ANDERSON, JR. 1973. PREDATOR-PREY PATTERNS (VOLTERRA-LOTKA EQUATIONS). PROC NATL. ACAD. SCI. U.S.A. 70(7), 2069-2071. [STABILITY]

ZIMKA, J. R. 1974. PREDATION OF FROGS, RANA ARVALIS NILSS., IN DIFFERENT FOREST SITE CONDITIONS. EKOLOGIYA POLSKA-SERIA A 22(1), 31-63. [ACTIVITY INDEX; PRESSURE INDEX; PREFERENCE INDEX]

COMPUTERS AND DATA-PROCESSING

ABRAMSON, N. J. 1963. COMPUTER PROGRAMS FOR FISHERIES PROBLEMS. TRANS. AMER. FISH. SOC. 92(3), 310.

ABRAMSON, N. J. 1965. VON BERTALANFFY GROWTH CURVE II, IBM 7094, UNIVAC 1107, FORTRAN IV. TRANS. AMER. FISH. SOC. 94(2), 195-196.

ABRAMSON, N. J. 1971. COMPUTER PROGRAMS FOR FISH STOCK ASSESSMENT. FAO FISHERIES TECHNICAL PAPER NO. 101, III, VARIOUS PAGES. [GROWTH CURVES; POPULATION MODEL]

ADAMS, L. 1969. USING COMPUTERS IN WILDLIFE MANAGEMENT. PP. 83-94. IN, WILDLIFE MANAGEMENT TECHNIQUES. (3RD REVISED EDITION). (R. H. GILES, JR., ED.). THE WILDLIFE SOCIETY, WASHINGTON, D. C.

ADAMS, R. P. 1970. CONTOUR MAPPING AND DIFFERENTIAL SYSTEMATICS OF GEOGRAPHIC VARIATION. SYST. ZOOL. 19(4), 385-390. [ALGORITHM; COMPUTER]

ALLEN, K. R. 1966. FITTING OF VON BERTALANFFY GROWTH -CURVES, IBM 709, 7094, FORTRAN IV. TRANS. AMER. FISH. SOC. 95(2), 231-232.

ALLEN, K. R. 1967. COMPUTER PROGRAMMES AVAILABLE AT ST. ANDREWS BIOLOGICAL STATION. FISH. RES. BOARD CAN. TECH. REPT. NO. 20.

ALLEN, K. R. 1969. AN APPLICATION OF COMPUTERS TO THE ESTIMATION OF EXPLOITED POPULATIONS. J. FISH. RES. BOARD CAN. 26(1), 179-189.

AMOS, M. H. 1966. COMPUTER USE IN FISHERIES RESEARCH. (ABSTRACT). BIOMETRICS 22(4), 961.

ANDERSON, D. R., C. F. KIMBALL AND F. R. FIEHRER. 1974. A COMPUTER PROGRAM FOR ESTIMATING SURVIVAL AND RECOVERY RATES. J. WILDL. MANAGE. 38(2), 369-370. [STOCHASTIC MODEL]

ANGELTON, G. M., C. D. BONHAM AND L. L. SHANNON. 1966. DIRECT PROCESSING OF FIELD DATA. J. RANGE MANAGE. 19(3), 143-144. [COMPUTER]

ANONYMOUS. 1965. DESK-TOP COMPUTER KEEPS TRACK OF OUR FISH POPULATION. COMPUTERS AND AUTOMATION, 14(2), 29.

ANONYMOUS. 1965. A HOST - PARASITE PROGRAM. INSTRUM. CONT. SYST. 38(6), 129. [COMPUTER; TIME LAG MODEL]

ANONYMOUS. 1971. ANNOUNCING FORTRAN IV PROGRAM FOR COMPUTING AND GRAPHING TREE GROWTH PARAMETERS FROM STEM ANALYSIS. FOREST SCI. 17(1), 102.

ANSCOMBE, F. J. 1971. KEYNOTE ADDRESS: A CENTURY OF STATISTICAL SCIENCE, 1890-1990. PP. XI-XVI. IN, STATISTICAL ECOLOGY, VOLUME 1. SPATIAL PATTERNS AND STATISTICAL DISTRIBUTIONS. (G. P. PATIL, E. C. PIELOU AND W. E. WATERS, EDS.). PENNSYLVANIA STATE UNIVERSITY PRESS, UNIVERSITY PARK, PENNSYLVANIA. [COMPUTER; HISTORY]

ARMSTRONG, J. S. 1971. MODELLING OF A GRAZING SYSTEM. PROC. ECOL. SOC. AUSTRALIA 6, 194-202. [COMPUTER]

ASKEW, A. J., W. W-G. YEH AND W. A. HALL. 1971. USE OF MONTE CARLO TECHNIQUES IN THE DESIGN AND OPERATION OF A MULTIPURPOSE RESERVOIR SYSTEM. WATER RESOUR. RES. 7(4), 819-826. [COMPUTER]

BEAUCHAMP, J. J. AND J. S. OLSON. 1973. CORRECTIONS FOR BIAS IN REGRESSION ESTIMATES AFTER LOGARITHMIC TRANSFORMATION. ECOLOGY 54(6), 1403-1407. [COMPUTER]

BENSON, D. A. 1964. A WILDLIFE BIOLOGIST LOOKS AT SAMPLING, DATA PROCESSING, AND COMPUTERS. CANADIAN WILDLIFE SERVICE OCCASSIONAL PAPER NO. 6. 16 PP.

BERUDE, C. L. AND N. J. ABRAMSON. 1972. RELATIVE FISHING POWER, CDC 6600, FORTRAN IV. TRANS. AMER. FISH. SOC. 101(1), 133.

BEVAN, D. E. 1969. THE USE OF COMPUTERS IN THE COLLEGE OF FISHERIES, UNIVERSITY OF WASHINGTON. PROBLEMS OF ICHTHYOLOGY (VOPROSY IKHTIOLOGII). TRANS. AMER. FISH. SOC. 9(6), 810-820. (ENGLISH TRANSLATION).

BLEDSOE, L. J. AND G. M. VAN DYNE. 1969. EVALUATION OF A DIGITAL COMPUTER METHOD FOR ANALYSIS OF COMPARTMENTAL MODELS OF ECOLOGICAL SYSTEMS. AEC REPT. ORNL/TM-2414. OAK RIDGE NATL. LAB., OAK RIDGE, TENNESSEE. 60 PP.

BONHAM, C. D. 1971. ECMAP A COMPUTER PROGRAM FOR MAPPING ECOLOGICAL DATA. COLORADO STATE UNIVERSITY, FT. COLLINS, COLORADO. RANGE SCIENCE SERIES NO. 9. 45 PP.

BOSSERT, W. 1969. COMPUTER TECHNIQUES IN SYSTEMATICS. PP. 595-605. IN, SYSTEMATIC BIOLOGY. NATIONAL ACADEMY OF SCIENCES, WASHINGTON, D. C.

BRENNAN, R. D., C. T. DE WIT, W. A. WILLIAMS AND E. V. QUATTRIN. 1970. THE UTILITY OF A DIGITAL SIMULATION LANGUAGE FOR ECOLOGICAL MODELLING. OECOLOGIA 4, 113-132. [COMPUTER]

BROGDEN, W. B., J. J. CECH, JR. AND C. H. OPPENHEIMER. 1974. A COMPUTERIZED SYSTEM FOR THE ORGANIZED RETRIEVAL OF LIFE HISTORY INFORMATION. ON. CHESAPEAKE SCIENCE 15(4), 250-254. [COMPUTER]

BROTZMAN, R. L. AND R. H. GILES, JR. 1966. ELECTRONIC DATA PROCESSING OF CAPTURE-RECAPTURE AND RELATED ECOLOGICAL DATA. J. WILDL. MANAGE. 30(2), 286-292. [COMPUTER; SCHNABEL METHOD OF ESTIMATION; SCHUMACHER AND ESCHMEYER METHOD ; LINCOLN-PETERSEN INDEX]

BROWN, D. J. AND L. L. LOW. 1974. THREE-DIMENSIONAL COMPUTER GRAPHICS IN FISHERIES SCIENCE. J. FISH. RES. BOARD CAN. 31(12), 1927-1935.

BUCKLEY, P. A. AND J. T. HANCOCK, JR. 1968. EQUATIONS FOR ESTIMATING AND A SIMPLE COMPUTER PROGRAM FOR GENERATING UNIQUE COLOR- AND ALUMINUM BAND SEQUENCES. BIRD-BANDING 39(2), 123-129.

BUFFINGTON, C. D. 1967. A COMPUTERIZED SYSTEM OF HARVEST ANALYSIS FOR IDAHO BIG GAME MANAGEMENT. M. S. THESIS, UNIVERSITY OF IDAHO, MOSCOW, IDAHO. 213 PP.

CADBURY, D. A., J. G. HAWKES AND R. C. READETT. 1971. METHODS OF RECORDING AND PROCESSING THE DATA. PP. 61-82. IN, A COMPUTER MAPPED FLORA: A STUDY OF THE COUNTY OF WARWICKSHIRE. (D. A. CADBURY ET AL., EDS.). ACADEMIC PRESS, NEW YORK, NEW YORK.

CANCELA DA FONSECA, J. P. AND P. DE BAILLIENCOURT. 1974. DIGITAL COMPUTER ANALYSIS OF ECOLOGICAL DATA STORED IN A RECORD CENTRE: STATING THE ACTUAL WORK. INTECOL BULLETIN 1974(4), 48-51.

CAPRIO, J. M., R. J. HOPP AND J. S. WILLIAMS. 1974. COMPUTER MAPPING IN PHENOLOGICAL ANALYSIS. PP. 77-82. IN, PHENOLOGY AND SEASONALITY MODELING. (H. LIETH, ED.). SPRINGER-VERLAG NEW YORK INC., NEW YORK, NEW YORK.

CASTRUCCIO, P. 1972. ECOLOGY AND THE COMPUTER. PP. 317-334. IN, ECOLOGY IN THEORY AND PRACTICE. (J. BENTHOLL, ED.). VIKING PRESS, NEW YORK, NEW YORK.

CESKA, A. AND H. ROEMER. 1971. A COMPUTER PROGRAM FOR IDENTIFYING SPECIES-RELEVE GROUPS IN VEGETATION STUDIES. VEGETATIO 23(3-4), 255-277.

CHADWICK, H. K. 1966. AGE AND GROWTH PROGRAM, UNIVAC 1107, FORTRAN IV. TRANS. AMER. FISH. SOC. 95(2), 229-230.

CHAPPELLE, D. E. 1966. ECONOMIC MODEL BUILDING AND COMPUTERS IN FORESTRY RESEARCH. J. FORESTRY 64(5), 329-333.

CHAPPELLE, D. E. 1966. TIMBER MANAGEMENT PLANNING. USER'S MANUAL FOR ARVOL COMPUTER PROGRAMMING (VERSION 5).

CIBULA, W. G. 1973. APPLICATIONS OF REMOTELY SENSED MULTISPECTRAL DATA TO AUTOMATED ANALYSIS OF MARSHLAND VEGETATION. (ABSTR.). BULL. ECOL. SOC. AMER. 54(1), 17. [COMPUTER]

CONWAY, G. R., N. R. GLASS AND J. C. WILCOX. 1970. FITTING NONLINEAR MODELS TO BIOLOGICAL DATA BY MARQUARDT'S ALGORITHM. ECOLOGY 51(3), 503-507. [COMPUTER; LOGISTIC]

COWARDIN, L. M. AND D. A. DAVENPORT. 1973. COMPUTERIZED SYSTEM FOR ORGANIZING AND MAINTAINING FILES OF BANDING DATA. BIRD-BANDING 44(3), 187-195. [COMPUTER]

DAHLBERG, M. L. 1966. LENGTH-FREQUENCY ANALYSIS IBM 7090/7094, FORTRAN II OR FORTRAN IV. TRANS. AMER. FISH. SOC. 95(3), 331. [COMPUTER]

DALLWITZ, M. J. 1974. A FLEXIBLE COMPUTER PROGRAM FOR GENERATING IDENTIFICATION KEYS. SYST. ZOOL. 23(1), 50-57.

DEAN, F. C. AND G. A. GALLAWAY. 1965. A FORTRAN PROGRAM FOR POPULATION STUDY WITH MINIMAL COMPUTER TRAINING. J. WILDL. MANAGE. 29(4), 892-894.

DOI, T. 1957. AN ANALOGUE COMPUTER FOR ANALYZING THE PROPERTIES OF MARINE RESOURCES AND PREDICTING CATCHES. JOINT SCI. MEET., INTERN. COMM. NORTHWEST ATLANTIC FISHERIES INTERN. COUNC. EXPLOR. SEA, FOOD AGR. ORGAN. UN, LISBON, PORTUGAL, DOCUMENT P 12, 14 PP.

DUNCAN, K. W. 1971. A GENERALIZED COMPUTER PROGRAM IN FORTRAN IV FOR LISTING ALL POSSIBLE COLOR BAND PERMUTATIONS. BIRD-BANDING 42(4), 279-287.

DURFEE, R. C. 1974. ORRMIS - OAK RIDGE REGIONAL MODELING INFORMATION SYSTEM PART I. AEC REPT. ORNL-NSF-EP-73, OAK RIDGE NATL. LAB., OAK RIDGE, TENNESSEE. IX, 95 PP. [COMPUTER]

ELLIS, D. V. 1968. A SERIES OF COMPUTER PROGRAMS FOR SUMMARISING DATA FROM QUANTITATIVE BENTHIC INVESTIGATIONS. J. FISH. RES. BOARD CAN. 25(8), 1737-1738.

ELLIS, E. J. 1970. A COMPUTER ANALYSIS OF FAWN SURVIVAL IN THE PRONGHORN ANTELOPE. DISSERTATION. UNIVERSITY OF CALIFORNIA, DAVIS, CALIFORNIA.

FAST, A. W. 1967. CORRELATION - REGRESSION WITH SCATTERGRAM, IBM 1620, FORTRAN II-D. TRANS. AMER. FISH. SOC. 96(2), 230.

FEE, E. J. 1973. A DIGITAL COMPUTER PROGRAM FOR CALCULATING INTEGRAL PRIMARY PRODUCTION IN VERTICALLY STRATIFIED WATERBODIES. FISH. RES. BOARD CAN. TECH. REPT. NO. 376. 14 PP.

FERNALD, R. D. AND P. HEINECKE. 1974. A COMPUTER COMPATIBLE MULTI-PURPOSE EVENT RECORDER. BEHAVIOUR 48(3/4), 268-275.

FERRIS, K. H. AND J. GY. FABOS. 1974. THE UTILITY OF COMPUTERS IN LANDSCAPE PLANNING: THE SELECTION AND APPLICATION OF A COMPUTER MAPPING AND ASSESSMENT SYSTEM FOR THE METROPOLITAN LANDSCAPE PLANNING MODEL (METLAND). MASS. AGRIC. EXPT. STN. BULL. NO. 617. UNIVERSITY OF MASSACHUSETTS, AMHERST, MASSACHUSETTS. 116 PP.

FLACK, J. E. AND D. A. SUMMERS. 1971. COMPUTER-AIDED CONFLICT RESOLUTION IN WATER RESOURCE PLANNING: AN ILLUSTRATION. WATER RESOUR. RES. 7(6), 1410-1414. [COGNOGRAPH]

FOX, W. W., JR. 1971. USER'S GUIDE TO FPW: A COMPUTER PROGRAM FOR ESTIMATING RELATIVE FISHING POWER AND RELATIVE POPULATION DENSITY BY THE METHOD OF ANALYSIS OF VARIANCE. QUANTITATIVE SCIENCE PAPER NO. 27. CENTER FOR QUANTITATIVE SCIENCE IN FORESTRY, FISHERES AND WILDLIFE, UNIVERSITY OF WASHINGTON, SEATTLE, WASHINGTON. MIMEO. 6 PP.

FOX, W. W., JR. 1971. USER'S GUIDE TO GSPFIT: A PROGRAM FOR ESTIMATING THE PARAMETERS OF THE GENERALIZED STOCK PRODUCTION MODEL USING GULLAND'S METHOD OF EQUILIBRIUM APPROXIMATION. QUANTITATIVE SCIENCE PAPER NO. 25. CENTER FOR QUANTITATIVE SCIENCE IN FORESTRY, FISHERIES AND WILDLIFE, UNIVERSITY OF WASHINGTON, SEATTLE, WASHINGTON. MIMEO. 5 PP.

FOX, W. W., JR. 1971. USER'S GUIDE TO GSPFIT: A COMPUTER PROGRAM FOR ESTIMATING THE PARAMETERS OF THE CHAPMAN-RICHARDS GROWTH FUNCTION. QUANTITATIVE SCIENCE PAPER NO. 24. CENTER FOR QUANTITATIVE SCIENCE IN FORESTRY, FISHERIES AND WILDLIFE, UNIVERSITY OF WASHINGTON, SEATTLE, WASHINGTON. MIMEO. 4 PP.

FRAYER, W. E. (ED.) 1968. SNOOP: A COMPUTER PROGRAM FOR 2-AND 3-DIMENSIONAL PLOTTING. U. S. FOR. SERV. NORTHEASTERN FOREST EXPT. STN. RES. PAPER NE-91. 24 PP.

FREESE, F. 1959. DESK CALCULATOR OR ELECTRONIC COMPUTER?. PP. 127-132. IN. TECHNIQUES AND METHODS OF MEASURING UNDERSTORY VEGETATION. (ANONYMOUS, ED.). U. S. FOREST SERVICE, WASHINGTON, D. C.

FRITTS, H. C. 1962. AN APPROACH TO DENDROCLIMATOLOGY: SCREENING BY MEANS OF MULTIPLE REGRESSION TECHNIQUES. J. GEOPHYS. RES. 67(4), 1413-1420. [MODEL; COMPUTER]

FRITTS, H. C. 1963. COMPUTER PROGRAMS FOR TREE RING RESEARCH. TREE-RING BULL. 25(3/4), 2-7.

FRITTS, H. C., J. E. MOSIMANN AND C. P. BOTTORFF. 1969. A REVISED COMPUTER PROGRAM FOR STANDARDIZING TREE-RING SERIES. TREE-RING BULL. 29(1-2), 15-20. [EXPONENTIAL MODEL]

GALLOPIN, G. C. 1971. A GENERALIZED MODEL OF A RESOURCE-POPULATION SYSTEM. I. GENERAL PROPERTIES. OECOLOGIA 7(4), 382-413. [COMPUTER]

GERKING, S. D. 1965. TWO COMPUTER PROGRAMS FOR AGE AND GROWTH STUDIES. PROG. FISH-CULT. 27(2), 59-66.

GERMERAAD, J. H. AND J. MULLER. 1970. A COMPUTER-BASED NUMERICAL CODING SYSTEM FOR THE DESCRIPTION OF POLLEN GRAINS AND SPORES. REV. PALAEOBOT. PALYNOL. 10(3), 175-202.

GILMER, D. S., S. E. MILLER AND L. M. COWARDIN. 1973. ANALYSIS OF RADIOTRACKING DATA USING DIGITIZED HABITAT MAPS. J. WILDL. MANAGE. 37(3), 404-409. [COMPUTER]

GLASS, N. R. 1970. A COMPARISON OF TWO MODELS OF THE FUNCTIONAL RESPONSE WITH EMPHASIS ON PARAMETER ESTIMATION PROCEDURES. CAN. ENTOMOL. 102(9), 1094-1101. [COMPUTER]

GOFF, F. G. 1974. RBAD: RELATIVE BASAL AREA DETERMINATION. A FORTRAN PROGRAM TO DETERMINE RELATIVE BASAL AREA BY SPECIES AND PLOT FROM IBP STANDARD FORMAT FOREST SERVICE PLOT TAPES. AEC REPT. EDFB-IBP-74-6. OAK RIDGE NATL. LAB., OAK RIDGE, TENNESSEE. III, 31 PP. [COMPUTER]

GOLDSTEIN, R. A. AND D. F. GRIGAL. 1972. COMPUTER PROGRAMS FOR THE ORDINATION AND CLASSIFICATION OF ECOSYSTEMS. AEC REPT. ORNL-IBP-71-10. OAK RIDGE NATL. LAB., OAK RIDGE, TENNESSEE. IV, 125 PP.

GRAF, R. L. 1973. METHODS FOR DELINEATING WILDLIFE AND OTHER ENVIRONMENTAL MANAGEMENT REGIONS. M. S. THESIS, VIRGINIA POLYTECHNIC INSTITUTE, BLACKSBURG, VIRGINIA. 164 PP [COMPUTER MAPPED; G-VALUE ALGORITHM]

GREGSON, K. AND P. V. BISCOE. 1975. BARLEY AND ITS ENVIRONMENT. II. STRATEGY FOR COMPUTING. J. APPL. ECOL. 12(1), 259-267. [COMPUTER]

GROSENBAUGH, L. R. 1958. THE ELUSIVE FORMULA OF BEST FIT: A COMPREHENSIVE NEW MACHINE PROGRAM. U. S. FOR. SERV. SOUTHERN FOREST EXPT. STN. OCCAS. PAPER 158. 9 PP.

GROSENBAUGH, L. R. 1964. STX-FORTRAN 4 PROGRAM FOR ESTIMATES OF TREE POPULATIONS FROM 3P SAMPLE-TREE-MEASUREMENTS. U. S. FOR. SERV. PACIFIC SOUTHWEST FOREST EXPT. STN. RES. PAPER PSW-13. 49 PP. [COMPUTER]

HACKER, C. S., D. W. SCOTT AND J. R. THOMPSON. 1973. TIME SERIES ANALYSIS OF MOSQUITO POPULATION DATA. J. MED. ENTOMOL. 10(6), 533-543. [COMPUTER]

HALL, A. V. 1970. A COMPUTER-BASED METHOD FOR SHOWING CONTINUA AND COMMUNITIES IN ECOLOGY. J. ECOL. 58(3), 591-602.

HOGMAN, W. J. 1970. A COMPUTER PROGRAM FOR STOCK ANALYSIS IN FORTRAN IV, UNIVAC 1108. TRANS. AMER. FISH. SOC. 99(2), 426-427.

HOPE, E. L. V. AND C. D. SAUER. 1967. A SYSTEM TO DIGITIZE BATHYTHERMOGRAPH APERTURE CARDS. J. FISH. RES. BOARD CAN. 24(5), 1155-1164. [COMPUTER]

HOURSTON, A. S. 1970. COMPUTER PROGRAMS FOR PROCESSING NEWFOUNDLAND HERRING DATA. FISH. RES. BOARD CAN. TECH. REPT. NO. 166. 143 PP.

HOURSTON, A. S. AND F. W. NASH. 1973. COMPUTER PROGRAMS FOR POPULATION ANALYSIS OF B. C. HERRING FROM CATCH, SAMPLING AND SPAWN DEPOSITION DATA. FISH. RES. BOARD CAN. TECH. REPT. NO. 399, I, 334 PP. [COMPUTER]

HULL, N. C., J. J. BEAUCHAMP AND D. E. REICHLE. 1970. PITFALL 1, A GENERAL PURPOSE DATA PROCESSING PROGRAM FOR ENVIRONMENTAL DATA. AEC REPT. ORNL-IBP-70-2. OAK RIDGE NATL. LAB., OAK RIDGE, TENNESSEE. 67 PP.

HUNT, R. AND I. T. PARSONS. 1974. A COMPUTER PROGRAM FOR DERIVING GROWTH-FUNCTIONS IN PLANT GROWTH-ANALYSIS. J. APPL. ECOL. 11(1), 297-307. [LOGARITHMIC TRANSFORMATION; POLYNOMIAL REGRESSION]

JAHODA, J. C. 1972. USING A COMPUTER TO MEASURE ACTIVITY OF SMALL MAMMALS. J. MAMMAL. 53(1), 212-213. [ANALYSIS OF VARIANCE; LSD; CORRELATION]

KESSELL, S. R. 1973. A MODEL FOR WILDERNESS FIRE MANAGEMENT. (ABSTR.). BULL. ECOL. SOC. AMER. 54(1), 17. [COMPUTER]

KIMBALL, D. C. AND W. T. HELM. 1971. A METHOD OF ESTIMATING FISH STOMACH CAPACITY. TRANS. AMER. FISH. SOC. 100(3), 572-575. [MODEL; COMPUTER]

KING, W. B., G. E. WATSON AND P. J. GOULD. 1967. AN APPLICATION OF AUTOMATIC DATA PROCESSING TO THE STUDY OF SEABIRDS, I. NUMERICAL CODING. PROC. U. S. NATL. MUS. 123(3609), 1-29.

KOJIMA, M. AND J. A. WAGAR. 1972. COMPUTER-GENERATED DRAWINGS OF GROUND FORM AND VEGETATION. J. FORESTRY 70(5), 282-285.

KRAL, E. 1972. THE USE OF COMPUTERS FOR THE DETERMINATION OF THE CORRESPONDING SIZE OF RECREATION AND AIR-CLEANING AREAS. PP. 83-90. IN. 3RD CONFERENCE ADVISORY GROUP OF FOREST STATISTICIANS. INSTITUT NATIONAL DE LA RECHERCHE AGRONOMIQUE, PARIS, FRANCE, I.N.R.A. PUBL. 72-3.

KWAN, Q. Y. 1972. ANNOUNCING PROGRAM POOD-DETERMINING ORIGIN AND SPREAD OF WILDFIRES. FOREST SCI. 18(4), 320. [COMPUTER]

LASSITER, R. R. 1966. USE OF COMPUTERS IN ECOLOGICAL RESEARCH. (ABSTRACT). BIOMETRICS 22(4), 961.

LEE, P. J. 1971. MULTIVARIATE ANALYSIS FOR THE FISHERIES BIOLOGIST. FISH. RES. BOARD CAN. TECH. REPT. NO. 244. 182 PP. [COMPUTER; SYSTEMS ANALYSIS]

LENARZ, W. H. AND J. H. GREEN 1971. ELECTRONIC PROCESSING OF ACOUSTICAL DATA FOR FISHERY RESEARCH. J. FISH. RES. BOARD CAN. 28(3), 446-447. [COMPUTER]

LIETH, H. AND J. S. RADFORD. 1971. PHENOLOGY, RESOURCE MANAGEMENT, AND SYNAGRAPHIC COMPUTER MAPPING. BIOSCIENCE 21(2), 62-70.

LINDSEY, J. K. AND A. M. SANDNES. 1970. PROGRAM FOR THE ANALYSIS OF NON-LINEAR RESPONSE SURFACES (EXTENDED VERSION). FISH. RES. BOARD CAN. TECH. REPT. NO. 173. 94 PP.

LINDSEY, J. K., D. F. ALDERDICE AND L. V. PIENAAR. 1970. ANALYSIS OF NONLINEAR MODELS - THE NONLINEAR RESPONSE SURFACE. J. FISH. RES. BOARD CAN. 27(4), 765-791. [LIKELIHOOD FUNCTION; COMPUTER]

LLOYD, P. S., V. MOFFETT AND D. W. WINDLE. 1972. COMPUTER STORAGE AND RETRIEVAL OF BOTANICAL SURVEY DATA. J. APPL. ECOL. 9(1), 1-10.

LOHREY, R. E. AND T. R. DELL. 1969. COMPUTER PROGRAMS USING HEIGHT ACCUMULATION FOR TREE VOLUMES AND PLOT SUMMARIES. J. FORESTRY 67(8), 554-555.

MAWSON, J. C. AND R. J. REED. 1970. THREE COMPUTER PROGRAMS: BACK-CALCULATION, CONDITION FACTOR, AND STOMACH CONTENTS, CDC 3600 FORTRAN/FORMAT. J. FISH. RES. BOARD CAN. 27(1), 156-157.

MCCANN, J. A. AND D. F. CRUSE. 1969. COMPUTER PROGRAM FOR MARK AND RECOVERY POPULATION ESTIMATES (CDC 3600 COMPUTER). TRANS. AMER. FISH. SOC. 98(2), 332-334.

MCCARTHY, M. M. ET AL. 1974. REGIONAL ENVIRONMENTAL SYSTEMS ANALYSIS: AN APPROACH FOR MANAGEMENT. PP. 130-135. IN, PROC. FIRST INTERN. CONGR. ECOL. CENTRE FOR AGRICULTURAL PUBLISHING AND DOCUMENTATION, WAGENINGEN, NETHERLANDS. [COMPUTER GRAPHICS]

MERRIAM, D. E. (ED.). 1966. COMPUTER APPLICATIONS IN THE EARTH SCIENCES: COLLOQUIM ON CLASSIFICATION PROCEDURES. COMPUTER CONTRIBUTION 7, STATE GEOLOGICAL SURVEY, UNIVERSITY OF KANSAS, LAWRENCE, KANSAS. II, 79 PP. [MULTIVARIATE; DESCRIMINANT FUNCTIONS]

MOGK, M. 1974. AUTOMATIC DATA PROCESSING IN ANALYSIS OF EPIDEMICS. PP. 55-77. IN, EPIDEMICS OF PLANT DISEASES: MATHEMATICAL ANALYSIS AND MODELING. (J. KRANZ, ED.). SPRINGER-VERLAG NEW YORK INC., NEW YORK, NEW YORK. [COMPUTER]

O'NEILL, R. V. 1970. AN INTRODUCTION TO THE NUMERICAL SOLUTION OF DIFFERENTIAL EQUATIONS IN ECOSYSTEM MODELS. AEC REPT. ORNL-IBP-70-4. OAK RIDGE NATL. LAB., OAK RIDGE, TENNESSEE. III, 29 PP. [COMPUTER]

O'REGAN, W. G. AND M. N. PALLEY. 1965. A COMPUTER TECHNIQUE FOR THE STUDY OF FOREST SAMPLING METHODS. FOREST SCI. 11(1), 99-114.

OSSIANDER, F. J. AND G. WEDEMEYER. 1973. COMPUTER PROGRAM FOR SAMPLE SIZES REQUIRED TO DETERMINE DISEASE INCIDENCE IN FISH POPULATIONS. J. FISH. RES. BOARD CAN. 30(9), 1383-1384. [HYPERGEOMETRIC DISTRIBUTION]

PATTEN, B. C. 1966. THE BIOCOENETIC PROCESS IN AN ESTUARINE PHYTOPLANKTON COMMUNITY. AEC REPT. ORNL-3946. OAK RIDGE NATL. LAB., OAK RIDGE, TENNESSEE. IV, 95 PP. [ANALOG COMPUTER MODEL; DIVERSITY INDEX; INFORMATION THEORY]

PAULIK, G. J. AND L. E. GALES. 1965. WEIGHTED LINEAR REGRESSION FOR TWO VARIABLES, IBM 709, FORTRAN II. TRANS. AMER. FISH. SOC. 94(2), 196.

PAULIK, G. J. AND W. H. BAYLIFF. 1967. A GENERALIZED COMPUTER PROGRAM FOR THE RICKER MODEL OF EQUILIBRIUM YIELD PER RECRUITMENT. J. FISH. RES. BOARD CAN. 24(2), 249-259.

PETERS, J. A. 1968. A COMPUTER PROGRAM FOR CALCULATING DEGREE OF BIOGEOGRAPHICAL RESEMBLANCE BETWEEN AREAS. SYST. ZOOL. 17(1), 64-69. [RESEMBLANCE EQUATION OF PRESTON; BURT COEFFICIENT; KULCZYNSKI COEFFICIENT; OTSUKA'S COEFFICIENT; SIMPSON COEFFICIENT; JACCARD'S COEFFICIENT; COEFFICIENT OF DIFFERENCE]

PETERS, J. A. 1971. BIOSTATISTICAL PROGRAMS IN BASIC LANGUAGE FOR TIME-SHARED COMPUTERS: COORDINATED WITH THE BOOK "QUANTITATIVE ZOOLOGY". SMITHSON. CONTRIB. ZOOL. NO. 69. 46 PP.

PETERS, R. C. AND R. W. WILSON, JR. 1967. THE NORTHEASTERN FOREST-INVENTORY DATA-PROCESSING SYSTEM. III. OPERATION OF SUBSYSTEM EDIT. U. S. FOR. SERV. NORTHEASTERN FOREST EXPT. STN. RES. PAPER NE-71. 31 PP.

PETERS, R. C. AND R. W. WILSON, JR. 1967. THE NORTHEASTERN FOREST-INVENTORY DATA-PROCESSING SYSTEM. IV. INFORMATION FOR PROGRAMMERS - SUBSYSTEM EDIT. U. S. FOR. SERV. NORTHEASTERN FOREST EXPT. STN. RES. PAPER NE-72. 9 PP.

PETERS, R. C. AND R. W. WILSON, JR. 1967. THE NORTHEASTERN FOREST-INVENTORY DATA-PROCESSING SYSTEM. IX. OPERATION OF SUBSYSTEM OUTPUT. U. S. FOR. SERV. NORTHEASTERN FOREST EXPT. STN. RES. PAPER NE-77. 18 PP.

PETERS, R. C. AND R. W. WILSON, JR. 1967. THE NORTHEASTERN FOREST-INVENTORY DATA-PROCESSING SYSTEM. VI. OPERATION OF SUBSYSTEM TABLE. U. S. FOR. SERV. NORTHEASTERN FOREST EXPT. STN. RES. PAPER NE-74. 25 PP.

PETERS, R. C. AND R. W. WILSON, JR. 1967. THE NORTHEASTERN FOREST-INVENTORY DATA-PROCESSING SYSTEM. VII. INFORMATION FOR PROGRAMMERS - SUBSYSTEM TABLE. U. S. FOR. SERV. NORTHEASTERN FOREST EXPT. STN. RES. PAPER NE-75. 22 PP.

PETERS, R. C. AND R. W. WILSON, JR. 1967. THE NORTHEASTERN FOREST-INVENTORY DATA-PROCESSING SYSTEM. X. INFORMATION FOR PROGRAMMERS - SUBSYSTEM OUTPUT. U. S. FOR. SERV. NORTHEASTERN FOREST EXPT. STN. RES. PAPER NE-78. 15 PP.

PIENAAR, L. V. AND J. A. THOMSON. 1973. THREE PROGRAMS USED IN POPULATION DYNAMICS WVONB-ALOMA-BHYLD (FORTRAN 1130). FISH. RES. BOARD CAN. TECH. REPT. NO. 367. 33 PP. [COMPUTER]

PIERCE, F. AND C. J. DICOSTANZO. 1967. INSTANTANEOUS MORTALITY AND MATURATION SCHEDULES IBM 1620, FORTRAN II. TRANS. AMER. FISH. SOC. 96(3), 366. [COMPUTER]

PINTER, M. AND I. BENDE. 1967. COMPUTER ANALYSIS OF ACINETOBACTER IWOFFII (MORAXELLA IWOFFII) AND ACINETOBACTER ANITRATUS (MORAXELLA GLUCIDOLYTICA) STRAINS. J. GEN. MICROBIOL. 46(2), 267-272.

POLIKARPOV, G. G. AND V. M. EGOROV. 1971. COMPUTERS IN EXPEDITION HYDROBIOLOGICAL INVESTIGATIONS. REPORTS OF ACADEMY OF SCIENCES OF UKRANIAN RSR. 5, 72-79.

PORTEOUS, D. AND S. A. POULET. 1973. COMPUTER PROGRAMS FOR PROCESSING DATA OF SUSPENDED MATERIAL COLLECTED BY THE COULTER COUNTER. FISH. RES. BOARD CAN. TECH. REPT. NO. 403. 126 PP.

PORTER, W. P. AND D. M. GATES. 1969. THERMODYNAMIC EQUILIBRIA OF ANIMALS WITH ENVIRONMENT. ECOL. MONOGR. 39(3), 227-244. [COMPUTER; MODEL]

REEVES, M., III. 1971. A CODE FOR LINEAR MODELING OF AN ECOSYSTEM. AEC REPT. ORNL-IBP-71-2. OAK RIDGE NATL. LAB., OAK RIDGE, TENNESSEE. V, 22 PP. [COMPUTER]

RICHARDS, C. E. 1968. ANALOG COMPUTER TECHNIQUES FOR AGE-GROWTH STUDIES OF FISHES. PROC. ANN. CONF. S. E. ASSOC. GAME AND FISH COMM. 21, 273-276. [VON BERTALANFFY]

SAILA, S. B. AND R. A. SHAPPY. 1964. ELECTRONIC DATA PROCESSING, INFORMATION RETRIEVAL AND TRANSLATION IN FISHERY SCIENCE. FAO FISHERIES TECHNICAL PAPER NO. 42. 9 PP. [COMPUTER]

SAVAGE, J. C. 1966. TELEMETRY AND AUTOMATIC DATA ACQUISITION SYSTEMS. PP. 69-98. IN, SYSTEMS ANALYSIS IN ECOLOGY. (K. E. F. WATT, ED.). ACADEMIC PRESS, NEW YORK, NEW YORK. [COMPUTER]

SCHREIBER, R. K., R. L. STEPHENSON, F. G. GOFF, D. C. WEST AND G. MUSE. 1974. GEOECOLOGY INFORMATION SYSTEM. PART I. BIOGEOGRAPHIC MAPPING OF SPECIES RANGES: DOCUMENTATION OF INPUT AND DATA CHECKING PROCEDURE FOR COMPUTER STORAGE AND RETRIEVAL OF INFORMATION. AEC REPT. EDFB-IBP-74-5. OAK RIDGE NATL LAB., OAK RIDGE, TENNESSEE. VII, 44 PP.

SHANKS, R. E. AND E. E. C. CLEBSCH. 1962. COMPUTER PROGRAMS FOR THE ESTIMATION OF FOREST STAND WEIGHT AND MINERAL POOL. ECOLOGY 43(2), 339-341.

SICCAMA, T. G. 1972. A COMPUTER TECHNIQUE FOR ILLUSTRATING THREE VARIABLES IN A PICTOGRAM. ECOLOGY 53(1), 177-181.

SILLIMAN, R. P. 1969. POPULATION MODELS AND TEST POPULATIONS AS RESEARCH TOOLS. BIOSCIENCE 19(6), 524-528. [COMPUTER]

SILVERSIDES, C. R., ET AL. 1968. THE USE OF COMPUTERS IN WOODLANDS MANAGEMENT RESEARCH IN CANADA. PULP PAPER MANAGE. CAN. 69(11, JUNE 7), 101-110.

SIMPSON, K. AND C. GROOT. 1972. PROGRAMS FOR ANALYSIS OF ORIENTATION DATA. FISH. RES. BOARD CAN. TECH. REPT. NO. 312. I, 61 PP. PLUS APPENDIX. [COMPUTER; FISH BEHAVIOUR]

SINIFF, D. B. AND J. R. TESTER. 1965. COMPUTER ANALYSIS OF ANIMAL-MOVEMENT DATA OBTAINED BY TELEMETRY. BIOSCIENCE 15(2), 104-108.

SMILLIE, K. W. 1963. ELECTRONIC DIGITAL COMPUTERS AND THEIR USE IN ENTOMOLOGY. MEM. ENTOMOL. SOC. CAN. 32, 11-15.

SOLLINS, P. 1971. CSS: A COMPUTER PROGRAM FOR MODELING ECOLOGICAL SYSTEMS. AEC REPT. ORNL-IBP-71-5. OAK RIDGE NATL. LAB., OAK RIDGE, TENNESSEE. IV, 96 PP.

SOLOMON, S. I., J. P. DENOUVILLIEZ, E. J. CHART, J. A. WOOLLEY AND C. CADOU. 1970. THE USE OF A SQUARE GRID SYSTEM FOR COMPUTER ESTIMATION OF PRECIPITATION, TEMPERATURE, AND RUNOFF. WATER RESOUR. RES. 4(5), 919-930.

SOUCEK, B. AND F. VENCL. 1975. BIRD COMMUNICATION STUDY USING DIGITAL COMPUTER. J. THEOR. BIOL. 49(1), 147-172. [PROBABILITY]

STARK, A. E. AND M. C. HALDEN. 1967. ESTIMATION OF CATCH COMPOSITION FROM SAMPLE SURVEYS BY STEPWISE APPLICATION OF RAISING FACTORS, CDC 3200, FORTRAN IV. TRANS. AMER. FISH. SOC. 96(2), 230.

STEPHENSON, W., W. T. WILLIAMS AND S. D. COOK. 1972. COMPUTER ANALYSES OF PETERSEN'S ORIGINAL DATA ON BOTTOM COMMUNITIES. ECOL. MONOGR. 42(4), 387-415. [MEASURE OF SIMILARITY; DENDOGRAM]

SWINGLE, W. E. 1964. INSTRUCTIONS FOR LENGTH-WEIGHT PROGRAMS FOR IBM 1620 IN FORTRAN-FORMAT (FORTRAN I). AUBURN UNIV. ZOOLOGY-ENTOMOLOGY DEPT. SERIES - FISHERIES NO. 1. 19 PP.

SWINGLE, W. E. AND H. S. SWINGLE. 1966. POPULATION ANALYSIS PROGRAMS, IBM 1620, FORTRAN/FORMAT. TRANS. AMER. FISH. SOC. 95(2), 232-233.

TYLER, A. V. 1974. USER'S MANUAL FOR PISES: A GENERAL FISH POPULATION SIMULATOR AND FISHERIES PROGRAM. FISH. RES. BOARD CAN. TECH. REPT. NO. 480. 30 PP. [COMPUTER]

VAN DYNE, G. M. 1965. PROBABILISTIC ESTIMATES OF RANGE FORAGE INTAKE. PROC. WEST. SCI. AMER. SOC. ANIMAL SCI. 16(77), 1-6. [REGRESSION; COMPUTER; GENERATION]

VAN DYNE, G. M. 1966. ECOSYSTEMS, SYSTEMS ECOLOGY, AND SYSTEMS ECOLOGISTS. AEC REPT. ORNL-3957. OAK RIDGE NATL. LAB., OAK RIDGE, TENNESSEE. III, 31 PP. [COMPUTER]

VARGA, L. P. AND C. P. FALLS. 1972. CONTINUOUS SYSTEM MODELS OF OXYGEN DEPLETION IN A EUTROPHIC RESERVOIR. ENVIRON. SCI. TECHNOL. 6(2), 135-142. [FORTRAN; COMPUTER]

VICHNEVETSKY, R. AND A. W. TOMALESKY. 1970. HYBRID COMPUTER PROGRAMS FOR THE STUDY OF POLLUTION IN RIVERS AND ESTUARIES. PP. 922-924. IN, PROCEEDINGS OF THE 1970 SUMMER COMPUTER SIMULATION CONFERENCE, JUNE 10, 11, 12, 1970, DENVER, COLORADO, VOL. II. [SIMULATION; MODEL]

VIEIRA, M. 1965. COMPUTERIZED BIRD CONSERVATION. INPUT 2(1), 21-23.

VOIGTLANDER, C. W. AND A. C. ROOCHVARG. 1967. AGE AND GROWTH PROGRAM, CONTROL DATA 3600, FORTRAN 63-FORTRAN IV. TRANS. AMER. FISH. SOC. 96(3), 364-366. [COMPUTER]

WALLIS, J. R. 1965. MULTIVARIATE STATISTICAL METHODS IN HYDROLOGY - A COMPARISON USING DATA OF KNOWN FUNCTIONAL RELATIONSHIP. WATER RESOUR. RES. 1(4), 447-461. [TRANSFORMATION; COMPUTER; MULTIPLE REGRESSION; PRINCIPAL COMPONENT ANALYSIS]

WATSON, C. R. 1973. INTERACTIVE COMPUTER GRAPHICS SYSTEM APPLIED TO THE LIFE SCIENCES. PH.D. DISSERTATION, OREGON STATE UNIVERSITY, CORVALLIS, OREGON. 164 PP.

WATT, K. E. F. 1961. USE OF A COMPUTER TO EVALUATE ALTERNATIVE INSECTICIDAL PROGRAMS. SCIENCE 133(3454), 706-707.

WATT, K. E. F. 1968. A COMPUTER APPROACH TO ANALYSIS OF DATA ON WEATHER, POPULATION FLUCTUATIONS, AND DISEASE. PP. 145-159. IN, BIOMETEROLOGY. (W. P. LOWRY, ED.). OREGON STATE UNIVERSITY PRESS, CORVALLIS, OREGON.

WEST, D., C. T. KELSEY, R. L. STEPHENSON AND F. G. GOFF. 1974. SAFT: STRUCTURAL ANALYSIS OF FOREST TREES. AEC REPT. EDFB-IBP-74-4. OAK RIDGE NATL. LAB., OAK RIDGE, TENNESSEE. V, 32 PP. [COMPUTER]

WEST, D., R. L. STEPHENSON, F. G. GOFF AND W. C. JOHNSON. 1974. REFORMATTING FOREST SURVEY TREE DETAIL DATA AT ORNL INCLUDING A DOCUMENTATION OF PROGRAM REFORM. AEC REPT. EDFB-IBP-74-3, OAK RIDGE NATL. LAB., OAK RIDGE, TENNESSEE. VII, 47 PP. [COMPUTER]

WHITEHEAD, C. J., JR. AND W. L. TURNER. 1970. COMPUTER APPLICATION TO DRAWINGS FOR SPECIAL PERMITS. PROC. ANN. CONF. S. E. ASSOC. GAME AND FISH COMMISSIONERS 23, 32-36.

WHYTE, A. G. D. 1968. THE USE OF ELECTRONIC COMPUTERS AS A TOOL OF FOREST MANAGEMENT. N. Z. J. FORESTRY 13(2), 184-190.

WILLIAMS, W. T. AND G. N. LANCE. 1969. APPLICATION OF COMPUTER CLASSIFICATION TECHNIQUES TO LAND SURVEYS. BULL. INTERN. STAT. INST. 42(1), 345-352.

WILSON, R. W., JR. AND R. C. PETERS. 1967. THE NORTHEASTERN FOREST-INVENTORY DATA-PROCESSING SYSTEM. I. INTRODUCTION. U. S. FOR. SERV. NORTHEASTERN FOREST EXPT. STN. RES. PAPER NE-61. 20 PP.

WILSON, R. W., JR. AND R. C. PETERS. 1967. THE NORTHEASTERN FOREST-INVENTORY DATA-PROCESSING SYSTEM. II. DESCRIPTION OF SUBSYSTEM EDIT. U. S. FOR. SERV. NORTHEASTERN FOREST EXPT. STN. RES. PAPER NE-70. 12 PP.

WILSON, R. W., JR. AND R. C. PETERS. 1967. THE NORTHEASTERN FOREST-INVENTORY DATA-PROCESSING SYSTEM. V. DESCRIPTION OF SUBSYSTEM TABLE. U. S. FOR. SERV. NORTHEASTERN FOREST EXPT. STN. RES. PAPER NE-73. 11 PP.

WILSON, R. W., JR. AND R. C. PETERS. 1967. THE NORTHEASTERN FOREST-INVENTORY DATA-PROCESSING SYSTEM. VIII. DESCRIPTION OF SUBSYSTEM OUTPUT. U. S. FOR. SERV. NORTHEASTERN FOREST EXPT. STN. RES. PAPER NE-76. 7 PP.

WRIGHT, R. G., JR. 1972. COMPUTER PROCESSING OF CHART QUADRAT MAPS AND THEIR USE IN PLANT DEMOGRAPHIC STUDIES. J. RANGE MANAGE. 25(6), 476-478.

YANDLE, D. O. AND E. R. ROTH. 1971. RATIO ESTIMATION OF FUSIFORM RUST INCIDENCE IN SOUTHERN PINE PLANTATIONS. FOREST SCI. 17(1), 135-142. [COMPUTER MAPPED]

ZAHNER, R. AND A. R. STAGE. 1966. A PROCEDURE FOR CALCULATING DAILY MOISTURE STRESS AND ITS UTILITY IN REGRESSIONS OF TREE GROWTH ON WEATHER. ECOLOGY 47(1), 64-74. [MODEL; COMPUTER; POLYNOMIAL EQUATION]

ZUBOY, J. R., R. T. LACKEY, N. S. PROSSER AND R. V. CORNING. 1974. COMPUTERIZED CREEL CENSUS SYSTEM FOR USE IN FISHERIES MANAGEMENT. PROC. S. E. ASSOC. GAME AND FISH COMMISSIONERS 27, 570-574. [COMPUTER]

COMPUTER SIMULATION

ANDERSON, F. M., G. E. CONNOLLY, A. N. HALTER AND W. M. LONGHURST. 1974. A COMPUTER SIMULATION STUDY OF DEER IN MENDOCINO COUNTY, CALIFORNIA. OREGON AGRIC. EXPT. STN. TECH. BULL. 130. 71 PP. [MODEL]

ANDERSON, F., ET AL. 1973. SEASONAL MODEL OF TEMPERATE DECIDUOUS FOREST. PP. 296-305. IN, MODELING FOREST ECOSYSTEMS. (L. KERN, ED.). AEC REPT. EDFB-IBP-73-7. OAK RIDGE NATL. LAB., OAK RIDGE, TENNESSEE. [COMPUTER]

ANDERSON, J. M. 1972. SYSTEM SIMULATION TO TEST ENVIRONMENTAL POLICY: THE EUTROPHICATION OF LAKES. ENVIRON. LETT. 3(3), 203-228. [MODEL; COMPUTER]

ARES, J. AND J. S. SINGH. 1974. A MODEL OF THE ROOT BIOMASS DYNAMICS OF A SHORTGRASS PRAIRIE DOMINATED BY BLUE GRAMA (BOUTELOUA GRACILIS). J. APPL. ECOL. 11(2), 727-743. [COMPUTER; SIMULATION]

ARVANITIS, L. G. AND W. G. O'REGAN. 1968. COMPUTER SIMULATION AND ECONOMIC EFFICIENCY IN FOREST SAMPLING. HILGARDIA 38(2), 133-164.

ATTIWILL, P. ET AL. 1973. SEASONAL MODEL OF THE BROADLEAVED EVERGREEN FOREST. PP. 313-319. IN, MODELING FOREST ECOSYSTEMS. (L. KERN, ED.). AEC REPT. EDFB-IBP-73-7. OAK RIDGE NATL. LAB., OAK RIDGE, TENNESSEE. [COMPUTER]

AUSTIN, M. P. AND B. G. COOK. 1974. ECOSYSTEM STABILITY: A RESULT FROM AN ABSTRACT SIMULATION. J. THEOR. BIOL. 45(2), 435-458. [MODEL; ALGORITHM; COMPUTER]

BANDU, D. ET AL. 1973. SEASONAL MODEL OF THE TROPICAL FORESTS. PP. 285-295. IN, MODELING FOREST ECOSYSTEMS. (L. KERN, ED.). AEC REPT. EDFB-IBP-73-7. OAK RIDGE NATL. LAB., OAK RIDGE, TENNESSEE. [COMPUTER]

BECKMAN, W. A., J. W. MITCHELL AND W. P. PORTER. 1973. THERMAL MODEL FOR PREDICTION OF A DESERT IGUANA'S DAILY AND SEASONAL BEHAVIOR. J. HEAT TRANSFER 95(2), 257-262. [COMPUTER]

BERRYMAN, A. A. AND L. V. PIENAAR. 1973. SIMULATION OF INTRASPECIFIC COMPETITION AND SURVIVAL OF SCOLYTUS VENTRALIS BROODS (COLEOPTERA: SCOLYTIDAE). ENVIRON. ENTOMOL. 2(3), 447-459. [MONTE CARLO; NEAREST NEIGHBOR; COMPUTER]

BERRYMAN, A. A., T. P. BOGYO AND L. C. DICKMANN. 1973. COMPUTER SIMULATION OF POPULATION REDUCTION BY RELEASE OF STERILE INSECTS. II. THE EFFECTS OF DYNAMIC SURVIVAL AND MULTIPLE MATING. PP. 31-43. IN, COMPUTER MODELS AND APPLICATION OF THE STERILE-MALE TECHNIQUE. INTERNATIONAL ATOMIC ENERGY AGENCY, VIENNA, AUSTRIA, PUBLICATION STI/PUB/340. [MODEL]

BERTHOUEX, P. M., A. M. ASCE AND L. C. BROWN. 1969. MONTE CARLO SIMULATION OF INDUSTRIAL WASTE DISCHARGES. AMER. SOC. CIVIL ENG., SAN. ENG. DIV. PROC. 95(SA5), 887-906. [SIMULATION; COMPUTER; PROBABILITY; NORMAL DISTRIBUTION; GAMMA DISTRIBUTION]

BISHOP, J. A. 1972. AN EXPERIMENTAL STUDY OF THE CLINE OF INDUSTRIAL MELANISM IN BISTON BETULARIA (L.) (LEPIDOPTERA) BETWEEN URBAN LIVERPOOL AND RURAL NORTH WALES. J. ANIM. ECOL. 41(1), 209-243. [DETERMINISTIC MODEL; SIMULATION; COMPUTER; SELECTIVE COEFFICIENT]

BLACK, M. L. AND I. D. GAY. 1965. SOME KINETIC PROPERTIES OF A DETERMINISTIC EPIDEMIC CONFIRMED BY COMPUTER SIMULATION. SCIENCE 148(3672), 981-985.

BLEDSOE, L. J., R. C. FRANCIS, G. L. SWARTZMAN AND J. D. GUSTAFSON. 1971. PWNEE: A GRASSLAND ECOSYSTEM MODEL. U. S. IBP GRASSLAND BIOME, COLORADO STATE UNIVERSITY, FT. COLLINS, COLORADO. TECH. REPT. 64. 179 PP. [COMPUTER]

BOLING, R. H., JR. AND J. A. VAN SICKLE. 1975. CONTROL THEORY IN ECOSYSTEM MANAGEMENT. PP. 219-229, IN, ECOSYSTEM ANALYSIS AND PREDICTION, (S. A. LEVIN, ED.), PROCEEDINGS OF A SIAM-SIMS CONFERENCE HELD AT ALTA, UTAH, JULY 1-5, 1974. [MODEL]

BOTKIN, D. B., J. F. JANAK AND J. R. WALLIS. 1972. SOME ECOLOGICAL CONSEQUENCES OF COMPUTER MODEL OF FOREST GROWTH. J. ECOL. 60(3), 849-872. [SIMULATION]

BROWN, G. W., R. H. BURGY, R. D. HARR AND J. P. RILEY. 1972. HYDROLOGIC MODELING IN THE CONIFEROUS FOREST BIOME. PP. 49-70. IN, PROCEEDINGS-RESEARCH ON CONIFEROUS FOREST ECOSYSTEMS-A SYMPOSIUM, (J. F. FRANKLIN, L. J. DEMPSTER AND R. H. WARING, EDS.), U. S. FOR. SERV. PACIFIC NORTHWEST FOREST AND RANGE EXPT. STN., PORTLAND, OREGON. [COMPUTER]

BUNNELL, F. AND D. RUSSELL. 1973. A REINDEER'S VIEW OF THE BOREAL FOREST ECOSYSTEM. PP. 722-728, IN, PROCEEDINGS OF THE 1973 SUMMER COMPUTER SIMULATION CONFERENCE, JULY 17, 18, 19, 1973. MONTREAL, P. Q., CANADA. VOL. II. [SIMULATION; COMPUTER; MODEL]

BUNNELL, F. L. 1972. LEMMINGS-MODELS AND THE REAL WORLD. PP. 1183-1197. IN, PROCEEDINGS OF THE 1972 SUMMER COMPUTER SIMULATION CONFERENCE, JUNE 14, 15, 16, 1972. SAN DIEGO, CALIFORNIA, VOL. II. [SIMULATION; COMPUTER]

BUSTARD, H. R. AND K. P. TOGNETTI. 1969. GREEN SEA TURTLES. A DISCRETE SIMULATION OF DENSITY-DEPENDENT POPULATION REGULATION. SCIENCE 163(3870), 939-941. [STOCHASTIC MODEL]

CASWELL, H. 1972 A SIMULATION STUDY OF A TIME LAG POPULATION MODEL. J. THEOR. BIOL. 34(3), 419-439. [VOLTERRA; GAUSE MODEL; COMPUTER]

CLARK, R. D., JR. AND R. T. LACKEY. 1975. COMPUTER-IMPLEMENTED SIMULATION AS A PLANNING AID FOR STATE FISHERIES MANAGEMENT AGENCIES. PUBL. FWS-3-75, VIRGINIA POLYTECHNIC INSTITUTE AND STATE UNIVERSITY, BLACKSBURG, VIRGINIA. V, 179 PP. [MODEL]

CLUTTER, J. L. AND J. H. BAMPING. 1966. COMPUTER SIMULATION OF AN INDUSTRIAL FORESTRY ENTERPRISE. PROC. SOC. AMER. FOR. (1965), 180-185.

CLYMER, A. B. 1972. NEXT-GENERATION MODELS IN ECOLOGY. PP. 533-569. IN, SYSTEMS ANALYSIS AND SIMULATION IN ECOLOGY, VOLUME II. (B. C. PATTEN, ED.), ACADEMIC PRESS, NEW YORK, NEW YORK.

CODY, M. L. 1971. FINCH FLOCKS IN THE MOHAVE DESERT. THEOR. POPUL. BIOL. 2(2), 142-158. [COMPUTER; RANDOM WALK]

COLE, C. R. 1967. A LOOK AT SIMULATION THROUGH A STUDY ON PLANKTON POPULATION DYNAMICS. AEC REPT. BNWL-485. BATTELLE-NORTHWEST LAB., RICHLAND, WASHINGTON. 19 PP.

CONRAD, M. AND H. H. PATTEE. 1970. EVOLUTION EXPERIMENTS WITH AN ARTIFICIAL ECOSYSTEM. J. THEOR. BIOL. 28(3), 393-409. [COMPUTER; MODEL]

CONWAY, G. R. 1970. COMPUTER SIMULATION AS AN AID TO DEVELOPING STRATEGIES FOR ANOPHELINE CONTROL. MISC. PUBL. ENTOMOL. SOC. AMER. 7(1), 181-191. [MODEL]

COOPERRIDER, A. Y. 1974. COMPUTER SIMULATION OF THE INTERACTION OF A DEER POPULATION WITH NORTHERN FOREST VEGETATION. PH.D. DISSERTATION. STATE UNIVERSITY OF NEW YORK. 229 PP. (DIS ABSTR. INTERN. B. 35(5), 2006-B).

CUELLAR, C. B. 1973. A VIEW OF COMPUTER SIMULATION IN STERILE MALE EXPERIMENTS. PP. 5-15, IN COMPUTER MODELS AND APPLICATION OF THE STERILE-MALE TECHNIQUE. INTERN. ATOMIC ENERGY AGENCY, VIENNA, AUSTRIA, PUBLICATION STI/PUB/340. [MODEL]

CULVER, D. C. 1975. THE RELATIONSHIP BETWEEN THEORY AND EXPERIMENT IN COMMUNITY ECOLOGY. PP. 103-107. IN, ECOSYSTEM ANALYSIS AND PREDICTION. (S. A. LEVIN, ED.). PROCEEDINGS OF A SIAM-SIMS CONFERENCE HELD AT ALTA, UTAH, JULY 1-5, 1974. [BROKEN-STICK MODEL; COMMUNITY MATRIX THEORY; COMPETITION]

CURDS, C. R. 1971. COMPUTER SIMULATIONS OF MICROBIAL POPULATION DYNAMICS IN THE ACTIVATED-SLUDGE PROCESS. WATER RES. 5(11), 1049-1066.

CURDS, C. R. 1971. A COMPUTER-SIMULATION STUDY OF PREDATOR-PREY RELATIONSHIPS IN A SINGLE-STAGE CONTINUOUS-CULTURE SYSTEM. WATER RES. 5(10), 793-812.

CURRY, R. B. 1970. DYNAMIC SIMULATION OF VEGETATIVE GROWTH IN A PLANT CANOPY. PP. 737-745. IN, PROCEEDINGS OF THE 1970 SUMMER COMPUTER SIMULATION CONFERENCE, JUNE 10, 11, 12, 1970. DENVER, COLORADO, VOL. II. [COMPUTER; MODEL]

CURTIS, C. F. 1973. THE USE OF SIMULATION IN THE CHOICE OF AN OPTIMUM RADIATION DOSE FOR CONTROL OF GLOSSINA MORSITANS. PP. 177-181. IN, COMPUTER MODELS AND APPLICATION OF THE STERILE-MALE TECHNIQUE. INTERNATIONAL ATOMIC ENERGY AGENCY, VIENNA, AUSTRIA. PUBLICATION STI/PUB/340. [COMPUTER; MODEL]

DAVIDSON, R. S. AND A. B. CLYMER. 1966. THE DESIRABILITY AND APPLICABILITY OF SIMULATING ECOSYSTEMS. ANN. N. Y. ACAD. SCI. 128(3), 790-794. [COMPUTER; MODEL]

DAVIS, R. K. AND J. J. SENECA. 1972. MODELS FOR SUPPLY AND DEMAND ANALYSIS IN STATE FISH AND GAME PLANNING. TRANS. N. AMER. WILDL. NAT. RESOUR. CONF. 37, 234-245. [SIMULATION]

DAWDY, D. R. AND T. H. THOMPSON. 1967. DIGITAL COMPUTER SIMULATION IN HYDROLOGY. J. AMER. WATER WORKS ASSOC. 59(6), 685-688.

DE ANELIS, D. L. 1975. STABILITY AND CONNECTANCE IN FOOD WEB MODELS. ECOLOGY 56(1), 238-243. [MONTE CARLO; COMPUTER]

DE WIT, C. T., R. BROUWER AND F. W. T. PENNING DE VRIES. 1970. THE SIMULATION OF PHOTOSYNTHETIC SYSTEMS. PP. 47-70. IN, PREDICTION AND MEASUREMENT OF PHOTOSYNTHETIC PRODUCTIVITY. (I. SETLIK, ED.). CENTRE FOR AGRICULTURAL PUBLISHING AND DOCUMENTATION, WAGENINGEN, NETHERLANDS. [MODEL; COMPUTER]

DEAN, F. C. 1972. POPSID, A FORTRAN IV PROGRAM FOR DETERMINISTIC POPULATION STUDIES. AEC REPT. ORNL-IBP-71-11. OAK RIDGE NATL. LAB., OAK RIDGE, TENNESSEE. IV, 42 PP.

DETTMANN, E. H. 1973. A MODEL OF SEASONAL CHANGES IN THE NITROGEN CONTENT OF LAKE WATER. PP. 753-761. IN, PROCEEDINGS OF THE 1973 SUMMER COMPUTER SIMULATION CONFERENCE, JULY 17, 18, 19, 1973. MONTREAL, P. Q., CANADA, VOL. II. [SIMULATION; COMPUTER]

DIXON, K. R. 1968. COMPUTER SIMULATION AND ANALYSIS OF A PREDATOR-PREY INTERACTION. M. S. THESIS, UNIVERSITY OF FLORIDA, GAINESVILLE, FLORIDA. 55 PP.

DOS SANTOS, E. P. 1968. POPULATIONAL DISTRIBUTION I: THE DIGITAL SIMULATION. MARINE BIOL. 1(4), 348-350. [COMPUTER; SPATIAL DISTRIBUTION; PSEUDORANDOM NUMBER GENERATION]

DUGDALE, R. C. AND T. WHITLEDGE. 1970. COMPUTER SIMULATION OF PHYTOPLANKTON GROWTH NEAR A MARINE SEWAGE OUTFALL. REV. INTERN. OCEANOGR. MED. 17, 201-210. [MODEL]

DUNCAN, W. G., R. S. LOOMIS, W. A. WILLIAMS AND R. HANAU. 1967. A MODEL FOR SIMULATING PHOTOSYNTHESIS IN PLANT COMMUNITIES. HILGARDIA 38(4), 181-205. [COMPUTER]

DUNHAM, R. J. AND P. H. NYE. 1974. THE INFLUENCE OF SOIL WATER CONTENT ON THE UPTAKE OF IONS BY ROOTS. II. CHLORIDE UPTAKE AND CONCENTRATION GRADIENTS IN SOIL. J. APPL. ECOL. 11(2), 581-595.

EBERHARDT, L. L. AND W. C. HANSON. 1969. A SIMULATION MODEL FOR AN ARCTIC FOOD CHAIN. HEALTH PHYSICS 17(6), 793-806. [COMPUTER; LOG TRANSFORMATION]

EBERHARDT, L. L., R. L. MEEKS AND T. J. PETERLE. 1970. DDT IN A FRESHWATER MARSH -- A SIMULATION STUDY. AEC REPT. BNWL-1297. BATTELLE-NORTHWEST LAB., RICHLAND, WASHINGTON. 62 PP.

ESTBERG, G. N. AND B. C. PATTEN. 1975. THE RELATION BETWEEN SENSITIVITY AND PERSISTENCE UNDER SMALL PERTURBATIONS. PP. 151-154. IN, ECOSYSTEM ANALYSIS AND PREDICTION. (S. A. LEVIN, ED.). PROCEEDINGS OF A SIAM-SIMS CONFERENCE HELD AT ALTA, UTAH, JULY 1-5, 1974. [STATE FUNCTION]

EWING, B., P. RAUCH AND J. F. BARBIERI. 1974. SIMULATING THE DYNAMICS AND STRUCTURE OF POPULATIONS. U. S. AEC REPT. UCRL-76046(REV. 1). UNIVERSITY OF CALIFORNIA, LIVERMORE, CALIFORNIA. 61 PP. [COMPUTER; MODEL]

FIELD, D. ET AL. 1973. SEASONAL MODEL OF THE CONIFEROUS ECOSYSTEM. PP. 306-312. IN, MODELING FOREST ECOSYSTEMS. (L. KERN, ED). AEC REPT. EDFB-IBP-73-7. OAK RIDGE NATL. LAB., OAK RIDGE, TENNESSEE.

FOX, W. W., JR. 1971. RANDOM VARIABILITY AND PARAMETER ESTIMATION FOR THE GENERALIZED PRODUCTION MODEL. U. S. NATL. MAR. FISH. SERV. FISH. BULL. 69(3), 569-580. [MONTE CARLO SIMULATION; STOCHASTIC; RANDOM NUMBER GENERATOR; COMPUTER]

FOX, W. W., JR. 1973. A GENERAL LIFE HISTORY EXPLOITED POPULATION SIMULATOR WITH PANDALID SHRIMP AS AN EXAMPLE. U. S. NATL. MAR. FISH. SERV. FISH. BULL. 71(4), 1019-1028. [COMPUTER; MODEL; BEVERTON-HOLT EQUATION; VON BERTALANFFY'S EQUATION]

FRANCIS, R. C. 1974. TUNPOP, A COMPUTER SIMULATION MODEL OF THE YELLOWFIN TUNA POPULATION AND THE SURFACE TUNA FISHERY OF THE EASTERN PACIFIC OCEAN. BULL. INTER-AMER. TROP. TUNA COMM. 16(3), 235-258.

FRENCH, N. AND R. H. SAUER. 1974. PHENOLOGICAL STUDIES AND MODELING IN GRASSLANDS. PP. 227-236. IN, PHENOLOGY AND SEASONALITY MODELING. (H. LIETH, ED.). SPRINGER-VERLAG NEW YORK INC., NEW YORK, NEW YORK. [SIMULATION; COMPUTER]

FRENCH, N. R. 1971. SIMULATION OF DISPERSAL IN DESERT RODENTS. PP. 367-374. IN, STATISTICAL ECOLOGY, VOLUME 3. MANY SPECIES POPULATIONS, ECOSYSTEMS, AND SYSTEMS ANALYSIS. (G. P. PATIL, E. C. PIELOU AND W. E. WATERS, EDS.). PENNSYLVANIA STATE UNIVERSITY PRESS, UNIVERSITY PARK, PENNSYLVANIA.

FRERE, M. H. AND M. E. JENSEN. 1970. MODELING WATER AND NITROGEN BEHAVIOR IN THE SOIL-PLANT SYSTEM. PP. 746-750. IN, PROCEEDINGS OF THE 1970 SUMMER COMPUTER SIMULATION CONFERENCE, JUNE 10, 11, 12, 1970, DENVER, COLORADO, VOL. II. [SIMULATION; COMPUTER; MODEL]

FRISSEL, M. J. AND K. J. A. WIJNANDS-STAB. 1973. COMPUTER MODELLING OF THE DYNAMICS OF INSECT POPULATIONS. PP. 23-29. IN, COMPUTER MODELS AND APPLICATION OF THE STERILE-MALE TECHNIQUE. INTERNATIONAL ATOMIC ENERGY AGENCY, VIENNA, AUSTRIA, PUBLICATION STI/PUB/340. [SIMULATION]

GALES, L. E. 1968. A PRELIMINARY REPORT ON A COMPUTER PROGRAM TO SIMULATE THE DYNAMICS OF A GROUP OF INTERRELATED ANIMAL POPULATIONS. QUANTITATIVE SCIENCE PAPER NO. 3. CENTER FOR QUANTITATIVE SCIENCE IN FORESTRY, FISHERIES AND WILDLIFE, UNIVERSITY OF WASHINGTON, SEATTLE, WASHINGTON. MIMEO. 16 PP. (PLUS SUPPLEMENT 3 PP.).

GALES, L. E. 1969. A COMPUTER PROGRAM TO SIMULATE THE LIFE CYCLE OF SOCKEYE SALMON (SIMF). QUANTITATIVE SCIENCE PAPER NO. 5. CENTER FOR QUANTITATIVE SCIENCE IN FORESTRY, FISHERIES AND WILDLIFE, UNIVERSITY OF WASHINGTON, SEATTLE, WASHINGTON. MIMEO. 5 PP.

GARFINKEL, D. 1962. DIGITAL COMPUTER SIMULATION OF ECOLOGICAL SYSTEMS. NATURE 194(4831), 856-857.

GARFINKEL, D. 1965. COMPUTER SIMULATION IN BIOCHEMISTRY AND ECOLOGY. PP. 292-310. IN, THEORETICAL AND MATHEMATICAL BIOLOGY. (H. J. MOROWITZ AND T. H. WATERMAN, EDS.). BLAISDELL PUBLISHING CO., NEW YORK, NEW YORK. [LOTKA-VOLTERRA; PREDATOR-PREY MODEL; MODEL]

GARFINKEL, D. 1965. SIMULATION OF ECOLOGICAL SYSTEMS. PP. 205-216. IN, COMPUTERS IN BIOMEDICAL RESEARCH, VOLUME 2. (R. W. STACEY AND B. D. WAXMAN, EDS.). ACADEMIC PRESS, NEW YORK, NEW YORK. [PREDATOR-PREY MODEL; COMPUTER]

GARFINKEL, D. A. 1967. A SIMULATION STUDY OF THE EFFECTS ON SIMPLE ECOLOGICAL SYSTEMS OF MAKING RATE OF INCREASE OF POPULATION DENSITY-DEPENDENT. J. THEOR. BIOL. 14(1), 46-58. [LOTKA-VOLTERRA; COMPUTER; MODEL]

GARFINKEL, D. AND R. SACK. 1964. DIGITAL COMPUTER SIMULATION OF AN ECOLOGICAL SYSTEM, BASED ON A MODIFIED MASS ACTION LAW. ECOLOGY 45(3), 502-507. [LOTKA-VOLTERRA MODEL]

GARFINKEL, D., R. H. MACARTHUR AND R. SACK. 1964. COMPUTER SIMULATION AND ANALYSIS OF SIMPLE ECOLOGICAL SYSTEMS. ANN. N. Y. ACAD. SCI. 115(2), 943-951.

GILBERT, N. AND R. D. HUGHES. 1971. A MODEL OF AN APHID POPULATION-THREE ADVENTURES. J. ANIM. ECOL. 40(2), 525-534. [COMPUTER; SIMULATION]

GILES, R. H., JR., C. D. BUFFINGTON AND J. A. DAVIS. 1969. A TOPOGRAPHIC MODEL OF POPULATION STABILITY. J. WILDL. MANAGE. 33(4), 1042-1045. [MODIFIED QUICK'S EQUATION; LIFE TABLE; COMPUTER]

GLASS, N. R. 1968. USE OF A COMPUTER MODEL TO DETERMINE ENERGY REQUIREMENTS OF A PREDATORY FISH, THE LARGEMOUTH BLACK BASS (MICROPTERUS SALMOIDES). PP. 333-336. IN, SECOND CONFERENCE ON APPLICATIONS OF SIMULATION, DECEMBER 2-4, 1968. INSTITUTE OF ELECTRICAL AND ELECTRONIC ENGINEERS, NEW YORK, NEW YORK. CAT. NO. 68C60-SIM. [SIMULATION]

GOLDSTEIN, R. A. AND J. B. MANKIN. 1972. PROSPER: A MODEL OF ATMOSPHERE-SOIL-PLANT WATER FLOW. PP. 1176-1181. IN, PROCEEDINGS OF THE 1972 SUMMER COMPUTER SIMULATION CONFERENCE, JUNE 14, 15, 16, 1972, SAN DIEGO, CALIFORNIA, VOL. II. [SIMULATION; COMPUTER]

GOLDSTEIN, R. A. AND W. F. HARRIS. 1973. SERENDIPITY-A WATERSHED LEVEL SIMULATION MODEL OF TREE BIOMASS DYNAMICS. PP. 691-696. IN, PROCEEDINGS OF THE 1973 SUMMER COMPUTER SIMULATION CONFERENCE, JULY 17, 18, 19, 1973, MONTREAL, P. Q., CANADA, VOL. II. [COMPUTER]

GOLDSTEIN, R. A., J. B. MANKIN AND R. J. LUXMOORE. 1974. DOCUMENTATION OF PROSPER: A MODEL OF ATMOSPHERE-SOIL-PLANT WATER FLOW. AEC REPT. EDFB-IBP-73-9. OAK RIDGE NATL. LAB., OAK RIDGE, TENNESSEE. 75 PP. [COMPUTER]

GOLOFF, A. A., JR. AND F. A. BAZZAZ. 1973. COMPUTER SIMULATION OF SEED GERMINATION. (ABSTR.). BULL. ECOL. SOC. AMER. 54(1), 15.

GOODALL, D. W. 1967. COMPUTER SIMULATION OF CHANGES IN VEGETATION SUBJECT TO GRAZING. J. INDIAN BOT. SOC. 46(4), 356-362.

GOODALL, D. W. 1969. SIMULATING THE GRAZING SITUATION. PP. 211-236. IN, CONCEPTS AND MODELS OF BIOMATHEMATICS: SIMULATION TECHNIQUES AND METHODS, VOLUME 1. (F. HEINMETS, ED.). MARCEL DEKKER, NEW YORK, NEW YORK. [COMPUTER; MODEL]

GOODALL, D. W. 1973. ECOSYSTEM SIMULATION IN THE US/IBP DESERT BIOME. PP. 777-780. IN, PROCEEDINGS OF THE 1973 SUMMER COMPUTER SIMULATION CONFERENCE, JULY 17, 18, 19, 1973, MONTREAL, P. Q., CANADA, VOL. II. [COMPUTER; MODEL]

GOODALL, D. W. 1974. THE HIERARCHICAL APPROACH TO MODEL BUILDING. PROC. INTERN. CONGR. ECOLOGY 1, 244-249. [SYSTEMS ANALYSIS]

GORE, A. J. P. 1972. A FIELD EXPERIMENT, A SMALL COMPUTER AND MODEL SIMULATION. PP. 309-325. IN, MATHEMATICAL MODELS IN ECOLOGY. (J. N. R. JEFFERS, ED.). BLACKWELL SCIENTIFIC PUBLICATIONS, OXFORD, ENGLAND.

GOUDRIAAN, J. AND C. T. DE WIT. 1973. A RE-INTERPRETATION OF GAUSE'S POPULATION EXPERIMENTS BY MEANS OF SIMULATION. J. ANIM. ECOL. 42(3), 521-530. [LOTKA-VOLTERRA EQUATION; COMPUTER; POISSON DISTRIBUTION; STOCHASTIC; MODEL; LOGISTIC EQUATION]

GOULD, E. M., JR. AND W. G. O'REGAN. 1965. SIMULATION: A STEP TOWARD BETTER PLANNING. HARVARD FOREST PAPERS, NO. 13. 86 PP.

GREENBERG, M. R., G. W. CAREY, L. ZOBLER AND R. M. HORDON. 1973. A STATISTICAL DISSOLVED OXYGEN MODEL FOR A FREE-FLOWING RIVER SYSTEM. J. AMER. STAT. ASSOC. 68(342), 279-283. [HARMONIC MODEL; SIMULATION]

GREENOUGH, J. W. 1967. A SIMULATION MODEL OF A HYPOTHETICAL INTRASEASONAL GANTLET FISHERY. M. S. THESIS, UNIVERSITY OF WASHINGTON, SEATTLE, WASHINGTON.

GUSTAFSON, J. AND G. INNIS. 1972. SIMCOMP: A SIMULATION COMPILER FOR BIOLOGICAL MODELLING. PP. 1090-1096. IN, PROCEEDINGS OF THE 1972 SUMMER COMPUTER SIMULATION CONFERENCE, JUNE 14, 15, 16, 1972, SAN DIEGO, CALIFORNIA, VOL. II. [COMPUTER; MODEL]

GUTIERREZ, A. P., D. E. HAVENSTEIN, H. A. NIX AND P. A. MOORE. 1974. THE ECOLOGY OF APHIS CRACCIVORA KOCH AND SUBTERRANEAN CLOVER STUNT VIRUS IN SOUTH-EAST AUSTRALIA. II. A MODEL OF COWPEA APHID POPULATIONS IN TEMPERATE PASTURES. J. APPL. ECOL. 11(1), 1-20. [COMPUTER; SIMULATION]

GUTIERREZ, A. P., W. H. DENTON, R. SHADE, H. MALTBY, T. BURGER AND G. MOOREHEAD. 1974. THE WITHIN-FIELD DYNAMICS OF THE CEREAL LEAF BEETLE (OULEMA MELANOPUS (L.)) IN WHEAT AND OATS. J. ANIM. ECOL. 43(3), 627-640. [SIMULATION MODEL]

HACKER, C. S., D. W. SCOTT AND J. R. THOMPSON. 1973. A FORECASTING MODEL FOR MOSQUITO POPULATION DENSITIES. J. MED. ENTOMOL. 10(6), 544-551. [HEURISTIC MODEL; AUTOREGRESSIVE INTEGRATED MOVING AVERAGE MODEL SIMULATION]

HACKER, C. S., D. W. SCOTT AND J. R. THOMPSON. 1975. A TRANSFER FUNCTION FORECASTING MODEL FOR MOSQUITO POPULATIONS. CAN. ENTOMOL. 107(3), 243-249. [COMPUTER; TIME LAG PARAMETER; VARIANCE; REGRESSION MODEL; SIMULATION]

HACKNEY, P. A. AND C. K. MINNS. 1974. A COMPUTER MODEL OF BIOMASS DYNAMICS AND FOOD COMPETITION WITH IMPLICATIONS FOR ITS USE IN FISHERY MANAGEMENT. TRANS. AMER. FISH. SOC. 103(2), 215-225. [DIFFERENTIAL EQUATION; SIMULATION; PEARL-VERHULST LOGISTIC EQUATION]

HALBACH, U. AND H. J. BURKHARDT. 1972. ARE SIMPLE TIME LAGS RESPONSIBLE FOR CYCLIC VARIATION OF POPULATION DENSITY? OECOLOGIA 9(3), 215-222. [LOGISTIC WITH TIME LAG; SIMULATION; COMPUTER]

HARBAUGH, J. W. 1966. MATHEMATICAL SIMULATION OF MARINE SEDIMENTATION WITH IBM 7090/7094 COMPUTERS. COMPUTER CONTRIBUTION 1, STATE GEOLOGICAL SURVEY, UNIVERSITY OF KANSAS, LAWRENCE, KANSAS. 52 PP.

HARBAUGH, J. W. AND W. J. WAHLSTEDT. 1967. FORTRAN IV PROGRAM FOR MATHEMATICAL SIMULATION OF MARINE SEDIMENTATION WITH IBM 7040 OR 7094 COMPUTERS. COMPUTER CONTRIBUTION 9, STATE GEOLOGICAL SURVEY, UNIVERSITY OF KANSAS, LAWRENCE, KANSAS. 40 PP. [MODEL; RESPONSE SURFACE]

HARRIS, D. R. AND W. A. BEYER. 1972. BIOTA2, A PROGRAM FOR MONTE CARLO SIMULATION OF POPULATION INTERACTIONS IN A BIOME. AEC REPT. LA-4865. LOS ALAMOS SCI. LAB., LOS ALAMOS, NEW MEXICO. 60 PP. [COMPUTER]

HARRIS, G. 1972. THE ECOLOGY OF CORTICOLOUS LICHENS III. A SIMULATION MODEL OF PRODUCTIVITY AS A FUNCTION OF LIGHT INTENSITY AND WATER AVAILABILITY. J. ECOL. 60(1), 19-40. [COMPUTER]

HARRIS, L. D. AND R. C. FRANCIS. 1972. AFCONS; A DYNAMIC SIMULATION MODEL OF AN INTERACTIVE HERBIVORE COMMUNITY. GRASSLAND BIOME, INTERN. BIOL. PROG., FT. COLLINS, COLORADO. IV, 88 PP. [NON-LINEAR DIFFERENTIAL EQUATION MODEL; FORTRAN; COMPUTER]

HATHEWAY, W. H., P. MACHNO AND E. HAMERLY. 1972. MODELING WATER MOVEMENT WITHIN THE UPPER ROOTING ZONE OF A CEDAR RIVER SOIL. PP. 95-101. IN, PROCEEDINGS-RESEARCH ON CONIFEROUS FOREST ECOSYSTEMS-A SYMPOSIUM. (J. F. FRANKLIN, L. J. DEMPSTER AND R. H. WARING, EDS.). U. S. FOR. SERV. PACIFIC NORTHWEST FOREST EXPT. STN., PORTLAND, OREGON. [COMPUTER]

HEALEY, M. C. 1973. EXPERIMENTAL GROUPING OF LAKES: 1. OUTLINE OF THE EXPERIMENT AND SIMULATION RESULTS. FISH. RES. BOARD CAN. TECH. REPT. NO. 383. 25 PP. [COMPUTER]

HETT, J. M. 1971. LAND-USE CHANGES IN EAST TENNESSEE AND A SIMULATION MODEL WHICH DESCRIBES THESE CHANGES FOR THREE COUNTIES. AEC REPT. ORNL-IBP-71-8. OAK RIDGE NATL. LAB., OAK RIDGE, TENNESSEE. V, 56 PP. [COMPUTER]

HILLMAN, P. E. 1974. SIMULATION AND MODELING OF TRANSIENT THERMAL RESPONSES OF THE WESTERN FENCE LIZARD, SCELOPORUS OCCIDENTALIS. THESIS, WASHINGTON STATE UNIVERSITY, PULLMAN, WASHINGTON. XI, 109 PP.

HOLLING, C. S. 1973. DEVELOPMENT AND USE OF ECOLOGICAL MODULES IN RESOURCE DEVELOPMENT SIMULATION. INTERN. INST. FOR APPLIED SYSTEMS ANALYSIS, RESEARCH MEMORANDUM RM-73-3. 10 PP. [SIMULATION MODEL]

HOLLING, C. S. AND S. EWING. 1971. BLIND MAN'S BUFF, EXPLORING THE RESPONSE SPACE GENERATED BY REALISTIC ECOLOGICAL SIMULATION MODELS. PP. 207-223. IN, STATISTICAL ECOLOGY, VOLUME 2, SAMPLING AND MODELING BIOLOGICAL POPULATIONS AND POPULATION DYNAMICS. (G. P. PATIL, E. C. PIELOU AND W. E. WATERS, EDS.). PENNSYLVANIA STATE UNIVERSITY PRESS, UNIVERSITY PARK, PENNSYLVANIA.

HOLLING, C. S., (ED.). 1974. MODELLING AND SIMULATION FOR ENVIRONMENTAL IMPACT ANALYSIS. INTERN. INST. FOR APPLIED SYSTEMS ANALYSIS, RESEARCH MEMORANDUM RM-74-4. 55 PP.

HUANG, B. K. 1970. COMPUTER SIMULATION OF PLANT GROWTH DYNAMICS. PP. 751-760. IN, PROCEEDINGS OF THE 1970 SUMMER COMPUTER SIMULATION CONFERENCE. JUNE 10, 11, 12, 1970, DENVER, COLORADO, VOL. II. [SIMULATION; MODEL]

HUGGINS, L. F. AND E. J. MONKE. 1968. A MATHEMATICAL MODEL FOR SIMULATING THE HYDROLOGIC RESPONSE OF A WATERSHED. WATER RESOUR. RES. 4(3), 529-539.

HUNT, H. W. AND E. W. HUDDLESTON. 1973. A SIMULATION MODEL FOR POPULATIONS OF THE MOSQUITO, CULEX TARSALIS. PP. 733-738. IN, PROCEEDINGS OF THE 1973 SUMMER COMPUTER SIMULATION CONFERENCE, JULY 17, 18, 19, 1973, MONTREAL, P. O., CANADA, VOL. II. [COMPUTER]

INNIS, G. 1972. SIMULATION OF BIOLOGICAL SYSTEMS: SOME PROBLEMS AND PROGRESS. PP. 1085-1090. IN, PROCEEDINGS OF THE 1972 SUMMER COMPUTER SIMULATION CONFERENCE, JUNE 14, 15, 16, 1972, SAN DIEGO, CALIFORNIA, VOL. II. [COMPUTER; MODEL]

INNIS, G. 1975. STABILITY, SENSITIVITY, RESILIENCE, PERSISTENCE. WHAT IS OF INTEREST? PP. 131-139. IN, ECOSYSTEM ANALYSIS AND PREDICTION. (S. A. LEVIN, ED.). PROCEEDINGS OF A SIAM-SIMS CONFERENCE HELD AT ALTA, UTAH, JULY 1-5, 1974. [EQUATIONS]

IVERSON, R. L., H. C. CURL, JR. AND J. L. SAUGEN. 1974. SIMULATION MODEL FOR WIND-DRIVEN SUMMER PHYTOPLANKTON DYNAMICS IN AUKE BAY, ALASKA. MARINE BIOL. 28(3), 169-177.

JAIN, S. K. 1970. SIMULATION OF POPULATION BIOLOGY MODELS IN THE THEORY OF EVOLUTION. PP. 769-777. IN, PROCEEDINGS OF THE 1970 SUMMER COMPUTER SIMULATION CONFERENCE, JUNE 10, 11, 12, 1970, DENVER, COLORADO, VOL. II. [COMPUTER]

JAMESON, D. A. 1972. A MARRIAGE OF SIMULATION AND OPTIMIZATION IN GRASSLAND ECOSYSTEM MANAGEMENT.(ABSTRACT). BIOMETRICS 28(1), 267.

JANSSEN, J. G. M. 1974. SIMULATION OF GERMINATION OF WINTER ANNUALS IN RELATION TO MICROCLIMATE AND MICRODISTRIBUTION. OECOLOGIA 14, 197-228. [MODEL; COMPUTER]

JAWORSKI, N. A. AND A. F. BARTSCH 1973. ENVIRONMENTAL MODELING--ECOSYSTEMS. PP. 81-92. IN PROC. INTERAGENCY CONFERENCE ON THE ENVIRONMENT. (G. D. SAUTER, ED.). U. S. AEC AND U. S. EPA REPT. CONF-72002. [SIMULATION]

JONES, D. D. 1974. BIOLOGY OF THE BUDWORM MODEL. INTERN. INST. FOR APPLIED SYSTEMS ANALYSIS, RESEARCH MEMORANDUM RM-74-3. 42 PP. [SIMULATION MODEL]

JONES, R. AND W. B. HALL. 1973. A SIMULATION MODEL FOR STUDYING THE POPULATION DYNAMICS OF SOME FISH SPECIES. PP. 35-59. IN, THE MATHEMATICAL THEORY OF THE DYNAMICS OF BIOLOGICAL POPULATIONS. (M. S. BARTLETT AND R. W. HIORNS, EDS.). ACADEMIC PRESS, NEW YORK, NEW YORK. [LIFE HISTORY MODEL; LOG-NORMAL DISTRIBUTION; VARIANCE; MORTALITY RATE]

JORGENSEN, C. D., D. T. SCOTT AND H. D. SMITH. 1972. SMALL-MAMMAL TRAPPING SIMULATOR. PP. 1154-1168. IN, PROCEEDINGS OF THE 1972 SUMMER COMPUTER SIMULATION CONFERENCE, JUNE 14, 15, 16, 1972, SAN DIEGO, CALIFORNIA, VOL II. [SIMULATION; COMPUTER; MODEL]

KERR, S. R. 1971. A SIMULATION MODEL OF LAKE TROUT GROWTH. J. FISH. RES. BOARD CAN. 28(6), 815-819. [COMPUTER]

KERSHAW, K. A. AND G. P. HARRIS. 1971. SIMULATION STUDIES AND ECOLOGY. USE OF THE MODEL. PP. 23-39. IN, STATISTICAL ECOLOGY, VOLUME 3. MANY SPECIES POPULATIONS, ECOSYSTEMS, AND SYSTEMS ANALYSIS. (G. P. PATIL, E. C. PIELOU AND W. E. WATERS, EDS.). PENNSYLVANIA STATE UNIVERSITY PRESS, UNIVERSITY PARK, PENNSYLVANIA.

KINERSON, R., JR. AND L. J. FRITSCHEN. 1971. MODELING A CONIFEROUS FOREST CANOPY. AGRIC. METEOROL. 8(6), 439-445. [COMPUTER]

KING, C. E. AND G. J. PAULIK. 1967. DYNAMIC MODELS AND THE SIMULATION OF ECOLOGICAL SYSTEMS. J. THEOR. BIOL. 16(2), 251-267. [DYNAMO; COMPUTER]

KITCHELL, J. F., J. F. KOONCE, R. V. O'NEILL, H. H. SHUGART, JR., J. J. MAGNUSON AND R. S. BOOTH. 1974. MODEL OF FISH BIOMASS DYNAMICS. TRANS. AMER. FISH. SOC. 103(4), 786-798. [SIMULATION]

KITCHING, R. 1971. A SIMPLE SIMULATION MODEL OF DISPERSAL OF ANIMALS AMONG UNITS OF DISCRETE HABITATS. OECOLOGIA 7(2), 95-116. [COMPUTER]

KLINE, J. R. 1973. MATHEMATICAL SIMULATION OF SOIL-PLANT RELATIONSHIPS AND SOIL GENESIS. SOIL SCI. 115(3), 240-249. [SYSTEMS ANALYSIS; MODEL; COMPUTER]

KNOX, J. B., D. M. HARDY AND C. A. SHERMAN. 1973. AQUATIC AND ATMOSPHERIC SIMULATION. AEC REPT. UCRL-51405. UNIVERSITY OF CALIFORNIA RADIATION LABORATORY, LIVERMORE, CALIFORNIA. 13 PP. [MODEL; COMPUTER]

KOURTZ, P. H. AND W. G. O'REGAN. 1971. A MODEL FOR A SMALL FOREST FIRE... TO SIMULATE BURNED AND BURNING AREAS FOR USE IN A DETECTION MODEL. FOREST SCI. 17(2), 163-169. [COMPUTER; MONTE CARLO]

KROH, G. C. AND S. H. WEISS. 1973. THE EFFECT OF SPATIAL PATTERN ON PRODUCTIVITY OF ANNUAL PLANTS--A DYNAMIC MODELING APPROACH. (ABSTRACT) BULL. ECOL. SOC. AMER. 54(1), 31. [MODEL DEVELOPED; SIMULATION]

LABINE, P. A. AND D. H. WILSON. 1973. A TEACHING MODEL OF POPULATION INTERACTIONS: AN ALGAE-DAPHNIA-PREDATOR SYSTEM. BIOSCIENCE 23(3), 162-167. [DIFFERENTIAL EQUATION; SIMULATION; COMPUTER; PREDATOR-PREY MODEL]

LAINE, P. A., G. H. LAUFF AND R. LEVINS. 1975. THE FEASIBILITY OF USING A HOLISTIC APPROACH IN ECOSYSTEM ANALYSIS. PP. 111-128. IN, ECOSYSTEM ANALYSIS AND PREDICTION. (S. A. LEVIN, ED.). PROCEEDINGS OF A SIAM-SIMS CONFERENCE HELD AT ALTA, UTAH, JULY 1-5, 1974. [LEVINS' THEORY OF NICHE; COMPETITION; MODEL; CARRYING CAPACITY; DIVERSITY]

LARKIN, P. A. 1971. SIMULATION STUDIES OF THE ADAMS RIVER SOCKEYE SALMON. J. FISH. RES. BOARD CAN. 28(10), 1493-1502.

LARKIN, P. A. AND A. S. HOURSTON. 1964. A MODEL FOR SIMULATION OF THE POPULATION BIOLOGY OF PACIFIC SALMON. J. FISH. RES. BOARD CAN. 21(5), 1245-1265. [COMPUTER]

LASSITER, R. R. 1970. A FINITE DIFFERENCE MODEL FOR SIMULATION OF DYNAMIC PROCESSES OF ECOSYSTEMS. DISSERTATION. UNIVERSITY OF NORTH CAROLINA, RALEIGH, NORTH CAROLINA.

LASSITER, R. R. AND D. W. HAYNE. 1971. A FINITE DIFFERENCE MODEL FOR SIMULATION OF DYNAMIC PROCESSES IN ECOSYSTEMS. PP. 387-440. IN, SYSTEMS ANALYSIS AND SIMULATION IN ECOLOGY, VOLUME I. (B. C. PATTEN, ED.). ACADEMIC PRESS, NEW YORK, NEW YORK.

LOBDELL, C. H., K. E. CASE AND H. S. MOSBY. 1972. EVALUATION OF HARVEST STRATEGIES FOR A SIMULATED WILD TURKEY POPULATION. J. WILDL. MANAGE. 36(2), 493-497. [COMPUTER; MODEL]

LONG, G. E. 1974. MODEL STABILITY, RESILIENCE, AND MANAGEMENT OF AN AQUATIC COMMUNITY. OECOLOGIA 17(1), 65-85. [VERHULST-PEARL LOGISTIC EQUATION; CARRYING CAPACITY; MATRIX; SIMULATION; EIGENVALUE; SENSITIVITY ANALYSIS]

MACCORMICK, A. J. A., O. L. LOUCKS, J. F. KOONCE, J. F. KITCHELL AND P. R. WEILER. 1974. AN ECOSYSTEM MODEL FOR THE PELAGIC ZONE OF LAKE WINGRA. AEC REPT. EDFB-IBP-74-7. OAK RIDGE NATL. LAB., OAK RIDGE, TENNESSEE. VII, 93 PP. [COMPUTER; SIMULATION]

MANKIN, J. B. AND A. A. BROOKS. 1971. NUMERICAL METHODS FOR ECOSYSTEMS ANALYSIS. AEC REPT. ORNL-IBP-71-1. OAK RIDGE NATL. LAB., OAK RIDGE, TENNESSEE. IV, 99 PP.

MANKIN, J. B., H. H. SHUGART AND R. I. VAN HOOK. 1973. COMPARISON OF MODELS OF AN OLD FIELD ARTHROPOD FOOD CHAIN. PP. 729-732. IN, PROCEEDINGS OF THE 1973 SUMMER COMPUTER SIMULATION CONFERENCE, JULY 17, 18, 19, 1973, MONTREAL, P. Q., CANADA, VOL. II. [SIMULATION; COMPUTER]

MATHEWS, S. B. 1967. THE ECONOMIC CONSEQUENCES OF FORECASTING SOCKEYE SALMON RUNS (ONCORHYNCHUS NERKA WALBAUM) TO BRISTOL BAY, ALASKA: A COMPUTER SIMULATION STUDY OF THE POTENTIAL BENEFITS TO A SALMON CANNING INDUSTRY FROM ACCURATE FORECASTS OF THE RUNS. DISSERTATION. UNIVERSITY OF WASHINGTON, SEATTLE, WASHINGTON.

MAYNARD SMITH, J. AND M. SLATKIN. 1973. THE STABILITY OF PREDATOR-PREY SYSTEMS. ECOLOGY 54(2), 384-391. [COMPUTER SIMULATION; MODEL; VOLTERRA'S EQUATIONS; GRAPHICAL]

MAZANOV, A. AND J. A. HARRIS. 1971. SIMULATION OF A CAVE MICROCOSM: A TRIP IN ECO-MATHEMATICAL REALITY. PROC. ECOL. SOC. AUSTRALIA 6, 116-134. [COMPUTER]

MCINTIRE, C. D. 1973. PERIPHYTON DYNAMICS IN LABORATORY STREAMS: A SIMULATION MODEL AND ITS IMPLICATIONS. ECOL. MONOGR. 43(3), 399-420. [MODEL; COMPUTER]

MEATS, A. 1974. SIMULATION OF POPULATION TRENDS OF TIPULA PALUDOSA USING A MODEL FED WITH CLIMATOLOGICAL DATA. OECOLOGIA 16(2), 139-147.

MEIER, R. L., E. H. BLAKELOCK AND H. HINOMOTO. 1964. SIMULATION OF ECOLOGICAL RELATIONSHIPS. BEHAV. SCI. 9(1), 67-76. [MODEL; COMPUTER]

MENSHUTKIN, V. V. 1971. THE MODELING OF THE INTERACTION BETWEEN A FISHERY AND THE ECOLOGICAL SYSTEM OF A BODY OF WATER. J. ICHTHYOLOGY 11(2), 158-163. (ENGLISH TRANSL.) [COMPUTER]

MENSHUTKIN, V. V. AND YU. YA. KISLYAKOV. 1968. THE SIMULATION OF THE INFLUENCE OF THE FEEDING BASE UPON THE DYNAMICS OF FISH POPULATIONS. ZOOLOGISCHESKII ZHURNAL 47(3), 325-330. [COMPUTER]

MILLER, D. R. 1974. SENSITIVITY ANALYSIS AND VALIDATION OF SIMULATION MODELS. J. THEOR. BIOL. 48(2), 345-360. [MOSQUITO POPULATION; COMPUTER]

MILLER, D. R. AND D. E. WEIDHAAS. 1973. PARAMETER SENSITIVITY IN INSECT POPULATION MODELING. J. THEOR. BIOL. 42(2), 263-274. [MODEL; STEADY-STATE ANALYSIS; SENSITIVITY ANALYSIS]

MILLER, R. S., G. S. HOCHBAUM AND D. B. BOTKIN 1972. A SIMULATION MODEL FOR THE MANAGEMENT OF SANDHILL CRANES. YALE UNIV. SCH. FOR. ENVIRON. STUD. BULL. NO. 80. 49 PP. [COMPUTER]

MILNER, C. 1972. THE USE OF COMPUTER SIMULATION IN CONSERVATION MANAGEMENT. PP. 249-275. IN, MATHEMATICAL MODELS IN ECOLOGY. (J. N. R. JEFFERS, ED.). BLACKWELL SCIENTIFIC PUBLICATIONS, OXFORD, ENGLAND.

MITCHELL, K. J. 1969. SIMULATION OF THE GROWTH OF EVEN-AGED STANDS OF WHITE SPRUCE. YALE UNIV. SCH. FOR. BULL. NO. 75. 48 PP. [COMPUTER]

MOBLEY, C. D. 1973. A SYSTEMATIC APPROACH TO ECOSYSTEMS ANALYSIS. J. THEOR. BIOL. 42(1), 119-136. [MODEL; ALGORITHM; LOTKA-VOLTERRA MODEL; MATRIX; SIMULATION; COMPUTER]

MONRO, J. 1973. SOME APPLICATIONS OF COMPUTER MODELLING IN POPULATION SUPPRESSION BY STERILE MALES. PP. 81-94. IN, COMPUTER MODELS AND APPLICATION OF THE STERILE-MALE TECHNIQUE. INTERN. ATOMIC ENERGY AGENCY, VIENNA, AUSTRIA, PUBLICATION STI/PUB/340. [MODEL]

MONTGOMERY, G. G. 1973. COMMUNICATION IN RED FOX DYADS: A COMPUTER SIMULATION STUDY. DISSERTATION. UNIVERSITY OF MINNESOTA, MINNEAPOLIS, MINNESOTA. 140 PP. [MODEL]

MULHOLLAND, R. J. 1975. STABILITY ANALYSIS OF THE RESPONSE OF ECOSYSTEMS TO PERTURBATIONS. PP. 166-181. IN, ECOSYSTEM ANALYSIS AND PREDICTION. (S. A. LEVIN, ED.), PROCEEDINGS OF A SIAM-SIMS CONFERENCE HELD AT ALTA, UTAH, JULY 1-5, 1974. [MODEL; AVERAGE MUTUAL INFORMATION; MATRIX; VECTOR]

MURAI, M. 1974. STUDIES ON THE INTERFERENCE AMONG LARVAE OF THE CITRUS LEAF MINER, PHYLLOCNISTIS CITRELLA STAINTON (LEPIDOPTERA: PHYLLOCNISTIDAE). RES. POPUL. ECOL.-KYOTO 16(1), 80-111. [MODEL OF MUTUAL CONTACT AMONG LARVAE; PROBABILITY; SIMULATION; COMPUTER]

MURDIE, G. 1971. SIMULATION OF THE EFFECTS OF PREDATOR/PARASITE MODELS ON PREY/LOST SPATIAL DISTRIBUTION. PP. 215-229. IN, STATISTICAL ECOLOGY, VOLUME 1, SPATIAL PATTERNS AND STATISTICAL DISTRIBUTIONS. (G. P. PATIL, E. C. PIELOU AND W. E. WATERS, EDS.), PENNSYLVANIA STATE UNIVERSITY PRESS, UNIVERSITY PARK, PENNSYLVANNIA.

MURPHY, G. I. 1967. VITAL STATISTICS OF THE PACIFIC SARDINE (SARDINOPS CAERULEA) AND THE POPULATION CONSEQUENCES. ECOLOGY 48(5), 731-736. [SIMULATION; COMPUTER; RICKER'S STOCK RECRUITMENT EQUATION; INTRINSIC RATE OF INCREASE; LOGISTIC]

NAKASUJI, F., K. KIRITANI AND E. TOMIDA. 1975. A COMPUTER SIMULATION OF THE EPIDEMIOLOGY OF THE RICE DWARF VIRUS. RES. POPUL. ECOL.-KYOTO 16(2), 245-251. [SYSTEMS MODEL]

NEEL, R. B. AND J. S. OLSON. 1962. USE OF ANALOG COMPUTERS FOR SIMULATING THE MOVEMENT OF ISOTOPES IN ECOLOGICAL SYSTEMS. AEC REPT. ORNL-3172. OAK RIDGE NATL. LAB., OAK RIDGE, TENNESSEE. XII, 108 PP. (M. S. THESIS: R. B. NEEL, VANDERBILT UNIVERSITY, NASHVILLE, TENNESSEE).

NEWELL, W. T. AND J. NEWTON. 1968. COMPUTER SIMULATION GAME MODELS FOR ECOLOGY AND NATURAL RESOURCES MANAGEMENT. QUANTITATIVE SCIENCE PAPER NO. 2, CENTER FOR QUANTITATIVE SCIENCE IN FORESTRY, FISHERIES AND WILDLIFE, UNIVERSITY OF WASHINGTON, SEATTLE, WASHINGTON. MIMEO, 30 PP.

NEWNHAM, R. M. 1968. SIMULATION MODELS IN FOREST MANAGEMENT AND HARVESTING. FORESTRY CHRON. 44(1), 7-13.

NIVEN, B. S. 1967. THE STOCHASTIC SIMULATION OF TRIBOLIUM POPULATIONS. PHYSIOL. ZOOL. 40(1), 67-82. [COMPUTER]

NIVEN, B. S. 1969. SIMULATION OF TWO INTERACTING SPECIES OF TRIBOLIUM. PHYSIOL. ZOOL. 42(2), 248-255. [FORTRAN]

O'REGAN, W. G., L. ARVANITIS AND E. M. GOULD, JR. 1966. SYSTEMS, SIMULATION, AND FOREST MANAGEMENT. PROC. SOC. AMER. FOR. (1965), 194-198. [SAMPLING; COMPUTER; FORTRAN]

OAK RIDGE SYSTEMS ECOLOGY GROUP. 1975. DYNAMIC ECOSYSTEM MODELS: PROGRESS AND CHALLENGES. PP. 280-293. IN, ECOSYSTEM ANALYSIS AND PREDICTION. (S. A. LEVIN, ED.). PROCEEDINGS OF A SIAM-SIMS CONFERENCE HELD AT ALTA, UTAH, JULY 1-5, 1974. [LINEAR AND NON-LINEAR MODELS; BIBLIOGRAPHY]

OLSON, J. S. 1963. ANALOG COMPUTER MODELS FOR MOVEMENT OF NUCLIDES THROUGH ECOSYSTEMS. PP. 121-125. IN, RADIOECOLOGY. (V. SCHULTZ, AND A. W. KLEMENT, JR., EDS.), REINHOLD PUBLISHING CORP., NEW YORK, NEW YORK.

OPIE, J. E. 1972. STANDSIM - A GENERAL MODEL FOR SIMULATION OF THE GROWTH OF EVENAGED STANDS. PP. 217-240. IN, 3RD CONFERENCE ADVISORY GROUP OF FOREST STATISTICIANS. INSTUT NATIONAL DE LA RECHERCHE AGRONOMIQUE, PARIS FRANCE. I.N.R.A. PUB. 72-3. [COMPUTER; SIMULATION]

OSTER, G. 1975. STOCHASTIC BEHAVIOR OF DETERMINISTIC MODELS. PP. 24-33. IN, ECOSYSTEM ANALYSIS AND PREDICTION. (S. A. LEVIN, ED.). PROCEEDINGS OF A SIAM-SIMS CONFERENCE HELD AT ALTA, UTAH, JULY 1-5, 1974.

OVERTON, W. S. 1975. DECOMPOSABILITY: A UNIFYING CONCEPT? PP. 297-298. IN, ECOSYSTEM ANALYSIS AND PREDICTION. (S. A. LEVIN, ED.). PROCEEDINGS OF A SIAM-SIMS CONFERENCE HELD AT ALTA, UTAH, JULY 1-5, 1974. [CANONICAL FORM; MODEL; MODAL FORM]

PARK, R. A. ET AL. 1974. A GENERALIZED MODEL FOR SIMULATING LAKE ECOSYSTEMS. SIMULATION 23(2), 33-50. [COMPUTER]

PARKER, R. A. 1968. SIMULATION OF AN AQUATIC ECOSYSTEM. BIOMETRICS 24(4), 803-821. [COMPUTER; MODEL]

PARKER, R. A. 1973. SOME PROBLEMS ASSOCIATED WITH COMPUTER SIMULATION OF AN ECOLOGICAL SYSTEM. PP. 269-288. IN, THE MATHEMATICAL THEORY OF THE DYNAMICS OF BIOLOGICAL POPULATIONS. (M. S. BARTLETT AND R. W. HIORNS EDS.). ACADEMIC PRESS, NEW YORK, NEW YORK. [MODEL; AQUATIC; MORTALITY RATE; COMPUTER]

PARTON, W. J. AND J. K. MARSHALL. 1973. MODENV: A GRASSLAND ECOSYSTEM MODEL. PP. 769-776. IN, PROCEEDINGS OF THE 1973 SUMMER COMPUTER SIMULATION CONFERENCE, JULY 17, 18, 19, 1973, MONTREAL, P. Q., CANADA, VOL. II. [SIMULATION; COMPUTER]

PASK, G. 1969. THE COMPUTER-SIMULATED DEVELOPMENT OF POPULATIONS OF AUTOMATA. MATH. BIOSCI. 4, 101-127.

PATTEN, B. C. 1971. A PRIMER FOR ECOLOGICAL MODELING AND SIMULATION WITH ANALOG AND DIGITAL COMPUTERS. PP. 3-121. IN, SYSTEMS ANALYSIS AND SIMULATION IN ECOLOGY, VOLUME 1. (B. C. PATTEN, ED.). ACADEMIC PRESS, NEW YORK, NEW YORK.

PATTEN, B. C. 1975. ECOSYSTEM LINEARIZATION: AN EVOLUTIONARY DESIGN PROBLEM. PP. 182-197. IN, ECOSYSTEM ANALYSIS AND PREDICTION. (S. A. LEVIN, ED.). PROCEEDINGS OF A SIAM-SIMS CONFERENCE HELD AT ALTA, UTAH, JULY 1-5, 1974. [MODEL; VECTOR; MATRIX]

PATTEN, B. C. 1975. THE RELATION BETWEEN SENSITIVITY AND STABILITY. PP. 141-143. IN, ECOSYSTEM ANALYSIS AND PREDICTION. (S. A. LEVIN, ED.). PROCEEDINGS OF A SIAM-SIMS CONFERENCE HELD AT ALTA, UTAH, JULY 1-5, 1974. [VECTOR; STATE DYNAMICS]

PAULIK, G. J., A. S. HOURSTON AND P. A. LARKIN. 1967. EXPLOITATION OF MULTIPLE STOCKS BY A COMMON FISHERY. J. FISH. RES. BOARD CAN. 24(12), 2527-2537. [COMPUTER]

PAVLIDIS, T. 1969. AN EXPLANATION OF THE OSCILLATORY FREE-RUNS IN CIRCADIAN RHYTHMS. AMER. NAT. 103(929), 31-42. [SIMULATION; COMPUTER; EQUATIONS]

PENNING DE VRIES, F. W. T. 1972. A MODEL FOR SIMULATING TRANSPIRATION OF LEAVES WITH SPECIAL ATTENTION TO STOMATAL FUNCTIONING. J. APPL. ECOL. 9(1), 57-77.

PENNYCUICK, C. J., R. M. COMPTON AND L. BECKINGHAM. 1968. A COMPUTER MODEL FOR SIMULATING THE GROWTH OF A POPULATION, OR OF TWO INTERACTING POPULATIONS. J. THEOR. BIOL. 18(3), 316-329. [LESLIE MATRIX; TIME LAG MODEL]

PENNYCUICK, L. 1969. A COMPUTER MODEL OF THE OXFORD GREAT TIT POPULATION. J. THEOR. BIOL. 22(3), 381-400. [FECUNDITY; LESLIE MATRIX]

PETERSEN, R. 1975. THE PARADOX OF THE PLANKTON: AN EQUILIBRIUM HYPOTHESIS. AMER. NAT. 109(965), 35-49. [MODEL; COMPUTER; SIMULATION]

PICARDI, A. C. 1973. GYPSY MOTH POPULATION SIMULATION: SYSTEM POSTULATION, VALIDATION, ANALYSIS. PP. 1069-1074. IN, PROCEEDINGS OF THE 1973 SUMMER COMPUTER SIMULATION CONFERENCE, JULY 17, 18, 19, 1973, MONTREAL, P. Q., CANADA, VOL. II. [COMPUTER; MODEL]

POWELL, R. A. 1973. A MODEL FOR RAPTOR PREDATION ON WEASELS. J. MAMMAL. 54(1), 259-263. [PREDATOR-PREY MODEL; SIMULATION; COMPUTER]

PRESTON, E. M. 1973. COMPUTER SIMULATED DYNAMICS OF A RABIES-CONTROLLED FOX POPULATION. J. WILDL. MANAGE. 37(4), 501-512. [SIMULATION; MODEL]

PRESTON, E. M. 1973. A COMPUTER SIMULATION OF COMPETITION AMONG FIVE SYMPATRIC CONGENERIC SPECIES OF XANTHID CRABS. ECOLOGY 54(3), 469-483.

PROKHOROV, V. M. AND R. GINZBURG. 1971. MODELING THE PROCESS OF MIGRATION OF RADIONUCLIDE IN FOREST ECOSYSTEMS AND DESCRIPTION OF THE MODEL. EKOLOGIYA 2(5), 11-19. (ENGLISH TRANSL. PP. 396-402). [COMPUTER; MATRIX]

RADFORD, P. J. 1972. THE SIMULATION LANGUAGE AS AN AID TO ECOLOGICAL MODELLING. PP. 277-295. IN, MATHEMATICAL MODELS IN ECOLOGY. (J. N. R. JEFFERS, ED.). BLACKWELL SCIENTIFIC PUBLICATIONS, OXFORD, ENGLAND.

RAINES, G. E., S. G. BLOOM, P. A. MCKEE AND J. C. BELL. 1971. MATHEMATICAL SIMULATION OF SEA OTTER POPULATION DYNAMICS AMCHITKA ISLAND, ALASKA. BIOSCIENCE 21(12), 686-691. [MODEL; DIFFERENTIAL EQUATION]

RARIDON, R. J., M. T. MILLS AND J. W. HUCKABEE. 1974. COMPUTER MODEL FOR CHEMICAL EXCHANGE IN THE STREAM SYSTEM. PP. 284-291. IN, PROCEEDINGS OF THE FIRST ANNUAL NSF TRACE CONTAMINANTS CONFERENCE. (W. FULKERSON, W. D. SHULTZ AND R. I. VAN HOOK, COMPS.). AEC REPT. CONF-730802. OAK RIDGE NATL. LAB., OACK RIDGE, TENNESSEE.

REDDINGIUS, J. AND P. J. DEN BOER. 1970. SIMULATION EXPERIMENTS ILLUSTRATING STABILIZATION OF ANIMAL NUMBERS BY SPREADING OF RISK. OECOLOGIA 5(3), 240-284. [COMPUTER; MODEL]

REICHLE, D. E. AND S. I. AUERBACH. 1972. ANALYSIS OF ECOSYSTEMS. PP. 260-280. IN, CHALLENGING BIOLOGICAL PROBLEMS. (J. A. BEHNKE, ED.). OXFORD UNIVERSITY PRESS, NEW YORK, NEW YORK. [MODEL; SIMULATION; COMPUTER; SYSTEMS ANALYSIS]

REICHLE, D. E., R. V. O'NEILL, S. V. KAYE, P. SOLLINS AND R. S. BOOTH. 1973. SYSTEMS ANALYSIS AS APPLIED TO MODELING ECOLOGICAL PROCESSES. OIKOS 24(3), 337-343.

REID, W. H. 1972. SIMULATION OF PREDATION EMPLOYING A STOCHASTIC TECHNIQUE. PP. 1148-1153A. IN, PROCEEDINGS OF THE 1972 SUMMER COMPUTER SIMULATION CONFERENCE, JUNE 14, 15, 16, 1972, SAN DIEGO, CALIFORNIA, VOL. II. [COMPUTER; MODEL; VOLTERRA; LOTKA'S EQUATION; GAUSE MODEL]

RICHARDS, C. E. 1970. ANALOG SIMULATION IN FISH POPULATION STUDIES. TRANS. ANALOG / HYBRID COMPUTER EDUCATIONAL USERS GROUP 11(7), 203-206.

RICKLEFS, R. E. 1970. THE ESTIMATION OF A TIME FUNCTION OF ECOLOGICAL USE. ECOLOGY 51(3), 508-513. [SIMULATION; PROBABILITY; COMPUTER; MODEL]

ROHLF, F. J. AND D. DAVENPORT. 1969. SIMULATION OF SIMPLE MODELS OF ANIMAL BEHAVIOR WITH A DIGITAL COMPUTER. J. THEOR. BIOL. 23(3), 400-424.

ROONEY, D. W. 1971. SIMULATED ALGAL CELL CULTURES: CORRELATION OF A GENERALIZED SYNCHRONY INDEX WITH DIVISION TIME. MATH. BIOSCI. 10, 149-155. [COMPUTER]

ROSE, D. W. 1973. SIMULATION OF JACK PINE BUDWORM ATTACKS. J. ENVIRON. MANAGE. 1(3), 259-276. [DETERMINISTIC MODEL; STOCHASTIC; RANDOM NUMBER GENERATOR; MODEL]

ROSEBERRY, J. L. 1974. RELATIONSHIPS BETWEEN SELECTED POPULATION PHENOMENA AND ANNUAL BOBWHITE AGE RATIOS. J. WILDL. MANAGE. 38(4), 665-673. [SIMULATION; COMPUTER; DETERMINISTIC MODEL]

RUESS, J. O. AND C. V. COLE. 1973. SIMULATION OF NITROGEN FLOW IN A GRASSLAND ECOSYSTEM. PP. 762-768. IN, PROCEEDINGS OF THE 1973 SUMMER COMPUTER SIMULATION CONFERENCE, JULY 17, 18, 19, 1973, MONTREAL, P. Q., CANADA, VOL. II. [COMPUTER; MODEL]

RYKIEL, E. J., JR. AND N. T. KUENZEL. 1971. ANALOG COMPUTER MODELS OF "THE WOLVES OF ISLE ROYALE," PP. 513-541. IN, SYSTEMS ANALYSIS AND SIMULATION IN ECOLOGY, VOLUME I. (B.C. PATTEN, ED.). ACADEMIC PRESS, NEW YORK, NEW YORK.

SASABA, T. AND K. KIRITANI. 1975. A SYSTEMS MODEL AND COMPUTER SIMULATION OF THE GREEN RICE LEAFHOPPER POPULATIONS IN CONTROL PROGRAMMES. RES. POPUL. ECOL.-KYOTO 16(2), 231-244. [MULTIPLE REGRESSION MODEL; LOGISTIC EQUATION]

SASABA, T., K. KIRITANI AND T. URABE. 1973. A PRELIMINARY MODEL TO SIMULATE THE EFFECT OF INSECTICIDES ON A SPIDER-LEAFHOPPER SYSTEM IN THE PADDY FIELD. RES. POPUL. ECOL.-KYOTO 15(1), 9-22. [COMPUTER; LOGISTIC]

SAUER, R. H. 1973. PHEN: A PHENOLOGICAL SIMULATION MODEL. PP. 830-834. IN, PROCEEDINGS OF THE 1973 SUMMER COMPUTER SIMULATION CONFERENCE, JULY 17, 18, 19, 1973, MONTREAL, P. Q., CANADA, VOL. II. [COMPUTER]

SCAVIA, D., J. A. BLOOMIELD, J. S. FISHER, J. NAGY AND R. A. PARK. 1974. DOCUMENTATION OF CLEANX: A GENERALIZED MODEL FOR SIMULATING THE OPEN-WATER ECOSYSTEMS OF LAKES. SIMULATION 23(2), 51-56. [COMPUTER]

SCHNEIDER, J. C. 1973. RESPONSE OF THE BLUEGILL POPULATION AND FISHERY OF A MILL LAKE TO EXPLOITATION RATE AND MINIMUM SIZE LIMIT: A SIMULATION MODEL. MICH. DEPT. NAT. RESOUR., FISH. RES. REPT. NO. 1804. 18 PP.

SCHNEIDER, J. C. 1973. RESPONSE OF THE BLUEGILL POPULATION AND FISHERY OF MILL LAKE TO INCREASED GROWTH: A SIMULATION MODEL. MICH. DEPT. NAT. RESOUR., FISH. RES. REPT. NO. 1805. 17 PP. (SFA 19(1): 17144).

SEIGER, M. B. 1967. A COMPUTER SIMULATION STUDY OF THE INFLUENCE OF IMPRINTING ON POPULATION STRUCTURE. AMER. NAT. 101(917), 47-57.

SHIH, S.-F., B. K. HUANG AND G. J. KRIZ. 1970. A SIMULATION ANALYSIS OF PLANT GROWTH DYNAMICS AS AFFECTED BY WATER TABLE, DEPTH TO DRAIN AND SOIL TYPES. PP. 980-989. IN, PROCEEDINGS OF THE 1970 SUMMER COMPUTER SIMULATION CONFERENCE, JUNE 10, 11, 12, 1970, DENVER, COLORADO. VOL II. [COMPUTER; MODEL]

SHIYOMI, M. 1974. A MODEL OF PLANT-TO-PLANT MOVEMENT OF APHIDS III. STUDIES OF ACTUAL MOVEMENT AND APPARENT MOVEMENT BY SIMULATION TECHNIQUE. RES. POPUL. ECOL.-KYOTO 15(2), 148-162. [DISPERSAL; PROBABILITY; MAXIMUM LIKELIHOOD; POISSON DISTRIBUTION; RANDOM NUMBER GENERATOR; SKELLAM'S MODEL; COMPUTER]

SHUGART, H. H., JR., T. R. CROW AND J. M. HETT. 1973. FOREST SUCCESSION MODELS: A RATIONALE AND METHODOLOGY FOR MODELING FOREST SUCCESSION OVER LARGE REGIONS. FOREST SCI. 19(3), 203-212. [DIFFERENTIAL EQUATION; SIMULATION; COMPUTER]

SILLIMAN, R. P. 1967. ANALOG COMPUTER MODELS OF FISH POPULATIONS. U. S. FISH WILDL. SERV. FISH. BULL. 66(1), 31-46. [SIMULATION; PLOTTER; GOMPERTZ EQUATION; BEVERTON-HOLT EQUATION; SURVIVAL CURVES; YIELD]

SILLIMAN, R. P. 1969. ANALOG COMPUTER SIMULATION AND CATCH FORECASTING IN COMMERCIALLY FISHED POPULATIONS. TRANS. AMER. FISH. SOC. 98(3), 560-569.

SINIFF, D. B. AND C. R. JESSEN. 1969. A SIMULATION MODEL OF ANIMAL MOVEMENT PATTERNS. PP. 185-219. IN, ADVANCES IN ECOLOGICAL RESEARCH, VOLUME 6. (J. B. CRAGG, ED.). ACADEMIC PRESS, NEW YORK, NEW YORK.

SLOBODKIN, L. B. 1975. COMMENTS FROM A BIOLOGIST TO A MATHEMATICIAN. PP. 318-327. IN, ECOSYSTEM ANALYSIS AND PREDICTION, (S. A. LEVIN, ED.). PROCEEDINGS OF A SIAM-SIMS CONFERENCE HELD AT ALTA, UTAH, JULY 1-5, 1974. [MODEL]

SMART, C. W. AND R. H. GILES, JR. 1973. A COMPUTER MODEL OF WILDLIFE RABIES EPIZOOTICS AND AN ANALYSIS OF INCIDENCE PATTERNS. WILDL. DIS. 61, 89 PP.

SMART, J. S. AND V. L. MORUZZI. 1971. COMPUTER SIMULATION OF CLINCH MOUNTAIN DRAINAGE NETWORKS. J. GEOL. 79(5), 572-584. [MODEL; RANDOM WALK]

SONLEITNER, F. J. 1971. COMPUTER SIMULATION OF A LABORATORY POPULATION OF TRIBOLIUM (COLEOPTERA). PROC. 13TH INTERN. CONGR. ENTOMOL. 1, 560-561. [MODEL; COMPUTER]

SOUTHWARD, G. M. 1966. A SIMULATION STUDY OF MANAGEMENT REGULATORY POLICIES IN THE PACIFIC HALIBUT FISHERY. DISSERTATION. UNIVERSITY OF WASHINGTON, SEATTLE, WASHINGTON.

SOUTHWARD, G. M. 1968. A SIMULATION OF MANAGEMENT STRATEGIES IN THE PACIFIC HALIBUT FISHERY. INTERN. PACIFIC HALIBUT COMM. REPT. NO. 47. 70 PP. [COEFFICIENT OF CATCHABILITY; GOMPERTZ EQUATION; COMPUTER; MODEL; STOCHASTIC; DETERMINISTIC COEFFICIENT OF MORTALITY]

STANISLAV, J. AND M. F. MOHTADI. 1971. MATHEMATICAL SIMULATION OF DISPERSION OF POLLUTANTS IN A LAKE WITH ICE COVER. WATER RES. 5(7), 401-412. [COMPUTER]

STITELER, W. M., III AND F. Y. BORDEN. 1967. THE GENERATION OF FOREST MODELS BY SIMULATION ON A COMPUTER. RESEARCH BRIEFS, SCHOOL OF FOREST RESOURCES, PENNSYLVANIA STATE UNIVERSITY, UNIVERSITY PARK, PENNSYLVANIA 2(1), 14-16. [COMPUTER]

STRONG, F. E. 1971. A COMPUTER-GENERATED MODEL TO SIMULATE MATING BEHAVIOR OF LYGUS BUGS. J. ECON. ENTOMOL. 64(1), 46-50. [RANDOM NUMBER GENERATOR]

SWARTZMAN, G. L. AND G. M. VAN DYNE. 1972. AN ECOLOGICALLY BASED SIMULATION-OPTIMIZATION APPROACH TO NATURAL RESOURCE PLANNING. PP. 347-398. IN, ANNUAL REVIEW OF ECOLOGY AND SYSTEMATICS, VOLUME 3. (R. F. JOHNSTON, P. W. FRANK AND C. D. MICHENER, EDS.). ANNUAL REVIEWS INC., PALO ALTO, CALIFORNIA.

TAKAHASHI, M., K. FUJII AND T. R. PARSONS 1973. SIMULATION STUDY OF PHYTOPLANKTON PHOTOSYNTHESIS AND GROWTH IN THE FRASER RIVER ESTUARY. MARINE BIOL. 19(2), 102-116. [MODEL]

TAYLOR, N. W. 1971. SIMULATION OF TRIBOLIUM POPULATIONS. PROC. ECOL. SOC. AUSTRALIA 6, 105-115. [COMPUTER]

THINGSTAD, T. F. AND T. I. LANGELAND. 1974. DYNAMICS OF A CHEMOSTAT CULTURE: THE EFFECT OF DELAY IN CELL RESPONSE. J. THEOR. BIOL. 48(1), 149-159. [MODEL; SIMULATION; LIMITING NUTRIENT]

THOMPSON, W. A., I. VERTINSKY AND J. R. KREBS. 1974. THE SURVIVAL VALUE OF FLOCKING IN BIRDS: A SIMULATION MODEL. J. ANIM. ECOL. 43(3), 785-820. [COMPUTER; PROBABILITY]

TIMIN, M. E. 1973. A MULTISPECIES CONSUMPTION MODEL. MATH. BIOSCI. 16, 59-66. [SIMULATION; COMPUTER]

TIMIN, M. E. AND B. D. COLLIER, 1972. SIMULATING THE ARCTIC TUNDRA ECOSYSTEM NEAR BARROW, ALASKA. PP. 1198-1204, IN, PROCEEDINGS OF THE 1972 SUMMER COMPUTER SIMULATION CONFERENCE, JUNE 14, 15, 16, 1972, SAN DIEGO, CALIFORNIA, VOL. II. [SIMULATION; COMPUTER; MODEL]

TIMIN, M. E., B. D. COLLIER, J. ZICH AND D. WALTERS. 1973. A COMPUTER SIMULATION OF THE ARCTIC ECOSYSTEM NEAR BARROW, ALASKA. U. S. IBP TUNDRA BIOME REPT. 73-1. III, 82 PP. [MODEL]

TOGNETTI, K. P. 1968. A DISCRETE BIO SIMULATION - THE POPULATION REGULATION OF TURTLES. PP. 364-363. IN, SECOND CONFERENCE ON APPLICATIONS OF SIMULATION, DECEMBER 2-4, 1968. NEW YORK, NEW YORK INSTITUTE OF ELECTRICAL AND ELECTRONIC ENGINEERS, NEW YORK, NEW YORK. CAT. NO. 68C60-SIM. [COMPUTER; MODEL]

UMNOV, A. A. 1971. APPLICATION OF THE MATHEMATICAL SIMULATION METHOD TO THE STUDY OF THE ROLE OF PHOTOSYNTHETIC AERATION OF LAKES. EKOLOGIYA 2(6), 5-12. (ENGLISH TRANSL. PP. 489-494). [COMPUTER; MODEL]

VILKITIS, J. R. AND R. H. GILES, JR. 1970. VIOLATION SIMULATION AS A TECHNIQUE FOR ESTIMATING ILLEGAL CLOSED-SEASON BIG GAME KILL. TRANS. N. E. FISH AND WILDLIFE CONF. 27, 83-87.

VINOGRADOV, M. E., V. V. MENSHUTKIN AND E. A. SHUSHKINA. 1972. ON MATHEMATICAL SIMULATION OF A PELAGIC ECOSYSTEM IN TROPICAL WATERS OF THE OCEAN. MARINE BIOL. 16(4), 261-268. [FINITE DIFFERENCE MODEL; COMPUTER]

WAGGONER, P. E. 1974. SIMULATION OF EPIDEMICS. PP. 137-160. IN, EPIDEMICS OF PLANT DISEASES: MATHEMATICAL ANALYSIS AND MODELING. (J. KRANZ, ED.). SPRINGER-VERLAG NEW YORK INC., NEW YORK. [COMPUTER]

WAGGONER, P. E. AND J. F. HORSFALL. 1969. EPIDEM - A SIMULATOR OF PLANT DISEASE WRITTEN FOR A COMPUTER. CONN. AGRIC. EXPT. STN. BULL. (NEW HAVEN) NO. 698. 80 PP.

WAGGONER, P. E., G. M. FURNIVAL AND W. E. REIFSNYDER. 1969. SIMULATION OF THE MICROCLIMATE IN A FOREST. FOREST SCI. 15(1), 37-45. [COMPUTER]

WALLER, W. T., M. L. DAHLBERG, R. E. SPARKS AND J. CAIRNS, JR. 1971. A COMPUTER SIMULATION OF THE EFFECTS OF SUPERIMPOSED MORTALITY DUE TO POLLUTANTS ON POPULATIONS OF FATHEAD MINNOWS (PIMEPHALES PROMELAS). J. FISH. RES. BOARD CAN. 28(8), 1107-1112. [STOCHASTIC MODEL]

WALSH, J. J. AND R. C. DUGDALE 1971. A SIMULATION MODEL OF THE NITROGEN FLOW IN THE PERUVIAN UPWELLING SYSTEM. INVESTIGACION PESQUERA 35(1), 309-330. [COMPUTER]

WALTERS, C. J. 1969. A GENERALIZED COMPUTER SIMULATION MODEL FOR FISH POPULATION STUDIES. TRANS. AMER. FISH. SOC. 98(3), 505-512.

WALTERS, C. J. 1975. DYNAMIC MODELS AND EVOLUTIONARY STRATEGIES. PP. 68-82. IN, ECOSYSTEM ANALYSIS AND PREDICTION. (S. A. LEVIN, ED.). PROCEEDINGS OF A SIAM-SIMS CONFERENCE HELD AT ALTA, UTAH, JULY 1-5, 1974.

WALTERS, C. J. AND F. BUNNELL. 1971. A COMPUTER MANAGEMENT GAME OF LAND USE IN BRITISH COLUMBIA. J. WILDL. MANAGE. 35(4), 644-657.

WALTERS, C. J. AND J. E. GROSS. 1972. DEVELOPMENT OF BIG GAME MANAGEMENT PLANS THROUGH SIMULATION MODELING. J. WILDL. MANAGE. 36(1), 119-128. [COMPUTER]

WALTERS, C. J. AND P. J. BANDY. 1972. PERIODIC HARVEST AS A METHOD OF INCREASING BIG GAME YIELDS. J. WILDL. MANAGE. 36(1), 128-134. [COMPUTER; SIMULATION MODEL]

WALTERS, C. J., R. HILLBORN, E. OGUSS, R. M. PETERMAN AND J. M. STANDER. 1974. DEVELOPMENT OF A SIMULATION MODEL OF MALLARD DUCK POPULATIONS. CANADIAN WILDLIFE SERVICE OCCASSIONAL PAPER NO. 20. 34 PP. [COMPUTER]

WALTON, G. S. 1965. A STUDY TO DEVELOP A COMPUTER PROGRAM FOR FOREST MANAGEMENT SIMULATIONS. M. S. THESIS, HARVARD UNIVERSITY, CAMBRIDGE, MASSACHUSETTS. THESIS.

WEBBER, M. I. 1974. FOOD WEB LINKAGE COMPLEXITY AND STABILITY IN A MODEL ECOSYSTEM. PP. 165-176. IN, ECOLOGICAL STABILITY. (M. B. USHER AND M. H. WILLIAMSON, EDS.). HALSTED PRESS. NEW YORK, NEW YORK. [STABILITY ANALYSIS; MATRIX; COMPUTER; SIMULATION]

WEHRHAHN, C. F. 1973. AN APPROACH TO MODELLING SPATIALLY HETEROGENEOUS POPULATIONS AND THE SIMULATION OF POPULATIONS SUBJECT TO STERILE INSECT RELEASE PROGRAMS. PP. 45-64. IN, COMPUTER MODELS AND APPLICATION OF THE STERILE-MALE TECHNIQUE. INTERNATIONAL ATOMIC ENERGY AGENCY. VIENNA, AUSTRIA. PUBLICATION STI/PUB/340. [FINITE DIFFERENCE MODEL; COMPUTER]

WIEBE, P. H. 1971. A COMPUTER MODEL STUDY OF ZOOPLANKTON PATCHINESS AND ITS EFFECTS ON SAMPLING ERROR. LIMNOL. OCEANOGR. 16(1), 29-38.

WIEGERT, R. G. 1973. A GENERAL ECOLOGICAL MODEL AND ITS USE IN SIMULATING ALGAL-FLY ENERGETICS IN A THERMAL SPRING COMMUNITY. PP. 85-102. IN, INSECTS: STUDIES IN POPULATION MANAGEMENT. (P. W. GEIER, L.R. CLARK, D. J. ANDERSON AND H. A. NIX, EDS.). ECOLOGICAL SOCIETY OF AUSTRALIA MEMOIRS 1. [SYSTEMS ANALYSIS; INTRINSIC RATE OF INCREASE; SIMULATION; EULER METHOD; COMPUTER]

WIENS, J. A. AND M. I. DYER. 1975. SIMULATION MODELLING OF RED-WINGED BLACKBIRD IMPACT ON GRAIN CROPS. J. APPL. ECOL. 12(1), 63-82. [MODEL; COMPUTER]

WIJNANDS-STAB, K. J. A. AND M. J. FRISSEL. 1973. COMPUTER SIMULATION FOR GENETIC CONTROL OF THE ONION FLY HYLEMYA ANTIQUA (MEIGEN). PP. 95-111. IN, COMPUTER MODELS AND APPLICATION OF THE STERILE-MALE TECHNIQUE. INTERNATIONAL ATOMIC ENERGY AGENCY. VIENNA, AUSTRIA. PUBLICATION STI/PUB/340. [MODEL]

WILLIAMS, R. B. 1971. COMPUTER SIMULATION OF ENERGY FLOW IN CEDAR BOG LAKE, MINNESOTA BASED ON THE CLASSICAL STUDIES OF LINDEMAN. PP. 543-582. IN, SYSTEMS ANALYSIS AND SIMULATION IN ECOLOGY, VOLUME I. (B. C. PATTEN, ED.). ACADEMIC PRESS, NEW YORK, NEW YORK. [MODEL]

WILSON, B. F. AND R. A. HOWARD. 1968. A COMPUTER MODEL FOR CAMBIAL ACTIVITY. FOREST SCI. 14(1), 77-90.

WINFREE, A. T. 1967. BIOLOGICAL RHYTHMS AND THE BEHAVIOR OF POPULATIONS OF COUPLED OSCILLATORS. J. THEOR. BIOL. 16(1), 15-42. [SIMULATION; COMPUTER]

WOODMANSEE, R. G. AND G. S. INNIS. 1973. A SIMULATION MODEL OF FOREST GROWTH AND NUTRIENT CYCLING. PP. 697-721. IN, PROCEEDINGS OF THE 1973 SUMMER COMPUTER SIMULATION CONFERENCE, JULY 17, 18, 19, 1973. MONTREAL, P. Q., CANADA, VOL. II. [COMPUTER]

WRIGHT, R. G. AND G. M. VAN DYNE (EDS.) 1970. SIMULATION AND ANALYSIS OF DYNAMICS OF A SEMI-DESERT GRASSLAND. COLORADO STATE UNIVERSITY, FT. COLLINS, COLORADO. RANGE SCIENCE SERIES NO. 6, III, 306 PP.

CYCLES

BALTENSWEILER, W., R. L. GIESE AND C. AUER. 1971. THE GREY LARCH BUD MOTH, ITS POPULATION FLUCTUATION IN OPTIMUM AND SUBOPTIMUM AREAS. PP. 401-415. IN, STATISTICAL ECOLOGY, VOLUME 2. SAMPLING AND MODELING BIOLOGICAL POPULATIONS AND POPULATION DYNAMICS. (G. P. PATIL, E. C. PIELOU AND W. E. WATERS, EDS.). PENNSYLVANIA STATE UNIVERSITY PRESS, UNIVERSITY PARK, PENNSYLVANIA.

BATSCHELET, E. 1965. STATISTICAL METHODS FOR THE ANALYSIS OF PROBLEMS IN ANIMAL ORIENTATION AND CERTAIN BIOLOGICAL RHYTHMS. AMERICAN INSTITUTE OF BIOLOGICAL SCIENCES, WASHINGTON, D. C. 57 PP.

BULMER, M. G. 1974. A STATISTICAL ANALYSIS OF THE 10-YEAR CYCLE IN CANADA. J. ANIM. ECOL. 43(3), 701-718. [CORRELOGRAM; PERIODOGRAM; SINE FUNCTION MODEL; SECOND-ORDER AUTOREGRESSIVE MODEL; SINE FUNCTION + FIRST-ORDER AUTOREGRESSIVE MODEL]

BULMER, M. G. 1975. PHASE RELATIONS IN THE TEN-YEAR CYCLE. J. ANIM. ECOL. 44(2), 609-621. [CONTINUOUS TIME MODEL; DISCRETE TIME MODEL; PREDATOR-PREY MODEL]

COLE, L. C. 1951. POPULATION CYCLES AND RANDOM OSCILLATIONS. J. WILDL. MANAGE. 15(3), 233-252. [PROBABILITY]

COLE, L. C. 1954. SOME FEATURES OF RANDOM POPULATION CYCLES. J. WILDL. MANAGE. 18(1), 2-24. [PROBABILITY]

COLE, L. C. 1957. BIOLOGICAL CLOCK IN THE UNICORN. SCIENCE 125(3253), 874-876.

COLWELL, R. K. 1974. PREDICTABILITY, CONSTANCY, AND CONTINGENCY OF PERIODIC PHENOMENA. ECOLOGY 55(5), 1148-1153. [MATRIX; MODEL]

COOK, L. M. 1965. OSCILLATION IN THE SIMPLE LOGISTIC GROWTH MODEL. NATURE 207(4994), 316.

CURLIN, J. W. 1970. MODELS OF THE HYDROLOGIC CYCLE. PP. 268-285. IN, ANALYSIS OF TEMPERATE FOREST ECOSYSTEMS. (D. E. REICHLE, ED.). SPRINGER-VERLAG, NEW YORK, NEW YORK.

DAVIDSON, J. AND H. G. ANDREWARTHA. 1948. ANNUAL TRENDS IN A NATURAL POPULATION OF THRIPS IMAGINIS (THYSANOPTERA). J. ANIM. ECOL. 17(2), 193-199. [LOGISTIC; GEOMETRIC MEAN; LOG TRANSFORMATION]

DE BACH, P. AND H. S. SMITH. 1941. ARE POPULATION OSCILLATIONS INHERENT IN THE HOST-PARASITE RELATION? ECOLOGY 22(4), 363-369. [NICHOLSON'S MODEL; LOTKA-VOLTERRA MODEL]

ENRIGHT, J. T. 1965. THE SEARCH FOR RHYTHMICITY IN BIOLOGICAL TIME-SERIES. J. THEOR. BIOL. 8(3), 426-468. [ESTIMATION OF FORM; PERIOD AND AMPLITUDE; PERIODOGRAM]

FINN, R. K. 1954. ACCOUNTING FOR PERIODICITIES IN BIOLOGY. BULL. MATH. BIOPHYS. 16, 181-182. [LINEAR CHEMICAL KINETICS; LINEAR DIFFERENTIAL EQUATIONS]

FRITTS, H. C. 1974. RELATIONSHIP OF RING WIDTHS IN ARID-SITE CONIFERS TO VARIATIONS IN MONTHLY TEMPERATURE AND PRECIPITATION. ECOL. MONOGR. 44(4), 411-440. [MULTIVARIATE ANALYSIS; CLUSTER ANALYSIS; PRINCIPAL COMPONENTS; COMPUTER]

FRITTS, H. C., T. J. BLASING, B. P. HAYDEN AND J. E. KUTZBACH 1971. MULTIVARIATE TECHNIQUES FOR SPECIFYING TREE-GROWTH AND CLIMATE RELATIONSHIPS AND FOR RECONSTRUCTING ANOMALIES IN PALEOCLIMATE. J. APPL. METEOROL. 10(5), 845-864. [PRINCIPAL COMPONENT ANALYSIS; MATRIX; MODEL; CANONICAL ANALYSIS]

FULLER, F. C., JR. AND C. P. TSOKOS. 1971. TIME SERIES ANALYSIS OF WATER POLLUTION DATA. BIOMETRICS 27(4), 1017-1034.

HARRIS, E. K. 1974. COMMENTS ON STATISTICAL METHODS FOR ANALYZING BIOLOGICAL RHYTHMS. PP. 757-760. IN, CHRONOBIOLOGY. (L. E. SCHEVING, F. HALBERG AND J. E. PAULY, EDS.). IGAKU SHOIN LTD., TOKYO, JAPAN.

HICKEY, J. J. 1954. MEAN INTERVALS IN INDICES OF WILDLIFE POPULATIONS. J. WILDL. MANAGE. 18(1), 90-106.

HOLGATE, P. 1966. TIME SERIES ANALYSIS APPLIED TO WILDFOWL COUNTS. APPL. STAT. 15(1), 15-23.

HUTCHINSON, G. E. 1954. THEORETICAL NOTES ON OSCILLATORY POPULATIONS. J. WILDL. MANAGE. 18(1), 107-109.

KOMIN, G. E., YU. A. P'YANKOV AND S. G. SHITYATOV. 1973. ASSESSMENT OF SIMILARITIES OF DENDROCHRONOLOGICAL SERIES. EKOLOGIYA 4(4), 29-34. (ENGLISH TRANSL. PP. 298-301). [TIME SERIES; COEFFICIENT OF TENDENCY; COEFFICIENT OF SIMILARITY; COMPUTER]

MORAN, P. A. P. 1949. THE STATISTICAL ANALYSIS OF THE SUNSPOT AND LYNX CYCLES. J. ANIM. ECOL. 18(1), 115-116. [CORRELATION COEFFICIENT; SERIAL CORRELATION COEFFICIENT; COEFFICIENTS; TIME SERIES]

MORAN, P. A. P. 1952. THE STATISTICAL ANALYSIS OF GAME-BIRD RECORDS. J. ANIM. ECOL. 21(1), 154-158. [CORRELATION COEFFICIENT; SERIAL CORRELATION COEFFICIENT]

MORAN, P. A. P. 1953. THE STATISTICAL ANALYSIS OF THE CANADIAN LYNX CYCLE I. STRUCTURE AND PREDICTION. AUST. J. ZOOL. 1(2), 163-173. [LOG TRANSFORMATION; MODELS OF STATIONARY STOCHASTIC PROCESSES; SERIAL CORRELATION COEFFICIENT]

MORAN, P. A. P. 1953. THE STATISTICAL ANALYSIS OF THE CANADIAN LYNX CYCLE II. SYNCHRONIZATION AND METEOROLOGY. AUST. J. ZOOL. 1(3), 291-298. [STOCHASTIC PROCESS; CORRELATION COEFFICIENT; SERIAL CORRELATION COEFFICIENT]

MORAN, P. A. P. 1954. THE STATISTICAL ANALYSIS OF GAME-BIRD RECORDS. II. J. ANIM. ECOL. 23(1), 35-37. [CORRELATION; SERIAL CORRELATION COEFFICIENT]

PAVLIDIS, T. 1967. A MODEL FOR CIRCADIAN CLOCKS. BULL. MATH. BIOPHYS. 29, 781-791.

PAVLIDIS, T. 1969. POPULATIONS OF INTERACTING OSCILLATORS AND CIRCADIAN RHYTHMS. J. THEOR. BIOL. 22(3), 418-436. [MODEL; SIMULATION; COMPUTER]

PAVLIDIS, T. 1973. THE FREE-RUN PERIOD OF CIRCADIAN RHYTHMS AND PHASE RESPONSE CURVES. AMER. NAT. 107(956), 524-530. [COMPUTER; SIMULATION]

PAVLIDIS, T., W. F. ZIMMERMAN AND J. OSBORN. 1968. A MATHEMATICAL MODEL FOR THE TEMPERATURE EFFECTS ON CIRCADIAN RHYTHMS. J. THEOR. BIOL. 18(2), 210-221.

READ, K. L. Q. AND J. R. ASHFORD. 1968. A SYSTEM OF MODELS FOR THE LIFE CYCLE OF A BIOLOGICAL ORGANISM. BIOMETRIKA 55(1), 211-221. [COMPUTER]

SCHUMACHER, F. X. AND B. B. DAY. 1939. THE INFLUENCE OF PRECIPITATION UPON THE WIDTH OF ANNUAL RINGS OF CERTAIN TIMBER TREES. ECOL. MONOGR. 9(4), 387-429. [MULTIPLE REGRESSION MODEL; ORTHOGONAL FUNCTIONS]

SKELLAM, J. G. 1967. SEASONAL PERIODICITY IN THEORETICAL POPULATION ECOLOGY. PP. 179-205. IN, BIOLOGY AND PROBLEMS OF HEALTH. PROC. FIFTH BERKELEY SYMP. MATH. STAT. PROBAB. VOL. 4. UNIVERSITY OF CALIFORNIA PRESS, BERKELEY, CALIFONIA.

SKELLAM, J. G. 1969. THE METHOD OF MINIMUM FLUCTUATIONS AND ITS APPLICATION TO THE STANDARDIZATION OF POPULATION INDICES. HANDBOOK OF 6TH INTERN. BIOMETRICS CONFERENCE (SYDNEY, AUSTRALIA), PP. 1.1-1.22.

SLOBODKIN, L. B. 1954. CYCLES IN ANIMAL POPULATIONS. AMER. SCI. 42(4), 658-660, 666.

STOCKTON, C. W. AND H. C. FRITTS. 1971. CONDITIONAL PROBABILITY OF OCCURRENCE FOR VARIATIONS IN CLIMATE BASED ON WIDTH OF ANNUAL TREE-RINGS IN ARIZONA. TREE-RING BULL. 31, 3-24. [COMPUTER; EXPONENTIAL]

SWADE, R. H. 1969. CIRCADIAN RHYTHMS IN FLUCTUATING LIGHT CYCLES: TOWARD A NEW MODEL OF ENTRAINMENT. J. THEOR. BIOL. 24(2), 227-239.

TAKAHASHI, F. 1964. REPRODUCTION CURVE WITH TWO EQUILIBRIUM POINTS: A CONSIDERATION OF THE FLUCTUATION OF INSECT POPULATIONS. RES. POPUL. ECOL.-KYOTO 6(1), 28-36.

TALBOT, G. B. 1954. FACTORS ASSOCIATED WITH FLUCTUATIONS IN ABUNDANCE OF HUDSON RIVER SHAD. U. S. FISH WILDL. SERV. FISH. BULL. 56, 373-413. [BINOMIAL DISTRIBUTION; TAG-RECAPTURE ESTIMATION; GEOMETRIC SERIES; MULTIPLE REGRESSION]

WASTLER, T. A. AND C. M. WALTER. 1968. STATISTICAL APPROACH TO ESTUARINE BEHAVIOR. AMER. SOC. CIVIL ENG., SAN. ENG. DIV. PROC. 94(6), 1175-1194. [TIME SERIES; POWER SPECTRUM; AUTOCORRELATION]

WHITTLE, P. 1954. THE STATISTICAL ANALYSIS OF A SEICHE RECORD. J. MARINE RES. 13(1), 76-100. [AUTOCOVARIANCE; CORRELATION COEFFICIENT; CORRELOGRAM; COVARIANCE; PERIODOGRAM; AUTOREGRESSION; MODEL; VECTOR; MATRIX]

WILLIAMSON, M. 1974. THE ANALYSIS OF DISCRETE TIME CYCLES. PP. 17-33. IN. ECOLOGICAL STABILITY. (M. B. USHER AND M. H. WILLIAMSON, EDS.). HALSTED PRESS, NEW YORK, NEW YORK. [MODEL; MATRIX; SIMULATION; 20 POINT CYCLE; CORRELOGRAM; MORAN-RICKER DIAGRAM]

DIVERSITY

BACKMANN, R. W. AND B. C. PATTEN. 1964. THE PLANKTON COMMUNITY. SCIENCE 144(3618), 556-558. [INFORMATION THEORY; SHANNON-WIENER MEASURE OF INFORMATION]

BHARGAVA, T. N. AND P. H. DOYLE. 1974. A GEOMETRIC STUDY OF DIVERSITY. J. THEOR. BIOL. 43(2), 241-251. [GINI'S DIVERSITY FUNCTION]

BLISS, C. I. 1965. AN ANALYSIS OF SOME INSECT TRAP RECORDS. PP. 385-397. IN, CLASSICAL AND CONTAGIOUS DISCRETE DISTRIBUTIONS. (G. P. PATIL, ED.). STATISTICAL PUBLISHING SOCIETY, CALCUTTA, INDIA. [LOGARTHMIC AND LOG NORMAL MODELS; FISHER'S LOGARTHMIC SERIES; POISSON DISTRIBUTION; DIVERSITY INDEX; TRUNCATED LOG NORMAL DISTRIBUTION; PRESTON'S OCTAVE]

BLISS, C. I. 1966. AN ANALYSIS OF SOME INSECT TRAP RECORDS. SANKHYA 28A(2/3), 123-136. [LOGARITHMIC SERIES; PRESTON'S TRUNCATED LOGNORMAL]

BOND, T. E. T. 1947. SOME CEYLON EXAMPLES OF THE LOGARITHMIC SERIES AND THE INDEX OF DIVERSITY OF PLANT AND ANIMAL POPULATIONS. CEYLON J. SCI. SECT. A BOT. 12(4), 195-202.

BONHAM, C. D. 1974. CLASSIFYING GRASSLAND VEGETATION WITH A DIVERSITY INDEX. J. RANGE MANAGE. 27(3), 240-243. [MCINTOSH'S INDEX]

BOTKIN, D. B. AND M. J. SOBEL. 1975. THE COMPLEXITY OF ECOSYSTEM STABILITY. PP. 144-150. IN, ECOSYSTEM ANALYSIS AND PREDICTION. (S. A. LEVIN, ED.). PROCEEDINGS OF A SIAM-SIMS CONFERENCE HELD AT ALTA, UTAH, JULY 1-5, 1974. [MODEL; DETERMINISTIC MODEL; STOCHASTIC]

BOWMAN, K. O., K. HUTCHESON, E. P. ODUM AND L. R. SHENTON. 1971. COMMENTS ON THE DISTRIBUTION OF INDICES OF DIVERSITY. PP. 315-359. IN, STATISTICAL ECOLOGY, VOLUME 3. MANY SPECIES POPULATIONS, ECOSYSTEMS, AND SYSTEMS ANALYSIS. (G. P. PATIL, E. C. PIELOU AND W. E. WATERS, EDS.). PENNSYLVANIA STATE UNIVERSITY PRESS, UNIVERSITY PARK, PENNSYLVANIA.

BRIAN, M. V. 1953. SPECIES FREQUENCIES IN RANDOM SAMPLES FROM ANIMAL POPULATIONS. J. ANIM. ECOL. 22(1), 57-64. [MODEL]

BULLOCK, J. A. 1971. THE INVESTIGATION OF SAMPLES CONTAINING MANY SPECIES I. SAMPLE DESCRIPTION. BIOL. J. LINN. SOC. 3(1), 1-21. [DIVERSITY INDEX; MODEL; INFORMATION; LOG-SERIES; LOG-NORMAL DISTRIBUTION; BROKEN-STICK MODEL; SHANNON-WIENER; MCINTOSH]

BULLOCK, J. A. 1971. THE INVESTIGATION OF SAMPLES CONTAINING MANY SPECIES II. SAMPLE COMPARISON. BIOL. J. LINN. SOC. 3(1), 23-56. [INDEX OF SIMILARITY; TRANSFORMATION; MOUNTFORD'S INDEX; KENDALL'S RANK CORRELATION COEFFICIENT; LOGARITHMIC SERIES; SORENSEN'S QUOTIENT; PRESTON'S OCTAVE; TRUNCATED NORMAL DISTRIBUTION]

BULMER, M. G. 1974. ON FITTING THE POISSON LOGNORMAL DISTRIBUTION TO SPECIES-ABUNDANCE DATA. BIOMETRICS 30(1), 101-110. [MACARTHUR'S BROKEN-STICK MODEL; MAXIMUM LIKELIHOOD; GAMMA DISTRIBUTION; PROBABILITY; DIVERSITY MEASURE; INFORMATION THEORY]

CAIRNS, J., JR. AND K. L. DICKSON. 1971. A SIMPLE METHOD FOR THE BIOLOGICAL ASSESSMENT OF THE EFFECTS OF WASTE DISCHARGES ON AQUATIC BOTTOM-DWELLING ORGANISMS. J. WATER POLLUT. CONTROL FED. 43(5), 755-772. [INFORMATION THEORY; COMPUTER; SEQUENTIAL SAMPLING]

CAIRNS, J., JR., D. W. ALBAUGH, F. BUSEY AND M. D. CHANAY. 1968. THE SEQUENTIAL COMPARISON INDEX- A SIMPLIFIED METHOD FOR NON-BIOLOGISTS TO ESTIMATE RELATIVE DIFFERENCES IN BIOLOGICAL DIVERSITY IN STREAM POLLUTION STUDIES. J. WATER POLLUT. CONTROL FED. 40(9), 1607-1613. [RUN TEST]

COHEN, J. E. 1968. ALTERNATE DERIVATIONS OF A SPECIES-ABUNDANCE RELATION. AMER. NAT. 102(924), 165-172. [BROKEN-STICK MODEL; EXPONENTIAL MODEL; BALL'S AND BOX'S MODEL]

CONRAD, M. 1972. STABILITY OF FOODWEBS AND ITS RELATION TO SPECIES DIVERSITY. J. THEOR. BIOL. 34(2), 325-335. [INDEX OF STABILITY; DIVERSITY INDEX]

CURTIS, J. T. AND R. P. MCINTOSH. 1951. ADDITIONAL NOTE ON THE INTERRELATIONS OF PHYTOSOCIOLOGICAL CHARACTERS. ECOLOGY 32(2), 345. [PRESTON'S LOG-NORMAL; POISSON SERIES]

DAWDY, D. R. 1964. STATISTICAL MODELS FOR PREDICTING NUMBERS OF PLANT SPECIES. SCIENCE 146(3647), 1074-1075.

DAWSON, G. W. P. 1951. A METHOD FOR INVESTIGATING THE RELATIONSHIP BETWEEN THE DISTRIBUTION OF INDIVIDUALS OF DIFFERENT SPECIES IN A PLANT COMMUNITY. ECOLOGY 32(2), 332-334. [DISCUSSION]

DEBENEDICTIS, P. A. 1973. ON THE CORRELATIONS BETWEEN CERTAIN DIVERSITY INDICES. AMER. NAT. 107(954), 295-302. [INFORMATION THEORY; MARGALEF'S DIVERSITY INDEX]

DICKMAN, M. 1968. SOME INDICES OF DIVERSITY. ECOLOGY 49(6), 1191-1193. [SHANNON-WEAVER FORMULA; MARGALEF'S INDEX]

EBERHARDT, L. L. 1969. SOME ASPECTS OF SPECIES DIVERSITY MODELS. ECOLOGY 50(3), 503-505.

EDDEN, A. C. 1971. A MEASURE OF SPECIES DIVERSITY RELATED TO THE LOGNORMAL DISTRIBUTION OF INDIVIDUALS AMONG SPECIES. J. EXPTL. MAR. BIOL. ECOL. 6(3), 199-209. [DIVERSITY INDEX; SHANNON-WEAVER INDEX; PRESTON'S INDEX]

ENGEN, S. 1974. ON SPECIES FREQUENCY MODELS. BIOMETRIKA 61(2), 263-270. [LOGARITHMIC SERIES; DIVERSITY; NEGATIVE BINOMIAL DISTRIBUTION]

ENGEN, S. 1975. THE COVERAGE OF A RANDOM SAMPLE FROM A BIOLOGICAL COMMUNITY. BIOMETRICS 31(1), 201-208. [TAXONOMIC GROUP; INFORMATION INDEX OF DIVERSITY; NEGATIVE BINOMIAL DISTRIBUTION; MODEL]

EVANS, F. C. 1950. RELATIVE ABUNDANCE OF SPECIES AND THE PYRAMID OF NUMBERS. ECOLOGY 31(4), 631-632.

FAGER, E. W. 1972. DIVERSITY: A SAMPLING STUDY. AMER. NAT. 106(949), 293-310. [DIVERSITY INDEX; RAREFACTION METHOD; SIMPSON'S INDEX; SHANNON-WIENER INDEX; SIMULATION; COMPUTER]

FINDLEY, J. S. 1973. PHENETIC PACKING AS A MEASURE OF FAUNAL DIVERSITY. AMER. NAT. 107(956), 580-584. [TAXONOMIC DISTANCE]

FISHER, R. A., A. S. CORBET AND C. B. WILLIAMS. 1943. THE RELATION BETWEEN THE NUMBER OF SPECIES AND THE NUMBER OF INDIVIDUALS IN A RANDOM SAMPLE OF AN ANIMAL POPULATION. J. ANIM. ECOL. 12(1), 42-58.

GAGE, J. AND P. B. TETT. 1973. THE USE OF LOG-NORMAL STATISTICS TO DESCRIBE THE BENTHOS OF LOCHS ETIVE AND CRERAN. J. ANIM. ECOL. 42(2), 373-382. [LOG-NORMAL DISTRIBUTION; ORDINATION; COMPONENTS OF DIVERSITY]

GALLOPIN, G. C. 1972. TROPHIC SIMILARITY BETWEEN SPECIES IN A FOOD WEB. AMER. MIDL. NAT. 87(2), 336-345. [INDICES]

GHENT, A. W. AND B. P. HANNA, 1968. APPLICATION OF THE "BROKEN STICK" FORMULA TO THE PREDICTION OF RANDOM TIME INTERVALS. AMER. MIDL. NAT. 79(2), 273-288.

GLEASON, H. A. 1955. ESTIMATION OF THE NUMBER OF SPECIES PRESENT ON A GIVEN AREA. ECOLOGY 36(2), 342-343.

GLIME, J. M. AND R. M. CLEMONS. 1972. SPECIES DIVERSITY OF STREAM INSECTS ON FONTINALIS SPP. COMPARED TO DIVERSITY ON ARTIFICIAL SUBSTRATES. ECOLOGY 53(3), 458-464. [SPEARMAN'S RANK CORRELATION; COMMUNITY COEFFICIENTS; SORENSEN'S K; INFORMATION THEORY]

GOOD, I. J. 1953. THE POPULATION FREQUENCIES OF SPECIES AND ESTIMATION OF POPULATION PARAMETERS. BIOMETRIKA 40(3/4), 237-264. [SHANNON'S ENTROPY; LOGARITHMIC SERIES; YULE'S CHARACTERISTIC]

GOODALL, D. W. 1970. STATISTICAL PLANT ECOLOGY. PP. 99-124. IN, ANNUAL REVIEW OF ECOLOGY AND SYSTEMATICS, VOLUME 1. (R. F. JOHNSTON, P. W. FRANK AND C. D. MICHENER, EDS.). ANNUAL REVIEWS, INC., PALO ALTO, CALIFORNIA. [SPATIAL PATTERN; SAMPLING; DIVERSITY; SIMILARITY; DISTANCE ORDINATION; NEAREST NEIGHBOR; INFORMATION ANALYSIS; MODEL ANALYSIS; MODEL]

GOULDEN, C. E. 1969. DEVELOPMENTAL PHASES OF THE BIOCOENOSIS. PROC. NATL. ACAD. SCI. U. S. A. 62(4), 1066-1073. [SHANNON-WIENER INFORMATION THEORY; DIVERSITY INDEX]

GREEN, R. H. 1971. A MULTIVARIATE STATISTICAL APPROACH TO THE HUTCHINSONIAN NICHE: BIVALVE MOLLUSCS OF CENTRAL CANADA. ECOLOGY 52(4), 543-556. [DISCRIMINANT FUNCTION]

GRUNDY, P. M. 1951. THE EXPECTED FREQUENCIES IN A SAMPLE OF AN ANIMAL POPULATION IN WHICH THE ABUNDANCES OF SPECIES ARE LOG-NORMALLY DISTRIBUTED. BIOMETRIKA 38(3/4), 427-434.

HAIRSTON, N. G. 1959. SPECIES ABUNDANCE AND COMMUNITY ORGANIZATION. ECOLOGY 40(3), 404-416. [LOGARITHMIC SERIES; OCTAVE; MODEL; INFORMATION THEORY; INDEX OF HETEROGENEITY]

HILL, M. O. 1973. DIVERSITY AND EVENNESS: A UNIFYING NOTATION AND ITS CONSEQUENCES. ECOLOGY 54(2), 427-432. [SIMPSON'S INDEX; SHANNON'S ENTROPY; SPECIES ABUNDANCE; RENYI'S GENERALIZED ENTROPY; SPECIES ABUNDANCE]

HORN, H. S. 1966. MEASUREMENT OF "OVERLAP" IN COMPARATIVE ECOLOGICAL STUDIES. AMER. NAT. 100(914), 419-424. [PROBABILITY; MODEL; MORISITA-I; SHANNON-WIENER MEASURE OF INFORMATION; SIMPSON'S INDEX OF DIVERSITY]

HUMMON, W. D. 1974. SH': A SIMILARITY INDEX BASED ON SHARED SPECIES DIVERSITY, USED TO ASSESS TEMPORAL AND SPATIAL RELATIONS AMONG INTERTIDAL MARINE GASTROTRICHA. OECOLOGIA 17(3), 203-220. [DENDOGRAM]

HURLBERT, S. H. 1971. THE NONCONCEPT OF SPECIES DIVERSITY: A CRITIQUE AND ALTERNATIVE PARAMETERS. ECOLOGY 52(4), 577-586. [DIVERSITY INDEX; INFORMATION; LOGARITHMIC SERIES; SHANNON'S FORMULA; PROBABILITY]

HURTUBIA, J. 1973. TROPHIC DIVERSITY MEASUREMENT IN SYMPATRIC PREDATORY SPECIES. ECOLOGY 54(4), 885-890. [PIELOU'S METHOD]

HUTCHESON, K. 1970. A TEST FOR COMPARING DIVERSITIES BASED ON THE SHANNON FORMULA. J. THEOR. BIOL. 29(1), 151-154.

JUMARS, P. A. 1974. TWO PITFALLS IN COMPARING COMMUNITIES OF DIFFERING DIVERSITIES. AMER. NAT. 108(961), 389-391.

KEMPTON, R. A. AND L. R. TAYLOR. 1974. LOG-SERIES AND LOG-NORMAL PARAMETERS AS DIVERSITY DISCRIMINANTS FOR THE LEPIDOPTERA. J. ANIM. ECOL. 43(2), 381-399. [MODEL; GAMMA DISTRIBUTION; WILLIAM'S MODEL]

KERNER, E. H. 1974. WHY ARE THERE SO MANY SPECIES? BULL. MATH. BIOL. 36(5/6), 477-488. [VOLTERRA'S MODEL; PRESTON'S MODEL; NICHE]

KING, C. E. 1964. RELATIVE ABUNDANCE OF SPECIES AND MACARTHUR'S MODEL. ECOLOGY 45(4), 716-727.

KOHN, A. J. 1959. THE ECOLOGY OF CONUS IN HAWAII. ECOL. MONOGR. 29(1), 47-90. [MACARTHUR'S ABUNDANCE EQUATION; KOCH'S DIVERSITY INDEX; MARGALEF'S HETEROGENEITY INDEX; INFORMATION THEORY]

KRYLOV, V. V. 1971. ON THE STATION-SPECIES CURVE. PP. 233-235. IN, STATISTICAL ECOLOGY, VOLUME 3. MANY SPECIES POPULATIONS, ECOSYSTEMS, AND SYSTEMS ANALYSIS. (G. P. PATIL, E. C. PIELOU AND W. E. WATERS, EDS.). PENNSYLVANIA STATE UNIVERSITY PRESS, UNIVERSITY PARK, PENNSYLVANIA.

LEGENDRE, L. 1973. PHYTOPLANKTON ORGANIZATION IN BAIE DES CHALEURS (GULF OF ST. LAWRENCE). J. ECOL. 61(1), 135-149. [BRILLOUIN'S FORMULA; DIVERSITY; MATRIX; SIMILARITY COEFFICIENTS]

LEIGH, E. G., JR. 1965. ON THE RELATION BETWEEN THE PRODUCTIVITY, BIOMASS, DIVERSITY, AND STABILITY OF A COMMUNITY. PROC. NATL. ACAD. SCI. U. S. A. 53(4), 777-783.

LEWONTIN, R. C. 1969. THE MEANING OF STABILITY. PP. 13-23. IN, DIVERSITY AND STABILITY IN ECOLOGICAL SYSTEMS. BROOKHAVEN SYMPOSIUM IN BIOLOGY NO. 22. [SPACE; VECTOR FIELD; MATRIX; EIGENVALUE; LOTKA-VOLTERRA MODEL]

LLOYD, M. 1964. WEIGHTING INDIVIDUALS BY REPRODUCTIVE VALUE IN CALCULATING SPECIES DIVERSITY. AMER. NAT. 98(900), 190-192.

LLOYD, M. 1967. 'MEAN CROWDING'. J. ANIM. ECOL. 36(1), 1-30.

LLOYD, M. AND R. J. GHELARDI. 1964. A TABLE FOR CALCULATING THE 'EQUITABILITY' COMPONENT OF SPECIES DIVERSITY. J. ANIM. ECOL. 33(2), 217-225.

LLOYD, M., J. H. ZAR AND J. R. KARR. 1968. ON THE CALCULATION OF INFORMATION-THEORETICAL MEASURES OF DIVERSITY. AMER. MIDL. NAT. 79(2), 257-272. [SHANNON'S FORMULA; BRILLOUIN'S DIVERSITY; STANDARD ERROR]

LLOYD, M., R. F. INGER AND F. W. KING. 1968. ON THE DIVERSITY OF REPTILE AND AMPHIBIAN SPECIES IN A BORNEAN RAIN FOREST. AMER. NAT. 102(928), 497-515. [SHANNON'S FUNCTION; BRILLOUIN MEASURE OF DIVERSITY; PIELOU'S METHOD]

LONGUET-HIGGINS, M. S. 1971. ON THE SHANNON-WEAVER INDEX OF DIVERSITY, IN RELATION TO THE DISTRIBUTION OF SPECIES IN BIRD CENSUSES. THEOR. POPUL. BIOL. 2(3), 271-289. [LOG EXPONENTIAL DISTRIBUTION; LOG UNIFORM DISTRIBUTION; BROKEN-STICK MODEL; GEOMETRIC SERIES; DIRICHLET SERIES]

MACARTHUR, R. 1960. ON THE RELATIVE ABUNDANCE OF SPECIES. AMER. NAT. 94(874), 25-36.

MACARTHUR, R. H. 1957. ON THE RELATIVE ABUNDANCE OF BIRD SPECIES. PROC. NATL. ACAD. SCI. U. S. A. 43(3), 293-295. [PROBABILITY; BROKEN-STICK MODEL]

MACARTHUR, R. H. 1964. ENVIRONMENTAL FACTORS AFFECTING BIRD SPECIES DIVERSITY. AMER. NAT. 98(903), 387-397. [DIVERSITY INDEX]

MACARTHUR, R. H. 1965. PATTERNS OF SPECIES DIVERSITY. BIOL. REV. 40(4), 510-533. [LOG-NORMAL DISTRIBUTION; POISSON DISTRIBUTION; PRESTON'S OCTAVE; INFORMATION THEORY]

MACARTHUR, R. H. AND J. W. MACARTHUR. 1961. ON BIRD DIVERSITY. ECOLOGY 42(3), 594-598. [DIVERSITY INDEX; POISSON DISTRIBUTION; INFORMATION]

MACARTHUR, R., H. RECHER, AND M. CODY. 1966. ON THE RELATION BETWEEN HABITAT SELECTION AND SPECIES DIVERSITY. AMER. NAT. 100(913), 319-332.

MARGALEF, D. R. 1958. INFORMATION THEORY IN ECOLOGY. MEMORIAS DE LA REAL ACADEMIA DE CIENCIAS Y ARTES DE BARCELONA 23, 373-449. (ENGLISH TRANSL., 1958, GENERAL SYSTEMS 3, 36-71.).

MARGALEF, R. 1961. COMMUNICATION OF STRUCTURE IN PLANKTONIC POPULATIONS. LIMNOL. OCEANOGR. 6(2), 124-128. [INFORMATION THEORY; ENTROPY; ANALOG CIRCUIT]

MARGALEF, R. 1969. DIVERSITY AND STABILITY: A PRACTICAL PROPOSAL AND A MODEL OF INTERDEPENDENCE. PP. 25-37. IN. DIVERSITY AND STABILITY IN ECOLOGICAL SYSTEMS. BROOKHAVEN SYMPOSIUM IN BIOLOGY NO. 22. [SHANNON-WEAVER FUNCTION; MATRIX]

MAY, R. M. 1974. GENERAL INTRODUCTION. PP. 1-14. IN. ECOLOGICAL STABILITY. (M. B. USHER AND M. H. WILLIAMSON, EDS.). HALSTED PRESS, NEW YORK, NEW YORK. [THEORY; HISTORY; MODEL; SPECIES PATTERNS; LOG-NORMAL DISTRIBUTION; EQUITABILITY INDICES]

MCCLOSKEY, J. W. 1965. A MODEL FOR THE DISTRIBUTION OF INDIVIDUALS BY SPECIES IN AN ENVIRONMENT. MICHIGAN STATE UNIVERSITY REPORT 7, DEPARTMENT OF STATISTICS, MICHIGAN STATE UNIVERSITY, EAST LANSING, MICHIGAN.

MCINTOSH, R. P. 1967. AN INDEX OF DIVERSITY AND THE RELATION OF CERTAIN CONCEPTS TO DIVERSITY. ECOLOGY 48(3), 392-404.

MENHINICK, E. F. 1964. A COMPARISON OF SOME SPECIES - INDIVIDUALS DIVERSITY INDICES APPLIED TO SAMPLES OF FIELD INSECTS. ECOLOGY 45(4), 859-861. [MONTE CARLO ANALYSIS; GLEASON'S INDEX; MARGALEF'S INDEX]

MILSUM, J. H. 1973. A SHORT NOTE ON "STABILITY IN MULTISPECIES COMMUNITY MODELS" BY ROBERT M. MAY, MATHEMATICAL BIOSCIENCES 12, 59-79 (1971). MATH. BIOSCI. 17, 189-190. [MODEL; PREDATOR-PREY MODEL]

MITCHELL, R. 1965. ANALYSIS OF SPECIES ABUNDANCE IN A WATER MITE GENUS. AMER. NAT. 99(905), 117-124. [MACARTHUR'S INDEX]

MOSS, B. 1973. DIVERSITY IN FRESH-WATER PHYTOPLANKTON. AMER. MIDL. NAT. 90(2), 341-355. [SHANNON'S FORMULA; GROWTH AND DIVERSITY MODEL]

ODUM, H. T., J. E. CANTLON AND L. S. KORNICKER. 1960. AN ORGANIZATIONAL HIERARCHY POSTULATE FOR THE INTERPRETATION OF SPECIES-INDIVIDUAL DISTRIBUTIONS, SPECIES ENTROPY, ECOSYSTEM EVOLUTION, AND THE MEANING OF A SPECIES-VARIETY INDEX. ECOLOGY 41(2), 395-399.

ORLOCI, L. 1968. INFORMATION ANALYSIS IN PHYTOSOCIOLOGY, PARTITION, CLASSIFICATION AND PREDICTION. J. THEOR. BIOL. 20(3), 271-284. [CLUSTER ANALYSIS; BRILLOUIN'S DIVERSITY; MATRIX; SHANNON'S FORMULA; ENTROPY; KULLBECK-LEIBLER INFORMATION STATISTIC]

ORLOCI, L. 1970. ANALYSIS OF VEGETATION SAMPLES BASED ON THE USE OF INFORMATION. J. THEOR. BIOL. 29(2), 173-189. [I-DIVERGENCES; ALGORITHM; INFORMATION FUNCTION]

PARRISH, J. D. AND S. B. SAILA. 1970. INTERSPECIFIC COMPETITION, PREDATION AND SPECIES DIVERSITY. J. THEOR. BIOL. 27(2), 207-220. [MODEL; LOTKA-VOLTERRA; LOGISTIC; COMPUTER]

PATRICK, R. AND D. STRAWBRIDGE. 1963. VARIATION IN THE STRUCTURE OF NATURAL DIATOM COMMUNITIES. AMER. NAT. 97(892), 51-57. [TRUNCATED NORMAL DISTRIBUTION; BIVARIATE NORMAL DISTRIBUTION; VARIANCE]

PATTEN, B. C. 1961. NEGENTROPY FLOW IN COMMUNITIES OF PLANKTON. LIMNOL. OCEANOGR. 6(1), 26-30. [MODEL; ENTROPY; INFORMATION THEORY]

PATTEN, B. C. 1962. SPECIES DIVERSITY IN NET PHYTOPLANKTON OF RARITAN BAY. J. MARINE RES. 20(1), 57-75. [INFORMATION THEORY; DIVERSITY INDEX]

PATTEN, B. C. 1963. INFORMATION PROCESSING BEHAVIOR OF A NATURAL PLANKTON COMMUNITY. AMER. BIOL. TEACH. 25(7), 489-501. [INFORMATION THEORY; ENTROPY]

PATTEN, B. C. 1963. PLANKTON: OPTIMUM DIVERSITY STRUCTURE OF A SUMMER COMMUNITY. SCIENCE 140(3569), 894-898. [LINEAR PROGRAMMING; DIVERSITY]

PEET, R. K. 1974. THE MEASUREMENT OF SPECIES DIVERSITY. ANNU. REV. ECOL. SYST. 5, 285-307. [SIMPSON'S INDEX; INFORMATION THEORY; SHANNON'S FORMULA; PIELOU'S J; HILL RATIOS; LLOYD AND GHELARDI'S E]

PIELOU, E. C. 1966. COMMENTS ON A REPORT BY J. H. VANDERMEER AND R. H. MACARTHUR CONCERNING THE BROKEN STICK MODEL OF SPECIES ABUNDANCE. ECOLOGY 47(6), 1073-1074.

PIELOU, E. C. 1966. THE MEASUREMENT OF DIVERSITY IN DIFFERENT BIOLOGICAL COLLECTIONS. J. THEOR. BIOL. 13, 131-144. [INFORMATION; BRILLOUIN'S DIVERSITY; SHANNON'S FORMULA]

PIELOU, E. C. 1966. SHANNON'S FORMULA AS A MEASURE OF SPECIFIC DIVERSITY: ITS USE AND MISUSE. AMER. NAT. 100(914), 463-465. [VARIANCE; BRILLOUIN'S FORMULA; INFORMATION; SHANNON AND WEAVER]

PIELOU, E. C. 1966. SPECIES-DIVERSITY AND PATTERN-DIVERSITY IN THE STUDY OF ECOLOGICAL SUCCESSION. J. THEOR. BIOL. 10(2), 370-383. [BRILLOUIN'S DIVERSITY; INFORMATION THEORY; SHANNON'S FORMULA]

PIELOU, E. C. 1967. THE USE OF INFORMATION THEORY IN THE STUDY OF THE DIVERSITY OF BIOLOGICAL POPULATIONS. PP. 163-177. IN, BIOLOGY AND PROBLEMS OF HEALTH. PROC. FIFTH BERKELEY SYMP. MATH. STAT. PROBAB. VOL. 4. UNIVERSITY OF CALIFORNIA PRESS, BERKELEY, CALIFONIA.

PIELOU, E. C. AND A. N. ARNASON. 1966. CORRECTION TO ONE OF MACARTHUR'S SPECIES-ABUNDANCE FORMULAS. SCIENCE 151(3710), 592.

POWER, D. M. 1971. WARBLER ECOLOGY: DIVERSITY, SIMILARITY, AND SEASONAL DIFFERENCES IN HABITAT SEGREGATION. ECOLOGY 52(3), 434-443. [DIVERSITY INDEX; EUCLIDEAN DISTANCE; NICHE OVERLAP]

PRESTON, F. W. 1948. THE COMMONNESS, AND RARITY, OF SPECIES. ECOLOGY 29(3), 254-283.

PRESTON, F. W. 1958. ANALYSIS OF THE AUDUBON CHRISTMAS COUNTS IN TERMS OF THE LOGNORMAL CURVE. ECOLOGY 39(4), 620-624. [OCTAVE]

PRESTON, F. W. 1960. TIME AND SPACE AND THE VARIATION OF SPECIES. ECOLOGY 41(4), 611-627. [LOG-NORMAL DISTRIBUTION; LOG TRANSFORMATION; ARRHENIUS'S EQUATION]

PRESTON, F. W. 1962. THE CANONICAL DISTRIBUTION OF COMMONNESS AND RARITY: PART I. ECOLOGY 43(2), 185-215.

PRESTON, F. W. 1962. THE CANONICAL DISTRIBUTION OF COMMONNESS AND RARITY: PART II. ECOLOGY 43(3), 410-432.

PRESTON, F. W. 1969. DIVERSITY AND STABILITY IN THE BIOLOGICAL WORLD. PP. 1-12. IN, DIVERSITY AND STABILITY IN ECOLOGICAL SYSTEMS. BROOKHAVEN SYMPOSIUM IN BIOLOGY NO. 22.

RECHER, H. F. 1971. BIRD SPECIES DIVERSITY: A REVIEW OF THE RELATION BETWEEN SPECIES NUMBER AND ENVIRONMENT. PROC. ECOL. SOC. AUSTRALIA 6, 135-152. [DIVERSITY INDEX; SHANNON INFORMATION FUNCTION]

REGIER, H. A. AND H. F. HENDERSON. 1973. TOWARDS A BROAD ECOLOGICAL MODEL OF FISH COMMUNITIES AND FISHERIES. TRANS. AMER. FISH. SOC. 102(1), 56-72. [DIVERSITY; MARGALEF'S INDEX; SHANNON-WIENER INDEX; MACARTHUR'S INDEX; RYDER'S INDEX]

RICE, E. L. AND R. W. KELTING. 1955. THE SPECIES-AREA CURVE. ECOLOGY 36(1), 7-11.

ROSENZWEIG, M. L. AND J. WINAKUR. 1969. POPULATION ECOLOGY OF DESERT RODENT COMMUNITIES: HABITATS AND ENVIRONMENTAL COMPLEXITY. ECOLOGY 50(4), 558-572. [DIVERSITY INDEX; MACARTHUR'S MODEL; MODEL OF HABITAT COMPLEXITY]

SANDERS, H. L. 1968. MARINE BENTHIC DIVERSITY: A COMPARATIVE STUDY. AMER. NAT. 102(925), 243-282. [MACARTHUR'S MODEL; RAREFACTION METHOD; FAUNAL INDICES; MARGALEF'S INDEX; DIVERSITY INDEX; SHANNON-WIENER INFORMATION; FUNCTION]

SHELDON, A. L. 1969. EQUITABILITY INDICES: DEPENDENCE ON THE SPECIES COUNT. ECOLOGY 50(3), 466-467. [DIVERSITY INDEX]

SHINOZAKI, K. AND N. URATA. 1953. APPARENT ABUNDANCE OF DIFFERENT SPECIES AND HETEROGENEITY. RES. POPUL. ECOL.-KYOTO 2, 8-21. [MODEL]

SIMBERLOFF, D. 1972. PROPERTIES OF THE RAREFACTION DIVERSITY MEASURE. AMER. NAT. 106(949), 414-418. [DIVERSITY INDEX; SAMPLE SIZE]

SIMPSON, E. H. 1949. MEASUREMENT OF DIVERSITY. NATURE 163(4148), 688.

SIROMONEY, G. 1962. ENTROPY OF LOGARITHMIC SERIES DISTRIBUTIONS. SANKHYA 24A(4), 419-420. [FLEA; BUTTERFLY]

SPIGHT, T. M. 1967. SPECIES DIVERSITY: A COMMENT ON THE ROLE OF THE PREDATOR. AMER. NAT. 101(922), 467-474. [PAINE'S HYPOTHESIS; MODEL]

TERBORGH, J. 1973. ON THE NOTION OF FAVORABLENESS IN PLANT ECOLOGY. AMER. NAT. 107(956), 481-501.

TRAVERS, M. 1971. DIVERSITE DU MICROPLANCTON DU GOLFE DE MARSEILLE EN 1964. MARINE BIOL. 8(4), 308-343. [DIVERSITY INDEX]

VAN EMDEN, H. F. AND G. F. WILLIAMS. 1974. INSECT STABILITY AND DIVERSITY IN AGRO-ECOSYSTEMS. ANNU. REV. ENTOMOL. 19, 455-475. [MODEL]

VANDERMEER, J. H. AND R. H. MACARTHUR. 1966. A REFORMULATION OF ALTERNATIVE (B) OF THE BROKEN STICK MODEL OF SPECIES ABUNDANCE. ECOLOGY 47(1), 139-140.

WATTERSON, G. A. 1974. MODELS FOR THE LOGARITHMIC SPECIES ABUNDANCE DISTRIBUTIONS. THEOR. POPUL. BIOL. 6(2), 217-250.

WEBB, D. J. 1974. THE STATISTICS OF RELATIVE ABUNDANCE AND DIVERSITY. J. THEOR. BIOL. 43(2), 277-291. [MACARTHUR AND COHEN MODELS]

WHITTAKER, R. H. 1965. DOMINANCE AND DIVERSITY IN LAND PLANT COMMUNITIES. SCIENCE 147(3655), 250-260. [SIMPSON'S INDEX OF DIVERSITY; OCTAVE; MACARTHUR'S INDEX; MARGALEF'S INDEX; FISHER, CORBET AND WILLIAMS INDEX; LOG-NORMAL DISTRIBUTION]

WHITTAKER, R. H. 1972. EVOLUTION AND MEASUREMENT OF SPECIES DIVERSITY. TAXON 21(2/3), 213-251. [DIVERSITY INDEX]

WHITTAKER, R. H. 1974. DIRECT GRADIENT ANALYSIS: RESULTS. PP. 33-51. IN, ORDINATION AND CLASSIFICATION OF COMMUNITIES. (R. H. WHITTAKER, ED.). DR. JUNK PUBLISHERS, THE HAGUE, NETHERLANDS. [DIVERSITY]

WHITTAKER, R. H. AND G. M. WOODWELL. 1974. RETROGRESSION AND COENOCLINE DISTANCE. PP. 53-73. IN, ORDINATION AND CLASSIFICATION OF COMMUNITIES. (R. H. WHITTAKER, ED.). DR. JUNK PUBLISHERS, THE HAGUE, NETHERLANDS. [DIVERSITY; SIMILARITY; WEIGHTED AVERAGE; COEFFICIENT OF COMMUNITY]

WILHM, J. 1972. GRAPHIC AND MATHEMATICAL ANALYSES OF BIOTIC COMMUNITIES IN POLLUTED STREAMS ANNU. REV. ENTOMOL. 17, 223-252. [INDEX OF POLLUTION; SHANNON'S FORMULA; BIOTIC INDEX; DIVERSITY INDEX; REGRESSION MODEL; ORDINATION]

WILHM, J. L. 1968. USE OF BIOMASS UNITS IN SHANNON'S FORMULA. ECOLOGY 49(1), 153-156. [DIVERSITY INDEX; INFORMATION; PIELOU'S CLARIFICATION]

WILHM, J. L. AND T. C. DORRIS. 1968. BIOLOGICAL PARAMETERS FOR WATER QUALITY CRITERIA. BIOSCIENCE 18(6), 477-481. [DIVERSITY INDEX; INFORMATION THEORY]

WILLIAMS, C. B. 1944. SOME APPLICATIONS OF THE LOGARITHMIC SERIES AND THE INDEX OF DIVERSITY TO ECOLOGICAL PROBLEMS. J. ECOL. 32(1), 1-44.

WILLIAMS, C. B. 1945. RECENT LIGHT TRAP CATCHES OF LEPIDOPTERA IN U.S.A. ANALYSED IN RELATION TO THE LOGARITHMIC SERIES AND THE INDEX OF DIVERSITY. ANN. ENTOMOL. SOC. AMER. 38(3), 357-364.

WILLIAMS, C. B. 1946. YULE'S 'CHARACTERISTIC' AND THE 'INDEX OF DIVERSITY'. NATURE 157, 482.

WILLIAMS, C. B. 1947. THE GENERIC RELATIONS OF SPECIES IN SMALL ECOLOGICAL COMMUNITIES. J. ANIM. ECOL. 16(1), 11-18. [LOGARITHMIC SERIES]

WILLIAMS, C. B. 1949. JACCARD'S GENERIC COEFFICIENT AND COEFFICIENT OF FLORAL COMMUNITY, IN RELATION TO THE LOGARITHMIC SERIES AND THE INDEX OF DIVERSITY. ANN. BOT. (LONDON) NEW SERIES 13(49), 53-58.

WILLIAMS, C. B. 1950. THE APPLICATION OF THE LOGARITHMIC SERIES TO THE FREQUENCY OF OCCURRENCE OF PLANT SPECIES IN QUADRATS. J. ECOL. 38(1), 107-138.

WILLIAMS, C. B. 1953. THE RELATIVE ABUNDANCE OF DIFFERENT SPECIES IN A WILD ANIMAL POPULATION. J. ANIM. ECOL. 22(1), 14-31. [LOG-NORMAL DISTRIBUTION; LOGARITHMIC SERIES]

WILLIAMS, C. B. 1954. THE STATISTICAL OUTLOOK IN RELATION TO ECOLOGY. J. ECOL. 42(1), 1-13. [HISTORY]

WILLIAMS, C. B. 1960. THE RANGE AND PATTERN OF INSECT ABUNDANCE. AMER. NAT. 94(875), 137-151. [LOGARITHMIC SERIES; LOG-NORMAL DISTRIBUTION]

WILLIAMS, W. T., G. N. LANCE, L. J. WEBB AND J. G. TRACY. 1973. STUDIES IN THE NUMERICAL ANALYSIS OF COMPLEX RAIN-FOREST COMMUNITIES. VI. MODELS FOR THE CLASSIFICATION OF QUANTITATIVE DATA. J. ECOL. 61(1), 47-70. [NUMERICAL MODELS; EUCLIDEAN METRIC; GOWER METRIC; BRAY-CURTIS MEASURE; CANBERRA METRIC; INFORMATION STATISTICS MEASURES; BRILLOUIN'S DIVERSITY; SHANNON'S DIVERSITY; INFORMATION ANALYSIS; COMPUTER]

WILLIAMSON, M. 1973. SPECIES DIVERSITY IN ECOLOGICAL COMMUNITIES. PP. 325-335. IN, THE MATHEMATICAL THEORY OF THE DYNAMICS OF BIOLOGICAL POPULATIONS. (M. S. BARTLETT AND R. W. HIORNS, EDS.). ACADEMIC PRESS, NEW YORK, NEW YORK. [DIVERSITY INDEX; INFORMATION THEORY; PATTERNS OF SPECIES ABUNDANCE; LOG-NORMAL CURVE; LOGARITHMIC CURVE; MACARTHUR'S BROKEN-STICK MODEL; SPECIES OR SPECIES-AREA RELATIONSHIPS; MATRIX]

WILSON, E. O. 1969. THE SPECIES EQUILIBRIUM. PP. 38-47. IN, DIVERSITY AND STABILITY IN ECOLOGICAL SYSTEMS. BROOKHAVEN SYMPOSIUM IN BIOLOGY NO. 22. [MODEL]

WOODIN, S. A. AND J. A. YORKE. 1975. DISTURBANCE, FLUCTUATING RATES OF RESOURCE RECRUITMENT, AND INCREASED DIVERSITY. PP. 38-41. IN, ECOSYSTEM ANALYSIS AND PREDICTION. (S. A. LEVIN, ED.). PROCEEDINGS OF A SIAM-SIMS CONFERENCE HELD AT ALTA, UTAH, JULY 1-5, 1974. [MODEL; COMPETITION]

ZAIKA, V. E. AND A. A. ANDRYUSHCHENKO. 1969. "TAXONOMIC DIVERSITY" OF PHYTO- AND ZOOPLANKTON OF THE BLACK SEA. HYDROBIOL. J. 5(3), 8-14. (ENGLISH TRANSL.). [DIVERSITY INDEX; INFORMATION THEORY]

ENERGETICS AND PRODUCTIVITY

ABRAMI, G. 1972. OPTIMUM MEAN TEMPERATURE FOR PLANT GROWTH CALCULATED BY A NEW METHOD OF SUMMATION. ECOLOGY 53(5), 893-900. [LOGISTIC CURVE]

ACOCK, B., J. H. M. THORNLEY AND J. W. WILSON. 1970. SPATIAL VARIATION OF LIGHT IN THE CANOPY, PP. 91-102. IN, PREDICTION AND MEASUREMENT OF PHOTOSYNTHETIC PRODUCTIVITY. (I. SETLIK, ED.), CENTRE FOR AGRICULTURAL PUBLISHING AND DOCUMENTATION, WAGENINGEN, NETHERLANDS. [MODEL]

ALDERFER, R. G. AND D. M. GATES. 1971. ENERGY EXCHANGE IN PLANT CANOPIES. ECOLOGY 52(5), 855-861. [MODEL; COMPUTER]

ALIMOV, A. F. 1970. THE ENERGY FLOW IN A MOLLUSK POPULATION (USING SPHAERIDAE AS AN EXAMPLE). HYDROBIOL. J. 6(2), 48-56. (ENGLISH TRANSL.). [GROWTH CURVES; ENERGY BALANCE]

ALLEN, K. R. 1950. THE COMPUTATION OF PRODUCTION IN FISH POPULATIONS. N. Z. SCI. REV. 8, 89. [MORTALITY CURVE; SIGMOID GROWTH CURVE]

ANDERSON, M. C. 1970. RADIATION CLIMATE, CROP ARCHITECTURE AND PHOTOSYNTHESIS. PP. 71-78. IN, PREDICTION AND MEASUREMENT OF PHOTOSYNTHETIC PRODUCTIVITY. (I. SETLIK, ED.). CENTRE FOR AGRICULTURAL PUBLISHING AND DOCUMENTATION, WAGENINGEN, NETHERLANDS. [MODEL]

BANNISTER, T. T. 1974. PRODUCTION EQUATIONS IN TERMS OF CHLOROPHYLL CONCENTRATION, QUANTUM YIELD, AND UPPER LIMIT TO PRODUCTION. LIMNOL. OCEANOGR. 19(1), 1-12.

BASKERVILLE, G. L. AND P. EMIN. 1969. RAPID ESTIMATION OF HEAT ACCUMULATION FROM MAXIMUM AND MINIMUM TEMPERATURES. ECOLOGY 50(3), 514-517.

BATZLI, G. O. 1974. PRODUCTION, ASSIMILATION AND ACCUMULATION OF ORGANIC MATTER IN ECOSYSTEMS. J. THEOR. BIOL. 45(1), 205-217. [EQUATIONS; MODEL]

BAZZAZ, F. A. AND J. S. BOYER. 1972. A COMPENSATING METHOD FOR MEASURING CARBON DIOXIDE EXCHANGE, TRANSPIRATION, AND DIFFUSIVE RESISTANCES OF PLANTS UNDER CONTROLLED ENVIRONMENTAL CONDITIONS. ECOLOGY 53(2), 343-349. [EQUATIONS]

BOCOCK, K. L., M. D. MOUNTFORD AND J. HEATH. 1967. ESTIMATION OF ANNUAL PRODUCTION OF A MILLIPEDE POPULATION. PP. 727-739. IN, SECONDARY PRODUCTIVITY OF TERRESTRIAL ECOSYSTEMS, VOLUME II, (K. PETRUSEWICZ, ED.) INSTITUTE OF ECOLOGY, POLISH ACAD. SCI., WARSAW, POLAND. [MAXIMUM LIKELIHOOD; PROBABILITY; VARIANCE]

BOLING, R. H., JR., E. D. GOODMAN, J. A. VAN SICKLE, J. O. ZIMMER, K. W. CUMMINS, R. C. PETERSEN AND S. R. REICE. 1975. TOWARD A MODEL OF DETRITUS PROCESSING IN A WOODLAND STREAM. ECOLOGY 56(1), 141-151. [SIMULATION; MATRIX]

BORMANN, F. H., G. E. LIKENS, T. G. SICCAMA, R. S. PIERCE AND J. S. EATON. 1974. THE EXPORT OF NUTRIENTS AND RECOVERY OF STABLE CONDITIONS FOLLOWING DEFORESTATION AT HUBBARD BROOK. ECOL. MONOGR. 44(3), 255-277. [MODEL]

BOTKIN, D. B. 1969. PREDICTION OF NET PHOTOSYNTHESIS OF TREES FROM LIGHT INTENSITY AND TEMPERATURE. ECOLOGY 50(5), 854-858. [MODEL]

BRYLINSKY, M. 1972. STEADY-STATE SENSITIVITY ANALYSIS OF ENERGY FLOW IN A MARINE ECOSYSTEM. PP. 81-101. IN, SYSTEMS ANALYSIS AND SIMULATION IN ECOLOGY, VOLUME II, (B. C. PATTEN, ED.), ACADEMIC PRESS, NEW YORK, NEW YORK. [MODEL; SYSTEMS]

BUJALSKA, G., R. ANDRZEJEWSKI AND K. PETRUSEWICZ. 1968. PRODUCTIVITY INVESTIGATION OF AN ISLAND POPULATION OF CLETHRIONOMYS GLAREOLUS (SCHREBER, 1780). II. NATALITY. ACTA THERIOL. 13(24), 415-425. [FORMULA]

BUTLER, E. I., E. D. S. CORNER AND S. M. MARSHALL. 1969. ON THE NUTRITION AND METABOLISM OF ZOOPLANKTON VI. FEEDING EFFICIENCY OF CALANUS IN TERMS OF NITROGEN AND PHOSPHORUS. J. MAR. BIOL. ASSOC. U. K. 49(3), 977-1001. [EQUATIONS]

BUTLER, E. I., E. D. S. CORNER AND S. M. MARSHALL. 1970. ON THE NUTRITION AND METABOLISM OF ZOOPLANKTON VII. SEASONAL SURVEY OF NITROGEN AND PHOSPHORUS EXCRETION BY CALANUS IN THE CLYDE SEA-AREA. J. MAR. BIOL. ASSOC. U. K. 50, 525-560. [EQUATIONS]

BYRAM, G. M. AND G. M. JEMISON. 1943. SOLAR RADIATION AND FOREST FUEL MOISTURE. J. AGRIC. RES. 67(4), 149-176.

CALDER, W. A., III. 1974. CONSEQUENCES OF BODY SIZE FOR AVIAN ENERGETICS. PP. 86-144. IN, AVIAN ENERGETICS. (R. A. PAYNTER, JR., ED.). PUBLICATION OF NUTTALL ORNITHOLOGICAL CLUB NO. 15., CAMBRIDGE, MASSACHUSETTS. [EQUATIONS]

CANFER, J. L. 1972. INTERRELATIONS AMONG PLANKTON, ATTACHED ALGAE, AND THE PHOSPHORUS CYCLE IN ARTIFICIAL OPEN SYSTEMS. ECOL. MONOGR. 42(1), 1-23. [CIRCULATION MODEL]

CERNUSCA, A. 1972. FREQUENCY OF MICROCLIMATIC MEASUREMENTS IN ECOSYSTEM ANALYSIS. OECOLOGIA 9(2), 112-122. [COMPUTER]

CHAPMAN, D. W. 1968. PRODUCTION. PP. 182-196. IN, METHODS FOR ASSESSMENT OF FISH PRODUCTION IN FRESH WATERS.(W. E. RICKER, ED.). BLACKWELL SCIENTIFIC PUBLICATIONS, OXFORD, ENGLAND. [MODEL; MORTALITY; GROWTH; GRAPHICAL METHOD; VARIANCE]

CLARKE, G. L. 1946. DYNAMICS OF PRODUCTION IN A MARINE AREA. ECOL. MONOGR. 16(4), 321-335. [MODEL]

CLARKE, G. L., W. T. EDMONDSON AND W. E. RICKER. 1946. MATHEMATICAL FORMULATION OF BIOLOGICAL PRODUCTIVITY. ECOL. MONOGR. 16(4), 336-337.

CONOVER, R. J. 1961. THE TURNOVER OF PHOSPHORUS BY CALANUS FINMARCHICUS. J. MAR. BIOL. ASSOC. U. K. 41(2), 484-488. [EQUATIONS]

CRISP, D. J. 1971. ENERGY FLOW MEASUREMENTS. PP. 197-279. IN, METHODS FOR THE STUDY OF MARINE BENTHOS. (N. A. HOLME AND A. D. MCINTYRE, EDS.). BLACKWELL SCIENTIFIC PUBLICATIONS, OXFORD, ENGLAND. [EQUATIONS]

CUSHING, D. H. 1953. STUDIES ON PLANKTON POPULATIONS. J. CONS. PERM. INT. EXPLOR. MER 19(1), 3-22. [MODEL; PRODUCTION MODEL]

CUSHING, D. H. 1968. GRAZING BY HERBIVOROUS COPEPODS IN THE SEA. J. CONS. PERM. INT. EXPLOR. MER 32(1), 70-82. [MODEL]

CUSHING, D. H. AND T. VUCETIC. 1963. STUDIES ON A CALANUS PATCH III. THE QUANTITY OF FOOD EATEN BY CALANUS FINMARCHICUS. J. MAR. BIOL. ASSOC. U. K. 43(2), 349-371. [EQUATIONS; GRAZING]

DAVIS, G. E. AND C. E. WARREN. 1968. ESTIMATION OF FOOD CONSUMPTION RATES. PP. 204-225. IN, METHODS FOR ASSESSMENT OF FISH PRODUCTION IN FRESH WATERS.(W. E. RICKER, ED.). BLACKWELL SCIENTIFIC PUBLICATIONS, OXFORD, ENGLAND. [MODEL]

DIAMOND, P. 1974. MODELS DESCRIBING ENERGY FLOW AND NUTRIENT CYCLING. PP. 16-21. IN, PROC. FIRST INTERN. CONGR. ECOL. CENTRE FOR AGRICULTURAL PUBLISHING AND DOCUMENTATION, WAGENINGEN, NETHERLANDS.

DILLON, P. J. AND F. H. RIGLER. 1974. A TEST OF A SIMPLE NUTRIENT BUDGET MODEL PREDICTING THE PHOSPHORUS CONCENTRATION IN LAKE WATER. J. FISH. RES. BOARD CAN. 31(11), 1771-1778. [VARIATION OF VOLLENWEIDER'S MODEL]

DOI, T, 1969, EXPERIMENTAL STUDY ON THE FEEDING OF THE GUPPY, POECILIA RETICULATA, JAPANESE J. ECOL, 19(2), 62-66, [FEEDING MODEL]

DROOP, M, R, 1974, THE NUTRIENT STATUS OF ALGAL CELLS IN CONTINUOUS CULTURE, J. MAR, BIOL, ASSOC, U, K, 54(4), 825-855, [GROWTH; EQUATIONS]

DUGDALE, R, C, 1967, NUTRIENT LIMITATION IN THE SEA: DYNAMICS, IDENTIFICATION, AND SIGNIFICANCE, LIMNOL, OCEANOGR, 12(4), 685-695, [EQUATIONS; MODEL]

EDMONDSON, W, T, 1971, SPECIAL COMPUTATIONS SYSTEMS, PP, 137-140, IN, A MANUAL ON METHODS FOR THE ASSESSMENT OF SECONDARY PRODUCTIVITY IN FRESH WATERS, (W, T, EDMONDSON AND G, G, WINBERG, EDS,), BLACKWELL SCIENTIFIC PUBLICATIONS, OXFORD, ENGLAND, [EQUATIONS; TOTAL ORGANISMS OR DISSOLVED MATERIAL]

EMLEN, J, M, 1966, THE ROLE OF TIME AND ENERGY IN FOOD PREFERENCE, AMER, NAT, 100(916), 611-617, [PREDATOR-PREY MODEL; PROBABILITY; MODEL]

FEDERER, C, A, AND C, B, TANNER, 1966, SENSORS FOR MEASURING LIGHT AVAILABLE FOR PHOTOSYNTHESIS, ECOLOGY 47(4), 654-657, [EQUATIONS]

FEDERER, C, A, AND C, B, TANNER, 1966, SPECTRAL DISTRIBUTION OF LIGHT IN THE FOREST, ECOLOGY 47(4), 555-560, [EQUATIONS]

FEE, E, J, 1969, A NUMERICAL MODEL FOR THE ESTIMATION OF PHOTOSYNTHETIC PRODUCTION, INTEGRATED OVER TIME AND DEPTH, IN NATURAL WATERS, LIMNOL, OCEANOGR, 14(6), 906-911,

FEE, E, J, 1973, MODELLING PRIMARY PRODUCTION IN WATER BODIES: A NUMERICAL APPROACH THAT ALLOWS VERTICAL INHOMOGENEITIES, J, FISH, RES, BOARD CAN, 30(10), 1469-1473, [MODEL; SIMULATION]

FEE, E, J, 1973, A NUMERICAL MODEL FOR DETERMINING INTEGRAL PRIMARY PRODUCTION AND ITS APPLICATION TO LAKE MICHIGAN, J, FISH, RES, BOARD CAN, 30(10), 1447-1468, [COMPUTER]

FISHER, S, G, AND G, E, LIKENS, 1973, ENERGY FLOW IN BEAR BROOK, NEW HAMPSHIRE: AN INTEGRATIVE APPROACH TO STREAM ECOSYSTEM METABOLISM, ECOL, MONOGR, 43(4), 421-439, [COMPARTMENT MODEL; OLSON'S MODEL]

FREDRICKSON, A, G, AND H, M, TSUCHIYA, 1970, UTILIZATION OF THE EFFECTS OF INTERMITTENT ILLUMINATION ON PHOTOSYNTHETIC MICROORGANISMS, PP, 519-541, IN, PREDICTION AND MEASUREMENT OF PHOTOSYNTHETIC PRODUCTIVITY, (I, SETLIK, ED,), CENTRE FOR AGRICULTURAL PUBLISHING AND DOCUMENTATION, WAGENINGEN, NETHERLANDS, [MODEL]

GATES, D, M, 1962, ENERGY EXCHANGE IN THE BIOSPHERE, HARPER AND ROW, PUBLISHERS, NEW YORK, NEW YORK, VIII, 151 PP, [BOOK]

GATES, D, M, 1965, ENERGY, PLANTS, AND ECOLOGY, ECOLOGY 46(1/2), 1-13, [EQUATIONS]

GAULD, D, T, 1951, THE GRAZING RATE OF PLANKTONIC COPEPODS, J, MAR, BIOL, ASSOC, U, K, 29, 695-706, [EQUATIONS]

GEORGE, D, G, AND R, W, EDWARDS, 1974, POPULATION DYNAMICS AND PRODUCTION OF DAPHNIA HYALINA IN A EUTROPHIC RESERVOIR, FRESHWATER BIOL, 4(5), 445-465, [POPULATION PARAMETERS]

GILMARTIN, M, 1964, THE PRIMARY PRODUCTION OF A BRITISH COLUMBIA FJORD, J, FISH, RES, BOARD CAN, 21(3), 505-538, [EQUATIONS]

GLASS, N, R, 1968, THE EFFECT OF TIME OF FOOD DEPRIVATION ON THE ROUTINE OXYGEN CONSUMPTION OF LARGEMOUTH BLACK BASS (MICROPTERUS SALMOIDES), ECOLOGY 49(2), 340-343, [MODEL]

GLASS, N. R. 1969. DISCUSSION OF CALCULATION OF POWER FUNCTION WITH SPECIAL REFERENCE TO RESPIRATORY METABOLISM IN FISH. J. FISH. RES. BOARD CAN. 26(10), 2643-2650. [ITERATIVE CURVE FITTING]

GLASS, N. R. 1971. COMPUTER ANALYSIS OF PREDATION ENERGETICS IN THE LARGEMOUTH BASS. PP. 325-363. IN, SYSTEMS ANALYSIS AND SIMULATION IN ECOLOGY, VOLUME I. (B. C. PATTEN, ED.). ACADEMIC PRESS, NEW YORK, NEW YORK.

GRACE, J. AND H. W. WOOLHOUSE. 1973. A PHYSIOLOGICAL AND MATHEMATICAL STUDY OF THE GROWTH AND PRODUCTIVITY OF A CALLUNA-SPHAGNUM COMMUNITY. II. LIGHT INTERCEPTION AND PHOTOSYNTHESIS IN CALLUNA. J. APPL. ECOL. 10(1), 63-76.

GUTEL'MAKHER, B. L. 1973. AUTORADIOGRAPHY AS A METHOD OF DETERMINING THE RELATIVE CONTRIBUTION OF INDIVIDUAL ALGAL SPECIES TO THE PRIMARY PRODUCTION OF PLANKTON. HYDROBIOL. J. 9(1), 61-64. (ENGLISH TRANSL.). [FILTERING RATE]

HELLER, H. C. AND D. M. GATES. 1971. ALTITUDINAL ZONATION IN CHIPMUNKS (EUTAMIAS): ENERGY BUDGETS. ECOLOGY 52(3), 424-433.

HUBBELL, S. P. 1971. OF SOWBUGS AND SYSTEMS: THE ECOLOGICAL BIOENERGETICS OF A TERRESTRIAL ISOPOD. PP. 269-324. IN, SYSTEMS ANALYSIS AND SIMULATION IN ECOLOGY, VOLUME I. (B. C. PATTEN, ED.). ACADEMIC PRESS, NEW YORK, NEW YORK.

HUBBELL, S. P. 1973. POPULATIONS AND SIMPLE FOOD WEBS AS ENERGY FILTERS. I. ONE-SPECIES SYSTEMS. AMER. NAT. 107(953), 94-121. [LINEAR MODEL; TIME LAG; INTRINSIC RATE]

HUBBELL, S. P. 1973. POPULATIONS AND SIMPLE FOOD WEBS AS ENERGY FILTERS. II TWO-SPECIES SYSTEMS. AMER. NAT. 107(953), 122-151. [LINEAR MODEL; COMPETITION; INTRINSIC RATE; MATRIX; PREDATOR-PREY MODEL; SIGNAL-FLOW GRAPH]

HUNT, R. 1973. A METHOD OF ESTIMATING ROOT EFFICIENCY. J. APPL. ECOL. 10(1), 157-164.

IDSO, S. B., R. D. JACKSON, W. L. EHRLER AND S. T. MITCHELL. 1969. A METHOD FOR DETERMINATION OF INFRARED EMITTANCE OF LEAVES. ECOLOGY 50(5), 899-902. [EQUATIONS]

IMBODEN, D. M. 1974. PHOSPHORUS MODEL OF LAKE EUTROPHICATION. LIMNOL. OCEANOGR. 19(2), 297-304.

IVANOVA, M. B. 1973. PRODUCTION OF ZOOPLANKTON POPULATIONS IN FRESH WATERS OF THE USSR. EKOLOGIYA 4(3), 52-62. (ENGLISH TRANSL. PP. 224-232). [EQUATIONS]

IVLEV, V. S. 1960. ON THE UTILIZATION OF FOOD BY PLANKTOPHAGE FISHES. BULL. MATH. BIOPHYS. 22, 371-389. [EQUATIONS]

IZAKOV, V. YA., S. M. RUTKEVICH, O. A. ZHIGAL'SHII AND V. S. KRUGLOV. 1973. POSSIBILITY OF EXISTENCE OF QUASI-HOMEOSTASIS OF CARDIAC ACTIVITY IN COLD-BLOODED ANIMALS WITH CHANGING TEMPERATURES IN THE ENVIRONMENT. EKOLOGIYA 4(1), 23-31. (ENGLISH TRANSL. PP. 14-20). [EQUATIONS]

JAHNKE, L. S. AND D. B. LAWRENCE. 1965. INFLUENCE OF PHOTOSYNTHETIC CROWN STRUCTURE ON POTENTIAL PRODUCTIVITY OF VEGETATION, BASED PRIMARILY ON MATHEMATICAL MODELS. ECOLOGY 46(3), 319-326.

JARVIS, P. G. 1970. CHARACTERISTICS OF THE PHOTOSYNTHETIC APPARATUS DERIVED FROM ITS RESPONSE TO NATURAL COMPLEXES OF ENVIRONMENTAL FACTORS. PP. 353-367. IN, PREDICTION AND MEASUREMENT OF PHOTOSYNTHETIC PRODUCTIVITY. (I. SETLIK, ED.). CENTRE FOR AGRICULTURAL PUBLISHING AND DOCUMENTATION, WAGENINGEN, NETHERLANDS. [MODEL]

JASSBY, A. D. AND C. R. GOLDMAN. 1974. A QUANTITATIVE MEASURE OF SUCCESSION RATE AND ITS APPLICATION TO THE PHYTOPLANKTON OF LAKES. AMER. NAT. 108(963), 688-693.

JOHNSON, M. G. AND R. O. BRINKHURST. 1971. BENTHIC COMMUNITY METABOLISM IN BAY OF QUINTE AND LAKE ONTARIO. J. FISH. RES. BOARD CAN. 28(11), 1715-1725. [EQUATIONS]

JONES, R. 1974. THE RATE OF ELIMINATION OF FOOD FROM THE STOMACHS OF HADDOCK MELANOGRAMMUS AEGLEFINUS, COD GADUS MORHUA AND WHITING MERLANGIUS MERLANGUS. J. CONS. PERM. INT. EXPLOR. MER 35(3), 225-243. [EQUATIONS]

JORDAN, P. A., D. B. BOTKIN AND M. L. WOLFE. 1971. BIOMASS DYNAMICS IN A MOOSE POPULATION. ECOLOGY 52(1), 147-152. [COMPUTER; MODEL]

KACZMAREK, W. 1967. METHODS OF PRODUCTION ESTIMATION IN VARIOUS TYPES OF ANIMAL POPULATION. PP. 413-446. IN, SECONDARY PRODUCTIVITY OF TERRESTRIAL ECOSYSTEMS, VOLUME II. (K. PETRUSEWICZ, ED.). INSTITUTE OF ECOLOGY, POLISH ACAD. SCI., WARSAW, POLAND.

KALANTYRENKO, I. I. 1972. CALCULATION OF THE CONCENTRATION OF THE BLUE-GREEN ALGA BIOMASS UPON SETTLING. HYDROBIOL. J. 8(5), 104-107. (ENGLISH TRANSL.). [ESTIMATION]

KELLY, J. M., P. A. OPSTRUP, J. S. OLSON, S. I. AUERBACH AND G. M. VAN DYNE 1969. MODELS OF SEASONAL PRIMARY PRODUCTIVITY IN EASTERN TENNESSEE FESTUCA AND ANDROPOGON ECOSYSTEMS. AEC REPT. ORNL-4310. OAK RIDGE NATL. LAB., OAK RIDGE, TENNESSEE. XXVIII, 296 PP.

KELLY, M. G., G. M. HORNBERGER AND B. J. COSBY. 1974. CONTINUOUS AUTOMATED MEASUREMENT OF RATES OF PHOTOSYNTHESIS AND RESPIRATION IN AN UNDISTURBED RIVER COMMUNITY. LIMNOL. OCEANOGR. 19(2), 305-312. [FOURIER SERIES; COMMUNITY PRODUCTION FUNCTION]

KERFOOT, W. B. 1970. BIOENERGETICS OF VERTICAL MIGRATION. AMER. NAT. 104(940), 529-546. [MODEL]

KERR, S. R. 1971. PREDICTION OF FISH GROWTH EFFICIENCY IN NATURE. J. FISH. RES. BOARD CAN. 28(6), 809-814. [MODEL]

KERR, S. R. 1974. THEORY OF SIZE DISTRIBUTION IN ECOLOGICAL COMMUNITIES. J. FISH. RES. BOARD CAN. 31(12), 1859-1862. [MODEL; WINBERG'S ENERGY BUDGET MODEL]

KERR, S. R. AND N. V. MARTIN. 1970. TROPHIC-DYNAMICS OF LAKE TROUT PRODUCTION SYSTEMS. PP. 365-376. IN, MARINE FOOD CHAINS. (J. H. STEELE, ED.). UNIVERSITY OF CALIFORNIA PRESS, BERKELEY, CALIFORNIA. [MODEL; ENERGY BUDGET]

KING, K. M. 1970. DISCUSSION SECTION 2. MASS AND ENERGY EXCHANGE BETWEEN PLANT STANDS AND ENVIRONMENT. PP. 197-198. IN, PREDICTION AND MEASUREMENT OF PHOTOSYNTHETIC PRODUCTIVITY.(I. SETLIK, ED.). CENTRE FOR AGRICULTURAL PUBLISHING AND DOCUMENTATION, WAGENINGEN, NETHERLANDS. [MODEL]

KLINE, J. R., J. R. MARTIN, C. F. JORDAN AND J. J. KORANDA. 1970. MEASUREMENT OF TRANSPIRATION IN TROPICAL TREES WITH TRITIATED WATER. ECOLOGY 51(6), 1068-1073. [EQUATIONS]

KNOERR, K. R. AND C. E. MURPHY, JR. 1972. THE EXCHANGE OF ENERGY, MASS AND MOMENTUM BETWEEN A VEGETATED SURFACE AND ITS ENVIRONMENT. PP. 13-36. IN, MODELING THE GROWTH OF TREES. (C. E. MURPHY, JR., J. D. HESKETH AND B. R. STRAIN, EDS.). AEC REPT. EDFB-IBP-72-11. OAK RIDGE NATL. LAB., OAK RIDGE, TENNESSEE.

KOLLER, D. 1970. CHARACTERISTICS OF THE PHOTOSYNTHETIC APPARATUS DERIVED FROM ITS RESPONSE TO NATURAL COMPLEXES OF ENVIRONMENTAL FACTORS. PP. 283-294. IN, PREDICTION AND MEASUREMENT OF PHOTOSYNTHETIC PRODUCTIVITY. (I. SETLIK, ED.). CENTRE FOR AGRICULTURAL PUBLISHING AND DOCUMENTATION, WAGENINGEN, NETHERLANDS. [MODEL]

KOWAL, N. E. AND D. A. CROSSLEY, JR. 1971. THE INGESTION RATES OF MICROARTHROPODS IN PINE MOR, ESTIMATED WITH RADIOACTIVE CALCIUM. ECOLOGY 52(3), 444-452. [MODEL]

KOZLOVA, I. V. 1972. CALCULATION OF THE TOTAL ZOOPLANKTON PRODUCTION WITH SPECIFIC REFERENCE TO URAL LAKES. HYDROBIOL. J. 8(3), 106-110. (ENGLISH TRANSL.). [ESTIMATION; PREDATOR-PREY MODEL]

KUROIWA, S. 1970. TOTAL PHOTOSYNTHESIS OF A FOLIAGE IN RELATION TO INCLINATION OF LEAVES. PP. 79-89. IN, PREDICTION AND MEASUREMENT OF PHOTOSYNTHETIC PRODUCTIVITY. (I. SETLIK, ED.). CENTRE FOR AGRICULTURAL PUBLISHING AND DOCUMENTATION, WAGENINGEN, NETHERLANDS. [MODEL]

LAISK, A. 1970. A MODEL OF LEAF PHOTOSYNTHESIS AND PHOTORESPIRATION. PP. 295-306. IN, PREDICTION AND MEASUREMENT OF PHOTOSYNTHETIC PRODUCTIVITY. (I. SETLIK, ED.). CENTRE FOR AGRICULTURAL PUBLISHING AND DOCUMENTATION, WAGENINGEN, NETHERLANDS.

LAKE, J. V. AND M. C. ANDERSON. 1970. DISCUSSION SECTION 1. DYNAMICS OF DEVELOPMENT OF PHOTOSYNTHETIC SYSTEMS. PP. 131-136. IN, PREDICTION AND MEASUREMENT OF PHOTOSYNTHETIC PRODUCTIVITY. (I. SETLIK, ED.). CENTRE FOR AGRICULTURAL PUBLISHING AND DOCUMENTATION, WAGENINGEN, NETHERLANDS. [MODEL]

LAKHANI, K. H. AND J. E. SATCHELL. 1970. PRODUCTION BY LUMBRICUS TERRESTRIS (L.). J. ANIM. ECOL. 39(2), 473-492. [MORTALITY AND SURVIVORSHIP CURVES; MODEL; RELATIVE PRODUCTIVITY EQUATION OF SKELLAM]

LANDSBERG, J. J. 1970. DISCUSSION SECTION 1. STRUCTURAL CHARACTERISTICS OF PHOTOSYNTHETIC SYSTEMS. PP. 143-144. IN, PREDICTION AND MEASUREMENT OF PHOTOSYNTHETIC PRODUCTIVITY.(I. SETLIK, ED.). CENTRE FOR AGRICULTURAL PUBLISHING AND DOCUMENTATION, WAGENINGEN, NETHERLANDS. [MODEL]

LANDSBERG, J. J. AND P. G. JARVIS. 1973. A NUMERICAL INVESTIGATION OF THE MOMENTUM BALANCE OF A SPRUCE FOREST. J. APPL. ECOL. 10(2), 645-655. [MODEL]

LEDIG, F. T. AND T. O. PERRY. 1969. NET ASSIMILATION RATE AND GROWTH IN LOBLOLLY PINE SEEDLINGS. FOREST SCI. 15(4), 431-438. [EQUATIONS]

LEMON, E. R. 1970. SUMMARY SECTION 2. MASS AND ENERGY EXCHANGE BETWEEN PLANT STANDS AND ENVIRONMENT. PP. 199-205. IN, PREDICTION AND MEASUREMENT OF PHOTOSYNTHETIC PRODUCTIVITY. (I. SETLIK, ED.). CENTRE FOR AGRICULTURAL PUBLISHING AND DOCUMENTATION, WAGENINGEN, NETHERLANDS. [MODEL]

LEVANDOWSKY, M. AND J. SIBERT. 1969. A CRITICISM OF PATTEN AND VAN DYNE'S PRODUCTIVITY MODEL. LIMNOL. OCEANOGR. 14(2), 311-312. [REPLY BY B. C. PATTEN, LIMNOL. OCEANOGR. 14(2), 312]

LEWIN, J. AND J. LOMAS. 1974. A COMPARISON OF STATISTICAL AND SOIL MOISTURE MODELING TECHNIQUES IN LONG-TERM STUDY OF WHEAT YIELD PERFORMANCE UNDER SEMI-ARID CONDITIONS. J. APPL. ECOL. 11(3), 1081-1090. [EXPONENTIAL CURVE; SIMULATION; COMPUTER; REGRESSION]

LEWIS, W. M., JR. 1974. PRIMARY PRODUCTION IN THE PLANKTON COMMUNITY OF A TROPICAL LAKE. ECOL. MONOGR. 44(4), 377-409. [ESTIMATION; METABOLISM EQUATION]

LIETH, H. 1970. PHENOLOGY IN PRODUCTIVITY STUDIES. PP. 29-46. IN, ANALYSIS OF TEMPERATE FOREST ECOSYSTEMS. (D. E. REICHLE, ED.) SPRINGER-VERLAG, NEW YORK, NEW YORK.

LINACRE, E. T. 1969. NET RADIATION TO VARIOUS SURFACES. J. APPL. ECOL. 6(1), 61-75. [EQUATIONS]

LINDEMAN, R. L. 1942. THE TROPHIC-DYNAMIC ASPECT OF ECOLOGY. ECOLOGY 23(4), 399-418. [ENERGY MODEL]

LOMNICKI, A., E. BANDOLA AND K. JANKOWSKA. 1968. MODIFICATION OF THE WEIGERT-EVANS METHOD FOR ESTIMATION OF NET PRIMARY PRODUCTION. ECOLOGY 49(1), 147-149.

LOOMIS, R. S. 1970. SUMMARY SECTION 1. DYNAMICS OF DEVELOPMENT OF PHOTOSYNTHETIC SYSTEMS. PP. 137-141. IN, PREDICTION AND MEASUREMENT OF PHOTOSYNTHETIC PRODUCTIVITY. (I. SETLIK, ED.). CENTRE FOR AGRICULTURAL PUBLISHING AND DOCUMENTATION, WAGENINGEN, NETHERLANDS. [MODEL]

LORENZEN, C. J. 1972. EXTINCTION OF LIGHT IN THE OCEAN BY PHYTOPLANKTON. J. CONS. PERM. INT. EXPLOR. MER 34(2), 262-267. [EQUATIONS]

LYUBARSKII, E. L. 1973. BIOMORPHOLOGICAL BOUNDARY BETWEEN LONG-RHIZOME AND SHORT-RHIZOME PLANTS. EKOLOGIYA 4(2), 94-95. (ENGLISH TRANSL. PP. 164-165). [MODEL]

MACFADYEN, A. 1948. THE MEANING OF PRODUCTIVITY IN BIOLOGICAL SYSTEMS. J. ANIM. ECOL. 17(1), 75-80. [EQUATIONS]

MACKINNON, J. C. 1973. ANALYSIS OF ENERGY FLOW AND PRODUCTION IN AN UNEXPLOITED MARINE FLATFISH POPULATION. J. FISH. RES. BOARD CAN. 30(11), 1717-1728. [MODEL; COMPUTER]

MADGWICK, H. A. I. 1970. BIOMASS AND PRODUCTIVITY MODELS OF FOREST CANOPIES. PP. 47-54. IN, ANALYSIS OF TEMPERATE FOREST ECOSYSTEMS. (D. E. REICHLE, ED.). SPRINGER-VERLAG, NEW YORK, NEW YORK.

MARGALEF, R. 1966. ECOLOGICAL CORRELATIONS AND THE RELATIONSHIP BETWEEN PRIMARY PRODUCTIVITY AND COMMUNITY STRUCTURE. PP. 355-364. IN, PRIMARY PRODUCTIVITY IN AQUATIC ENVIRONMENTS. (C. R. GOLDMAN, ED.). UNIVERSITY OF CALIFORNIA PRESS, BERKELEY, CALIFORNIA. [REGRESSION; AQUATIC; DIVERSITY INDEX]

MCALLISTER, C. D. 1969. ASPECTS OF ESTIMATING ZOOPLANKTON PRODUCTION FROM PHYTOPLANKTON PRODUCTION. J. FISH. RES. BOARD CAN. 26(2), 199-220. [GRAZING EQUATIONS]

MCALLISTER, C. D. 1970. ZOOPLANKTON RATIONS, PHYTOPLANKTON MORTALITY AND THE ESTIMATION OF MARINE PRODUCTION. PP. 419-457. IN, MARINE FOOD CHAINS. (J. H. STEELE, ED.). UNIVERSITY OF CALIFORNIA PRESS, BERKELEY, CALIFORNIA. [MODEL; COMPUTER]

MCBRAYER, J. F. 1974. ENERGY FLOW AND NUTRIENT CYCLING IN A CRYPTOZOAN FOOD-WEB. THESIS. UNIVERSITY OF TENNESSEE, KNOXVILLE, TENNESSEE. XII, 78 PP. [COMPUTER; MODEL; SIMULATION]

MCCREE, K. J. 1970. AN EQUATION FOR THE RATE OF RESPIRATION OF WHITE CLOVER PLANTS GROWN UNDER CONTROLLED CONDITIONS. PP. 221-229. IN, PREDICTION AND MEASUREMENT OF PHOTOSYNTHETIC PRODUCTIVITY. (I. SETLIK, ED.). CENTRE FOR AGRICULTURAL PUBLISHING AND DOCUMENTATION, WAGENINGEN, NETHERLANDS.

MCCULLOUGH, E. C. AND W. P. PORTER. 1971. COMPUTING CLEAR DAY SOLAR RADIATION SPECTRA FOR THE TERRESTRIAL ECOLOGICAL ENVIRONMENT. ECOLOGY 52(6), 1008-1015. [EQUATIONS]

MCLELLAN, H. J. 1958. ENERGY CONSIDERATIONS IN THE BAY OF FUNDY SYSTEM. J. FISH. RES. BOARD CAN. 15(2), 115-134. [EQUATIONS]

MCNAB, B. K. 1963. A MODEL OF THE ENERGY BUDGET OF A WILD MOUSE. ECOLOGY 44(3), 521-532.

MCNAB, B. K. 1974. THE BEHAVIOR OF TEMPERATURE CAVE BATS IN A SUBTROPICAL ENVIRONMENT. ECOLOGY 55(5), 943-958. [EQUATIONS]

MEGARD, R. O. 1972. PHYTOPLANKTON, PHOTOSYNTHESIS, AND PHOSPHORUS IN LAKE MINNETONKA, MINNESOTA. LIMNOL. OCEANOGR. 17(1), 68-87. [EQUATIONS]

MENSHUTKIN, V. V. AND A. A. UMNOV. 1971. ENERGY MODEL OF THE PELAGIC ECOSYSTEM OF LAKE DAL'NEYE. HYDROBIOL. J. 7(4), 7-11. (ENGLISH TRANSL.) [ALGORITHM; SIMULATION; MATRIX; COMPUTER]

MENSHUTKIN, V. V. AND T. I. PRIKHOD'KO. 1969. PRODUCTIVE PROPERTIES OF STABLE POPULATIONS WITH A PROLONGED PERIOD OF REPRODUCTION. HYDROBIOL. J. 5(1), 1-7. (ENGLISH TRANSL.). [POPULATION MODEL; GROWTH RATE]

MIGITA, M. AND T. HANAOKA. 1938. AMOUNT OF FOOD PROTEIN AND THE ACCUMULATION OF BODY PROTEIN IN FISH. II. FED ON DIFFERENT AMOUNTS OF DIET OF THE SAME COMPOSITION. BULL. JAPANESE SOC. SCI. FISH. 7(3), 171-175. [EQUATIONS]

MIGITA, M. AND T. HANAOKA. 1938. AMOUNT OF FOOD PROTEIN AND THE ACCUMULATION OF BODY PROTEIN IN FISH. III. ECONOMICAL AMOUNT OF FOOD PROTEIN. BULL. JAPANESE SOC. SCI. FISH. 7(4), 220-226. [EQUATIONS]

MIGITA, M. AND T. HANAOKA. 1939. AMOUNT OF FOOD PROTEIN AND THE ACCUMULATION OF BODY PROTEIN IN FISH. IV. INFLUENCE OF SOME FEEDING CONDITIONS ON THE GROWTH OF FISH. BULL. JAPANESE SOC. SCI. FISH. 7(5), 288-296. [EQUATIONS]

MILLER, P. C. 1971. SAMPLING TO ESTIMATE MEAN LEAF TEMPERATURES AND TRANSPIRATION RATES IN VEGETATION CANOPIES. ECOLOGY 52(5), 885-889. [EQUATIONS; SAMPLE SIZE]

MILLER, P. C. 1972. BIOCLIMATE, LEAF TEMPERATURE, AND PRIMARY PRODUCTION IN RED MANGROVE CANOPIES IN SOUTH FLORIDA. ECOLOGY 53(1), 22-45. [MODEL]

MOEN, A. N. 1968. ENERGY EXCHANGE OF WHITE-TAILED DEER, WESTERN MINNESOTA. ECOLOGY 49(4), 676-682. [EQUATIONS]

MOIR, W. H. 1972. LITTER, FOLIAGE, BRANCH, AND STEM PRODUCTION IN CONTRASTING LODGEPOLE PINE HABITATS OF THE COLORADO FRONT RANGE. PP. 189-198. IN, PROCEEDINGS-RESEARCH ON CONIFEROUS FOREST ECOSYSTEMS-A SYMPOSIUM. (J. F. FRANKLIN, L. J. DEMPSTER AND R. H. WARING, EDS.). U. S. FOR. SERV. PACIFIC NORTHWEST FOREST EXPT. STN., PORTLAND, OREGON. [MODEL; ALLOMETRIC EQUATION]

MONSI, M. AND Y. MURATA. 1970. DEVELOPMENT OF PHOTOSYNTHETIC SYSTEMS AS INFLUENCED BY DISTRIBUTION OF MATTER. PP. 115-129. IN, PREDICTION AND MEASUREMENT OF PHOTOSYNTHETIC PRODUCTIVITY. (I. SETLIK, ED.). CENTRE FOR AGRICULTURAL PUBLISHING AND DOCUMENTATION, WAGENINGEN, NETHERLANDS. [MODEL]

MONTEITH, J. L. 1970. SUMMARY SECTION 1. STRUCTURAL CHARACTERISTICS OF PHOTOSYNTHETIC SYSTEMS. PP. 145-146. IN, PREDICTION AND MEASUREMENT OF PHOTOSYNTHETIC PRODUCTIVITY. (I. SETLIK, ED.). CENTRE FOR AGRICULTURAL PUBLISHING AND DOCUMENTATION, WAGENINGEN, NETHERLANDS. [MODEL]

MORHARDT, S. S. AND D. M. GATES. 1974. ENERGY-EXCHANGE ANALYSIS OF THE BELDING GROUND SQUIRREL AND ITS HABITAT. ECOL. MONOGR. 44(1), 17-44. [MODEL; COMPUTER]

MOSZYNSKA, B. 1973. METHODS FOR ASSESSING PRODUCTION OF THE UPPER PARTS OF SHRUBS AND CERTAIN PERENNIAL PLANTS. EKOLOGIYA POLSKA-SERIA A 21(24), 359-367. ["W" INDEX; EQUATIONS]

MUNK, W. H. AND G. A. RILEY. 1952. ABSORPTION OF NUTRIENTS BY AQUATIC PLANTS. J. MARINE RES. 11(2), 215-240. [EQUATIONS]

MURPHY, C. E., JR. AND K. R. KNOERR. 1970. A GENERAL MODEL FOR THE ENERGY EXCHANGE AND MICROCLIMATE OF PLANT COMMUNITIES. PP. 786-797. IN, PROCEEDINGS OF THE 1970 SUMMER COMPUTER SIMULATION CONFERENCE, JUNE 10, 11, 12, 1970, DENVER, COLORADO, VOL. II. [SIMULATION; COMPUTER]

MURPHY, C. E., JR. AND K. R. KNOERR. 1972. MODELING THE ENERGY BALANCE PROCESSES OF NATURAL ECOSYSTEMS. AEC REPT. EDFB-IBP-72-10. OAK RIDGE NATL. LAB., OAK RIDGE, TENNESSEE. VII, 135 PP. [SIMULATION; COMPUTER]

MURPHY, C. E., JR., J. B. MANKIN, JR. AND K. R. KNOERR. 1972. A DYNAMIC ENERGY BALANCE AND MICROCLIMATE MODEL OF A FOREST ECOSYSTEM. PP. 1169-1175. IN, PROCEEDINGS OF THE 1972 SUMMER COMPUTER SIMULATION CONFERENCE, JUNE 14, 15, 16, 1972, SAN DIEGO, CALIFORNIA, VOL. II. [SIMULATION; COMPUTER]

MURPHY, C. E., JR., T. R. SINCLAIR, R. S. KINERSON, K. O. HIGGINBOTHAM, K. R. KNOERR AND B. R. STRAIN. 1973. MODELING THE PRIMARY PRODUCTIVITY PROCESS-A COORDINATED EFFORT AT AN INTENSIVE RESEARCH SITE. PP. 684-690. IN, PROCEEDINGS OF THE 1973 SUMMER COMPUTER SIMULATION CONFERENCE, JULY 17, 18, 19, 1973. MONTREAL, P. Q., CANADA. VOL. II. [SIMULATION; COMPUTER; MODEL]

MURPHY, C. E., T. R. SINCLAIR AND K. R. KNOERR. 1974. MODELING THE PHOTOSYNTHESIS OF PLANT STANDS. PP. 123-147. IN, VEGETATION AND ENVIRONMENT. (B. R. STRAIN AND W. D. BILLINGS, EDS.). DR. W. JUNK PUBLISHERS, THE HAGUE, NETHERLANDS. [MODEL; SIMULATION]

MURPHY, G. I. 1971. CLARIFYING A PRODUCTION MODEL. LIMNOL. OCEANOGR. 16(6), 981-982. [FEEDBACK EQUATION; SVERDRUP'S MODEL; STANDING CROP]

NAKAMURA, K., Y. ITO, M. NAKAMURA, T. MATSUMOTO AND K. HAYAKAWA. 1971. ESTIMATION OF POPULATION PRODUCTIVITY OF PARAPLEURUS ALLIACEUS GERMAR (ORTHOPTERA: ACRIDIIDAE) ON A MISCANTHUS SINENSIS ANDERS. GRASSLAND I. ESTIMATION OF POPULATION PARAMETERS. OECOLOGIA 7(1), 1-15. [TAG-RECAPTURE; LESLIE'S THREE POINT METHOD; BAILEY'S TRIPLE-CATCH METHOD]

NEESS, J. AND R. C. DUGDALE. 1959. COMPUTATION OF PRODUCTION FOR POPULATIONS OF AQUATIC MIDGE LARVAE. ECOLOGY 40(3), 425-430. [MODEL]

NEFF, E. L. 1973. WATER STORAGE CAPACITY OF CONTOUR FURROWS IN MONTANA. J. RANGE MANAGE. 26(4), 298-301. [EQUATIONS]

NEWMAN, E. I. 1969. RESISTANCE TO WATER FLOW IN SOIL AND PLANT I. SOIL RESISTANCE IN RELATION TO AMOUNTS OF ROOT: THEORETICAL ESTIMATES. J. APPL. ECOL. 6(1), 1-12. [EQUATIONS]

NIILISK, H., T. NILSON AND J. ROSS. 1970. RADIATION IN PLANT CANOPIES AND ITS MEASUREMENT. PP. 165-177. IN, PREDICTION AND MEASUREMENT OF PHOTOSYNTHETIC PRODUCTIVITY. (I. SETLIK, ED.). CENTRE FOR AGRICULTURAL PUBLISHING AND DOCUMENTATION, WAGENINGEN, NETHERLANDS. [MODEL]

O'BRIEN, W. J. 1974. THE DYNAMICS OF NUTRIENT LIMITATION OF PHYTOPLANKTON ALGAE: A MODEL RECONSIDERED. ECOLOGY 55(1), 135-141. [COMPUTER]

O'CONNOR, J. S. AND B. C. PATTEN. 1968. MATHEMATICAL MODELS OF PLANKTON PRODUCTIVITY. PP. 207-228. IN, RESERVOIR FISHERY RESOURCES SYMPOSIUM (F. F. FISH ET AL.,EDS.). AMERICAN FISHERIES SOCIETY, WASHINGTON, D. C.

O'NEILL, R. V. 1969. INDIRECT ESTIMATION OF ENERGY FLUXES IN ANIMAL FOOD WEBS. J. THEOR. BIOL. 22(2), 284-290. [MODEL]

ODUM, H. T. 1956. EFFICIENCIES, SIZE OF ORGANISMS, AND COMMUNITY STRUCTURE. ECOLOGY 37(3), 592-597. [HUTCHINSON'S ENERGY MODEL]

ODUM, H. T. 1962. THE USE OF A NETWORK ENERGY SIMULATOR TO SYNTHESIZE SYSTEMS AND DEVELOP ANALOGOUS THEORY: THE ECOSYSTEM EXAMPLE. PP. 291-297. IN, THE CULLOWHEE CONFERENCE ON TRAINING IN BIOMATHEMATICS. (H. L. LUCAS, ED.). TYPING SERVICE, RALEIGH, NORTH CAROLINA.

ODUM, H. T. 1972. AN ENERGY CIRCUIT LANGUAGE FOR ECOLOGICAL AND SOCIAL SYSTEMS: ITS PHYSICAL BASIS. PP. 139-211. IN, SYSTEMS ANALYSIS AND SIMULATION IN ECOLOGY, VOLUME II. (B. C. PATTEN, ED.) ACADEMIC PRESS, NEW YORK, NEW YORK. [SIMULATION; COMPUTER]

ODUM, H. T. 1974. ENERGY COST-BENEFIT MODELS FOR EVALUATING THERMAL PLUMES. PP. 628-649. IN, THERMAL ECOLOGY. (J. W. GIBBONS AND R. R. SHARITZ, EDS.). U.S. AEC REPT. CONF-730505. [COMPUTER]

ODUM, H. T. 1975. COMBINING ENERGY LAWS AND COROLLARIES OF THE MAXIMUM POWER PRINCIPLE WITH VISUAL SYSTEMS MATHEMATICS. PP. 239-262. IN, ECOSYSTEM ANALYSIS AND PREDICTION. (S. A. LEVIN, ED.). PROCEEDINGS OF A SIAM-SIMS CONFERENCE HELD AT ALTA, UTAH, JULY 1-5, 1974. [MODEL; ENERGY CIRCUIT LANGUAGE]

ODUM, H. T. AND R. C. PINKERTON. 1955. TIMES SPEED REGULATOR, THE OPTIMUM EFFICIENCY FOR MAXIMUM OUTPUT IN PHYSICAL AND BIOLOGICAL SYSTEMS. AMER. SCI. 43(2), 331-343.

OLECHOWICZ, E. 1971. PRODUCTIVITY INVESTIGATION OF TWO TYPES OF MEADOWS IN THE VISTULA VALLEY VIII. THE NUMBER OF EMERGED DIPTERA AND THEIR ELIMINATION. EKOLOGIYA POLSKA-SERIA A 19(14), 183-195. [MODEL; ESTIMATION]

OLSON, J. S. 1963. ENERGY STORAGE AND THE BALANCE OF PRODUCERS AND DECOMPOSERS IN ECOLOGICAL SYSTEMS. ECOLOGY 44(2), 322-331. [MODEL]

OSWALD, W. J. 1970. GROWTH CHARACTERISTICS OF MICROALGAE CULTURED IN DOMESTIC SEWAGE: ENVIRONMENTAL EFFECTS ON PRODUCTIVITY. PP. 473-487. IN, PREDICTION AND MEASUREMENT OF PHOTOSYNTHETIC PRODUCTIVITY. (I. SETLIK, ED.). CENTRE FOR AGRICULTURAL PUBLISHING AND DOCUMENTATION, WAGENINGEN, NETHERLANDS. [MODEL]

OTSUKI, A. AND T. HANYA. 1972. PRODUCTION OF DISSOLVED ORGANIC MATTER FROM GREEN ALGAL CELLS. I. AEROBIC MICROBIAL DECOMPOSITION. LIMNOL. OCEANOGR. 17(2), 248-257. [MODEL; PRODUCTION MODEL]

PALOHEIMO, J. E. AND L. M. DICKIE. 1970. PRODUCTION AND FOOD SUPPLY. PP. 499-527. IN, MARINE FOOD CHAINS. (J. H. STEELE, ED.). UNIVERSITY OF CALIFORNIA PRESS, BERKELEY, CALIFORNIA. [MODEL]

PALTRIDGE, G. W. AND J. V. DENHOLM. 1974. PLANT YIELD AND THE SWITCH FROM VEGETATIVE TO REPRODUCTIVE GROWTH. J. THEOR. BIOL. 44(1), 23-34. [MODEL]

PARK, P. K. 1969. OCEANIC CO2 SYSTEM: AN EVALUATION OF TEN METHODS OF INVESTIGATION. LIMNOL. OCEANOGR. 14(2), 179-186. [EQUATIONS]

PARKHURST, D. F. AND O. L. LOUCKS. 1972. OPTIMAL LEAF SIZE IN RELATION TO ENVIRONMENT. J. ECOL. 60(2), 505-537. [MODEL; SIMULATION]

PASCIAK, W. J. AND J. GAVIS. 1974. TRANSPORT LIMITATION OF NUTRIENT UPTAKE IN PHYTOPLANKTON. LIMNOL. OCEANOGR. 19(6), 881-888. [MONOD'S MODEL; MODEL]

PATTEN, B. 1961. PLANKTON ENERGETICS OF RARITAN BAY. LIMNOL. OCEANOGR. 6(4), 369-387. [MODEL]

PATTEN, B. C. 1965. COMMUNITY ORGANIZATION AND ENERGY RELATIONSHIPS IN PLANKTON. AEC REPT, ORNL-3634, OAK RIDGE NATL. LAB., OAK RIDGE, TENNESSEE. VI, 58 PP. [MODEL; GROWTH FORMULA; DIVERSITY; COMPUTER]

PATTEN, B. C. 1968. MATHEMATICAL MODELS OF PLANKTON PRODUCTION. BULL. INT. REV. GES. HYDROBIOL. 53(3), 357-408.

PATTEN, B. C. AND G. M. VAN DYNE. 1968. FACTORIAL PRODUCTIVITY EXPERIMENTS IN A SHALLOW ESTUARY: ENERGETICS OF INDIVIDUAL PLANKTON SPECIES IN MIXED POPULATIONS. LIMNOL. OCEANOGR. 13(2), 309-314.

PERRY, T. O. 1972. DRY MATTER PRODUCTION OF DECIDUOUS TREES - A NEW MODEL. PP. 167-170. IN, MODELING THE GROWTH OF TREES. (C. E. MURPHY, JR., J. D. HESKETH AND B. R. STRAIN, EDS.), AEC REPT. EDFB-IBP-72-11. OAK RIDGE NATL. LAB., OAK RIDGE, TENNESSEE.

PERRY, T. O., H. E. SELLERS AND C. O. BLANCHARD. 1969. ESTIMATION OF PHOTOSYNTHETICALLY ACTIVE RADIATION UNDER A FOREST CANOPY WITH CHLOROPHYLL EXTRACTS AND FROM BASAL AREA MEASUREMENTS. ECOLOGY 50(1), 39-44. [EQUATIONS; BEER'S LAW]

PETRUSEWICZ, K. 1967. CONCEPTS IN STUDIES ON THE SECONDARY PRODUCTIVITY OF TERRESTRIAL ECOSYSTEMS. PP. 17-49. IN, SECONDARY PRODUCTIVITY OF TERRESTRIAL ECOSYSTEMS, VOLUME I. (K. PETRUSEWICZ, ED.) INSTITUTE OF ECOLOGY, POLISH ACAD. SCI., WARSAW POLAND. [EQUATIONS]

PETRUSEWICZ, K. 1967. SUGGESTED LIST OF MORE IMPORTANT CONCEPTS IN PRODUCTIVITY STUDIES (DEFINITIONS AND SYMBOLS). PP. 51-58. IN, SECONDARY PRODUCTIVITY OF TERRESTRIAL ECOSYSTEMS, VOLUME I. (K. PETRUSEWICZ, ED.) INSTITUTE OF ECOLOGY, POLISH ACAD. SCI., WARSAW, POLAND.

PETRUSEWICZ, K., R. ANDRZEJEWSKI, G. BUJALSKA AND J. GLIWICZ. 1968. PRODUCTIVITY INVESTIGATION OF AN ISLAND POPULATION OF CLETHRIONOMYS GLAREOLUS (SCHREBER, 1780). IV. PRODUCTION. ACTA THERIOL. 13(26), 435-445. [MODEL; SURVIVORSHIP CURVE]

PHILLIPSON, J. 1967. SECONDARY PRODUCTIVITY IN INVERTEBRATES REPRODUCING MORE THAN ONCE IN A LIFETIME. PP. 459-475. IN, SECONDARY PRODUCTIVITY OF TERRESTRIAL ECOSYSTEMS, VOLUME II. (K. PETRUSEWICZ, ED.) INSTITUTE OF ECOLOGY, POLISH ACAD. SCI., WARSAW, POLAND.

PIERPONT, G. 1974. ESTIMATING RELATIVE WATER CONTENT OF LEAVES WITHOUT OVEN DRYING. FOREST SCI. 20(1), 101-103. [EQUATIONS]

PLATT, T. AND C. FILION. 1973. SPATIAL VARIABILITY OF THE PRODUCTIVITY: BIOMASS RATIO FOR PHYTOPLANKTON IN A SMALL MARINE BASIN. LIMNOL. OCEANOGR. 18(5), 743-749. [VARIANCE; ANALYSIS OF VARIANCE]

PLOTNIKOV, V. V. 1973. WAYS OF DEVELOPMENT OF WOODY PLANTS IN CONNECTION WITH THEIR DENSITY DYNAMICS IN A COMMUNITY. EKOLOGIYA 4(3), 44-51. (ENGLISH TRANSL. PP. 218-223). [SPECIFIC DENSITY; DENSITY INDEX]

PULLIAM, H. R. 1973. COMPARATIVE FEEDING ECOLOGY OF A TROPICAL GRASSLAND FINCH (TIARIS OLIVACEA) ECOLOGY 54(2), 284-299. [EXPECTED NUMBER OF SEEDS EATEN]

RAFAIL, S. Z. 1969. FURTHER ANALYSIS OF RATION AND GROWTH RELATIONSHIP OF PLAICE (PLEURONECTES PLATESSA). J. FISH. RES. BOARD CAN. 26(12), 3237-3241. [EQUATIONS]

RANDOLPH, J. C. 1973. ECOLOGICAL ENERGETICS OF A HOMEOTHERMIC PREDATOR, THE SHORT-TAILED SHREW. ECOLOGY 54(5), 1166-1187. [MODEL]

RASHEVSKY, N. 1959. SOME REMARKS ON THE MATHEMATICAL THEORY OF NUTRITION OF FISHES. BULL. MATH. BIOPHYS. 21, 161-183.

REED, K. L. AND R. H. WARING. 1974. COUPLING OF ENVIRONMENT TO PLANT RESPONSE: A SIMULATION MODEL OF TRANSPIRATION. ECOLOGY 55(1), 62-72. [STOCHASTIC MODEL; COMPUTER]

RICHARDS, F. A., J. D. CLINE, W. W. BROENKOW AND L. P. ATKINSON. 1965. SOME CONSEQUENCES OF THE DECOMPOSITION OF ORGANIC MATTER IN LAKE NITINAT, AN ANOXIC FJORD. LIMNOL. OCEANOGR. 10(SUPPL.), R185-R201. [EQUATIONS]

RICKER, W. E. 1958. PRODUCTION, REPRODUCTION AND YIELD. VERH. INTERNAT. VER LIMNOL. 13, 84-100. [MODEL]

RICKER, W. E. 1968. INTRODUCTION. PP. 1-6. IN, METHODS FOR ASSESSMENT OF FISH PRODUCTION IN FRESH WATERS. (W.E. RICKER, ED.). BLACKWELL SCIENTIFIC PUBLICATIONS, OXFORD, ENGLAND. [ECOTROPHIC COEFFICIENT; BIOMASS MODEL]

RIELEY, J. O., D. MACHIN AND A. MORTON. 1969. THE MEASUREMENT OF MICROCLIMATIC FACTORS UNDER A VEGETATION CANOPY - A REAPPRAISAL OF WILM'S METHOD. J. ECOL. 57(1), 101-108. [REGRESSION; COMPUTER]

RIJKS, D. A. 1968. WATER USE BY IRRIGATED COTTON IN SUDAN II. NET RADIATION AND SOIL HEAT FLUX. J. APPL. ECOL. 5(3), 685-706. [EQUATIONS; COMPUTER]

RIJKS, D. A. 1971. WATER USE BY IRRIGATED COTTON IN SUDAN III. BOWEN RATIOS AND ADVECTIVE ENERGY. J. APPL. ECOL. 8(3), 643-663. [EQUATIONS; COMPUTER]

ROBINS, P. C. 1974. A METHOD OF MEASURING THE AERODYNAMIC RESISTANCE TO THE TRANSPORT OF WATER VAPOUR FROM FOREST CANOPIES. J. APPL. ECOL. 11(1), 315-325. [MODEL]

RODHE, W. 1966. STANDARD CORRELATIONS BETWEEN PELAGIC PHOTOSYNTHESIS AND LIGHT. PP. 365-381. IN, PRIMARY PRODUCTIVITY IN AQUATIC ENVIRONMENTS. (C. R. GOLDMAN, ED.). UNIVERSITY OF CALIFORNIA PRESS, BERKELEY, CALIFORNIA. [TALLING'S PHOTOSYNTHETIC MODEL]

ROSS, J. 1970. MATHEMATICAL MODELS OF PHOTOSYNTHESIS IN A PLANT STAND. PP. 29-45. IN, PREDICTION AND MEASUREMENT OF PHOTOSYNTHETIC PRODUCTIVITY. (I. SETLIK, ED.). CENTRE FOR AGRICULTURAL PUBLISHING AND DOCUMENTATION, WAGENINGEN, NETHERLANDS.

RUNNING, S. W., R. H. WARING AND R. A. RYDELL. 1975. PHYSIOLOGICAL CONTROL OF WATER FLUX IN CONIFERS: A COMPUTER SIMULATION MODEL. OECOLOGIA 18(1), 1-16.

RYSZKOWSKI, L. AND K. PETRUSEWICZ 1967. ESTIMATION OF ENERGY FLOW THROUGH SMALL RODENT POPULATIONS. PP. 125-146. IN, SECONDARY PRODUCTIVITY OF TERRESTRIAL ECOSYSTEMS, VOLUME I. (K. PETRUSEWICZ, ED.) INSTITUTE OF ECOLOGY, POLISH ACAD. SCI., WARSAW, POLAND.

SAITO, S. 1969. ENERGETICS OF ISOPOD POPULATIONS IN A FOREST OF CENTRAL JAPAN. RES. POPUL. ECOL.-KYOTO 11(2), 229-258.

SAUER, J. F. T. AND E. R. ANDERSON. 1956. THE HEAT BUDGET OF A BODY OF WATER OF VARYING VOLUME. LIMNOL. OCEANOGR. 1(4), 247-251. [EQUATIONS]

SCHINDLER, D. W. 1971. A HYPOTHESIS TO EXPLAIN DIFFERENCES AND SIMILARITITES AMONG LAKES IN THE EXPERIMENTAL LAKES AREA, NORTHWESTERN ONTARIO. J. FISH. RES. BOARD CAN. 28(2), 295-301. [EQUATIONS]

SCHINDLER, D. W., V. E. FROST AND R. V. SCHMIDT. 1973. PRODUCTION OF EPILITHIPHYTON IN TWO LAKES OF THE EXPERIMENTAL LAKES AREA, NORTHWESTERN ONTARIO. J. FISH. RES. BOARD CAN. 30(10), 1511-1524. [EQUATIONS]

SCHULZE, E.-D., O. L. LANGE, M. EVENARI, L. KAPPEN AND V. BUSCHBOM. 1974. THE ROLE OF AIR HUMIDITY AND LEAF TEMPERATURE IN CONTROLLING STOMATAL RESISTANCE OF PRUNUS ARMENIACA L. UNDER DESERT CONDITIONS I. A SIMULATION OF THE DAILY COURSE OF STOMATAL RESISTANCE. OECOLOGIA 17(2), 159-170. [MODEL; COMPUTER]

SCOTT, D. 1965. THE DETERMINATION AND USE OF THERMODYNAMIC DATA IN ECOLOGY. ECOLOGY 46(5), 673-680. [ENTROPY; INFORMATION THEORY]

SCOTT, D. 1969. DETERMINING THE TYPE OF RELATIONSHIP BETWEEN PLANTS AND ENVIRONMENTAL FACTORS. PROC. N. Z. ECOL. SOC. 16, 29-31. [STEPWISE MULTIPLE REGRESSION]

SETLIK, I. (ED.). 1970. PREDICTION AND MEASUREMENT OF PHOTOSYNTHETIC PRODUCTIVITY. PROC. IBP/PP TECH. MEETING, TREBON, 14-21. 1969. CENTRE FOR AGRICULTURAL PUBLISHING AND DOCUMENTATION, WAGENINGEN, NETHERLANDS. 632 PP.

SHONTING, D. H. 1964. SOME OBSERVATIONS OF SHORT-TERM HEAT TRANSFER THROUGH THE SURFACE LAYERS OF THE OCEAN. LIMNOL. OCEANOGR. 9(4), 576-588. [EQUATIONS]

SHUSHKINA, E. A., S. I. ANISIMOV AND R. Z. KLEKOWSKI. 1968. CALCULATION OF PRODUCTION EFFICIENCY IN PLANKTON COPEPODS. POL. ARCH. HYDROBIOL. 15, 251-261. [EQUATIONS]

SINCLAIR, T. R. 1972. ERROR ANALYSIS OF LATENT, SENSIBLE AND PHOTOCHEMICAL HEAT FLUX DENSITIES CALCULATED FROM ENERGY BALANCE MEASUREMENTS ABOVE FORESTS. PP. 55-66. IN. MODELING THE GROWTH OF TREES. (C. E. MURPHY, JR., J. D. HESKETH AND B. R. STRAIN, EDS.). AEC REPT. EDFB-IBP-72-11. OAK RIDGE NATL. LAB., OAK RIDGE, TENNESSEE.

SKELLAM, J. G. 1967. PRODUCTIVE PROCESSES IN ANIMAL POPULATIONS CONSIDERED FROM THE BIOMETRICAL STANDPOINT. PP. 59-82. IN. SECONDARY PRODUCTIVITY OF TERRESTRIAL ECOSYSTEMS, VOLUME I. (K. PETRUSEWICZ, ED.). INSTITUTE OF ECOLOGY, POLISH ACAD. SCI., WARSAW, POLAND.

SLOBODKIN, L. B. 1960. ECOLOGICAL ENERGY RELATIONSHIPS AT THE POPULATION LEVEL. AMER. NAT. 94(876), 213-236.

SLOBODKIN, L. B. 1962. ENERGY IN ANIMAL ECOLOGY. PP. 69-101. IN. ADVANCES IN ECOLOGICAL RESEARCH VOLUME 1. (J. B. CRAGG, ED.). ACADEMIC PRESS, NEW YORK, NEW YORK. [MODEL; INFORMATION THEORY; ENTROPY; MACARTHUR]

SMAYDA, T. J. 1965. A QUANTITATIVE ANALYSIS OF THE PHYTOPLANKTON OF THE GULF OF PANAMA II. ON THE RELATIONSHIP BETWEEN C-14 ASSIMILATION AND THE DIATOM STANDING CROP. BULL. INTER-AMER. TROP. TUNA COMM. 9(7), 467-531. [ESTIMATION; REGRESSION]

SMITH, F. E. 1969. EFFECTS OF ENRICHMENT IN MATHEMATICAL MODELS. PP. 631-645. IN. EUTROPHICATION: CAUSES, CONSEQUENCES, CORRECTIVES. NATIONAL ACADEMY OF SCIENCES, WASHINGTON, D. C. [SIMULATION; SYSTEMS ANALYSIS; COMPUTER]

SMITH, K. G. AND H. H. PRINCE. 1973. THE FASTING METABOLISM OF SUBADULT MALLARDS ACCLIMATIZED TO LAW AMBIENT TEMPERATURES. CONDOR 75(3), 330-335. [ANALYSIS OF VARIANCE; ANALYSIS OF COVARIANCE; REGRESSION; COMPUTER]

SOHOLT, L. F. 1973. CONSUMPTION OF PRIMARY PRODUCTION BY A POPULATION OF KANGAROO RATS (DIPODOMYS MERRIAMI) IN THE MOJAVE DESERT. ECOL. MONOGR. 43(3), 357-376. [TAG-RECAPTURE; ENERGY EQUATION]

SOUTHWICK, E. E. AND D. M. GATES. 1975. ENERGETICS OF OCCUPIED HUMMINGBIRD NESTS. PP. 417-430. IN, PERSPECTIVE OF BIOPHYSICAL ECOLOGY. (D. M. GATES AND R. B. SCHMERL, EDS.). SPRINGER-VERLAG NEW YORK INC., NEW YORK, NEW YORK. [MODEL]

SPOTILA, J. R., O. H. SOULE AND D. M. GATES. 1972. THE BIOPHYSICAL ECOLOGY OF THE ALLIGATOR: HEAT ENERGY BUDGETS AND CLIMATE SPACES. ECOLOGY 53(6), 1094-1102.

STACHURSKI, A. 1974. STABILIZATION MECHANISMS OF ENERGY TRANSFER BY LIGIDIUM DYPNORUM (CUVIER) (ISOPODA) POPULATION IN ALDER WOOD (CARICI ELONGATAE - ALNETUM). EKOLOGIYA POLSKA-SERIA A 22(1), 3-29. [PRODUCTIVITY EQUATIONS; INSECT POPULATION MODEL]

STANDEN, V. 1973. THE PRODUCTION AND RESPIRATION OF AN ENCHYTRAEID POPULATION IN BLANKET BOG. J. ANIM. ECOL. 42(2), 219-245. [MODEL; VARIANCE OF DEATH RATES]

STEELE, J. H. 1962. ENVIRONMENTAL CONTROL OF PHOTOSYNTHESIS IN THE SEA. LIMNOL. OCEANOGR. 7(2), 137-150. [EQUATIONS]

STEELE, J. H. 1966. NOTES ON SOME THEORETICAL PROBLEMS IN PRODUCTION ECOLOGY. PP. 383-398. IN, PRIMARY PRODUCTIVITY IN AQUATIC ENVIRONMENTS. (C. R. GOLDMAN, ED.). UNIVERSITY OF CALIFORNIA PRESS, BERKELEY, CALIFORNIA. [PHOTOSYNTHESIS; AQUATIC]

STEPANOVA, L. A. 1971. PRODUCTION OF SOME COMMON PLANKTONIC CRUSTACEANS IN LAKE ILMEN. HYDROBIOL. J. 7(6), 13-23. (ENGLISH TRANSL.). [GROWTH EQUATIONS; POPULATION MODEL]

STEPANOVA, L. A. 1973. COMPARATIVE EVALUATION OF TWO METHODS OF CALCULATING ZOOPLANKTON PRODUCTION EXEMPLIFIED ON THE LAKE IL'MEN POPULATION. EKOLOGIYA 4(6), 18-25. (ENGLISH TRANSL. PP. 476-482). [EDMONDSON'S EQUATION]

STRICKLAND, J. D. H. 1960. MEASURING THE PRODUCTION OF MARINE PHYTOPLANKTON. FISH. RES. BOARD CAN. BULL. 122. VIII, 172 PP. [MODEL]

STROGONOV, A. A. 1969. CIRCADIAN RHYTHMS OF THE HYPONEUSTON IN THE BLACK SEA AND THE MEDITERRANEAN HYDROBIOL. J. 5(5), 78-83. (ENGLISH TRANSL.). [BIOMASS EQUATION; CONFIDENCE INTERVAL; ABSOLUTE ERROR]

SUSHCHENYA, L. M. 1970. FOOD RATIONS, METABOLISM AND GROWTH OF CRUSTACEANS. PP. 127-141. IN, MARINE FOOD CHAINS. (J. H. STEELE, ED.). UNIVERSITY OF CALIFORNIA PRESS, BERKELEY, CALIFORNIA. [MODEL]

SZEICZ, G. 1974. SOLAR RADIATION IN CROP CANOPIES. J. APPL. ECOL. 11(3), 1117-1156. [EQUATIONS; MODEL]

TAJCHMAN, S. J. 1972. THE RADIATION AND ENERGY BALANCES OF CONIFEROUS AND DECIDUOUS FORESTS. J. APPL. ECOL. 9(2), 359-375. [EQUATIONS]

TAKAHASHI, M. AND S. ICHIMURA. 1970. PHOTOSYNTHETIC PROPERTIES AND GROWTH OF PHOTOSYNTHETIC SULFUR BACTERIA IN LAKES. LIMNOL. OCEANOGR. 15(6), 929-944. [EQUATIONS]

TALLING, J. 1970. GENERALIZED AND SPECIALIZED FEATURES OF PHYTOPLANKTON AS A FORM OF PHOTOSYNTHETIC COVER. PP. 431-445. IN, PREDICTION AND MEASUREMENT OF PHOTOSYNTHETIC PRODUCTIVITY. (I. SETLIK, ED.). CENTRE FOR AGRICULTURAL PUBLISHING AND DOCUMENTATION, WAGENINGEN, NETHERLANDS. [MODEL]

THORKELSON, J. AND R. K. MAXWELL. 1974. DESIGN AND TESTING OF A HEAT TRANSFER MODEL OF A RACCOON (PROCYON LOTOR) IN A CLOSED TREE DEN. ECOLOGY 55(1), 29-39.

TIMIN, M. E. AND B. D. COLLIER 1971. A MODEL INCORPORATING ENERGY UTILIZATION FOR THE DYNAMICS OF SINGLE SPECIES POPULATIONS. THEOR. POPUL. BIOL. 2(2), 237-251.

TROJAN, P. 1967. INVESTIGATIONS ON PRODUCTION OF CULTIVATED FIELDS. PP. 545-561. IN, SECONDARY PRODUCTIVITY OF TERRESTRIAL ECOSYSTEMS, VOLUME II. (K. PETRUSEWICZ, ED.) INSTITUTE OF ECOLOGY, POLISH ACAD. SCI., WARSAW, POLAND.

TROJAN, P. 1968. ESTIMATED FOOD CONSUMPTION BY THE COLORADO BEETLE (LEPTINOTARSA DECEMLINEATA SAY) UNDER CONDITIONS OF NATURAL REDUCTION. EKOLOGIYA POLSKA-SERIA A 16(18), 385-393. [MODEL]

TUCKER, V. A. 1966. DIURNAL TORPOR AND ITS RELATION TO FOOD CONSUMPTION AND WEIGHT CHANGES IN THE CALIFORNIA POCKET MOUSE PEROGNATHUS CALIFORNICUS. ECOLOGY 47(2), 245-252. [EQUATIONS]

TUCKER, V. A. 1974. ENERGETICS OF NATURAL AVIAN FLIGHT. PP. 298-328. IN, AVIAN ENERGETICS. (R. A. PAYNTER, JR., ED.). PUBLICATION OF NUTTALL ORNITHOLOGICAL CLUB NO. 15, CAMBRIDGE, MASSACHUSETTS. [EQUATIONS]

TUCKER, V. A. AND K. SCHMIDT-KOENIG. 1971. FLIGHT SPEEDS OF BIRDS IN RELATION TO ENERGETICS AND WIND DIRECTIONS. AUK 88(1), 97-107. [MODEL]

UCHIJIMA, Z. 1970. CARBON DIOXIDE ENVIRONMENT AND FLUX WITHIN A CORN CROP CANOPY. PP. 179-196. IN, PREDICTION AND MEASUREMENT OF PHOTOSYNTHETIC PRODUCTIVITY. (I. SETLIK, ED.). CENTRE FOR AGRICULTURAL PUBLISHING AND DOCUMENTATION, WAGENINGEN, NETHERLANDS. [MODEL]

ULANOWICZ, R. E. 1972. MASS AND ENERGY FLOW IN CLOSED ECOSYSTEMS. J. THEOR. BIOL. 34(2), 239-253. [DIFFERENTIAL EQUATION ; ENTROPY]

UNGER, K. AND ST. CLAUS. 1970. MODEL INVESTIGATION FOR THE ENERGY BALANCE OF CULTIVATED PLANTS. PP. 567-570. IN, PREDICTION AND MEASUREMENT OF PHOTOSYNTHETIC PRODUCTIVITY. (I. SETLIK, ED.). CENTRE FOR AGRICULTURAL PUBLISHING AND DOCUMENTATION, WAGENINGEN, NETHERLANDS.

UTTER, J. M. AND E. A. LE FEBVRE. 1973. DAILY ENERGY EXPENDITURE OF PURPLE MARTINS (PROGNE SUBIS) DURING THE BREEDING SEASON; ESTIMATES USING D2O18 AND TIME BUDGET METHODS. ECOLOGY 54(3), 597-604. [EQUATIONS]

VAN DYNE, G. M. 1968. MEASURING QUANTITY AND QUALITY OF THE DIET OF LARGE HERBIVORES. PP. 54-94. IN, A PRACTICAL GUIDE TO THE PRODUCTIVITY OF LARGE HERBIVORES. (F. B. GOLLEY AND H. K. BUECHNER, EDS.). BLACKWELL SCIENTIFIC PUBLICATIONS, OXFORD, ENGLAND. [DIGESTIBILITY EQUATIONS]

VAN DYNE, G. M. AND J. H. MEYER. 1964. A METHOD FOR MEASUREMENT OF FORAGE INTAKE OF GRAZING LIVESTOCK USING MICRODIGESTION TECHNIQUES. J. RANGE MANAGE. 17(4), 204-208. [EQUATIONS]

VAN DYNE, G. M., R. G. WRIGHT AND J. F. DOLLAR. 1968. INFLUENCE OF SITE FACTORS ON VEGETATION PRODUCTIVITY: A REVIEW AND SUMMARIZATION OF MODELS, DATA, AND REFERENCES. AEC REPT. ORNL-TM-1974. OAK RIDGE NATL. LAB., OAK RIDGE, TENNESSEE. III, 235 PP.

VERMEIJ, G. J. 1973. MORPHOLOGICAL PATTERNS IN HIGH-INTERTIDAL GASTROPODS: ADAPTIVE STRATEGIES AND THEIR LIMITATIONS. MARINE BIOL. 20(4), 319-346. [ENERGY EQUATION]

VINBERG, G. G. 1972. INVESTIGATIONS OF THE BIOLOGICAL ENERGY BALANCE AND THE BIOLOGICAL PRODUCTIVITY OF LAKES IN THE SOVIET UNION. EKOLOGIYA 3(4), 5-18. (ENGLISH TRANSL. PP. 295-304). [GROWTH EQUATIONS; METABOLISM RATE]

VINBERG, G. G. 1973. DISCUSSION OF THE ARTICLE BY YU. A. KOLESNIK "SOME QUESTIONS RELATING TO GROWTH AND METABOLIC RATE IN ANIMALS." J. ICHTHYOLOGY 13(5), 775-777. [EQUATIONS]

VINBERG, G. G. ET AL. 1973. THE PROGRESS AND STATE OF RESEARCH ON THE METABOLISM, GROWTH, NUTRITION, AND PRODUCTION OF FRESH-WATER INVERTEBRATE ANIMALS. HYDROBIOL. J. 9(3), 77-84. (ENGLISH TRANSL.). [GROWTH EQUATIONS]

VINOGRADOV, G. A. 1973. FUNCTIONING OF OSMOTIC REGULATION SYSTEMS OF A FRESHWATER AMPHIPOD IN WATER OF VARYING SALINITY. EKOLOGIYA 4(3), 77-84. (ENGLISH TRANSL. PP. 244-249). [EQUATIONS]

VLYMEN, W. J. 1970. ENERGY EXPENDITURE OF SWIMMING COPEPODS. LIMNOL. OCEANOGR. 15(3), 348-356. [EQUATIONS]

VLYMEN, W. J., III. 1974. SWIMMING ENERGETICS OF THE LARVAL ANCHOVY ENGRAULIS MORDAX. U. S. FISH WILDL. SERV. FISH. BULL. 72(4), 885-899. [EQUATIONS]

VOLLENWEIDER, R. A. 1966. CALCULATION MODELS OF PHOTOSYNTHESIS-DEPTH CURVES AND SOME IMPLICATIONS REGARDING DAY RATE ESTIMATES IN PRIMARY PRODUCTIVITY MEASUREMENTS. PP. 425-457. IN, PRIMARY PRODUCTIVITY IN AQUATIC ENVIRONMENTS. (C. R. GOLDMAN, ED.). UNIVERSITY OF CALIFORNIA PRESS, BERKELEY, CALIFORNIA. [TALLING'S PHOTOSYNTHETIC MODEL; GRAPHICAL DETERMINATION; STEELE'S INTEGRAL; SMITH'S FORMULA; SIMULATION]

VOLLENWEIDER, R. A. 1970. MODELS FOR CALCULATING INTEGRAL PHOTOSYNTHESIS AND SOME IMPLICATIONS REGARDING STRUCTURAL PROPERTIES OF THE COMMUNITY METABOLISM OF AQUATIC SYSTEMS. PP. 455-472. IN, PREDICTION AND MEASUREMENT OF PHOTOSYNTHETIC PRODUCTIVITY. (I. SETLIK, ED.). CENTRE FOR AGRICULTURAL PUBLISHING AND DOCUMENTATION, WAGENINGEN, NETHERLANDS.

VOLOKHONSKII, A. G. 1973. STRUCTURAL AND ENERGETIC ASPECTS OF THE PROBLEM OF DEFICIENCY OF BIOGENIC ELEMENTS. EKOLOGIYA 4(2), 5-11. (ENGLISH TRANSL. PP. 95-100). [EQUATIONS; NEGENTROPY; PHYTOPLANKTON; FRESHWATER]

WALLENTINUS, H.-G. 1973. ABOVE-GROUND PRIMARY PRODUCTION OF A JUNCETUM GERARDI ON A BALTIC SEA-SHORE MEADOW. OIKOS 24(2), 200-219. [FORMULA]

WANG, J. Y. 1960. A CRITIQUE OF THE HEAT UNIT APPROACH TO PLANT RESPONSE STUDIES. ECOLOGY 41(4), 785-790. [EQUATIONS]

WATT, K. E. F. 1955. STUDIES ON POPULATION PRODUCTIVITY I. THREE APPROACHES TO THE OPTIMUM YIELD PROBLEM IN POPULATIONS OF TRIBOLIUM CONFUSUM. ECOL. MONOGR. 25(3), 269-290. [MODEL; HYPERSURFACE; GAUSE MODEL; GRAPHICAL REGRESSION ANALYSIS]

WATT, K. E. F. 1959. STUDIES ON POPULATION PRODUCTIVITY II. FACTORS GOVERNING PRODUCTIVITY IN A POPULATION OF SMALLMOUTH BASS. ECOL. MONOGR. 29(4), 367-392. [MODEL; TAG-RECAPTURE ESTIMATION; MULTIPLE REGRESSION; LOG TRANSFORMATION]

WEBB, W. L. 1972. A MODEL OF LIGHT AND TEMPERATURE CONTROLLED NET PHOTOSYNTHETIC RATES FOR TERRESTRIAL PLANTS. PP. 237-242. IN, PROCEEDINGS-RESEARCH ON CONIFEROUS FOREST ECOSYSTEMS-A SYMPOSIUM. (J. F. FRANKLIN, L. J. DEMPSTER AND R. H. WARING, EDS.). U. S. FOR. SERV. PACIFIC NORTHWEST FOREST EXPT. STN., PORTLAND, OREGON. [COMPUTER]

WEBB, W. L., M. NEWTON AND D. STARR. 1974. CARBON DIOXIDE EXCHANGE OF ALNUS RUBRA. A MATHEMATICAL MODEL. OECOLOGIA 17(4), 281-291.

WEIHS, D. 1974. ENERGETIC ADVANTAGES OF BURST SWIMMING OF FISH. J. THEOR. BIOL. 48(1), 215-229. [EQUATIONS]

WHITE, H. 1965. THE ENTROPY OF A CONTINUOUS DISTRIBUTION. BULL. MATH. BIOPHYS. 27(SPECIAL ISSUE), 135-143. [INFORMATION THEORY]

WHITTAKER, R. H. AND G. M. WOODWELL. 1967. SURFACE AREA RELATIONS OF WOODY PLANTS AND FOREST COMMUNITIES. AMER. J. BOT. 54(8), 931-939. [REGRESSION; LOG TRANSFORMATION]

WHITTAKER, R. H. AND G. M. WOODWELL. 1968. DIMENSION AND PRODUCTION RELATIONS OF TREES AND SHRUBS IN THE BROOKHAVEN FOREST, NEW YORK. J. ECOL. 56(1), 1-25. [REGRESSION; LOG TRANSFORMATION]

WIEGERT, R. G. 1968. THERMODYNAMIC CONSIDERATIONS IN ANIMAL NUTRITION. AMER. ZOOL. 8(1), 71-81.

WIENS, J. A. AND G. S. INNIS 1973. ESTIMATION OF ENERGY FLOW IN BIRD COMMUNITIES. II. A SIMULATION MODEL OF ACTIVITY BUDGETS AND POPULATION BIOENERGETICS. PP. 739-752. IN, PROCEEDINGS OF THE 1973 SUMMER COMPUTER SIMULATION CONFERENCE, JULY 17, 18, 19, 1973, MONTREAL, P. Q., CANADA, VOL. II. [COMPUTER]

WIENS, J. A. AND G. S. INNIS 1974. ESTIMATION OF ENERGY FLOW IN BIRD COMMUNITIES: A POPULATION BIOENERGETICS MODEL. ECOLOGY 55(4), 730-746. [SIMULATION; LOGISTIC FUNCTION; COMPUTER]

WILSON, K. J. AND A. K. LEE. 1974. ENERGY EXPENDITURE OF A LARGE HERBIVORUS LIZARD. COPEIA 1974(2), 338-348. [EQUATIONS]

WILSON, T. A. 1965. NATURAL MORTALITY AND REPRODUCTION FOR A FOOD SUPPLY AT MINIMUM METABOLISM. AMER. NAT. 99(908), 373-376.

WINBERG, G. G., K. PATALAS, J. C. WRIGHT, A. HILLBRICHT-ILKOWSKA, W. E. COOPER AND K. H. MANN. 1971. METHODS FOR CALCULATING PRODUCTIVITY. PP. 296-317. IN, A MANUAL ON METHODS FOR THE ASSESSMENT OF SECONDARY PRODUCTIVITY IN FRESH WATERS. (W. T. EDMONDSON AND G. G. WINBERG, EDS.) BLACKWELL SCIENTIFIC PUBLICATIONS, OXFORD, ENGLAND. [PRODUCTION MODEL; COHORT]

WRIGHT, J. C. AND R. M. HORRALL. 1967. HEAT BUDGET STUDIES ON THE MADISON RIVER, YELLOWSTONE NATIONAL PARK. LIMNOL. OCEANOGR. 12(4), 578-583. [EQUATIONS]

YOSHIDA, Y. 1970. STUDIES ON THE EFFICIENCY OF FOOD CONVERSION TO FISH BODY GROWTH-I. THE FORMULAE ON THE EFFICIENCY OF CONVERSION. BULL. JAPANESE SOC. SCI. FISH. 36(2), 156-159.

YOSHIDA, Y. 1970. STUDIES ON THE EFFICIENCY OF FOOD CONVERSION TO FISH BODY GROWTH-III. TOTAL UPTAKE OF FOOD AND EFFICIENCY OF TOTAL FOOD CONVERSION. BULL. JAPANESE SOC. SCI. FISH. 36(9), 914-916. [EQUATIONS]

EXPLOITATION, MANAGEMENT, AND CONTROL

AASEN, O. 1954. A METHOD FOR THEORETICAL CALCULATION OF THE NUMERICAL STOCK-STRENGTH IN FISH POPULATIONS SUBJECT TO SEASONAL FISHERIES. RAPP. PROCES-VERB. REUNIONS J. CONS. PERM. INT. EXPLOR. MER 136, 77-86.

ABRAMSON, N. J. AND P. K. THOMLINSON. 1972. AN APPLICATION OF YIELD MODELS TO A CALIFORNIA OCEAN SHRIMP POPULATION. U. S. NATL. MAR. FISH. SERV. FISH. BULL. 70(3), 1021-1041. [SCHAEFER'S MODEL; MURPHY MODEL]

ABROSOV, V. N. 1969. DETERMINATION OF COMMERCIAL TURNOVER IN NATURAL BODIES OF WATER. PROBLEMS OF ICHTHYOLOGY 9(4), 482-489. [CATCH EQUATION]

AHMED, N. U. AND N. D. GEORGANAS. 1973. OPTIMAL CONTROL THEORY APPLIED TO A DYNAMIC AQUATIC ECOSYSTEM. J. FISH. RES. BOARD CAN. 30(4), 576-579. [MODEL]

AIKAWA, H. 1937. RATE OF VARIATION IN THE STOCK OF HOKKAIDO HERRING. BULL. JAPANESE SOC. SCI. FISH. 6(2), 59-65. [RECRUITMENT CURVE; GROWTH CURVES; STOCK CURVE]

ALLEN, K. R. 1947. SOME ASPECTS OF THE PRODUCTION AND CROPPING OF FRESH WATERS. N. Z. SCI. CONGR. PP. 222-228. [GROWTH CURVES]

ALLEN, K. R. 1953. A METHOD FOR COMPUTING THE OPTIMUM SIZE-LIMIT FOR A FISHERY. NATURE 172(4370), 210.

ALLEN, K. R. 1954. FACTORS AFFECTING THE EFFICIENCY OF RESTRICTIVE REGULATIONS IN FISHERIES MANAGEMENT. I. SIZE LIMITS. N. Z. J. SCI. TECHNOL. SEC. B 35(6), 498-529.

ALLEN, K. R. 1955. FACTORS AFFECTING THE EFFICIENCY OF RESTRICTIVE REGULATIONS IN FISHERIES MANAGEMENT. II. BAG LIMITS. N. Z. J. SCI. TECHNOL. SEC. B. 36(4), 305-334. [MODEL; DISTRIBUTION; LOG TRANSFORMATION]

ALLEN, K. R. 1967. SOME QUICK METHODS FOR ESTIMATING THE EFFECT ON CATCH OF CHANGES IN THE SIZE LIMIT. J. CONS. PERM. INT. EXPLOR. MER 31(1), 111-126. [GROWTH CURVES; VON BERTALANFFY'S GROWTH CURVES; EXPONENTIAL GROWTH]

ALLEN, K. R. 1971. RELATION BETWEEN PRODUCTION AND BIOMASS. J. FISH. RES. BOARD CAN. 28(10), 1573-1581. [MODEL]

ALLEN, K. R. 1973. THE INFLUENCE OF RANDOM FLUCTUATIONS IN THE STOCK-RECRUITMENT RELATIONSHIP ON THE ECONOMIC RETURN FROM SALMON FISHERIES. RAPP. PROCES-VERB. REUNIONS J. CONS. PERM. INT. EXPLOR. MER 164, 350-359. [MODEL]

ALLEN, K. R. AND C. R. FORRESTER. 1966. APPROPRIATE SIZE LIMITS FOR LEMON SOLE (PAROPHRYS VETULUS) IN THE STRAIT OF GEORGIA. J. FISH. RES. BOARD CAN. 23(4), 511-520. [GROWTH CURVES]

ALVERSON, D. L. AND G. J. PAULIK. 1973. OBJECTIVES AND PROBLEMS OF MANAGING AQUATIC LIVING RESOURCES. J. FISH. RES. BOARD CAN. 30(12, PART 2), 1936-1947. [MODEL]

ANDERSON, K. P. AND O. BAGGE. 1963. THE BENEFIT OF PLAICE TRANSPLANTATION AS ESTIMATED BY TAGGING EXPERIMENTS. INTERN. COMM. NORTHWEST ATLANTIC FISH. SPEC. PUBL. NO. 4. [YIELD MODEL; MORTALITY COEFFICIENT; TAG-RECAPTURE]

ANDERSON, L. G. 1973. OPTIMUM ECONOMIC YIELD OF A FISHERY GIVEN A VARIABLE PRICE OF OUTPUT. J. FISH. RES. BOARD CAN. 30(4), 509-518. [MODEL]

ANDERSON, L. G. 1975. OPTIMUM ECONOMIC YIELD OF AN INTERNATIONALLY UTILIZED COMMON PROPERTY RESOURCE. U. S. NATL. MAR. FISH. SERV. FISH. BULL. 73(1), 51-66. [MODEL]

ARIMIZU, T. 1958. WORKING GROUP MATRIX IN DYNAMIC MODEL OF FOREST MANAGEMENT. J. JAPANESE FOR. SOC. 40(5), 185-190.

AULSTAD, D., T. GJEDREM AND H. SKJERVOLD. 1972. GENETIC AND ENVIRONMENTAL SOURCES OF VARIATION IN LENGTH AND WEIGHT OF RAINBOW TROUT (SALMO GAIRDNERI). J. FISH. RES. BOARD CAN. 29(3), 237-241. [ANALYSIS OF VARIANCE]

BAER, R. L. 1971. OPTIMAL CONTROL AND THE MANAGEMENT OF ECOLOGICAL SYSTEMS. DISSERTATION. UNIVERSITY OF PENNSYLVANIA, PHILADELPHIA, PENNSYLVANIA. 202 PP. [MODEL; COMPUTER]

BAMS, R. A. 1967. DIFFERENCES IN PERFORMANCE OF NATURALLY AND ARTIFICIALLY PROPAGATED SOCKEYE SALMON MIGRANT FRY, AS MEASURED WITH SWIMMING AND PREDATION TESTS. J. FISH. RES. BOARD CAN. 24(5), 1117-1153. [NON-PARAMETRIC STATISTICS]

BAYLIFF, W. H. 1967. GROWTH, MORTALITY, AND EXPLOITATION OF THE ENGRAULIDAE, WITH SPECIAL REFERENCE TO THE ANCHOVETA, CETENGRAULIS MYSTICETUS, AND THE COLORADO, ANCHOVA NASO, IN THE EASTERN PACIFIC OCEAN. BULL. INTER-AMER. TROP. TUNA COMM. 12(5), 367-408. [COEFFICIENT OF MORTALITY; VON BERTALANFFY'S EQUATION; BEVERTON YIELD EQUATION]

BECKER, N. G. 1970. CONTROL OF A PEST POPULATION. BIOMETRICS 26(3), 365-375.

BEDDINGTON, J. R. 1974. AGE STRUCTURE, SEX RATIO AND POPULATION DENSITY IN THE HARVESTING OF NATURAL ANIMAL POPULATIONS. J. APPL. ECOL. 11(3), 915-924. [MODEL; MATRIX; FIRST ORDER DIFFERENCE EQUATIONS; OPTIMAL CROPPING]

BEDDINGTON, J. R. AND D. B. TAYLOR 1973. OPTIMUM AGE SPECIFIC HARVESTING OF A POPULATION. BIOMETRICS 29(4), 801-809. [LESLIE MATRIX; ALGORITHM; RED DEER]

ELAND, P. 1974. ON PREDICTING HE YIELD FROM BROOK TROUT OPULATIONS. TRANS. AMER. FISH. OC. 103(2), 353-355. [JENSEN'S ATRIX; LESLIE MATRIX]

BERRYMAN, A. A. 1967. MATHEMATICAL DESCRIPTION OF THE STERILE MALE PRINCIPLE. CAN. ENTOMOL. 99(8), 858-865.

BEVERTON, R. J. H. 1953. SOME OBSERVATIONS ON THE PRINCIPLES OF FISHERY REGULATION. J. CONS. PERM. INT. EXPLOR. MER 19(1), 56-68. [EXPLOITATION MODEL; VON BERTALANFFY GROWTH FUNCTION]

BEVERTON, R. J. H. 1954. NOTES ON THE USE OF THEORETICAL MODELS IN THE STUDY OF THE DYNAMICS OF EXPLOITED FISH POPULATIONS. U. S. FISHERY LAB., BEAUFORT, NORTH CAROLINA. MISC. CONTR. NO. 2, 186 PP.

BEVERTON, R. J. H. 1963. MATURATION, GROWTH AND MORTALITY OF CLUPEID AND ENGRAULID STOCKS IN RELATION TO FISHING. RAPP. PROCES-VERB. REUNIONS J. CONS. PERM. INT. EXPLOR. MER 154, 44-67. [INDEX OF CATCH PER RECRUIT; MORTALITY RATE; GROWTH FUNCTION; VON BERTALANFFY'S EQUATION; BEVERTON-HOLT EQUATION]

BEVERTON, R. J. H. AND B. B. PARRISH. 1956. COMMERCIAL STATISTICS IN FISH POPULATION STUDIES. RAPP. PROCES-VERB. REUNIONS J. CONS. PERM. INT. EXPLOR. MER 140(1), 58-66. [SAMPLING; ESTIMATION; STRATIFICATION]

BEVERTON, R. J. H. AND S. J. HOLT. 1956. THE THEORY OF FISHING. PP. 372-441.(PLUS REFERENCES). IN, SEA FISHERIES: THEIR INVESTIGATION IN THE UNITED KINGDOM. (M. GRAHAM, ED.). EDWARD ARNOLD PUBLISHERS, LONDON, ENGLAND. [MODEL; VON BERTALANFFY'S EQUATION; MORTALITY; GROWTH]

BEVERTON, R. J. H. AND S. J. HOLT. 1964. TABLES OF YIELD FUNCTIONS FOR FISHERY ASSESSMENT. FAO FISHERIES TECHNICAL PAPER NO. 38. 49 PP. [BEVERTON-HOLT EQUATION; VON BERTALANFFY GROWTH EQUATION]

BEVERTON, R. J. H. AND S. J. HOLT. 1966. MANUAL OF METHODS FOR FISH STOCK ASSESSMENT PART II-TABLES OF YIELD FUNCTIONS. FAO FISHERIES TECHNICAL PAPER NO. 38(REV.1), VI, VARIOUS PAGES PLUS TABLES. [YIELD EQUATION]

BOOTH, D. E. 1972. A MODEL FOR OPTIMAL SALMON MANAGEMENT. U. S. NATL. MAR. FISH. SERV. FISH. BULL. 70(2), 497-506. [PROFIT MAXIMIZING MODEL]

BORGHI, C., A. DE MORAIS REGO, M. L. DE OLIVEIRA AND Z. M. DE AZEVEDO. 1971. ORGANIZING A PROGRAM FOR ERADICATION OF CULEX PIPIENS FATIGANS IN RECIFE, BRAZIL. PP. 373-378. IN, STERILITY PRINCIPLE FOR INSECT CONTROL OR ERADICATION. IAEEDITORS OR EDITOR* [MODEL]

BOWER, D. R. 1970. ROUNDING EFFECTS IN SIMPLE LINEAR REGRESSION. FOREST SCI. 16(3), 301-303. [FORESTRY]

BROWN, W. G., A. SINGH AND E. N. CASTLE. 1965. NET ECONOMIC VALUE OF THE OREGON SALMON-STEELHEAD SPORT FISHERY. J. WILDL. MANAGE. 29(2), 266-279. [CLAWSON ECONOMIC VALUE MODEL; REGRESSION MODEL]

BUCHANAN-WOLLASTON, H. J. 1927. ON THE SELECTIVE ACTION OF A TRAWL NET, WITH SOME REMARKS ON SELECTIVE ACTION OF DRIFT NETS. J. CONS. PERM. INT. EXPLOR. MER 2(3), 343-355. [PROBABILITY OF CAPTURE; CURVE OF ERROR; METHOD OF LEAST SQUARES; PARABOLA; NATURAL LOGARITHMS; CURVE FITTING]

BURD, A. C. 1963. ON SELECTION BY THE DRIFTER FLEETS IN THE EAST ANGLIAN HERRING FISHERY. J. CONS. PERM. INT. EXPLOR. MER 28(1), 91-120. [EQUATIONS]

BUTLER, R. L. AND D. P. BORGESON. 1965. CALIFORNIA "CATCHABLE" TROUT FISHERIES. CALIF. DEPT. FISH GAME FISH BULL. NO. 127, 47 PP. [MORTALITY RATE; RICKER'S CATCHABILITY EQUATION; RECRUITMENT EQUATION]

CALIFORNIA BUREAU OF MARINE FISHERIES. 1952. THE COMMERCIAL FISH CATCH OF CALIFORNIA FOR THE YEAR 1950 WITH A DESCRIPTION OF THE METHODS USED IN COLLECTING AND COMPILING THE STATISTICS. CALIF. DEPT. FISH GAME FISH BULL. NO. 86. 120 PP.

CARLANDER, K. D. 1958. SOME SIMPLE MATHEMATICAL MODELS AS AIDS IN INTERPRETING THE EFFECT OF FISHING. IOWA STATE J. SCI. 32(3), 395-418.

CHADWICK, H. K. 1969. AN EVALUATION OF STRIPED BASS ANGLING REGULATIONS BASED ON AN EQUILIBRIUM YIELD MODEL. CALIF. FISH GAME 55(1), 12-19. [COMPUTER; BEVERTON-HOLT EQUATION]

CHAPMAN, D. G. 1969. STATISTICAL PROBLEMS IN THE OPTIMUM UTILIZATION OF FISHERIES RESOURCES. BULL. INTERN. STAT. INST. 42(1), 268-290. [MODEL; LOGISTIC LAW; PEARL-VERHULST; GROWTH FUNCTION; LOTKA-VOLTERRA; SCHAEFER'S MODEL; MORTALITY RATE; VON BERTALANFFY'S EQUATION; RICKER'S MODEL; BEVERTON-HOLT MODEL; M LIKELIHOOD; TIME-SEQUENTIAL TEST; VARIANCE; POISSON DISTRIBUTION; SUBSAMPLING; ANALYSIS OF VARIANCE; VARIANCE COMPONENTS; SIMULATION; COMPUTER; TAG-RECAPTURE; HYPERGEOMETRIC MODEL; BIAS]

CHAPMAN, D. G. 1973. SPAWNER-RECRUIT MODELS AND ESTIMATION OF THE LEVEL OF MAXIMUM SUSTAINABLE CATCH. RAPP. PROCES-VERB. REUNIONS J. CONS. PERM. INT. EXPLOR. MER 164, 325-332.

CHAPMAN, D. G., R. J. MYHRE AND G. M. SOUTHWARD. 1962. UTILIZATION OF PACIFIC HALIBUT STOCKS: ESTIMATION OF MAXIMUM SUSTAINABLE YIELD, 1960. INTERN. PACIFIC HALIBUT COMM. REPT. NO. 31. 35 PP. [MODEL]

CHAPMAN, D. W. 1967. PRODUCTION IN FISH POPULATIONS. PP. 3-29. IN, THE BIOLOGICAL BASIS OF FRESHWATER FISH PRODUCTION. (S. D. GERKING, ED.). JOHN WILEY AND SONS, NEW YORK, NEW YORK.

CHAPPELLE, D. E, AND T. C, THOMAS. 1964. ESTIMATION OF OPTIMAL STOCKING LEVELS AND ROTATION AGES OF LOBLOLLY PINE. FOREST SCI. 10(4), 471-502. [MODEL; ECONOMIC MODEL]

CHATTERJEE, S. 1973. A MATHEMATICAL MODEL FOR PEST CONTROL. BIOMETRICS 29(4), 727-734. [COST FUNCTION; OPTIMIZATION]

CHRISTIE, J. M. 1972. THE CHARACTERIZATION OF THE RELATIONSHIP BETWEEN BASIC CROP PARAMETERS IN YIELD TABLE CONSTRUCTION. PP. 37-54. IN, 3RD CONFERENCE ADVISORY GROUP OF FOREST STATISTICIANS. INSTITUT NATIONAL DE LA RECHERCHE AGRONOMIQUE, PARIS, FRANCE. I.N.R.A. PUBL. 72-3. [ORTHOGONAL; YIELD MODEL; COMPUTER]

CLARK, C. W. 1971. ECONOMICALLY OPTIMAL POLICIES FOR THE UTILIZATION OF BIOLOGICALLY RENEWABLE RESOURCES. MATH. BIOSCI. 12, 245-260. [MODEL]

CLARK, C. W. 1972. THE DYNAMICS OF COMMERCIALLY EXPLOITED NATURAL ANIMAL POPULATIONS. MATH. BIOSCI. 13, 149-164. [MODEL; OPTIMIZATION]

CLARK, C. W. 1973. THE ECONOMICS OF OVEREXPLOITATION. SCIENCE 181(4100), 630-634. [COST FUNCTION; RECRUITMENT MODEL; MAXIMIZATION; ECONOMIC ANALYSIS]

CLARK, C. W. 1973. PROFIT MAXIMIZATION AND THE EXTINCTION OF ANIMAL SPECIES. J. POLITICAL ECON. 81(4), 950-961. [BIOECONOMIC MODEL]

CLARK, C. W. 1974. MATHEMATICAL BIOECONOMICS. PP. 29-45. IN, MATHEMATICAL PROBLEMS IN BIOLOGY, LECTURE NOTES IN BIOMATHEMATICS 2. (P. VAN DEN DRIESSCHE, ED.). SPRINGER-VERLAG NEW YORK INC., NEW YORK, NEW YORK. [MODEL; RESOURCE MANAGEMENT; LOGISTIC MODEL; BEVERTON-HOLT MODEL; ECONOMIC ANALYSIS]

CLARK, C., G. EDWARDS AND M. FRIEDLAENDER. 1973. BEVERTON-HOLT MODEL OF A COMMERCIAL FISHERY: OPTIMAL DYNAMICS. J. FISH. RES. BOARD CAN. 30(11), 1629-1640. [MODEL; OPTIMAL CONTROL THEORY; FISHER RULE]

CLARK, G. H. 1931. THE CALIFORNIA HALIBUT (PARALICHTHYS CALIFORNICUS) AND AN ANALYSIS OF THE BOAT CATCHES. CALIF. DIV. FISH GAME FISH BULL. NO. 32. 54 PP. [GEOMETRIC MEAN]

CONWAY, G. R. AND G. MURDIE. 1972. POPULATION MODELS AS A BASIS FOR PEST CONTROL. PP. 195-213. IN, MATHEMATICAL MODELS IN ECOLOGY. (J. N. R. JEFFERS, ED.). BLACKWELL SCIENTIFIC PUBLICATIONS, OXFORD, ENGLAND.

CRAWFORD-SIDEBOTHAM, T. J. 1970. DIFFERENTIAL SUSCEPTIBILITY OF SPECIES OF SLUGS TO METALDEHYDE/BRAN AND TO METHIOCARB BATES. OECOLOGIA 5, 303-324. [INFORMATION MATRIX; BARTLETT'S TEST FOR HOMOGENEITY; ANALYSIS OF CONTINGENCY TABLES]

CROWE, D. M. 1975. A MODEL FOR EXPLOITED BOBCAT POPULATIONS IN WYOMING. J. WILDL. MANAGE. 39(2), 408-415.

CUNIA, T. 1964. WEIGHTED LEAST SQUARES METHOD AND CONSTRUCTION OF VOLUME TABLES. FOREST SCI. 10(2), 180-191. [REGRESSION; VARIANCE]

CURTIN, R. A. 1970. DYNAMICS OF TREE AND CROWN STRUCTURE IN EUCALYPTUS OBLIQUA. FOREST SCI. 16(3), 321-328. [EQUATIONS]

CURTIS, R. O. 1967. HEIGHT-DIAMETER AND HEIGHT-DIAMETER-AGE EQUATIONS FOR SECOND-GROWTH DOUGLAS-FIR. FOREST SCI. 13(4), 365-375.

CURTIS, R. O. 1967. A METHOD OF ESTIMATION OF GROSS YIELD OF DOUGLAS FIR. FOREST SCI. MONOGR. 13. 24 PP. [MODEL; REGRESSION; AUTOCORRELATION]

CURTIS, R. O. 1970. STAND DENSITY MEASURES; AN INTERPRETATION. FOREST SCI. 16(4), 403-414.

CURTIS, R. O. 1971. A TREE AREA POWER FUNCTION AND RELATED STAND DENSITY MEASURES FOR DOUGLAS-FIR. FOREST SCI. 17(2), 146-159. [EQUATIONS; MEASURE OF STAND DENSITY]

CURTIS, R. O., D. J. DEMARS AND F. R. HERMAN. 1974. WHICH DEPENDENT VARIABLE IN SITE INDEX-HEIGHT-AGE REGRESSIONS? FOREST SCI. 20(1), 74-87.

CUSHING, D. H. 1959. ON THE EFFECT OF FISHING ON THE HERRING OF THE SOUTHERN NORTH SEA. J. CONS. PERM. INT. EXPLOR. MER 24(2), 283-307. [FISHING INTENSITY MODEL; EFFICIENCY MODEL; POPULATION MODEL]

CUSHING, D. H. 1973. RECRUITMENT AND PARENT STOCK IN FISHES. WASHINGTON SEA GRANT PUBLICATION WGS 73-1. UNIVERSITY OF WASHINGTON PRESS, SEATTLE, WASHINGTON. XI, 197 PP. [STOCK AND RECRUITMENT MODELS; RICKER; BEVERTON-HOLT EQUATION; CUSHING]

DAHLBERG, M. L. 1973. STOCK-AND-RECRUITMENT RELATIONSHIPS AND OPTIMUM ESCAPEMENTS OF SOCKEYE SALMON STOCKS OF THE CHIGNIK LAKES, ALASKA. RAPP. PROCES-VERB. REUNIONS J. CONS. PERM. INT. EXPLOR. MER 164, 98-105. [RICKER'S MODEL; MODEL]

DARWIN, J. H. AND R. M. WILLIAMS. 1964. THE EFFECT OF TIME OF HUNTING ON THE SIZE OF A RABBIT POPULATION. N. Z. J. SCI. 7(3), 341-352. [MODEL; LESLIE MATRIX; COMPUTER]

DAVENPORT, D. A., G. A. SHERWOOD AND H. W. MURDY. 1973. A METHOD TO DETERMINE WATERFOWL SHOOTING DISTANCES. WILDL. SOC. BULL. 1(2), 101-105. [TRIANGULATION]

DAVIDOFF, E. B. 1969. VARIATIONS IN YEAR-CLASS STRENGTH AND ESTIMATES OF THE CATCHABILITY COEFFICIENT OF YELLOWFIN TUNA, THUNNUS ALBACARES, IN THE EASTERN PACIFIC OCEAN. BULL. INTER-AMER. TROP. TUNA COMM. 14(1), 3-28. [HENNEMUTH'S METHOD; COEFFICIENT OF CATCHABILITY]

DAVIS, D. E. AND J. J. CHRISTIAN. 1958. POPULATION CONSEQUENCES OF A SUSTAINED YIELD PROGRAM FOR NORWAY RATS. ECOLOGY 39(2), 217-222. [LOGISTIC CURVE]

DEMAERSCHALK, J. P. 1972. CONVERTING VOLUME EQUATIONS TO COMPATIBLE TAPER EQUATIONS. FOREST SCI. 18(3), 241-245.

DICKIE, L. M. 1973. MANAGEMENT OF FISHERIES; ECOLOGICAL SUBDIVISIONS. TRANS. AMER. FISH. SOC. 102(2), 470-480. [MODEL; SYSTEMS ANALYSIS]

DICKIE, L. M. AND F. D. MCCRACKEN. 1955. ISOPLETH DIAGRAMS TO PREDICT EQUILIBRIUM YIELDS OF A SMALL FLOUNDER FISHERY. J. FISH. RES. BOARD CAN. 12(2), 187-209. [GROWTH FORMULA; BEVERTON GROWTH FORMULA; BRODY GROWTH FORMULA]

DODGE, W. E., C. M. LOVELESS AND N. B. KVERNO. 1967. DESIGN AND ANALYSIS OF FOREST-MAMMAL REPELLENT TESTS. FOREST SCI. 13(3), 333-336.

DOERNER, K., JR. 1964. SOME CAUSES AND EFFECTS OF HORIZONTAL DENSITY VARIATION IN TREE STEMS. FOREST SCI. 10(1), 24-27. [EQUATIONS]

DOERNER, K., JR. 1965. SOME DIMENSIONAL RELATIONSHIPS AND FORM DETERMINANTS OF TREES. FOREST SCI. 11(1), 50-54. [EQUATIONS]

DOI, T. 1949. ON THE STOCK OF BLUE SHARK FROM STATISTICS OF FIN (I). BULL. JAPANESE SOC. SCI. FISH. 15(7), 301-305. [MODEL]

DOI, T. 1950. A MATHEMATICAL CONSIDERATION ON THE ANALYSIS OF ANNUAL YIELD OF FISH AND ITS APPLICATION TO "BURI" (SERIOLA QUIUQUERADIATA). BULL. JAPANESE SOC. SCI. FISH. 15(11), 661-664. [CATCH MODEL]

DOI, T. 1951. A MATHEMATICAL CONSIDERATION ON THE ANALYSIS OF ANNUAL YIELD OF FISH AND ITS APPLICATION TO "BURI" (SERIOLA QUINQUERADIATA). CENTRAL FISHERIES STAT., JAPAN, CONTRIB. 1948-1949, (117), PP. 65, 100, 327, 329.

DOI, T. 1955. DYNAMICAL ANALYSIS OF PORGY PAGROSOMUS MAJOR (T. & S.) FISHERY IN THE ISLAND SEA OF JAPAN. BULL. JAPANESE SOC. SCI. FISH. 21(5), 320-334. [POPULATION MODEL; VECTOR; SURVIVAL RATE; MORTALITY RATE; OPTIMUM FISHING RATE]

DOI, T. 1956. DYNAMICAL TREATMENT OF EXPLOITATION OF AQUATIC RESOURCE-II. EFFECT ON THE ASPECT OF APPEARANCE OF RECRUIT BY THE FEED-BACK OF INFORMATION OF THE SIZE OF STOCK. BULL. TOKAI REG. FISH. RES. LAB. 13, 73-84. [MODEL]

DOI, T. 1956. DYNAMICAL TREATMENT OF EXPLOITATION OF AQUATIC RESOURCES - I. DYNAMIC CHARACTERISTICS OF A SINGLE FISH POPULATION. BULL. JAPANESE SOC. SCI. FISH. 21(11), 1121-1133. [STOCK-RECRUITMENT MODEL; RICKER'S MODEL]

DOI, T. 1957. SOME CONSIDERATIONS OF THE PERSONAL ERROR AND PRECISION IN OBSERVING BIOLOGICAL PROPERTY OF FISH IN RELATION WITH THE COMMUNICATION HEORY. BULL. TOKAI REG. FISH. RES. LAB. 15, 1-13. [INFORMATION; NTROPY]

DOI, T. 1962. THEORETICAL ONSIDERATIONS ON RATIONALITY OF FFSHORE SALMON FISHERIES FROM THE TANDPOINT OF EFFECTUAL UTILIZATION F THE POPULATION. BULL. TOKAI EG. FISH. RES. LAB. 34, 1-11. RICKER'S MODEL; BEVERTON MODEL]

DOI, T. 1971. A THEORETICAL TREATMENT ON REPRODUCTIVE RELATIONSHIP BETWEEN RECRUITMENT AND ADULT STOCK. BULL. TOKAI REG. FISH. RES. LAB. 64, 39-56. [MODEL; PRINCIPAL COMPONENT ANALYSIS]

DOI, T. 1973. A THEORETICAL TREATMENT OF THE REPRODUCTIVE RELATIONSHIP BETWEEN RECRUITMENT AND ADULT STOCK. RAPP. PROCES-VERB. REUNIONS J. CONS. PERM. INT. EXPLOR. MER 164, 341-349. [MODEL; MATRIX; PRINCIPAL COMPONENTS]

DOI, T. AND Y. SHIMADZU. 1972. A METHOD OF ESTIMATING EFFECTIVE OVERALL FISHING EFFORT RELEVANT TO SELECTION OF WHALING GROUND - I. ESTIMATES OF CATCHABILITY COEFFICIENT BY COUNTRY AND SPECIES. BULL. TOKAI REG. FISH. RES. LAB. 71, 15-35. [MODEL]

DOUBLEDAY, W. G. 1975. HARVESTING IN MATRIX POPULATION MODELS. BIOMETRICS 31(1), 189-200. [LESLIE'S MODEL; LEFKOVICH MODEL; LINEAR PROGRAMMING; COMPUTER]

DRAGESUND, O. AND J. JAKOBSSON. 1963. STOCK STRENGTHS AND RATES OF MORTALITY OF THE NORWEGIANS SPRING SPAWNERS AS INDICATED BY TAGGING EXPERIMENTS IN ICELANDIC WATERS. RAPP. PROCES-VERB. REUNIONS J. CONS. PERM. INT. EXPLOR. MER 154, 83-90. [SURVIVAL RATE; MORTALITY RATE; BEVERTON-HOLT EQUATION]

DRAGESUND, O. AND O. NAKKEN. 1973. RELATIONSHIP OF PARENT STOCK SIZE AND YEAR CLASS STRENGTH IN NORWEGIAN SPRING SPAWNING HERRING. RAPP. PROCES-VERB. REUNIONS J. CONS. PERM. INT. EXPLOR. MER 164, 15-29. [LARVAL MORTALITY MODEL]

ELSNER, G. H. 1970. CAMPING USE-AXLE COUNT RELATIONSHIP: ESTIMATION WITH DESIRABLE PROPERTIES. FOREST SCI. 16(4), 493-495. [NON-LINEAR REGRESSION EQUATION; GROSENBAUGH'S SIGMOID FUNCTION]

EVERT, F. 1969. ESTIMATING STAND VOLUME BY MEASURING FORM CLASS WITHOUT MEASURING DIAMETERS. FOREST SCI. 16(2), 145-148. [FORM FACTOR; EQUATIONS]

FAO, FISHERIES DIVISION, BIOLOGY BRANCH. 1965. REPORT ON THE EFFECTS OF WHALE STOCKS OF PELAGIC OPERATIONS IN THE ANTARCTIC DURING THE 1964/65 SEASON AND ON THE PRESENT STATUS OF THESE STOCKS. FAO FISHERIES TECHNICAL PAPER NO. 59, I, 36 PP. [YIELD MODEL]

FISHER, G. 1972. AN ENVIRONMENTAL CARRYING CAPACITY EQUATION. PP. 128-134. IN, ANNUAL REPT.-FISCAL YEAR 1972. (STAFF OF THE RADIOBIOLOGY LABORATORY). UNIV. OF CALIFORNIA AT DAVIS, U. S. AEC REPT. UCD-472-119.

FOSTER, J. J. 1969. THE INFLUENCE OF FISH BEHAVIOUR ON TRAWL DESIGN WITH SPECIAL REFERENCE TO MATHEMATICAL INTERPRETATIONS OF OBSERVATIONS ON THE SWIMMING SPEEDS OF FISH AND RESULTS OF C. F. EXPERIMENTS. PP. 731-773. IN, PROCEEDINGS OF THE FAO CONFERENCE ON FISH BEHAVIOUR IN RELATION TO FISHING TECHNIQUES AND TACTICS. (A. BEN-TUVIA AND W. DICKSON, EDS.). FAO FISHERIES REPORT NO. 62, VOL. 3. [BEHAVIOR MODEL; YIELD MODEL; ENDURANCE EQUATIONS]

FOX, W. W., JR. 1970. AN EXPONENTIAL SURPLUS-YIELD MODEL FOR OPTIMIZING EXPLOITED FISH POPULATIONS. TRANS. AMER. FISH. SOC. 99(1), 80-88. [GOMPERTZ GROWTH FUNCTION]

FOX, W. W., JR. 1970. RANDOM VARIABILITY AND PARAMETER ESTIMATION FOR THE GENERALIZED PRODUCTION MODEL. QUANTITATIVE SCIENCE PAPER NO. 16. CENTER FOR QUANTITATIVE SCIENCE IN FORESTRY, FISHERIES AND WILDLIFE, UNIVERSITY OF WASHINGTON, SEATTLE, WASHINGTON. MIMEO. 33 PP. [COMPUTER]

FOX, W. W., JR. 1975. FITTING THE GENERALIZED STOCK PRODUCTION MODEL BY LEAST-SQUARES AND EQUILIBRIUM APPROXIMATION. U. S. NATL. MAR. FISH. SERV. FISH. BULL. 73(1), 23-37. [COMPUTER; STOCHASTIC; DETERMINISTIC MODEL; ESTIMATION; MATRIX]

FREESE, F. 1960. TESTING ACCURACY. FOREST SCI. 6(2), 139-145.

FRENCH, R. R. AND J. R. DUNN. 1973. LOSS OF SALMON FROM HIGH-SEAS GILLNETTING WITH REFERENCE TO THE JAPANESE SALMON MOTHERSHIP FISHERY. U. S. NATL. MAR. FISH. SERV. FISH. BULL. 71(3), 845-875. [REGRESSION MODEL]

FRY, F. E. J. 1949. STATISTICS OF A LAKE TROUT FISHERY. BIOMETRICS 5(1), 27-67. [EQUATIONS]

FUKUDA, Y. 1953. AN APPLICATION OF LEAST SQUARES METHOD TO ANALYSIS OF SIZE COMPOSITIONS. BULL. JAPANESE SOC. SCI. FISH. 19(4), 262-272. [VECTOR; MATRIX; MORTALITY RATE; GROWTH PARAMETER]

FUKUDA, Y. 1973. A GAP BETWEEN THEORY AND PRACTICE. J. FISH. RES. BOARD CAN. 30(12, PART 2), 1986-1991. [WHALE; MODEL; EXPLOITATION MODEL]

FULLENBAUM, R. F. AND F. W. BELL. 1974. A SIMPLE BIOECONOMIC FISHERY MANAGEMENT MODEL: A CASE STUDY OF THE AMERICAN LOBSTER FISHERY. U. S. NATL. MAR. FISH. SERV. FISH. BULL. 72(1), 13-25. [SCHAEFER'S MODEL; COMPUTER; SIMULATION]

FURNIVAL, G. M. 1961. AN INDEX FOR COMPARING EQUATIONS USED IN CONSTRUCTING VOLUME TABLES. FOREST SCI. 7(4), 337-341.

FURUKAWA, I. 1961. STUDIES ON THE TUNA LONG LINE FISHERY IN THE EASTERN CHINA SEA AND OKINAWA REGION - II. SCHOOL COMPOSITION. BULL. JAPANESE SOC. SCI. FISH. 27(6), 566-577. [MODEL OF SCHOOL SIZE]

GALES, L. E. 1969. SPAWNER-RECRUIT RELATIONSHIPS WITH A VARIABLE EXPLOITATION RATE (ITER). QUANTITATIVE SCIENCE PAPER NO. 4. CENTER FOR QUANTITATIVE SCIENCE IN FORESTRY, FISHERIES, AND WILDLIFE, UNIVERSITY OF WASHINGTON, SEATTLE, WASHINGTON. MIMEO. 2 PP.

GARROD, D. J. 1964. EFFECTIVE FISHING EFFORT AND THE CATCHABILITY COEFFICIENT Q. RAPP. PROCES-VERB. REUNIONS J. CONS. PERM. INT. EXPLOR. MER 155, 66-70. [BEVERTON-HOLT EQUATION; REGRESSION]

GARROD, D. J. 1969. EMPIRICAL ASSESSMENTS OF CATCH/EFFORT RELATIONSHIPS IN NORTH ATLANTIC COD STOCKS. INTERN. COMM. NORTHWEST ATLANTIC FISH. RES. BULL. 6, 26-34. [YIELD MODEL]

GARROD, D. J. 1973. THE VARIATION OF REPLACEMENT AND SURVIVAL IN SOME FISH STOCKS. RAPP. PROCES-VERB. REUNIONS J. CONS. PERM. INT. EXPLOR. MER 164, 43-56. [MODEL]

GETTINBY, G. 1974. ASSESSMENT OF THE EFFECTIVENESS OF CONTROL TECHNIQUES FOR LIVER FLUKE INFECTION. PP. 89-97. IN, ECOLOGICAL STABILITY. (M. B. USHER AND M. H. WILLIAMSON, EDS.). HALSTED PRESS, NEW YORK, NEW YORK. [ENCOUNTER AND ABSORPTION MODEL; SIMULATION; SENSITIVITY ANALYSIS]

GILES, R. H., JR, AND R. F. SCOTT. 1969. A SYSTEMS APPROACH TO REFUGE MANAGEMENT. TRANS. N. AMER. WILDL. NAT. RESOUR. CONF. 34, 103-115.

OH, B. S. 1971. THE POTENTIAL TILITY OF CONTROL THEORY IN THE ANAGEMENT OF INSECT POPULATIONS. ROC. ECOL. SOC. AUSTRALIA 6, 4-90.

OLIKOV, A. N. AND V. V. ENSHUTKIN. 1973. ESTIMATION OF RODUCTION PROPERTIES OF MOLLUSK OPULATIONS. MARINE BIOL. 20(1) -13. [COMPUTER; ALGORITHM]

GORDON, H. S. 1953. AN ECONOMIC APPROACH TO THE OPTIMUM UTILIZATION OF FISHERY RESOURCES. J. FISH. RES. BOARD CAN. 10(7), 442-457. [GRAPHIC]

GORDON, H. S. 1954. THE ECONOMIC THEORY OF A COMMON PROPERTY RESOURCE: THE FISHERY. J. POLITICAL ECON. 62(2), 124-142. [COST FUNCTION; MAXIMIZATION; MODEL]

GOTSHALL, D. W. 1972. POPULATION SIZE, MORTALITY RATES, AND GROWTH RATES OF NORTHERN CALIFORNIA OCEAN SHRIMP, PANDALUS JORDANI, 1965 THROUGH 1968. CALIF. DEPT. FISH GAME FISH BULL. NO. 155. 47 PP. [COMPUTER; RICKER'S EQUATION; VARIANCE; VON BERTALANFFY GROWTH EQUATION]

GRAHAM, M. 1935. MODERN THEORY OF EXPLOITING A FISHERY, AND APPLICATION TO NORTH SEA TRAWLING. J. CONS. PERM. INT. EXPLOR. MER 10(3), 264-274. [RATE OF NATURAL INCREASE; RUSSELL'S EQUATION; MODEL]

GRAHAM, M. 1939. THE SIGMOID CURVE AND THE OVERFISHING PROBLEM. RAPP. PROCES-VERB. REUNIONS J. CONS. PERM. INT. EXPLOR. MER 110( ), 15-20.

GRAHAM, M. 1952. OVERFISHING AND OPTIMUM FISHING. RAPP. PROCES-VERB. REUNIONS J. CONS. PERM. INT. EXPLOR. MER 132, 72-78. [MODEL; BEVERTON-HOLT MODEL; BARANOV'S MODEL; RUSSELL'S MODEL]

GRAHAM, S. A. 1929. THE NEED FOR STANDARDIZED QUANTITATIVE METHODS IN FOREST BIOLOGY. ECOLOGY 10(2), 245-250.

GREENBANK, J. 1954. DEPLORABLE PRESENTATION OF RESULTS OF FISHERY RESEARCH. TRANS. AMER. FISH. SOC. 83, 115-119. [SIGNIFICANT FIGURES; GROWTH FORMULA]

GREGORY, G. R. 1955. AN ECONOMIC APPROACH TO MULTIPLE USE. FOREST SCI. 1(1), 6-13. [COST FUNCTION]

GRIEB, J. R. 1958. WILDLIFE STATISTICS. COLORADO GAME AND FISH DEPT., FED AID. DIV., DENVER, COLORADO. MIMEO II, 96 PP.

GULLAND, J. A. 1956. A NOTE ON THE STATISTICAL DISTRIBUTION OF TRAWL CATCHES. RAPP. PROCES-VERB. REUNIONS J. CONS. PERM. INT. EXPLOR. MER 140(1), 28-29. [VARIANCE; POISSON DISTRIBUTION]

GULLAND, J. A. 1956. THE STUDY OF FISH POPULATIONS BY THE ANALYSIS OF COMMERCIAL CATCHES. A STATISTICAL REVIEW. RAPP. PROCES-VERB. REUNIONS J. CONS. PERM. INT. EXPLOR. MER 140(1), 21-27. [SAMPLING; STRATIFICATION]

GULLAND, J. A. 1958. SAMPLING OF SEMI-OCEANIC STOCKS OF FISH. PP. 71-76. IN, SOME PROBLEMS FOR BIOLOGICAL FISHERY SURVEY AND TECHNIQUES FOR THEIR SOLUTION. INTERN. COMM. NORTHWEST ATLANTIC FISH. SPEC. PUBL. NO. 1. [CATCH MODEL; DIFFERENTIAL CAPTURE]

GULLAND, J. A. 1961. THE ESTIMATION OF THE EFFECT ON CATCHES OF CHANGES IN GEAR SELECTIVITY. J. CONS. PERM. INT. EXPLOR. MER 26(2), 204-214. [MODEL]

GULLAND, J. A. 1964. THE ABUNDANCE OF FISH STOCKS IN THE BARENTS SEA. RAPP. PROCES-VERB. REUNIONS J. CONS. PERM. INT. EXPLOR. MER 155, 126-137. [INDEX OF ABUNDANCE; STOCK FUNCTION]

GULLAND, J. A. 1965. MANUAL OF METHODS FOR FISH STOCK ASSESSMENT. PART I. FISH POPULATION ANALYSIS. FAO FISHERIES TECHNICAL PAPER NO. 40 (REV. 1), II, MISC PAGES.

GULLAND, J. A. 1968. APPRAISAL OF A FISHERY. PP. 236-245. IN, METHODS FOR ASSESSMENT OF FISH PRODUCTION IN FRESH WATERS.(W. E. RICKER, ED.). BLACKWELL SCIENTIFIC PUBLICATIONS, OXFORD, ENGLAND. [FISHING EFFORT; LOGISTIC CURVE; POPULATION GROWTH; VON BERTALANFFY; YIELD MODEL]

GULLAND, J. A. 1968. THE CONCEPT OF THE MARGINAL YIELD FROM EXPLOITED FISH STOCKS. J. CONS. PERM. INT. EXPLOR. MER 32(2), 256-261. [CATCH EQUATION; SCHAEFER'S MODEL; BEVERTON-HOLT MODEL]

GULLAND, J. A. 1968. THE CONCEPT OF THE MAXIMUM SUSTAINABLE YIELD AND FISHERY MANAGEMENT. FAO FISHERIES TECHNICAL PAPER NO. 70. I, 13 PP. [MODEL]

GULLAND, J. A. 1969. MANUAL OF METHODS FOR FISH STOCK ASSESSMENT PART 1. FISH POPULATION ANALYSIS. FAO MANUALS IN FISHERIES SCIENCE NO. 4, VII, 154 PP. [SAMPLING; GROWTH; CATCH-EFFORT MODELS; TAG-RECAPTURE; RECRUITMENT MODEL; YIELD CURVE; MORTALITY ESTIMATES]

GULLAND, J. A. 1970. FOOD CHAIN STUDIES AND SOME PROBLEMS IN WORLD FISHERIES. PP. 296-315. IN, MARINE FOOD CHAINS. (J. H. STEELE, EDS.), UNIVERSITY OF CALIFORNIA PRESS, BERKELEY, CALIFORNIA. [MODEL]

GULLAND, J. A. 1971. ECOLOGICAL ASPECTS OF FISHERY RESEARCH. PP. 115-176. IN, ADVANCES IN ECOLOGICAL RESEARCH, VOLUME 7, (J. B. CRAGG, ED.). ACADEMIC PRESS, NEW YORK, NEW YORK. [MODEL]

GULLAND, J. A. 1973. CAN A STUDY OF STOCK AND RECRUITMENT AID MANAGEMENT DECISIONS? RAPP. PROCES-VERB. REUNIONS J. CONS. PERM. INT. EXPLOR. MER 164, 368-372. [MODEL]

HALL, F. C. 1971. SOME USES AND LIMITATIONS OF MATHEMATICAL ANALYSIS IN PLANT ECOLOGY AND LAND MANAGEMENT. PP. 377-395. IN, STATISTICAL ECOLOGY, VOLUME 3, MAN SPECIES POPULATIONS, ECOSYSTEMS, AND SYSTEMS ANALYSIS, (G. P. PATIL E. C. PIELOU AND W. E. WATERS, EDS.), PENNSYLVANIA STATE UNIVERSITY PRESS, UNIVERSITY PARK, PENNSYLVANIA. [MULTIPLE REGRESSION]

HALL, W. B. 1968. MANUAL OF SAMPLING AND STATISTICAL METHODS FOR FISHERIES BIOLOGY PART II-STATISTICAL METHODS. CHAPTER 6: CONTINUOUS DISTRIBUTIONS. FAO FISHERIES TECHNICAL PAPER NO. 26(SUPPL. 2). V, 36 PP. [NORMAL DISTRIBUTION; BIVARIATE NORMAL DISTRIBUTION; MOMENTS; MEAN; VARIANCE; MEASURES OF LOCATION; MEASURE OF DISPERSION]

HAMLEY, J. M. AND H. A. REGIER. 1973. DIRECT ESTIMATES OF GILLNET SELECTIVITY TO WALLEYE (STIZOSTEDION VITREUM VITREUM). J. FISH. RES. BOARD CAN. 30(6), 817-830. [PEARSON'S TYPE I CURVE; TAG-RECAPTURE]

HANAMURA, N. 1953. ON THE HERRING RESOURCES OF HOKKAIDO AND THE SOUTH SAGHALIEN. BULL. JAPANESE SOC. SCI. FISH. 19(4), 283-291. [BARANOV'S MODEL; RICKER'S MODEL]

HANCOCK, D. A. 1965. YIELD ASSESSMENT IN THE NORFOLK FISHERY FOR CRABS (CANCER PAGURUS). RAPP. PROCES-VERB. REUNIONS J. CONS. PERM. INT. EXPLOR. MER 156, 81-93. [BEVERTON-HOLT EQUATION; MODEL; RATE OF EXPLOITATION; MORTALITY COEFFICIENT]

HAYNE, D. W. 1946. THE RELATION BETWEEN NUMBER OF EARS OPENED AND THE AMOUNT OF GRAIN TAKEN BY REDWINGS IN CORNFIELDS. J. AGRIC. RES. 72(8), 289-295.

HAYNE, D. W. 1969. THE USE OF MODELS IN RESOURCE MANAGEMENT. PP. 119-122. IN, WHITE-TAILED DEER N THE SOUTHERN FOREST HABITAT. ANONYMOUS, ED.) U. S. FOR. SERV. OUTHERN FOREST EXPT. STN., ACOGDOCHES, TEXAS.

EADLEY, J. C. 1972. ECONOMICS F AGRICULTURAL PEST CONTROL. NNU. REV. ENTOMOL. 17, 273-286. COST FUNCTION]

ENDERSON, F. 1972. THE YNAMICS OF THE MEAN-SIZE STATISTIC N A CHANGING FISHERY. FAO ISHERIES TECHNICAL PAPER NO. 116. 5 PP. [MODEL; SIMULATION; TIME ERIES; AUTOCOVARIANCE; POPULATION ODEL]

HENNEMUTH, R. C. 1961. YEAR CLASS ABUNDANCE, MORTALITY AND YIELD-PER-RECRUIT OF YELLOWFIN TUNA IN THE EASTERN PACIFIC OCEAN, 1954-1959. BULL. INTER-AMER. TROP. TUNA COMM. 6(1), 3-32. [REGRESSION; GROWTH EQUATIONS; MORTALITY RATE; COEFFICIENT OF MORTALITY; BEVERTON-HOLT EQUATION]

HERRINGTON, W. C. 1948. LIMITING FACTORS FOR FISH POPULATIONS. SOME THEORIES AND AN EXAMPLE. BULL. BINGHAM OCEANOGR. COLLECT. YALE UNIV. 11(ART. 4), 229-279. [MODEL]

HIRAYAMA, N. 1969. STUDIES ON THE FISHING MECHANISM OF TUNA LONG-LINE - III. THE DIFFERENCE OF CATCH BY RETRIEVING METHODS. BULL. JAPANESE SOC. SCI. FISH. 35(7), 629-634. [EQUATIONS OF SOAKING TIME]

HIRAYAMA, N. 1969. STUDIES ON THE FISHING MECHANISM OF TUNA LONG-LINE - IV. THEORETICAL ANALYSIS OF FISHING EFFECTIVENESS OF THE GEAR. BULL. JAPANESE SOC. SCI. FISH. 35(7), 635-643. [CATCH EQUATION]

HIRAYAMA, N. 1969. STUDIES ON THE FISHING MECHANISM OF TUNA LONG-LINE-I. RELATION BETWEEN CATCH AND SIZE OF GEAR. BULL. JAPANESE SOC. SCI. FISH. 35(6), 546-549. [EQUATIONS]

HIRAYAMA, N. 1969. STUDIES ON THE FISHING MECHANISM OF TUNA LONG-LINE-II. RELATION BETWEEN SETTING COURSE OF THE GEAR AND MOVING DIRECTION OF THE FISH. BULL. JAPANESE SOC. SCI. FISH. 35(6), 550-554. [EQUATIONS]

HIRSCHHORN, G. 1966. EFFECT OF GROWTH PATTERN AND EXPLOITATION FEATURES ON THE YIELDS OF PINK SALMON IN THE NORTHWESTERN NORTH PACIFIC. TRANS. AMER. FISH. SOC. 95(1), 39-51. [DOI'S YIELD EQUATIONS; DOI'S INDEX OF RATIONALITY OF OFFSHORE FISHING]

HOLT, S. J. 1958. THE EVALUATION OF FISHERIES RESOURCES BY THE DYNAMIC ANALYSIS OF STOCKS, AND NOTES ON THE TIME FACTORS INVOLVED. PP. 77-95. IN, SOME PROBLEMS FOR BIOLOGICAL FISHERY SURVEY AND TECHNIQUES FOR THEIR SOLUTION. INTERN. COMM. FOR THE NORTHWEST ATLANTIC FISHERIES, SPECIAL PUBL. NO. 1.

HOLT, S. J. 1958. THE EVALUATION OF FISHERIES RESOURCES BY THE DYNAMIC ANALYSIS OF STOCKS, AND NOTES ON THE TIME FACTORS INVOLVED. PP. 77-95. IN, SOME PROBLEMS FOR BIOLOGICAL FISHERY SURVEY AND TECHNIQUES FOR THEIR SOLUTION. INTERN. COMM. NORTHWEST ATLANTIC FISH. SPEC. PUBL. NO. 1. [BEVERTON MODEL]

ILES, T. D. 1973. DWARFING OR STUNTING IN THE GENUS TILAPIA (CICHLIDAE) A POSSIBLY UNIQUE RECRUITMENT MECHANISM. RAPP. PROCES-VERB. REUNIONS J. CONS. PERM. INT. EXPLOR. MER 164, 247-254. [BIOMASS MODEL]

ILES, T. D. 1973. INTERACTION OF ENVIRONMENT AND PARENT STOCK SIZE IN DETERMINING RECRUITMENT IN THE PACIFIC SARDINE AS REVEALED BY ANALYSIS OF DENSITY-DEPENDENT O-GROUP GROWTH. RAPP. PROCES-VERB. REUNIONS J. CONS. PERM. INT. EXPLOR. MER 164, 228-240. [MODEL]

ISHIDA, M., N. SANO, S. MISHIMA AND S. SAITO. 1969. ON MEASURING THE DROPPING RATE OF SALMON GILL NETS BY MEANS OF UNDERWATER TELEVISION TECHNIQUES. BULL. JAPANESE SOC. SCI. FISH. 35(12), 1157-1166. [EQUATIONS]

ISHIWATA, N. 1969. ECOLOGICAL STUDIES ON THE FEEDING OF FISHES-IX. MAINTENANCE REQUIREMENT. BULL. JAPANESE SOC. SCI. FISH. 35(11), 1049-1059. [EQUATIONS]

JAQUETTE, D. L. 1972. A DISCRETE TIME POPULATION CONTROL MODEL. MATH. BIOSCI. 15, 231-252. [DYNAMIC PROGRAMMING; STOCHASTIC MODEL; DETERMINISTIC MODEL; POISSON BRANCHING PROCESS]

JAQUETTE, D. L. 1972. MATHEMATICAL MODELS FOR CONTROLLING BIOLOGICAL POPULATIONS: A SURVEY. OPERATIONS RES. 20(6), 1142-1151. [DECISION THEORY; SYSTEMS ANALYSIS; DETERMINISTIC MODEL; STOCHASTIC MODEL]

JEFFERS, J. N. R. 1973. SYSTEMS MODELLING AND ANALYSIS IN RESOURCE MANAGEMENT. J. ENVIRON. MANAGE. 1(1), 13-28. [MULTIVARIATE ANALYSIS; STOCHASTIC; COMPUTER; MODEL]

JENSEN, A. J. C. 1939. ON THE LAWS OF DECREASE IN FISH STOCKS. RAPP. PROCES-VERB. REUNIONS J. CONS. PERM. INT. EXPLOR. MER 110( ), 85-96.

JENSEN, A. L. 1972. POPULATION BIOMASS, NUMBER OF INDIVIDUALS, AVERAGE INDIVIDUAL WEIGHT, AND THE LINEAR SURPLUS-PRODUCTION MODEL. J. FISH. RES. BOARD CAN. 29(11), 1651-1655. [LOGISTIC; MODEL]

JENSEN, A. L. 1973. RELATION BETWEEN SIMPLE DYNAMIC POOL AND SURPLUS PRODUCTION MODELS FOR YIELD FROM A FISHERY. J. FISH. RES. BOARD CAN. 30(7), 998-1002.

JOHNSON, L. 1966. CONSUMPTION BY THE RESIDENT POPULATION OF PIKE, ESOX LUCIUS, IN LAKE WINDERMERE. J. FISH. RES. BOARD CAN. 23(10), 1523-1535. [YIELD ESTIMATION; BEVERTON-HOLT EQUATION; VON BERTALANFFY; KOSTITZIN]

JONES, D. R., J. W. KICENIUK AND O S. BAMFORD. 1974. EVALUATION OF THE SWIMMING PERFORMANCE OF SEVERAL FISH SPECIES FROM THE MACKENZIE RIVER. J. FISH. RES. BOARD CAN. 31(10), 1641-1647. [EQUATIONS]

JONES, R. 1963. SOME INVESTIGATIONS OF HADDOCK MOVEMENTS. PP. 84-88. IN, NORTH ATLANTIC FISH MARKING SYMPOSIUM. INTERN. COMM. NORTHWEST ATLANTIC FISH. SPEC. PUBL. NO. 4. [MOVEMENT MODEL; VARIANCE; DIFFUSION PARAMETER]

JONES, R. 1966. MANUAL OF METHODS FOR FISH STOCK ASSESSMENT. PART IV. MARKING. FAO FISHERIES TECHNICAL PAPER NO. 51, SUPPLEMENT 1, IX. MISC. PAGES. [TAG-RECAPTURE ESTIMATION; VON BERTALANFFY GROWTH EQUATION; GROWTH FORMULA]

KABAK, I. W. 1970. WILDLIFE MANAGEMENT: AN APPLICATION OF FINITE MARKOV CHAIN. AMER. STATISTICIAN 24(5), 27-29.

KAMIYA, S. 1934. THE DISTRIBUTION OF TENSIONS IN THE NETS. I. BULL. JAPANESE SOC. SCI. FISH. 3(1), 5-7. [VARIANCE]

KAMIYA, S. 1936. THE VISIBLE DISTANCE OF OBJECTS IMMERSED IN SEA-WATER. I. BULL. JAPANESE SOC. SCI. FISH. 4(6), 365-373. [EQUATIONS]

KANDA, K. 1953. ON THE EXPERIMENTAL STUDY HOW THE FISHES PASS THROUGH THE MESH OF VARIOUS SHORTENING NETS - I. BULL. JAPANESE SOC. SCI. FISH. 18(8), 365-372. [EQUATIONS]

KAWADA, S. AND Y. TAWARA. 1957. AN ATTEMPT AT MEASURING DEFORMATION OF THE SET NET BY A FISH FINDER (PRELIMINARY REPORT). BULL. JAPANESE SOC. SCI. FISH. 22(10), 593-597. [EQUATIONS]

AWADA, S., Y. TAWARA AND C. OSHIMUTA. 1958. AN ATTEMPT FOR ETERMINING THE SWIMMING SPEED OF ISH SCHOOLS BY THE FISH FINDER PRELIMINARY REPORT). BULL. APANESE SOC. SCI. FISH. 24(1), -4. [EQUATIONS]

AWAI, T. 1969. ANALYSIS OF ELLOWTAIL ABUNDANCE CAUGHT BY SET ETS IN MIE PERFECTURE. BULL. OKAI REG. FISH. RES. LAB. 58, -17. [MODEL]

AWAKAMI, T. 1952. ON THE REDICTION OF FISHERY BASED ON THE ARIATION OF STOCK. MEM. COLL. GRIC. KYOTO UNIV. 62, 27-35. ODEL]

KAWAKAMI, T. 1953. MECHANICAL ACTION OF THE OTTER BOARD OF THE TRAWL NET. BULL. JAPANESE SOC. SCI. FISH. 19(4), 228-232. [EQUATIONS]

KAWAKAMI, T. 1956. ON THE PREDICTION OF FISHERY BASED ON THE VARIATION OF STOCK - II. BULL. JAPANESE SOC. SCI. FISH. 21(10), 1047-1048. [CATCH EQUATION]

KAWAKAMI, T. 1961. ON THE LAW OF MECHANICAL SIMILARITY FOR DRIFT GILL NET. BULL. JAPANESE SOC. SCI. FISH. 27(2), 124-127. [EQUATIONS]

KAWAKAMI, T. AND K. NAKASAI. 1962. ON THE MECHANICAL CHARACTER OF THE DRAG-NET-II. BULL. JAPANESE SOC. SCI. FISH. 28(7), 664-670. [EQUATIONS]

KAWAKAMI, T. AND T. KITAHARA. 1964. ON THE DECLINE IN THE CATCH PER UNIT OF FISHING EFFORT AND THE DIMINUTION OF AVERAGE BODY LENGTH DUE TO THE EXPLOITATION OF A VERGIN STOCK. BULL. JAPANESE SOC. SCI. FISH. 30(10), 821-827. [MODEL]

KAWAMURA, G. 1972. GILL-NET MESH SELECTIVITY CURVE DEVELOPED FROM LENGTH-GIRTH RELATIONSHIP. BULL. JAPANESE SOC. SCI. FISH. 38(10), 1119-1127. [EQUATIONS]

KAWASAKI, T. 1973. ON THE STOCK CONDITIONS OF THE PACIFIC SUBPOPULATION OF MACKEREL. BULL. TOKAI REG. FISH. RES. LAB. 74, 45-65. [MODEL]

KELLY, W. H. 1965. A STOCKING FORMULA FOR HEAVILY FISHED TROUT STREAMS. N. Y. FISH GAME J. 12(2), 170-179.

KIM, Y. M., T. DOI AND K. OKADA. 1972. A TRIAL OF POPULATION ANALYSIS BY A NEWLY DEVELOPED PROCESS ON YELLOW CROAKER IN THE EAST CHINA SEA AND THE YELLOW SEA-BASED UPON DATA TAKEN BY LARGE PAIR TRAWLERS OF KOREA. BULL. TOKAI REG. FISH. RES. LAB. 71, 37-50. [REGRESSION MODEL]

KITAHARA, T. 1969. ON SWEEPING TRAMMEL NET (KOGISASHIAMI) FISHERY ALONG COAST OF SAN'IN DISTRICT-IV. YEARLY CHANGE IN EFFICIENCY OF SWEEPING TRAMMEL NET TO BRANQUILLOS IN WAKASA BAY. BULL. JAPANESE SOC. SCI. FISH. 35(3), 258-264. [EQUATIONS]

KITAHARA, T. 1971. ON SELECTIVITY CURVE OF GILLNET. BULL. JAPANESE SOC. SCI. FISH. 37(4), 289-296. [REGIER AND ROBSON CURVE]

KITAHARA, T. 1973. ON SWEEPING TRAMMEL NET FISHERY ALONG THE COAST IN THE SAN'IN DISTRICT-V. EXPLOITATION RATE OF BRANQUILLOS WITH THE SWEEPING TRAMMEL NET IN WAKASA BAY BULL. JAPANESE SOC. SCI. FISH. 39(5), 471-476. [MODEL]

KNIPLING, E. F. 1968. POPULATION MODELS TO APPRAISE THE LIMITATIONS AND POTENTIALITIES OF TRICHOGRAMMA IN MANAGING HOST INSECT POPULATIONS. USDA TECH. BULL. NO. 1387, 44 PP.

KNIPLING, E. F. AND J. U. MCGUIRE. 1972. POTENTIAL ROLE OF STERILIZATION FOR SUPPRESSING RAT POPULATION: A THEORETICAL APPRAISAL. U. S. DEPT. AGRICULTURE, TECH. BULL. NO. 1455. 27 PP. [MODEL]

KOIKE, A. 1961. ON THE WEIGHT DISTRIBUTION OF FISHES CAUGHT BY THE SALMON DRIFT NETS IN THE REGION OF THE NORTHERN PACIFIC OCEAN - I. FUNDAMENTAL STUDY OF THE WEIGHT DISTRIBUTION OF RAINBOW TROUT CAUGHT BY THE GILL NETS IN THE OUTDOOR POOL. BULL. JAPANESE SOC. SCI. FISH 27(5), 372-376. [TAUTI'S THEORY; CATCH-CURVE]

KOJIMA, K. 1971. STOCHASTIC MODELS FOR EFFICIENT CONTROL OF INSECT POPULATIONS BY STERILE-INSECT RELEASE METHODS. PP. 477-487. IN, STERILITY PRINCIPLE FOR INSECT CONTROL OR ERADICATION. INTERNATIONAL ATOMIC ENERGY AGENCY, VIENNA, REPORT STI/PUB/265. [COMPUTER]

KONAGAYA, T. 1969. RESISTANCE OF PLANE NET SET PARALLEL TO STREAM - I. DRAG FORCES OF THE WIRES IN WAKE. BULL. JAPANESE SOC. SCI. FISH. 35(7), 644-647. [EQUATIONS]

KONAGAYA, T. 1971. RESISTANCE OF A PLANE NET SET PARALLEL TO A STREAM-III. THE RELATIONSHIP BETWEEN THE DRAG COEFFICIENT AND HANGING COEFFICIENT OR ANGLE OF INCIDENCE. BULL. JAPANESE SOC. SCI. FISH. 37(11), 1033-1036. [EQUATIONS]

KONAGAYA, T. 1971. STUDIES ON THE PURSE SEINE-II. EFFECT OF THE MESH AND THE SPECIFIC GRAVITY OF WEBBING. BULL. JAPANESE SOC. SCI. FISH. 37(1), 8-12. [EQUATIONS]

KONAGAYA, T. 1971. STUDIES ON THE PURSE SEINE-III. ON THE EFFECT OF SINKERS ON THE PERFORMANCE OF A PURSE SEINE. BULL. JAPANESE SOC. SCI. FISH. 37(9), 861-865. [EQUATIONS]

KONAGAYA, T. 1971. STUDIES ON THE PURSE SEINE-IV. THE INFLUENCE OF HANGING AND NET DEPTH. BULL. JAPANESE SOC. SCI. FISH. 37(9), 866-870. [EQUATIONS]

KONAGAYA, T. AND T. KAWAKAMI. 1971. RESISTANCE OF A PLANE NET SET PARALLEL TO A STREAM-II. DRAG FORCE OF A PLANE NET PULLED PARALLEL TO THE DIRECTION OF MOTION. BULL. JAPANESE SOC. SCI. FISH. 37(10), 944-947. [EQUATIONS]

KONTAR, V. A. 1971. THE BIOMATHEMATICAL THEORY OF COMMUNIT CONTROL AND ITS APPLICATION. PROC. 13TH INTERN. CONGR. ENTOMOL. 1, 512. [MODEL; VOLTERRA; LOTKA' EQUATION]

KRISHNAN KUTTY, M. 1968. ESTIMATION OF THE AGE OF EXPLOITATION AT A GIVEN FISHING MORTALITY. J. FISH. RES. BOARD CAN. 25(6), 1291-1294. [BEVERTON-HOLT EQUATION]

KRISHNAN KUTTY, M. 1968. SOME MODIFICATIONS IN THE BEVERTON AND HOLT MODEL FOR ESTIMATING THE YIELD OF EXPLOITED FISH POPULATIONS. PROC. NATL. INST. SCI. INDIA 34(6, PART B), 293-302.

KRISHNAN KUTTY, M. AND S. Z. QASIM. 1968. THE ESTIMATION OF OPTIMUM AGE OF EXPLOITATION AND POTENTIAL YIELD IN FISH POPULATIONS. J. CONS. PERM. INT. EXPLOR. MER 32(2), 249-255..

KROGIUS, F. V. 1968. CALCULATION OF THE PROPORTION OF LOCAL STOCKS IN THE TOTAL STOCK OF SOCKEYE SALMON (ONCORHYNCHUS NERKA (WALB.)) IN THE KAMCHATKA RIVER BASIN. PROBLEMS OF ICHTHYOLOGY 8(6), 779-783. [DETERMINANT; NORMAL EQUATIONS]

KURITA, S. 1957. CAUSES AFFECTING THE SIZE OF SARDINE STOCK IN THE WATERS OFF JAPAN AND ADJACENT REGIONS - WITH PARTICULAR REFERENCE TO THE CATCH DECLINING SINCE 1941-. BULL. TOKAI REG. FISH. RES. LAB. 18, 1-14. [MODEL]

KURITA, S. AND C. TANAKA. 1956. ESTIMATION OF ANNUAL CATCHES OF SARDINE AND ANCHOVY IN JAPAN, 1926-50 - USING THE AMOUNTS PROCESSED BY IWASHI PRODUCTS. BULL. JAPANESE SOC. SCI. FISH. 2(6), 338-347. [MULTIPLE REGRESSION MODEL; COEFFICIENT OF PARTIAL REGRESSION; COEFFICIENT OF MULTIPLE CORRELATION]

URITA, S., S. TANAKA AND M. MOGI. 973. ABUNDANCE INDEX AND YNAMICS OF THE SAURY POPULATION IN HE PACIFIC OCEAN OFF NORTHERN APAN. BULL. JAPANESE SOC. SCI. ISH. 39(1), 7-16. [EQUATIONS; OEFFICIENT OF DECREASE]

UROKI, T. 1953. A RELATION ETWEEN THE FLUCTUATIONS OF CATCHES D THE DIFFUSIVE MOVEMENTS OF ISH-SCHOOLS. BULL. JAPANESE SOC. I. FISH. 19(4), 258-261. QUATIONS]

LANDER, R. H. AND K. A. HENRY. 1973. SURVIVAL, MATURITY, ABUNDANCE, AND MARINE DISTRIBUTION OF 1965-66 BROOD COHO SALMON, ONCORHYNCHUS KISUTCH, FROM COLUMBIA RIVER HATCHERIES. U. S. NATL. MAR. FISH. SERV. FISH. BULL. 71(3), 679-695. [CLEAVER MODEL; MAXIMUM LIKELIHOOD; LIMIT-MEAN MODEL]

LARKIN, P. A. 1966. EXPLOITATION IN A TYPE OF PREDATOR-PREY RELATIONSHIP. J. FISH. RES. BOARD CAN. 23(3), 349-356. [LOTKA-VOLTERRA MODEL; KOSTITZIN MODEL]

LARKIN, P. A. 1973. SOME OBSERVATIONS ON MODELS OF STOCK AND RECRUITMENT RELATIONSHIPS FOR FISHES. RAPP. PROCES-VERB. REUNIONS J. CONS. PERM. INT. EXPLOR. MER 164, 316-324.

LARKIN, P. A. AND A. WALTON. 1969. FISH SCHOOL SIZE AND MIGRATION. J. FISH. RES. BOARD CAN. 26(5), 1372-1374. [EQUATIONS; CIRCULAR NORMAL DISTRIBUTION]

LARKIN, P. A. AND J. G. MCDONALD. 1968. FACTORS IN THE POPULATION BIOLOGY OF THE SOCKEYE SALMON OF THE SKEENA RIVER. J. ANIM. ECOL. 37(1), 229-258. [MODEL OF LARKIN AND HOURSTON; SIMULATION; RICKER-TYPE MODEL; COMPUTER]

LARKIN, P. A. AND W. E. RICKER. 1964. FURTHER INFORMATION ON SUSTAINED YIELDS FROM FLUCTUATING ENVIRONMENTS. J. FISH. RES. BOARD CAN. 21(1), 1-7. [SIMULATION; COMPUTER; RANDOM NORMAL DEVIATES]

LAVOV, M. A. 1972. THE MODELLING OF CHANGES IN ABUNDANCE AND COMPOSITION OF ROE DEER HERDS. BYULL. MOSK. O-VA. ISPYT. PRIR. OTD. BIOL. 76(4), 33-36.

LAWSON, F. R. 1967. THEORY OF CONTROL OF INSECT POPULATIONS BY SEXUALLY STERILE MALES. ANN. ENTOMOL. SOC. AMER. 60(4), 713-722. [MODEL; PROBABILITY; KNIPLING'S MODEL]

LE CREN, E. D. 1951. THE LENGTH-WEIGHT RELATIONSHIP AND SEASONAL CYCLE IN GONAD WEIGHT AND CONDITION IN THE PERCH (PERCA FLUVIATILIS). J. ANIM. ECOL. 20(2), 201-219. [LENGTH-WEIGHT DATA ANALYSIS REVIEWED]

LEE, P. J. 1971. THE PRINCIPLE AND OPERATION OF FISH CATCH AND POLLUTION INFORMATION SYSTEM. FISH. RES. BOARD CAN. TECH. REPT. NO. 250. III, 105 PP. [COMPUTER]

LENARZ, W. H. 1972. MESH RETENTION OF LARVAE OF SARDINOPS CAERULEA AND ENGRAULIS MORDAX BY PLANKTON NETS. U. S. NATL. MAR. FISH. SERV. FISH. BULL. 70(3), 839-848. [SAMPLING; MODEL]

LENARZ, W. H. 1973. DEPENDENCE OF CATCH RATES ON SIZE OF FISH LARVAE. RAPP. PROCES-VERB. REUNIONS J. CONS. PERM. INT. EXPLOR. MER 164, 270-275. [CATCH EQUATION]

LENARZ, W. H., W. W. FOX, JR., G. T. SAKAGAWA AND B. J. ROTHSCHILD. 1974. AN EXAMINATION OF THE YIELD PER RECRUIT BASIS FOR A MINIMUM SIZE REGULATION FOR ATLANTIC YELLOWFIN TUNA, THUNNUS ALBACARES. U. S. NATL. MAR. FISH. SERV. FISH. BULL. 72(1), 37-61. [COMPUTER; RICKER'S MODEL; BEVERTON-HOLT MODEL]

LORD, G. E. 1973. CHARACTERISATION OF THE OPTIMUM DATA ACQUISITION AND MANAGEMENT OF A SALMON FISHERY AS A STOCHASTIC DYNAMIC PROGRAM. U. S. NATL. MAR. FISH. SERV. FISH. BULL. 71(4), 1029-1037. [DECISION THEORY; BAYES RISK; RICKER'S EQUATION]

LOWES, A. L. AND C. C. BLACKWELL, JR. 1975. APPLICATIONS OF MODERN CONTROL THEORY TO ECOLOGICAL SYSTEMS. PP. 299-305. IN, ECOSYSTEM ANALYSIS AND PREDICTION. (S. A. LEVIN, ED.). PROCEEDINGS OF A SIAM-SIMS CONFERENCE HELD AT ALTA, UTAH, JULY 1-5, 1974. [MODEL]

LYNESS, F. K. AND E. H. M. BADGER. 1970. A MEASURE OF WINTER SEVERITY. APPL. STAT. 19(2), 119-134. [LOG-NORMAL DISTRIBUTION]

MACKETT, D. J. 1973. MANUAL OF METHODS FOR FISHERIES RESOURCE SURVEY AND APPRAISAL, PART 3- STANDARD METHODS AND TECHNIQUES FOR DEMERSAL FISHERIES RESOURCE SURVEYS. FAO FISHERIES TECHNICAL PAPER NO. 124. III, 39 PP. [PROBABILITY SAMPLING; SIMPLE RANDOM SAMPLING; STRATIFIED RANDOM SAMPLING; STANDARD ERROR; SAMPLE SIZE]

MANN, S. H. 1970. A MATHEMATICAL THEORY FOR THE EXPLOITATION OF BIOLOGICAL POPULATIONS. APPENDIX PP. 117. IN, PLANNING CHALLENGES OF THE 70'S IN THE PUBLIC DOMAIN. SCIENCE AND TECHNOLOGY SERIES OF THE AMERICAN ASTRONAUTICAL SOCIETY, WASHINGTON, D. C., VOLUME 22. [MARKOVIAN]

MANN, S. H. 1971. MATHEMATICAL MODELS FOR THE CONTROL OF PEST POPULATIONS. BIOMETRICS 27(2), 357-368.

MANN, S. H. 1971. A MATHEMATICAL THEORY FOR THE HARVEST OF NATURAL ANIMAL POPULATIONS IN THE CASE OF MALE- AND FEMALE-DEPENDENT BIRTH RATES. PP. 537-551. IN, STATISTICAL ECOLOGY, VOLUME 1, SPATIAL PATTERNS AND STATISTICAL DISTRIBUTIONS. (G. P. PATIL, E. C. PIELOU AND W. E. WATERS, EDS.). PENNSYLVANIA STATE UNIVERSITY PRESS, UNIVERSITY PARK, PENNSYLVANIA.

MARIAN, J. E. AND D. A. STUMBO. 1960. A NEW METHOD OF GROWTH RING ANALYSIS AND THE DETERMINATION OF DENSITY BY SURFACE TEXTURE MEASUREMENTS. FOREST SCI. 6(3), 276-291. [EQUATIONS]

MARR, J. C. 1951. ON THE USE OF THE TERMS ABUNDANCE, AVAILABILITY AND APPARENT ABUNDANCE IN FISHERY BIOLOGY. COPEIA 1951(2), 163-169 [EQUATIONS]

MATHEWS, C. P. 1970. ESTIMATES OF PRODUCTION WITH REFERENCE TO GENERAL SURVEYS. OIKOS 21(1), 129-133.

MATIS, J. H. AND D. CHILDERS, 1973. MULTIVARIABLE MULTILAG MODELS OF FISH BEHAVIOR IN AN OPEN FIELD UNDER CONTROL CONDITIONS. (ABSTRACT). BIOMETRICS 29(3), 611. [AUTOCORRELATION]

MATUDA, K. 1963. ON THE MECHANICAL CHARACTERS OF THE SWEEPING TRAMMEL NET-I. BULL. JAPANESE SOC. SCI. FISH. 29(2), 135-138. [EQUATIONS]

MATUDA, K. 1967. RELATIONSHIP BETWEEN CATCH PER UNIT EFFORT AND POWER OF ENGINE IN SWEEPING TRAMMEL NET FISHERY. BULL. JAPANESE SOC. SCI. FISH. 33(12), 1092-1095. [EQUATIONS]

MATUDA, K. AND T. KAWAKAMI. 1968. ON THE TENSION OF TUNA-LONG LINE RETRIEVING-I. BULL. JAPANESE SOC. SCI. FISH. 34(7), 594-598. [EQUATIONS]

MATUDA, K. AND T. KITAHARA. 1967. ON THE ESTIMATION OF CATCH EFFICIENCY OF SWEEPING TRAMMEL NET. BULL. JAPANESE SOC. SCI. FISH. 33(12), 1096-1098. [EQUATIONS]

MATUDA, K. AND T. KITAHARA. 1967. ON THE ESTIMATION OF SWEEP AREA OF SWEEPING TRAMMEL NET (KOGISASIAMI). BULL. JAPANESE SOC. SCI. FISH. 33(6), 524-530. [EQUATIONS]

MCCARL, B., D. RAPHAEL AND E. STAFFORD. 1975. THE IMPACT OF MAN ON THE WORLD NITROGEN CYCLE. J. ENVIRON. MANAGE. 3(1), 7-19. [MODEL; MATRIX]

MCCOMBIE, A. M. AND F. E. J. FRY. 1960. SELECTIVITY OF GILL NETS FOR LAKE WHITEFISH, COREGONUS CLUPEAFORMIS. TRANS. AMER. FISH. SOC. 89(2), 176-184. [NORMAL DISTRIBUTION; CATCH EQUATION]

MCMANUS, M. L. AND R. L. GIESE. 1968. THE COLUMBIAN TIMBER BEETLE, CORTHYLUS COLUMBIANUS. VII. THE EFFECT OF CLIMATE INTEGRANTS ON HISTORIC DENSITY FLUCTUATIONS. FOREST SCI. 14(3), 242-253. [MULTIPLE REGRESSION MODEL; MODEL]

MEIER, R., G. J. PAULIK, L. GALES AND W. CLARK. 1970. GAMES: A MULTI-PURPOSE INTERACTIVE RESOURCE MANAGEMENT GAME. QUANTITATIVE SCIENCE PAPER NO. 17. CENTER FOR QUANTITATIVE SCIENCE IN FORESTRY, FISHERIES AND WILDLIFE, UNIVERSITY OF WASHINGTON, SEATTLE, WASHINGTON. MIMEO 17 PP.

MILLAR, J. B. 1973. ESTIMATION OF AREA AND CIRCUMFERENCE OF SMALL WETLANDS. J. WILDL. MANAGE. 37(1), 30-38. [ELLIPSE FORMULAE]

MILNE, A. 1962. ON A THEORY OF NATURAL CONTROL OF INSECT POPULATION. J. THEOR. BIOL. 3(1), 19-50. [MODEL; LOGISTIC MODIFIED]

MIYAZAKI, Y. 1970. STUDIES ON APPROXIMATE FORMULAS FOR TENSION AND CONFIGURATION OF THE TOWING ROPE-I. METHOD OF APPROXIMATION. BULL. JAPANESE SOC. SCI. FISH. 36(1), 48-57. [EQUATIONS]

MIYAZAKI, Y. 1970. STUDIES ON APPROXIMATE FORMULAS FOR TENSION AND CONFIGURATION OF THE TOWING ROPE-II. METHOD OF CONFIGURATION. BULL. JAPANESE SOC. SCI. FISH. 36(1), 58-67. [EQUATIONS]

MIYAZAKI, Y. 1972. STUDIES ON APPROXIMATE FORMULAS FOR TENSION AND CONFIGURATION OF THE TOWING ROPE-III. METHOD OF APPROXIMATION OF THREE DIMENSIONAL PROBLEMS. BULL. JAPANESE SOC. SCI. FISH. 38(11), 1215-1222. [EQUATIONS]

MIYAZAKI, Y. 1972. STUDIES ON APPROXIMATE FORMULAS FOR TENSION AND CONFIGURATION OF THE TOWING ROPE-IV. METHOD OF CALCULATION OF THREE DIMENSIONAL PROBLEMS. BULL. JAPANESE SOC. SCI. FISH. 38(11), 1223-1228. [EQUATIONS]

MOTTLEY, C. M. 1942. EXPERIMENTAL DESIGNS FOR DEVELOPING AND TESTING A STOCKING POLICY. TRANS. N. AMER. WILDL. CONF. 7, 224-232. [RANDOMIZED BLOCK; LATIN SQUARE; YOUDEN SQUARE]

MOTTLEY, C. M. 1942. MODERN METHODS OF STUDYING FISH POPULATIONS. TRANS. N. AMER. WILDL. CONF. 7, 356-360. [ANALYSIS OF VARIANCE]

MULLER, G. R. 1972. A NEW METHOD TO FIND THE GROWTH AND YIELD OF FORESTS WITHOUT USE OF YIELD-TABLES. PP. 55-58. IN, 3RD CONFERENCE ADVISORY GROUP OF FOREST STATISTICIANS, INSTITUT NATIONAL DE LA RECHERCHE AGRONOMIQUE, PARIS, FRANCE. I.N.R.A. PUBL. 72-3. [EXPONENTIAL FUNCTION]

MUNRO, J. L. 1974. THE MODE OF OPERATION OF ANTILLEAN FISH TRAPS AND THE RELATIONSHIPS BETWEEN INGRESS, ESCAPEMENT, CATCH AND STOCK. J. CONS. PERM. INT. EXPLOR. MER 35(3), 337-350. [CATCH EQUATION; MODEL]

MYHRE, R. J. 1969. GEAR SELECTION AND PACIFIC HALIBUT. INTERN. PACIFIC HALIBUT COMM. REPT. NO. 51. 35 PP. [SELECTION CURVE]

MYHRE, R. J. 1974. MINIMUM SIZE AND OPTIMUM AGE OF ENTRY FOR PACIFIC HALIBUT. INTERN. PACIFIC HALIBUT COMM. SCI. REPT. NO. 55. 15 PP. [RICKER'S MODEL]

NAKAI, Z. AND S. HATTORI. 1962. QUANTITATIVE DISTRIBUTION OF EGGS AND LARVAE OF THE JAPANESE SARDINE BY YEAR, 1949 THROUGH 1951. BULL. TOKAI REG. FISH. RES. LAB. 9, 23-53. [EQUATIONS; SURVIVAL RATE]

NAKASAI, K. AND T. KAWAKAMI. 1965. ON A SIMPLE ESTIMATION OF WORKING DEPTH OF MID-WATER TRAWL. BULL. JAPANESE SOC. SCI. FISH. 31(4), 277-280. [EQUATIONS]

NAKASAI, K., O. SUZUKI AND T. KAWAKAMI. 1961. STUDIES ON THE CONFIGURATION OF DANISH SEINE. BULL. JAPANESE SOC. SCI. FISH. 27(7), 641-644. [EQUATIONS]

NASLUND, B. 1969. OPTIMAL ROTATION AND THINNING. FOREST SCI. 15(4), 446-451. [ECONOMICS PONTRYAGIN'S MAXIMUM PRINCIPLE; EQUATIONS]

NEU, C. W., C. R. BYERS AND J. M. PEEK. 1974. A TECHNIQUE FOR ANALYSIS OF UTILIZATION-AVAILABILITY DATA. J. WILDL. MANAGE. 38(3), 541-545. [CHI-SQUARE; BONFERRONI Z STATISTIC; MULTINOMIAL DISTRIBUTION]

NEWNHAM, R. M. 1968. A CLASSIFICATION OF CLIMATE BY PRINCIPAL COMPONENT ANALYSIS AND ITS RELATIONSHIP TO TREE SPECIES DISTRIBUTION. FOREST SCI. 14(3), 254-264. [MULTIVARIATE ANALYSIS]

NOAKES, R. R. AND J. J. J. PIGRAM. 1973. IMPACT MULTIPLIERS AND FOREST RESOURCE MANAGEMENT. J. ENVIRON. MANAGE. 1(3), 277-287. [ECONOMETRIC MODEL]

NONODA, T. 1967. STUDIES ON THE MECHANICAL CHARACTERS OF PATTI-AMI-I. ON THE TENSIONS ON HEAD LINE AND FOOT ROPE. BULL. JAPANESE SOC. SCI. FISH. 33(5), 385-391. [EQUATIONS]

NONODA, T. 1969. ON THE RESISTANCE OF PLANE MINNOW NETTING IN A CURRENT. BULL. JAPANESE SOC. SCI. FISH. 35(12), 1151-1156. [EQUATIONS]

OKA, M. 1960. STUDIES ON THE FORMATION OF SHOAL KIDAI (TAIUS TUMIFRONS) BY THE CATCH PER UNIT OF EFFORT. BULL. JAPANESE SOC. SCI. FISH. 23(3), 211-216. [POISSON DISTRIBUTION]

OKA, M. 1962. STUDIES ON THE ESTIMATION OF THE STOCK CAPACITY OF "GYOSHO". BULL. JAPANESE SOC. SCI. FISH. 28(5), 477-483. [FISHING EFFORT EQUATION; STOCK EQUATION; DIFFERENTIAL EQUATION; CATCH EQUATION]

OLSEN, S. 1959. MESH SELECTION IN HERRING GILL NETS. J. FISH. RES. BOARD CAN. 16(3), 339-349. [MODEL]

ONODERA, K. 1961. ON THE ESTIMATION OF CATCH AND FISHING EFFORT IN RIVER BY A NEW METHOD OF CREEL CENSUS. BULL. JAPANESE SOC. SCI. FISH. 27(6), 521-529. [MODEL; CORRECTIVE COEFFICIENT]

OTTESTAD, P. 1934. STATISTICAL ANALYSIS OF THE NORWEGIAN HERRING POPULATION. RAPP. PROCES-VERB. REUNIONS J. CONS. PERM. INT. EXPLOR. MER 88(3), 1-45. [VARIANCE; NORMAL DISTRIBUTION; EQUATIONS]

PAL, R. AND L. E. LACHANCE. 1974. THE OPERATIONAL FEASIBILITY OF GENETIC METHODS FOR CONTROL OF INSECTS OF MEDICAL AND VETERINARY IMPORTANCE. ANNU. REV. ENTOMOL. 19, 269-291. [MODEL]

PALOHEIMO, J. E. AND A. C. KOHLER. 1968. ANALYSIS OF THE SOUTHERN GULF OF ST. LAWRENCE COD POPULATION. J. FISH. RES. BOARD CAN. 25(3), 555-578. [ESTIMATION OF GROWTH; RECRUITMENT AND PRODUCTIVITY]

PALOHEIMO, J. E. AND L. M. DICKIE. 1964. ABUNDANCE AND FISHING SUCCESS. RAPP. PROCES-VERB. REUNIONS J. CONS. PERM. INT. EXPLOR. MER 155, 152-163. [CATCH EQUATION; RICKER'S EQUATION; VARANOV'S EQUATION; NICHOLSON-BAILEY MODEL; ASSUMPTIONS; SCHOOLING EQUATIONS]

PARKER, R. R. 1960. CRITICAL SIZE AND MAXIMUM YIELD FOR CHINOOK SALMON. (ONCORHYNCHUS TSHAWYTSCHA). J. FISH. RES. BOARD CAN. 17(2), 199-210. [NATURAL MORTALITY EQUATION; GROWTH RATE]

PARKER, R. R. 1963. ON THE PROBLEM OF MAXIMUM YIELD FROM NORTH PACIFIC SOCKEYE SALMON STOCKS. J. FISH. RES. BOARD CAN. 20(6), 1371-1396. [GROWTH EQUATIONS; MORTALITY ESTIMATES; TAG-RECAPTURE]

PARKER, R. R., E. C. BLACK AND P. . LARKIN. 1959. FATIGUE AND ORTALITY IN TROLL-CAUGHT PACIFIC ALMON (ONCORHYNCHUS). J. FISH. ES. BOARD CAN. 16(4), 429-448. VARIANCE OF COMBINED PROBABILITY STIMATES]

PAULIK, G. J. 1973. STUDIES OF THE POSSIBLE FORM OF THE STOCK-RECRUITMENT CURVE. RAPP. PROCES-VERB. REUNIONS J. CONS. PERM. INT. EXPLOR. MER 164, 302-315. [MODEL]

PAULIK, G. J. AND J. W. GREENOUGH, JR. 1966. MANAGEMENT ANALYSIS FOR A SALMON RESOURCE SYSTEM. PP. 215-252. IN, SYSTEMS ANALYSIS IN ECOLOGY. (K. E. F. WATT, ED.). ACADEMIC PRESS, NEW YORK, NEW YORK.

PELLA, J. J. AND P. K. TOMLINSON. 1969. A GENERALIZED STOCK PRODUCTION MODEL. BULL. INTER-AMER. TROP. TUNA COMM. 13(3), 421-458.

PETERSON, A. E. 1966. GILL NET MESH SELECTION CURVES FOR PACIFIC SALMON ON THE HIGH SEAS. U. S. FISH WILDL. SERV. FISH. BULL. 65(2), 381-390. [EQUATIONS]

PIENAAR, L. V. AND W. E. RICKER. 1968. ESTIMATING MEAN WEIGHT FROM LENGTH STATISTICS. J. FISH. RES. BOARD CAN. 25(12), 2743-2747. [NORMAL DISTRIBUTION]

PIKUSH, N. V. 1971. A PNEUMATIC METHOD FOR MEASURING THE VELOCITY OF A WATER CURRENT. HYDROBIOL. J. 7(4), 83-91. (ENGLISH TRANSL.). [EQUATIONS]

PITCHER, T. J. AND P. D. M. MACDONALD. 1973. A NUMERICAL INTEGRATION METHOD FOR FISH POPULATION FECUNDITY. J. FISH. BIOL. 5(4), 549-553. [VARIANCE; COMPUTER]

POLLARD, R. A. 1955. MEASURING SEEPAGE THROUGH SALMON SPAWNING GRAVEL. J. FISH. RES. BOARD CAN. 12(5), 706-741. [EQUATIONS]

POPE, J. A. 1966. MANUAL OF METHODS FOR FISH STOCK ASSESSMENT PART III. SELECTIVITY OF FISHING GEAR. FAO FISHERIES TECHNICAL PAPER NO. 41. I, 41 PP. PLUS FIGURES. [VARIANCE; GRAPHICAL METHOD; NUMERICAL METHOD; LOGISTIC CURVE]

POPE, J. A. AND B. B. PARRISH, 1964, THE IMPORTANCE OF FISHING POWER STUDIES IN ABUNDANCE ESTIMATION. RAPP. PROCES-VERB. REUNIONS J. CONS. PERM. INT. EXPLOR. MER 155, 81-89. [PROBABILITY DISTRIBUTION; CATCH EQUATION; WEIGHTING]

POPE, J. G. 1973. AN INVESTIGATION INTO THE EFFECTS OF VARIABLE RATES OF THE EXPLOITATION OF FISHERY RESOURCES. PP. 23-34. IN, THE MATHEMATICAL THEORY OF THE DYNAMICS OF BIOLOGICAL POPULATIONS. (M. S. BARTLETT AND R. W. HIRONS, EDS.), ACADEMIC PRESS, NEW YORK, NEW YORK. [POPULATION DYNAMICS; SIMULATION; COMPUTER]

QUICK, H. F. 1958. ESTIMATING THE EFFECTS OF EXPLOITATION BY LIFE TABLES. TRANS. N. AMER. WILDL. NAT. RESOUR. CONF. 23, 426-442.

RAPPORT, D. J. AND J. E. TURNER, 1974. ECOLOGICAL AND ECONOMIC MARKETS. PP. 206-210. IN, MATHEMATICAL PROBLEMS IN BIOLOGY. LECTURE NOTES IN BIOMATHEMATICS 2. (P. VAN DEN DRIESSCHE, ED.). SPRINGER-VERLAG NEW YORK INC., NEW YORK, NEW YORK. [HARVEST AND YIELD FUNCTIONS; LOTKA-VOLTERRA MODEL]

REGIER, H. A. AND D. S. ROBSON, 1966. SELECTIVITY OF GILL NETS, ESPECIALLY TO LAKE WHITEFISH. J. FISH. RES. BOARD CAN. 23(3), 423-454. [ARITHMETIC PROGRESSION; GEOMETRIC SERIES; COMPUTER; NORMAL DISTRIBUTION; COEFFICIENT OF SKEWNESS; MODEL; GRAPHIC]

RICH, W. H. 1943. AN APPLICATION OF THE CONTROL CHART METHOD TO THE ANALYSIS OF FISHERIES DATA. SCIENCE 97(2516), 269-270.

RICKER, W. E. 1945. A METHOD OF ESTIMATING MINIMUM SIZE LIMITS FOR OBTAINING MAXIMUM YIELD. COPEIA 1945 (2), 84-94.

RICKER, W. E. 1945. SOME APPLICATIONS OF STATISTICAL METHODS TO FISHERY PROBLEMS. BIOMETRICS 1(6), 73-79. [GRAPHIC; NON-GRAPHICAL; CORRELATION; REGRESSION; BINOMIAL DISTRIBUTION; POISSON DISTRIBUTION; CONFIDENCE INTERVAL; ANALYSIS OF VARIANCE; ANALYSIS OF COVARIANCE; MAXIMUM LIKELIHOOD; TAG-RECAPTURE ESTIMATION]

RICKER, W. E. 1954. STOCK AND RECRUITMENT. J. FISH. RES. BOARD CAN. 11(5), 559-623. [MODEL]

RICKER, W. E. 1958. MAXIMUM SUSTAINED YIELDS FROM FLUCTUATING ENVIRONMENTS AND MIXED STOCKS. J. FISH. RES. BOARD CAN. 15(5), 991-1006. [MODEL]

RICKER, W. E. 1962. REGULATION OF THE ABUNDANCE OF PINK SALMON POPULATIONS. PP. 155-200. IN, SYMPOSIUM ON PINK SALMON. INSTITUTE OF FISHERIES, UNIVERSITY OF BRITISH COLUMBIA, VANCOUVER, BRITISH COLUMBIA, CANADA. [MODEL]

RICKER, W. E. 1969. EFFECTS OF SIZE-SELECTIVE MORTALITY AND SAMPLING BIAS ON ESTIMATES OF GROWTH, MORTALITY, PRODUCTION AND YIELD. J. FISH. RES. BOARD CAN. 26(9), 479-541. [MODEL]

RICKER, W. E. 1973. CRITICAL STATISTICS FROM TWO REPRODUCTION CURVES. RAPP. PROCES-VERB. REUNIONS J. CONS. PERM. INT. EXPLOR. MER 164, 333-340. [RECRUITMENT MODEL; BEVERTON-HOLT MODEL; RICKER'S MODEL]

RICKER, W. E. 1973. LINEAR REGRESSIONS IN FISHERY RESEARCH. J. FISH. RES. BOARD CAN. 30(3), 409-434. [NORMAL DISTRIBUTION; GEOMETRIC MEAN; CORRELATION; NON-NORMAL DISTRIBUTION]

RICKER, W. E. 1973. TWO MECHANISMS THAT MAKE IT IMPOSSIBLE TO MAINTAIN PEAK-PERIOD YIELDS FROM STOCKS OF PACIFIC SALMON AND OTHER FISHES. J. FISH. RES. BOARD CAN. 30(9), 1275-1286. [MODEL; PRODUCTIVITY; CATCH RECRUITMENT; RICKER'S FORMULA; COMPUTER]

RICKER, W. E. AND R. E. FOERSTER. 1948. COMPUTATION OF FISH PRODUCTION. BULL. BINGHAM OCEANOGR. COLLECT. YALE UNIV. 11(ART. 4), 173-211.

ROBSON, D. S. 1966. ESTIMATION OF THE RELATIVE FISHING POWER OF INDIVIDUAL SHIPS. INTERN. COMM. NORTHWEST ATLANTIC FISH. RES. BULL. 3, 5-14. [CATCH RATE MODEL; METHOD OF LEAST SQUARES; MAXIMUM LIKELIHOOD; MATRIX; VARIANCE]

RORRES, C. AND W. FAIR. 1975. OPTIMAL HARVESTING POLICY FOR AN AGE-SPECIFIC POPULATION. MATH. BIOSCI. 24(1/2), 31-47. [LESLIE'S MODEL; MATRIX]

ROSENZWEIG, M. L. 1973. EXPLOITATION IN THREE TROPHIC LEVELS. AMER. NAT. 107(954), 275-294. [MODEL; TIME LAG; MATRIX]

ROTHSCHILD, B. J. 1967. COMPETITION FOR GEAR IN A MULTIPLE-SPECIES FISHERY. J. CONS. PERM. INT. EXPLOR. MER 31(1), 102-110. [MODEL; VARIANCE; COVARIANCE; MAXIMUM LIKELIHOOD]

ROTHSCHILD, B. J. 1970. SYSTEMS VIEW OF FISHERY MANAGEMENT WITH SOME NOTES ON THE TUNA FISHERIES. QUANTITATIVE SCIENCE PAPER NO. 14. CENTER FOR QUANTITATIVE SCIENCE IN FORESTRY, FISHERIES AND WILDLIFE. UNIVERSITY OF WASHINGTON, SEATTLE, WASHINGTON. MIMEO, 78 PP. [COMPUTER]

ROTHSCHILD, B. J. 1971. A SYSTEMS VIEW OF FISHERY MANAGEMENT WITH SOME NOTES ON THE TUNA FISHERIES. FAO FISHERIES TECHNICAL PAPER NO. 106. III, 33 PP..

ROTHSCHILD, B. J. 1973. QUESTIONS OF STRATEGY IN FISHERY MANAGEMENT AND DEVELOPMENT. J. FISH. RES. BOARD CAN. 30(12, PART 2), 2017-2030. [MODEL]

ROTHSCHILD, B. J. AND D. S. ROBSON. 1972. THE USE OF CONCENTRATION INDICES IN FISHERIES. U. S. NATL. MAR. FISH. SERV. FISH. BULL. 70(2), 511-514. [PREDATOR-PREY MODEL]

ROUNSEFELL, G. A. 1949. METHODS OF ESTIMATING TOTAL RUNS AND ESCAPEMENTS OF SALMON. BIOMETRICS 5(2), 115-126. [MULTIPLE REGRESSION; SIMPLE LINEAR REGRESSION; CORRELATION COEFFICIENT; T-TEST; ANALYSIS OF COVARIANCE]

ROUNSEFELL, G. A. 1958. FACTORS CAUSING DECLINE IN SOCKEYE SALMON OF KARLUK RIVER, ALASKA. U. S. FISH WILDL. SERV. FISH. BULL. 58, 83-169. [CURVILINEAR CORRELATION PRODUCTION CURVE; REGRESSION]

RUSSELL, E. S. 1931. SOME THEORETICAL CONSIDERATIONS ON THE "OVERFISHING" PROBLEM. J. CONS. PERM. INT. EXPLOR. MER 6(1), 3-20. [MODEL]

RYDER, R. A., S. R. KERR, K. H. LOFTUS AND H. A. REGIER. 1974. THE MORPHOEDAPHIC INDEX, A FISH YIELD ESTIMATOR-REVIEW AND EVALUATION. J. FISH. RES. BOARD CAN. 31(5), 663-688.

SAILA, S. B. 1962. PROPOSED HURRICANE BARRIERS RELATED TO WINTER FLOUNDER MOVEMENTS IN NARRAGANSETT BAY. TRANS. AMER. FISH. SOC. 91(2), 189-195. [BARRIER MODEL]

SAILA, S. B. AND J. M. FLOWERS. 1969. TOWARD A GENERALIZED MODEL OF FISH MIGRATION. TRANS. AMER. FISH. SOC. 98(3), 582-588.

SAILA, S. B., D. B. HORTON AND R. J. BERRY. 1965. ESTIMATES OF THE THEORETICAL BIOMASS OF JUVENILE WINTER FLOUNDER, PSEUDOPLEURONECTES AMERICANUS (WALBAUM) REQUIRED FOR A FISHERY IN RHODE ISLAND. J. FISH. RES. BOARD CAN. 22(4), 945-954. [RICKER'S MODEL; COMPUTER]

SANDEMAN, E. J. 1969. DIURNAL VARIATION IN AVAILABILITY OF DIFFERENT SIZES OF REDFISH, SEBASTES MENTELLA. INTERN. COMM. NORTHWEST ATLANTIC FISH. RES. BULL. 6, 35-46. [YIELD MODEL; POPULATION MODEL]

SCHAAF, W. E. AND G. R. HUNTSMAN. 1972. EFFECTS OF FISHING ON THE ATLANTIC MENHADEN STOCK: 1955-1969. TRANS. AMER. FISH. SOC. 101(2), 290-297. [MORTALITY RATE; CATCH EQUATION; BEVERTON-HOLT EQUATION; RICKER'S MODEL; COMPUTER; SIMULATION]

SCHAEFER, M. B. 1954. SOME ASPECTS OF THE DYNAMICS OF POPULATIONS IMPORTANT TO THE MANAGEMENT OF THE COMMERCIAL MARINE FISHERIES. BULL. INTER-AMER. TROP. TUNA COMM. 1(2), 25-56. [GROWTH CURVES; VERHULST-PEARL LOGISTIC EQUATION; CATCH-CURVE; PREDATOR-PREY MODEL; LOTKA-VOLTERRA]

SCHAEFER, M. B. 1957. SOME CONSIDERATIONS OF POPULATION DYNAMICS AND ECONOMICS IN RELATION TO THE MANAGEMENT OF THE COMMERCIAL MARINE FISHERIES. J. FISH. RES. BOARD CAN. 14(5), 669-681. [MODEL]

SCHAEFER, M. B. 1967. FISHERY DYNAMICS AND PRESENT STATUS OF THE YELLOWFIN TUNA POPULATION OF THE EASTERN PACIFIC OCEAN. BULL. INTER-AMER. TROP. TUNA COMM. 12(3), 89-112. [MODEL; LOGISTIC MODEL; BEVERTON-HOLT MODEL]

SCHAEFER, M. B. 1968. METHODS OF ESTIMATING EFFECTS OF FISHING ON FISH POPULATIONS. TRANS. AMER. FISH. SOC. 97(3), 231-241. [COMPUTER]

SCHAEFER, M. B. AND R. J. H. BEVERTON. 1963. FISHERY DYNAMICS-THEIR ANALYSIS AND INTERPRETATION. PP. 464-483. IN, THE SEA, VOLUME 2. (M. N. HILL, ED.). INTERSCIENCE PUBLISHERS, NEW YORK, NEW YORK.

SCHREUDER, G. F. 1971. THE SIMULTANEOUS DETERMINATION OF OPTIMAL THINNING SCHEDULE AND ROTATION FOR AN EVEN-AGED FOREST. FOREST SCI. 17(3), 333-339. [DYNAMIC PROGRAMMING; MODEL; COST FUNCTION]

SCHULTZ, A. M. 1956. THE USE OF REGRESSION IN RANGE RESEARCH. J. RANGE MANAGE. 9(1), 41-46.

SCOTT, J. R. 1972. REDUNDANT VARIABLES IN MULTIVARIATE ANALYSIS. PP. 153-159. IN, 3RD CONFERENCE ADVISORY GROUP OF FOREST STATISTICIANS. INSTITUT NATIONAL DE LA RECHERCHE AGRONOMIQUE, PARIS, FRANCE, I.N.R.A. PUBL. 72-3. [FORESTRY]

SECKEL, G. R. 1972. HAWAIIAN-CAUGHT SKIPJACK TUNA AND THEIR PHYSICAL ENVIRONMENT. U. S. NATL. MAR. FISH. SERV. FISH. BULL. 70(3), 763-787. [FISH DRIFT MODEL]

SHIMADA, B. M. AND M. B. SCHAEFER. 1956. A STUDY OF CHANGES IN FISHING EFFORT, ABUNDANCE, AND YIELD FOR YELLOWFIN AND SKIPJACK TUNA IN THE EASTERN TROPICAL PACIFIC OCEAN. BULL. INTER-AMER. TROP. TUNA COMM. 1(7), 351-421. [RUSSELL'S FORMULA; STANDARDIZED UNIT OF EFFORT]

SHIUE, C.-J. AND S. S. PAULEY. 1961. SOME CONSIDERATIONS ON THE STATISTICAL DESIGN FOR PROVENANCE AND PROGENY TESTS IN TREE IMPROVEMENT PROGRAMS. FOREST SCI. 7(2), 116-122. [PLOT SIZE; PLOT SHAPE]

SHOEMAKER, C. 1973. OPTIMIZATION OF AGRICULTURAL PEST MANAGEMENT II: FORMULATION OF A CONTROL MODEL. MATH. BIOSCI. 17, 357-366.

SHOEMAKER, C. 1973. OPTIMIZATION OF AGRICULTURAL PEST MANAGEMENT I: BIOLOGICAL AND MATHEMATICAL BACKGROUND. MATH. BIOSCI. 16, 143-175. [MODEL; ALGORITHM; LOTKA-VOLTERRA MODEL; COMPUTER; LESLIE MATRIX; LOGISTIC]

SHOEMAKER, C. 1973. OPTIMIZATION OF ARGICULTURAL PEST MANAGEMENT III: RESULTS AND EXTENSIONS OF MODEL. MATH. BIOSCI. 18, 1-22. [COMPUTER; OPTIMIZATION TECHNIQUES; DYNAMIC PROGRAMMING]

SILLIMAN, R. P. 1971. ADVANTAGES AND LIMITATIONS OF "SIMPLE" FISHERY MODELS IN LIGHT OF LABORATORY EXPERIMENTS. J. FISH. RES. BOARD CAN. 28(8), 1211-1214.

SISSENWINE, M. P. 1974. VARIABILITY IN RECRUITMENT AND EQUILIBRIUM CATCH OF THE SOUTHERN NEW ENGLAND YELLOWTAIL FLOUNDER FISHERY. J. CONS. PERM. INT. EXPLOR. MER 36(1), 15-26. [MULTIPLE CORRELATION; MODEL]

SLOBODKIN, L. B. 1973. SUMMARY AND DISCUSSION OF THE SYMPOSIUM. RAPP. PROCES-VERB. REUNIONS J. CONS. PERM. INT. EXPLOR. MER 164, 7-14. [FISH STOCKS AND RECRUITMENT; MODEL]

SMITH, P. E. 1972. THE INCREASE IN SPAWNING BIOMASS OF NORTHERN ANCHOVY, ENGRAULIS MORDAX. U. S. NATL. MAR. FISH. SERV. FISH. BULL. 70(3), 849-874. [ESTIMATION]

SMOLONOGOV, E. P. 1970. USE OF TREE SIZES IN TREE STANDS TO DIAGNOSE FOREST TYPES. EKOLOGIYA 1(1), 50-59. (ENGLISH TRANSL. PP. 37-44). [VOLUME INDEX]

SO KO, K., M. SUZUKI AND Y. KONDO. 1970. AN ELEMENTARY STUDY ON BEHAVIOUR OF COMMON SHRIMP TO MOVING NET. BULL. JAPANESE SOC. SCI. FISH. 36(6), 556-570. [EQUATIONS]

SPIVY, W. A. 1973. OPTIMIZATION IN COMPLEX MANAGEMENT SYSTEMS. TRANS. AMER. FISH. SOC. 102(2), 492-499. [MODEL; MATRIX; PROGRAMMING]

STACHURSKI, A. 1968. EMIGRATION AND MORTALITY RATES AND THE FOOD-SHELTER CONDITIONS OF LIGIDIUON HYPNORUM L. (ISOPODA). EKOLOGIYA POLSKA-SERIA A 16(21), 445-459. [LOSS INDEX; INDEX OF MORTALITY; EMIGRATION INDEX]

STAFF. 1960. UTILIZATION OF PACIFIC HALIBUT STOCKS: YIELD PER RECRUITMENT. INTERN. PACIFIC HALIBUT COMM. REPT. NO. 28. 52 PP. [MORTALITY RATE]

STAGE, A. R. 1963. A MATHEMATICAL APPROACH TO POLYMORPHIC SITE INDEX CURVES FOR GRAND FIR. FOREST SCI. 9(2), 167-180. [SIGMOID FUNCTION]

STRAND, L. 1964. NUMERICAL CONSTRUCTIONS OF SITE-INDEX CURVES. FOREST SCI. 10(4), 410-414.

STRUB, M. R. AND H. E. BURKHART. 1975. A CLASS-INTERVAL-FREE METHOD FOR OBTAINING EXPECTED YIELDS FROM DIAMETER DISTRIBUTIONS. FOREST SCI. 21(1), 67-69. [EQUATIONS]

SUND, O. 1934. COD-MEASUREMENTS DURING 21 YEARS ON THE WESTERN AND NORTHERN COASTS OF NORWAY. METHODS AND RESULTS. RAPP. PROCES-VERB. REUNIONS J. CONS. PERM. INT. EXPLOR. MER 88(7), 1-20. [MODEL]

SUZUKI, O. 1961. MECHANICAL ANALYSIS ON THE WORKING BEHAVIOUR OF MID-WATER TRAWL. BULL. JAPANESE SOC. SCI. FISH. 27(10), 903-907. [EQUATIONS]

SUZUKI, O. 1962. BEHAVIOUR OF SWEEP LINE IN PAIR-TRAWLING. BULL. JAPANESE SOC. SCI. FISH. 28(11), 1051-1055. [EQUATIONS]

SUZUKI, O. 1963. BEHAVIOUR OF SWEEP LINE IN DANISH SEINING - I. BULL. JAPANESE SOC. SCI. FISH. 29(12), 1071-1076. [EQUATIONS]

SUZUKI, O. 1964. BEHAVIOUR OF SWEEP LINE IN DANISH SEINING - II. BULL. JAPANESE SOC. SCI. FISH. 30(1), 21-28. [EQUATIONS]

SUZUKI, O. 1964. BEHAVIOUR OF SWEEP LINE IN DANISH SEINING - III. BULL. JAPANESE SOC. SCI. FISH. 30(1), 29-36. [EQUATIONS]

SUZUKI, O. 1965. BEHAVIOUR OF CHAIN FITTED SWEEP LINE IN PAIR-TRAWLING. BULL. JAPANESE SOC. SCI. FISH. 31(6), 409-413. [EQUATIONS]

SUZUKI, O. 1965. BEHAVIOUR OF SWEEP LINE WITH HEAVY FALL. BULL. JAPANESE SOC. SCI. FISH. 31(6), 403-408. [EQUATIONS]

SUZUKI, O. 1965. FIELD EXPERIMENT ON PAIR-TRAWL. BULL. JAPANESE SOC. SCI. FISH. 31(2), 493-499. [EQUATIONS]

SUZUKI, O. AND K. MATUDA. 1965. DRAG FORCE OF PLANE NET SET PARALLEL TO STREAM. BULL. JAPANESE SOC. SCI. FISH. 31(8), 579-584. [EQUATIONS]

TANAKA, S. 1954. THE EFFECT OF RESTRICTION OF FISHING EFFORT ON THE YIELD. BULL. JAPANESE SOC. SCI. FISH. 20(7), 599-603. [VARANOV'S MODEL; YIELD CURVE]

TANAKA, S. 1960. STUDIES ON THE DYNAMICS AND THE MANAGEMENT OF FISH POPULATIONS. BULL. TOKAI REG. FISH. RES. LAB. 29, 1-200. [MODEL]

TANIGUCHI, T. 1968. ON THE RESISTANCE OF VARIOUS COD ENDS FIXED IN A STREAM-V. BULL. JAPANESE SOC. SCI. FISH. 34(4), 295-299. [EQUATIONS]

TANIGUCHI, T., S. MINAMI AND Y. SUMIKAWA. 1968. FIELD EXPERIMENTS OF 100-FOOT TRAWL NET. BULL. JAPANESE SOC. SCI. FISH. 34(10), 889-894. [EQUATIONS]

TAUTI, M. 1928. ON THE INFLUENCE OF THE VARIATION OF TEMPERATURE OF WATER UPON THE HATCH RATE AND THE HATCHING DAYS OF FISH EGGS. J. IMPERIAL FISH. INST. 24(1), 13-18.

TAUTI, M. 1934. THE FORCE ACTING ON THE PLANE NET IN MOTION THROUGH THE WATER. BULL. JAPANESE SOC. SCI. FISH. 3(1), 1-4. [EQUATIONS]

TAUTI, M. 1934. A RELATION BETWEEN EXPERIMENTS ON MODEL AND ON FULL SCALE OF FISHING NETS. BULL. JAPANESE SOC. SCI. FISH. 3(4), 171-177. [EQUATIONS]

TAUTI, M. 1948. AN ESTIMATION OF OPTIMAL CATCH OF TRAWL FISHERY IN THE CHINESE SEAS. BULL. JAPANESE SOC. SCI. FISH. 14(5), 227-232. [MODEL]

TAUTI, M. 1951. AN DETERMINATION OF THE MAXIMUM YIELD ALLOWABLE TO CONSERVE A FISH POPULATION. BULL. JAPANESE SOC. SCI. FISH. 16(12), 197-200. [MODEL]

TAUTZ, A., P. A. LARKIN AND W. E. RICKER. 1969. SOME EFFECTS OF SIMULATED LONG-TERM ENVIRONMENTAL FLUCTUATIONS ON MAXIMUM SUSTAINED YIELD. J. FISH. RES. BOARD CAN. 26(10), 2715-2726. [RANDOM NORMAL DEVIATES; STOCHASTIC; DETERMINISTIC MODEL; COMPUTER]

TAYLOR, C. R. AND J. C. HEADLEY. 1975. INSECTICIDE RESISTANCE AND THE EVALUATION OF CONTROL STRATEGIES FOR AN INSECT POPULATION. CAN. ENTOMOL. 107(3), 237-242. [MONTE CARLO; DYNAMIC PROGRAMMING; MODEL; GENETICS; MARKOVIAN DENSITY FUNCTION]

TEEGUARDEN, D. E. AND H.-L. VON SPERBER. 1968. SCHEDULING DOUGLAS-FIR REFORESTATION INVESTMENTS: A COMPARISON OF METHODS. FOREST SCI. 14(4), 354-368. [MODEL; LINEAR PROGRAMMING]

TESTER, A. L. 1953. THEORETICAL YIELDS AT VARIOUS RATES OF NATURAL AND FISHING MORTALITY IN STABILIZED FISHERIES. TRANS. AMER. FISH. SOC. 82, 115-122. [MODEL; GOMPERTZ AND LOGISTIC EQUATION]

THOMAS, J. C. 1968. MANAGEMENT OF THE WHITE SEABASS (CYNOSCION NOBILIS) IN CALIFORNIA WATERS. CALIF. DEPT. FISH GAME FISH BULL. NO. 142, 34 PP. [BEVERTON-HOLT EQUATION; VON BERTALANFFY GROWTH EQUATION; SILLIMAN'S FISHING RATE METHOD]

THOMPSON, E. F. 1968. THE THEORY OF DECISION UNDER UNCERTAINTY AND POSSIBLE APPLICATIONS IN FOREST MANAGEMENT. FOREST SCI. 14(2), 156-163. [MODEL; DECISION THEORY]

THOMPSON, W. F. 1952. CONDITION OF STOCKS OF HALIBUT IN THE PACIFIC. J. CONS. PERM. INT. EXPLOR. MER 18(2), 141-166. [MODEL; RECIPROCAL RELATIONSHIP]

THOMPSON, W. F. 1962. THE RELATIONSHIP BETWEEN NUMBERS SPAWNING AND NUMBERS RETURNING IN PACIFIC SALMON. PP. 213-225. IN, SYMPOSIUM ON PINK SALMON. (N. J. WILIMOVSKY, ED.). INSTITUTE OF FISHERIES, UNIVERSITY OF BRITISH COLUMBIA, VANCOUVER, BRITISH COLUMBIA. [STOCK-RECRUITMENT]

THOMPSON, W. R. 1930. THE UTILITY OF MATHEMATICAL METHODS IN RELATION TO WORK ON BIOLOGICAL CONTROL. ANN. APPL. BIOL. 17(3), 641-648. [MODEL; TRANSFORMATION]

TOKAREV, A. S. 1974. DETERMINATION OF THE PERMISSIBLE ANNUAL FISH CATCH FOR LONG-TERM PLANNING PURPOSES. J. ICTHYOLOGY 14(1), 60-67. (ENGLISH TRANSL.). [GRAHAM'S MODEL; LOGISTIC; RUSSELL'S AXIOM]

TOMASSONE, M. R. 1972. INTRODUCTION. PP. 7-8. IN, 3RD CONFERENCE ADVISORY GROUP OF FOREST STATISTICIANS. INSTITUT NATIONAL DE LA RECHERCHE AGRONOMIQUE, PARIS, FRANCE. I.N.R.A. PUBL. 72-3.

TOOMING, H. 1970. MATHEMATICAL DESCRIPTION OF NET PHOTOSYNTHESIS AND ADAPTATION PROCESSES IN THE PHOTOSYNTHETIC APPARATUS OF PLANT COMMUNITIES. PP. 103-113. IN, PREDICTION AND MEASUREMENT OF PHOTOSYNTHETIC PRODUCTIVITY. (I. SETLIK, ED.). CENTRE FOR AGRICULTURAL PUBLISHING AND DOCUMENTATION, WAGENINGEN, NETHERLANDS. [MODEL]

TRESCHEV, A. I. 1964. ON FISHING INTENSITY. RAPP. PROCES-VERB. REUNIONS J. CONS. PERM. INT. EXPLOR. MER 155, 19-20. [INTENSITY OF FISHING FUNCTION]

TSUDA, R. AND N. INOUE. 1973. STUDY ON THE UNDERWATER VISIBILITY OF NET TWINES BY A HUMAN EYE-III. ESTIMATION OF THRESHOLD OF BRIGHTNESS-CONTRAST. BULL. JAPANESE SOC. SCI. FISH. 39(3), 253-264. [EQUATIONS]

UDA, M. 1939. ON THE CHARACTERISTICS OF THE FREQUENCY CURVE FOR THE CATCH OF "KATUO", EUTHYNNUS VAGANS (LESSON), REFERRED TO THE WATER TEMPERATURE. BULL. JAPANESE SOC. SCI. FISH. 8(4), 169-172.

USHER, M. B. 1966. A MATRIX APPROACH TO THE MANAGEMENT OF RENEWABLE RESOURCES, WITH SPECIAL REFERENCE TO SELECTION FORESTS. J. APPL. ECOL. 3(2), 355-367. [LESLIE MATRIX]

USHER, M. B. 1969. A MATRIX APPROACH TO THE MANAGEMENT OF RENEWABLE RESOURCES, WITH SPECIAL REFERENCE TO SELECTION FORESTS-TWO EXTENSIONS. J. APPL. ECOL. 6(2), 347-348. [COMPUTER]

USHER, M. B. 1969. A MATRIX MODEL FOR FOREST MANAGEMENT. BIOMETRICS 25(2), 309-315. [COMPUTER]

VERNON, E. H., A. S. HOURSTON AND G. A. HOLLAND. 1964. THE MIGRATION AND EXPLOITATION OF PINK SALMON RUNS IN AND ADJACENT TO THE FRASER RIVER CONVENTION AREA IN 1959. INTERN. PACIFIC SALMON FISH. COMM. BULL. 15, 296 PP. [MODEL; ESTIMATION; TAG-RECAPTURE]

VINBERG, G. G. 1971. COMPARISON OF SOME WIDELY USED METHODS FOR CALCULATING THE PRODUCTION OF AQUATIC BACTERIA. HYDROBIOL. J. 7(4), 72-82. (ENGLISH TRANSL.)

VON GELDERN, C. E., JR. AND P. K. TOMLINSON. 1973. ON THE ANALYSIS OF ANGLER CATCH RATE DATA FROM WARMWATER RESERVOIRS. CALIF. FISH GAME 59(4), 281-292. [SAMPLING; VARIANCE; COMPUTER; POISSON DISTRIBUTION; TWO-STAGE SAMPLING; STRATIFIED SAMPLING]

WADLEY, F. M. 1945. INCOMPLETE BLOCK EXPERIMENTAL DESIGNS IN INSECT POPULATION PROBLEMS. J. ECON. ENTOMOL. 38(6), 651-654.

WALKER, D. W. AND K. B. PEDERSEN. 1969. POPULATION MODELS FOR SUPPRESSION OF THE SUGARCANE BORER BY INHERITED PARTIAL STERILITY. ANN. ENTOMOL. SOC. AMER. 62(1), 21-26.

WATT, K. E. F. 1956. THE CHOICE AND SOLUTION OF MATHEMATICAL MODELS FOR PREDICTING AND MAXIMIZING THE YIELD OF A FISHERY. J. FISH. RES. BOARD CAN. 13(5), 613-645.

WATT, K. E. F. 1961. MATHEMATICAL MODELS FOR USE IN INSECT PEST CONTROL. CAN. ENTOMOL. 93(SUPPL. 19), 1-62. [COMPUTER]

WATT, K. E. F. 1964. COMPUTERS AND THE EVALUATION OF RESOURCE MANAGEMENT STRATEGIES. AMER. SCI. 52(4), 408-418.

WATT, K. E. F. 1964. THE USE OF MATHEMATICS AND COMPUTERS TO DETERMINE OPTIMAL STRATEGY AND TACTICS FOR A GIVEN INSECT PEST CONTROL PROBLEM. CAN. ENTOMOL. 96(1/2), 202-220.

WENNERGREN, E. B., H. H. FULLERTON AND J. C. WRIGLEY. 1973. ESTIMATION OF QUALITY AND LOCATION VALUES FOR RESIDENT DEER HUNTING IN UTAH . . UTAH AGRIC. EXPT. STAT. BULL. 488. IV, 24 PP. [MODEL; COST FUNCTION; MATRIX; REGRESSION]

WHITE, E. H. AND D. J. MEAD. 1971. DISCRIMINANT ANALYSIS IN TREE NUTRITION RESEARCH. FOREST SCI. 17(4), 425-427.

WHITNEY, R. R. AND K. D. CARLANDER. 1956. INTERPRETATION OF BODY-SCALE REGRESSION FOR COMPUTING BODY LENGTH OF FISH. J. WILDL. MANAGE. 20(1), 21-27.

WILIMOVSKY, N. J. AND E. C. WICKLUND. 1963. TABLES OF THE INCOMPLETE BETA FUNCTION FOR THE CALCULATION OF FISH POPULATION YIELD. INST. OF FISHERIES, UNIVERSITY OF BRITISH COLUMBIA, VANCOUVER, CANADA. III, 291 PP.

WORLUND, D. D. AND R. A. FREDIN. 1962. DIFFERENTIATION OF STOCKS. PP. 143-151. IN. SYMPOSIUM ON PINK SALMON. INSTITUTE OF FISHERIES, UNIVERSITY OF BRITISH COLUMBIA, VANCOUVER, BRITISH COLUMBIA, CANADA. [MODEL]

WORLUND, D. D., R. J. WAHLE AND P. D. ZIMMER. 1969. CONTRIBUTION OF COLUMBIA RIVER HATCHERIES TO HARVEST OF FALL CHINOOK SALMON (ONCORHYNCHUS TSHAWYTSCHA). U. S. FISH WILDL. SERV. FISH. BULL. 67(2), 361-391. [TAG-RECAPTURE; ESTIMATION]

YAMAMOTO, T. 1961. ON THE MECHANISM OF FISHING IN WHICH DISTRIBUTION CURVE OF CATCH PER UNIT EFFORT IS VERY LIKE LOGARITHMIC DISTRIBUTION. BULL. JAPANESE SOC. SCI. FISH. 27(2), 137-142. [POISSON DISTRIBUTION; VARIANCE]

YOSHIHARA, T. 1951. DISTRIBUTION OF FISHES CAUGHT BY THE LONG LINE II. VERTICAL DISTRIBUTION. BULL. JAPANESE SOC. SCI. FISH. 16(8), 370-374. [NORMAL DISTRIBUTION; EQUATIONS]

YOSHIHARA, T. 1952. ON BARANOV'S PAPER. BULL. JAPANESE SOC. SCI. FISH. 17(11), 363-366. [EXPLOITATION MODEL; LOGISTIC MODEL]

ZELLNER, A. 1962. ON SOME ASPECTS OF FISHERY CONSERVATION PROBLEMS. PP. 495-510. IN. ECONOMIC EFFECTS OF FISHERY REGULATION. (R. HAMLISCH. ED.). FAO FISHERIES REPORTS NO. 5. [SIGMOID CURVE; POPULATION MODEL; LOGISTIC FUNCTION; EXPLOITATION MODEL; ECONOMIC THEORY; COST FUNCTION]

FREQUENCY DISTRIBUTIONS

ABERDEEN, J. E. C. 1955. QUANTITATIVE METHODS FOR ESTIMATING THE DISTRIBUTION OF SOIL FUNGI. UNIV. QUEENSL. PAP. DEP. BOT. 3, 83-96.

ABRAHAMSEN, G. AND L. STRAND. 1970. STATISTICAL ANALYSIS OF POPULATION DENSITY DATA OF SOIL ANIMALS, WITH PARTICULAR REFERENCE TO ENCHYTRAEIDAE (OLIGOCHAETA). OIKOS 21(2), 276-284. [NEGATIVE BINOMIAL DISTRIBUTION; CONFIDENCE INTERVAL; COMPUTER; POISSON DISTRIBUTION; ANALYSIS OF VARIANCE; POLYA-AEPPLI DISTRIBUTION]

AHUJA, J. C. AND S. W. NASH. 1967. THE GENERALIZED GOMPERTZ-VERHULST FAMILY OF DISTRIBUTIONS. SANKHYA 29A(2), 141-156. [LOGISTIC]

ALLEN, K. R. 1959. THE DISTRIBUTION OF STREAM BOTTOM FAUNAS. PROC. N. Z. ECOL. SOC. 6, 5-8.

ANDERSEN, F. S. 1965. THE NEGATIVE BINOMIAL DISTRIBUTION AND THE SAMPLING OF INSECT POPULATIONS. PROC. INTERN. CONGR. ENTOMOL. 12, 395.

ANSCOMBE, F. J. 1949. THE STATISTICAL ANALYSIS OF INSECT COUNTS BASED ON THE NEGATIVE BINOMIAL DISTRIBUTION. BIOMETRICS 5(2), 165-173.

ANSCOMBE, F. J. 1950. SAMPLING THEORY OF THE NEGATIVE BINOMIAL AND LOGARITHMIC SERIES DISTRIBUTIONS. BIOMETRIKA 37(3/4), 358-382.

ARCHIBALD, E. E. A. 1948. PLANT POPULATIONS. I. A NEW APPLICATION OF NEYMAN'S CONTAGIOUS DISTRIBUTION. ANN. BOT. (LONDON) NEW SERIES 12(47), 221-235.

ASHBY, E. 1935. THE QUANTITATIVE ANALYSIS OF VEGETATION. ANN. BOT. (LONDON) 49(196), 779-802. [JACCARD'S COEFFICIENT; POISSON DISTRIBUTION; VARIANCE]

ASHBY, E. 1936. STATISTICAL ECOLOGY. BOT. REV. 2(5), 221-235. [POISSON DISTRIBUTION]

ASHBY, E. 1948. STATISTICAL ECOLOGY. II. A REASSESSMENT. BOT. REV. 14(4), 222-234.

BAGENAL, M. 1955. A NOTE ON THE RELATIONS OF CERTAIN PARAMETERS FOLLOWING A LOGARITHMIC TRANSFORMATION. J. MAR. BIOL. ASSOC. U. K. 34(2), 289-296. [LOG TRANSFORMATION; LOG-NORMAL DISTRIBUTION]

BAILEY, R. L. AND T. R. DELL. 1973. QUANTIFYING DIAMETER DISTRIBUTIONS WITH THE WEIBULL FUNCTION. FOREST SCI. 19(2), 97-104. [EXPONENTIAL FUNCTION; MAXIMUM LIKELIHOOD; SIMULATION]

BARNES, H. 1949. A STATISTICAL STUDY OF THE VARIATION IN VERTICAL PLANKTON HAULS, WITH SPECIAL REFERENCE TO THE LOSS OF THE CATCH WITH DIVIDED HAULS. J. MAR. BIOL. ASSOC. U. K. 28(2), 429-446. [CHI-SQUARE; COEFFICIENT OF VARIANCE; ANALYSIS OF VARIANCE; POISSON DISTRIBUTION]

BARNES, H. 1952. THE USE OF TRANSFORMATIONS IN MARINE BIOLOGICAL STATISTICS. J. CONS. PERM. INT. EXPLOR. MER 18(1), 61-71. [SQUARE ROOT; LOGARITHMIC; HYPERBOLIC; DISTRIBUTION; POISSON DISTRIBUTION; NEGATIVE BINOMIAL DISTRIBUTION; PEARSON'S TYPE III CURVE]

BARNES, H. 1953. A NOTE ON CUMULATIVE FREQUENCY CURVES. MEM. INST. ITAL. IDROBIOL. 7, 201-207. [NORMAL PROBABILITY PAPER]

BARNES, H. AND S. M. MARSHALL. 1951. ON THE VARIABILITY OF REPLICATE PLANKTON SAMPLES AND SOME APPLICATIONS OF 'CONTAGIOUS' SERIES TO THE STATISTICAL DISTRIBUTION OF CATCHES OVER RESTRICTED PERIODS. J. MAR. BIOL. ASSOC. U. K. 30(2), 233-263. [POISSON DISTRIBUTION; CHI-SQUARE; NEYMAN'S TYPE A DISTRIBUTION; THOMAS'S DISTRIBUTION; ANALYSIS OF VARIANCE; COEFFICIENT OF DISPERSION]

BARTON, D. E. AND F. N. DAVID, 1960. A CONGREGATING OF WORMS AND WOODLICE. TRABAJOS DE ESTADISTICA 11(3), 187-197. [INDEX OF DISPERSION; DISTRIBUTION; MULTINOMIAL DISTRIBUTION]

BEALL, G. 1940. THE FIT AND SIGNIFICANCE OF CONTAGIOUS DISTRIBUTIONS WHEN APPLIED TO OBSERVATIONS ON LARVAL INSECTS. ECOLOGY 21(4), 460-474. [NEYMAN'S TYPE A, B AND C DISTRIBUTIONS]

BEALL, G. 1942. THE TRANSFORMATION OF DATA FROM ENTOMOLOGICAL FIELD EXPERIMENTS SO THAT THE ANALYSIS OF VARIANCE BECOMES APPLICABLE. BIOMETRIKA 32(3/4), 243-262. [POISSON DISTRIBUTION; CHARLIER COEFFICIENT OF DISTURBANCE; SQUARE ROOT AND LOG TRANSFORMATION]

BEALL, G. 1954. DATA IN BINOMIAL OR NEAR-BINOMIAL DISTRIBUTION: WITH PARTICULAR APPLICATION TO PROBLEMS IN ENTOMOLOGICAL RESEARCH. PP. 295-302. IN, STATISTICS AND MATHEMATICS IN BIOLOGY. (O. KEMPTHORNE, T. A. BANCROFT, J. W. GOWEN AND J. L. LUSH, EDS.). IOWA STATE COLLEGE PRESS, AMES, IOWA.

BEALL, G. AND R. R. RESCIA. 1953. A GENERALIZATION OF NEYMAN"S CONTAGIOUS DISTRIBUTIONS. BIOMETRICS 9(3), 354-386.

BEAZLEY, R. AND SHIUE, C.-J. 1957. FURTHER APPLICATIONS OF SKEWNESS AND CENTRAL TENDENCY TESTS WITH THE RECTANGULAR DISTRIBUTION AS A CRITERION. FOREST SCI. 3(4), 321-328.

BERRYMAN, A. A. 1968. DISTRIBUTIONS OF SOLYTUS VENTRALIS ATTACKS, EMERGENCE, AND PARASITES IN GRAND FIR. CAN. ENTOMOL. 100(1), 57-68. [NEGATIVE BINOMIAL DISTRIBUTION; POISSON DISTRIBUTION; NORMAL; BINOMIAL]

BERTHET, P. AND G. GERARD. 1965. A STATISTICAL STUDY OF MICRODISTRIBUTION OF ORIBATEI (ACARI). PART I. THE DISTRIBUTION PATTERN. OIKOS 16(1/2), 214-227. [POISSON DISTRIBUTION; NEGATIVE BINOMIAL DISTRIBUTION]

BLACKMAN, G. E. 1935. A STUDY BY STATISTICAL METHODS OF THE DISTRIBUTION OF SPECIES IN GRASSLAND ASSOCIATIONS. ANN. BOT. (LONDON) 49(196), 749-778. [POISSON DISTRIBUTION; LOG TRANSFORMATION; REGRESSION; BINOMIAL DISTRIBUTION]

BLACKMAN, G. E. 1942. STATISTICAL AND ECOLOGICAL STUDIES IN THE DISTRIBUTION OF SPECIES IN PLANT COMMUNITIES. I; DISPERSION AS A FACTOR IN THE STUDY OF CHANGES IN PLANT POPULATIONS. ANN. BOT. (LONDON) NEW SERIES 6(22), 315-370. [POISSON DISTRIBUTION; COEFFICIENT OF DISPERSION]

BLISCHKE, W. R. 1965. MIXTURES OF DISCRETE DISTRIBUTIONS. PP. 351-372. IN, CLASSICAL AND CONTAGIOUS DISCRETE DISTRIBUTIONS. (G. P. PATIL, ED.). STATISTICAL PUBLISHING SOCIETY, CALCUTTA, INDIA. [POISSON DISTRIBUTION-NEGATIVE BINOMIAL DISTRIBUTION; MAXIMUM LIKELIHOOD; NEYMAN'S TYPE A, B AND C DISTRIBUTIONS; BINOMIAL DISTRIBUTION; PROBABILITY; POISSON DISTRIBUTION; NEGATIVE HYPERGEOMETRIC DISTRIBUTION]

BLISS, C. I. 1958. THE ANALYSIS OF INSECT COUNTS AS NEGATIVE BINOMIAL DISTRIBUTIONS. PROC. INTERN. CONGR. ENTOMOL., 10TH, 2, 1015-1032. [POISSON DISTRIBUTION]

BLISS, C. I. AND A. R. G. OWEN. 1958. NEGATIVE BINOMIAL DISTRIBUTIONS WITH A COMMON K. BIOMETRIKA 45(1/2), 37-58.

BLISS, C. I. AND K. A. REINKER. 1964. A LOGNORMAL APPROACH TO DIAMETER DISTRIBUTIONS IN EVEN-AGED STANDS. FOREST SCI. 10(3), 350-360.

BLISS, C. I. AND R. A. FISHER. 1953. FITTING THE NEGATIVE BINOMIAL DISTRIBUTION TO BIOLOGICAL DATA AND NOTE ON THE EFFICIENT FITTING OF THE NEGATIVE BINOMIAL. BIOMETRICS 9(2), 176-200.

BOSWELL, M. T. AND G. P. PATIL. 1971. CHANCE MECHANISMS GENERATING THE LOGARITHMIC SERIES DISTRIBUTION USED IN THE ANALYSIS OF NUMBER OF SPECIES AND INDIVIDUALS. PP. 99-125. IN, STATISTICAL ECOLOGY, VOLUME 1. SPATIAL PATTERNS AND STATISTICAL DISTRIBUTIONS. (G. P. PATIL, E. C. PIELOU AND W. E. WATERS, ED.). PENNSYLVANIA STATE UNIVERSITY PRESS, UNIVERSITY PARK, PENNSYLVANIA.

BOWDEN, D. C., A. E. ANDERSON AND D. E. MEDIN. 1969. FREQUENCY DISTRIBUTIONS OF MULE DEER FECAL GROUP COUNTS. J. WILDL. MANAGE. 33(4), 895-905. [NEYMAN'S TYPE A DISTRIBUTION; THOMAS'S DISTRIBUTION; NEGATIVE BINOMIAL DISTRIBUTION; MAXIMUM LIKELIHOOD; POISSON DISTRIBUTION]

BUCHANAN-WOLLASTON, H. J. 1934. LETTER TO THE EDITOR. J. CONS. PERM. INT. EXPLOR. MER 9(3), 394-396. [CONCERNING P. OTTESTAD'S COMMENTS; EXPONENTIAL DISTRIBUTION; NORMAL DISTRIBUTION; CURVE FITTING]

BUCHANAN-WOLLASTON, H. J. 1935. ON THE COMPONENT OF A FREQUENCY DISTRIBUTION ASCRIBABLE TO REGRESSION. J. CONS. PERM. INT. EXPLOR. MER 10(1), 81-98. [RANDOM SAMPLING; SKEWNESS; FREQUENCY CURVE; NORMAL DISTRIBUTION; CURVE FITTING]

BUCHANAN-WOLLASTON, H. J. 1935. THE PHILOSOPHIC BASIS OF STATISTICAL ANALYSIS. J. CONS. PERM. INT. EXPLOR. MER 10(3), 249-263. [METHOD OF LEAST SQUARES; NORMAL DISTRIBUTION; POISSON SERIES; BINOMIAL DISTRIBUTION; PEARSON'S TYPE III RANDOM SAMPLE; DEGREES FREEDOM; MAXIMUM LIKELIHOOD; TESTING]

BUCHANAN-WOLLASTON, H. J. 1936. THE PHILOSOPHIC BASIS OF STATISTICAL ANALYSIS. J. CONS. PERM. INT. EXPLOR. MER 11(1), 7-26. [CONTINUATION OF 10(3), 249-263; METHOD OF LEAST SQUARES; TESTING; SAMPLE SIZE; BINOMIAL DISTRIBUTION; POISSON DISTRIBUTION; MULTINOMIAL DISTRIBUTION; CORRELATION; REGRESSION; ANALYSIS OF VARIANCE]

BUCHANAN-WOLLASTON, H. J. AND W. C. HODGSON. 1929. A NEW METHOD OF TREATING FREQUENCY CURVES IN FISHERY STATISTICS, WITH SOME RESULTS. J. CONS. PERM. INT. EXPLOR. MER 4(2), 207-225. [PARABOLA; MODE; CURVE OF ERROR; CURVE FITTING]

CASSIE, R. M. 1950. THE ANALYSIS OF POLYMODAL FREQUENCY DISTRIBUTIONS BY THE PROBABILITY PAPER METHOD. N. Z. SCI. REV. 8, 89-91.

CASSIE, R. M. 1954. SOME USES OF PROBABILITY PAPER IN THE ANALYSIS OF SIZE FREQUENCY DISTRIBUTIONS. AUST. J. MAR. FRESHWATER RES. 5(3), 513-522.

CASSIE, R. M. 1961. STATISTICAL AND SAMPLING PROBLEMS IN PRIMARY PRODUCTION. PP. 163-171. IN, PROCEEDINGS OF THE CONFERENCE ON PRIMARY PRODUCTIVITY MEASUREMENT, MARINE AND FRESHWATER. (M. S. DOTY, ED.). UNIVERSITY OF HAWAII, HONOLULU, HAWAII. AUGUST 21-SEPTEMBER 6, 1961. U. S. AEC REPT. TID-7633. [SERIAL CORRELATION; TRANSFORMATION; REGRESSION; POISSON DISTRIBUTION; NORMAL DISTRIBUTION; LOG-NORMAL DISTRIBUTION; GEOMETRIC MEAN; NEGATIVE BINOMIAL DISTRIBUTION; MIXED POISSON DISTRIBUTION]

CASSIE, R. M. 1962. FREQUENCY DISTRIBUTION MODELS IN THE ECOLOGY OF PLANKTON AND OTHER ORGANISMS. J. ANIM. ECOL. 31(1), 65-92. [BINOMIAL DISTRIBUTION; CONTAGIOUS; POISSON DISTRIBUTION; LOG-NORMAL DISTRIBUTION; THOMAS'S DISTRIBUTION; NEYMAN'S DISTRIBUTION; NEGATIVE BINOMIAL DISTRIBUTION]

CASSIE, R. M. 1967. MATHEMATICAL MODELS FOR THE INTERPRETATION OF INSHORE PLANKTON COMMUNITIES. PP. 509-514. IN, ESTUARIES. (G. H. LAUFF, ED.). AMER. ASSOC. ADV. SCI., WASHINGTON, D. C. [POISSON DISTRIBUTION; LOG TRANSFORMATION; MATRIX]

CHAPMAN, D. G. 1951. SOME PROPERTIES OF THE HYPERGEOMETRIC DISTRIBUTION WITH APPLICATIONS TO ZOOLOGICAL SAMPLE CENSUSES. UNIV. CALIF. PUBL. STAT. 1(7), 131-160.

COLE, L. C. 1946. A THEORY FOR ANALYSING CONTAGIOUSLY DISTRIBUTED POPULATIONS. ECOLOGY 27(4), 329-341. [POISSON DISTRIBUTION; MIXTURE OF POISSONS]

COLEBROOK, J. M. 1960. PLANKTON AND WATER MOVEMENTS IN WINDERMERE. J. ANIM. ECOL. 29(2), 217-240. [NEGATIVE BINOMIAL DISTRIBUTION; MOVEMENT EQUATIONS; STATISTICAL PROPERTIES]

CUNIA, T. 1971. ON FREQUENCIES AND FREQUENCY FUNCTIONS. FOREST SCI. 17(3), 339-340. [FORESTRY]

CUSHING, D. H. 1962. PATCHINESS. RAPP. PROCES-VERB. REUNIONS J. CONS. PERM. INT. EXPLOR. MER 153, 152-163. [PLANKTON; DENSITY FUNCTION; DIFFUSION]

DARWIN, J. H. 1960. AN ECOLOGICAL DISTRIBUTION AKIN TO FISHER'S LOGARITHMIC DISTRIBUTION. BIOMETRICS 16(1), 51-60.

DAVID, F. N. AND R. G. MOORE. 1954. NOTES ON CONTAGIOUS DISTRIBUTIONS IN PLANT POPULATIONS. ANN. BOT. (LONDON) NEW SERIES 18(69), 47-53. [POISSON DISTRIBUTION; CLUMPING MODEL; NEGATIVE BINOMIAL DISTRIBUTION; THOMAS'S DISTRIBUTION]

DEBAUCHE, H. R. 1962. THE STRUCTURAL ANALYSIS OF ANIMAL COMMUNITIES OF THE SOIL. PP. 10-25. IN, PROGRESS IN SOIL ZOOLOGY. (P. W. MURPHY, ED.). BUTTERWORTHS, LONDON, ENGLAND. [INDEX OF CLUMPING; INDEX OF DISTURBANCE OF LEXIS; SAMPLING; DENSITY; NEGATIVE BINOMIAL DISTRIBUTION; NEYMAN'S DISTRIBUTION; DISTRIBUTION IN SPACE; POISSON DISTRIBUTION; AGGREGATION; DOMINANCE CORRELATION]

DICE, L. R. 1948. RELATIONSHIP BETWEEN FREQUENCY INDEX AND POPULATION DENSITY. ECOLOGY 29(3), 389-391. [POISSON DISTRIBUTION]

EVANS, D. A. 1953. EXPERIMENTAL EVIDENCE CONCERNING CONTAGIOUS DISTRIBUTIONS IN ECOLOGY. BIOMETRIKA 40(1/2), 186-211.

GAUCH, H. G., JR. AND G. B. CHASE. 1974. FITTING THE GAUSSIAN CURVE TO ECOLOGICAL DATA. ECOLOGY 55(6), 1377-1381. [PROBIT; LOG-NORMAL DISTRIBUTION; COMPUTER]

GHENT, A. W. 1966. BINOMIAL CORNER-ASSOCICATION ASSESSMENT OF CONTAGION: A RESPONSE TO CERTAIN CRITICISMS OF THE METHOD. TRANS. AMER. FISH. SOC. 95(4), 437-441.

GHENT, A. W. 1972. A METHOD FOR EXACT TESTING OF 2X2, 2X3, 3X3, AND OTHER CONTINGENCY TABLES, EMPLOYING BINOMIAL COEFFICIENTS. AMER. MIDL. NAT. 88(1), 15-27. [FISHER'S EXACT TEST; PROBABILITY]

GHENT, A. W. AND B. GRINSTEAD. 1965. A NEW METHOD OF ASSESSING CONTAGION, APPLIED TO A DISTRIBUTION OF REDEAR SUNFISH. TRANS. AMER. FISH. SOC. 94(2), 135-142.

GOODALL, D. W. 1952. QUANTITATIVE ASPECTS OF PLANT DISTRIBUTION. BIOL. REV. 27(2), 194-245. [POISSON DISTRIBUTION; POLYA DISTRIBUTION; NEYMAN'S DISTRIBUTION]

GUPTA, V. L. 1970. SELECTION OF FREQUENCY DISTRIBUTION MODELS. WATER RESOUR. RES. 6(4), 1193-1198.

GURLAND, J. AND P. HINZ. 1971. ESTIMATING PARAMETERS, TESTING FIT, AND ANALYZING UNTRANSFORMED DATA PERTAINING TO THE NEGATIVE BINOMIAL AND OTHER DISTRIBUTIONS. PP. 143-178. IN, STATISTICAL ECOLOGY, VOLUME 1. SPATIAL PATTERNS AND STATISTICAL DISTRIBUTIONS. (G. P. PATIL, E. C. PIELOU AND W. E. WATERS, EDS.). PENNSYLVANIA STATE UNIVERSITY PRESS, UNIVERSITY PARK, PENNSYLVANIA.

HARDING, J. P. 1949. THE USE OF PROBABILITY PAPER FOR GRAPHICAL ANALYSIS OF POLYMODAL FREQUENCY DISTRIBUTIONS. J. MAR. BIOL. ASSOC. U. K. 28(1), 141-153.

HARRIS, D. 1968. A METHOD OF SEPARATING TWO SUPERIMPOSED NORMAL DISTRIBUTIONS USING ARITHMETIC PROBABILITY PAPER. J. ANIM. ECOL. 37(2), 315-320. [GRAPHIC]

HOEL, P. G. 1943. ON INDICES OF DISPERSION. ANN. MATH. STAT. 14(2), 155-162. [POISSON DISTRIBUTION; BINOMIAL DISTRIBUTION; MOMENTS]

HOLGATE, P. 1964. A MODIFIED GEOMETRIC DISTRIBUTION ARISING IN TRAPPING STUDIES. ACTA THERIOL. 9(20), 353-356.

HOLMES, R. W. AND T. M. WIDRIG. 1956. THE ENUMERATION AND COLLECTION OF MARINE PHYTOPLANKTON. J. CONS. PERM. INT. EXPLOR. MER 22(1), 21-32. [NOMOGRAM; BINOMIAL DISTRIBUTION; POISSON DISTRIBUTION; EXPECTED NUMBER; NEGATIVE BINOMIAL DISTRIBUTION; PRECISION; SAMPLE SIZE]

HOPKINS, B. 1954. A NEW METHOD FOR DETERMINING THE TYPE OF DISTRIBUTION OF PLANT INDIVIDUALS. ANN. BOT. (LONDON) NEW SERIES 18(70), 213-226. [POISSON DISTRIBUTION]

HOZUMI, K. 1971. STUDIES ON THE FREQUENCY DISTRIBUTION OF THE WEIGHT OF INDIVIDUAL TREES IN A FOREST STAND III. A BETA-TYPE DISTRIBUTION. JAPANESE J. ECOL. 21(3/4), 152-167. [EQUATIONS]

HOZUMI, K. AND K. SHINOZAKI. 1970. STUDIES ON THE FREQUENCY DISTRIBUTION OF THE WEIGHT OF INDIVIDUAL TREES IN A FOREST STAND II. EXPONENTIAL DISTRIBUTION. JAPANESE J. ECOL. 20(1), 1-9.

HOZUMI, K. AND K. SHINOZAKI. 1974. STUDIES ON THE FREQUENCY DISTRIBUTION OF THE WEIGHT OF INDIVIDUAL TREES IN A FOREST STAND IV. ESTIMATION OF THE TOTAL FUNCTION OF A FOREST STAND AND A GENERALIZED MEAN PLANT. JAPANESE J. ECOL. 24(3), 207-212. [EQUATIONS]

HOZUMI, K., K. SHINOZAKI AND Y. TADAKI. 1968. STUDIES ON THE FREQUENCY DISTRIBUTION OF THE WEIGHT OF INDIVIDUAL TREES IN A FOREST STAND I. A NEW APPROACH TOWARD THE ANALYSIS OF THE DISTRIBUTION FUNCTION AND THE -3/2TH POWER DISTRIBUTION. JAPANESE J. ECOL. 18(1), 10-20.

HUGHES, R. D. 1962. THE STUDY OF AGGREGATED POPULATIONS. PP. 51-55. IN. PROGRESS IN SOIL ZOOLOGY. (P. W. MURPHY, ED.). BUTTERWORTHS, LONDON, ENGLAND. [RANDOM SAMPLING; PROBABILITY]

IPATOV, V. S., L. A. KIRIKOVA AND YU. I. SAMOILOV. 1974. SOME METHODOLOGICAL ASPECTS OF CONSTRUCTING ECOLOGICAL AMPLITUDES OF SPECIES. EKOLOGIYA 5(1), 13-23. (ENGLISH TRANSL. PP. 8-16). [QUARTILE; NORMAL DISTRIBUTION; CONFIDENCE INTERVAL]

JENSEN, P. 1959. FIT OF CERTAIN DISTRIBUTION FUNCTIONS TO COUNTS OF TWO SPECIES OF CRYPTOZOA ECOLOGY 40(3), 447-453. [NEGATIVE BINOMIAL DISTRIBUTION; POLYA DISTRIBUTION; NEYMAN'S TYPE A DISTRIBUTION; COLE'S THEORY]

KATTI, S. K. AND A. VIJAYA RAO. 1970. THE LOG-ZERO-POISSON DISTRIBUTION. BIOMETRICS 26(4), 801-813. [LIKELIHOOD FUNCTION AND EQUATIONS; ANIMAL AND PLANT EXAMPLES]

KEMP, A. W. 1973. THE EVALUATION OF SECOND-ORDER ENTROPY AND INACCURACY FOR CERTAIN DISCRETE DISTRIBUTIONS. BULL. INTERN. STAT. INST. 45(2), 45-51. [BINOMIAL DISTRIBUTION; POISSON DISTRIBUTION; NEGATIVE BINOMIAL DISTRIBUTION; SHANNON'S ENTROPY]

KEMP, C. D. 1971. PROPERTIES OF SOME DISCRETE ECOLOGICAL DISTRIBUTIONS. PP. 1-16. IN, STATISTICAL ECOLOGY, VOLUME 1. SPATIAL PATTERNS AND STATISTICAL DISTRIBUTIONS. (G. P. PATIL, E. C. PIELOU AND W. E. WATERS, EDS.). PENNSYLVANIA STATE UNIVERSITY PRESS, UNIVERSITY PARK, PENNSYLVANIA.

KOBAYASHI, S. 1968. POPULATION DENSITY AND DEGREE OF AGGREGATION WITH SPECIAL REFERENCE TO A COMMON K IN THE NEGATIVE BINOMIAL DISTRIBUTIONS. JAPANESE J. ECOL. 18(3), 120-124.

KOCH, A. L. 1966. THE LOGARITHM IN BIOLOGY. 1. MECHANISMS GENERATING THE LOG-NORMAL DISTRIBUTION EXACTLY. J. THEOR. BIOL. 12(2), 276-290.

KOCH, A. L. 1969. THE LOGARITHM IN BIOLOGY II. DISTRIBUTIONS SIMULATING THE LOG-NORMAL. J. THEOR. BIOL. 23(2), 251-268.

KRISHNA IYER, P. V. 1950. THE THEORY OF PROBABILITY DISTRIBUTIONS OF POINTS ON A LATTICE. ANN. MATH. STAT. 21, 198-217.

LEAK, W. B. 1965. THE J-SHAPED PROBABILITY DISTRIBUTION. FOREST SCI. 11(4), 405-409. [VARIANCE]

LINDSEY, J. K. AND F. W. NASH. 1972. LIKELIHOOD ANALYSIS OF THREE-WAY CONTINGENCY TABLES. J. FISH. RES. BOARD CAN. 29(5), 590-591. [ANALYSIS OF VARIANCE; MULTINOMIAL DISTRIBUTION; MAXIMUM LIKELIHOOD]

LONG, C. A. 1970. AN ANALYSIS OF PATTERNS OF VARIATION IN SOME REPRESENTATIVE MAMMALIA. PART III. SOME EQUATIONS OF THE NATURE OF FREQUENCY DISTRIBUTIONS OF ESTIMATED VARIABILITIES. ACTA THERIOL. 15(30), 517-528.

MANTEL, N. AND J. C. BAILAR, III. 1970. A CLASS OF PERMUTATIONAL AND MULTINOMIAL TESTS ARISING IN EPIDEMIOLOGICAL RESEARCH. BIOMETRICS 26(4), 687-700.

MAYFIELD, H. 1965. CHANCE DISTRIBUTION OF COWBIRD EGGS. CONDOR 67(3), 257-263. [PROBABILITY; POISSON DISTRIBUTION]

MCCONNELL, B. R. AND J. G. SMITH. 1970. FREQUENCY DISTRIBUTIONS OF DEER AND ELK PELLET GROUPS. J. WILDL. MANAGE. 34(1), 29-36. [MODEL; POISSON DISTRIBUTION; NEYMAN'S TYPE A DISTRIBUTION; THOMAS'S DOUBLE POISSON DISTRIBUTION; NEGATIVE BINOMIAL DISTRIBUTION]

MCGUIRE, J. U., T. A. BRINDLEY AND T. A. BANCROFT. 1957. THE DISTRIBUTION OF EUROPEAN CORNBORER LARVAE PYRAUSTA NUBILALIS (HBN). IN FIELD CORN. BIOMETRICS 13(1), 65-78. (ERRATA BIOMETRICS 14(3), 432-434).

MCNEIL, W. J. 1967. RANDOMNESS IN DISTRIBUTION OF PINK SALMON REDDS. J. FISH. RES. BOARD CAN. 24(7), 1629-1634. [POISSON DISTRIBUTION; NEGATIVE BINOMIAL DISTRIBUTION; COEFFICIENT OF DISPERSION]

MOSIMANN, J. E. 1962. ON THE COMPOUND MULTINOMIAL DISTRIBUTION, THE MULTIVARIATE BETA-DISTRIBUTION, AND CORRELATIONS AMONG PROPORTIONS. BIOMETRIKA 49(1/2), 65-82.

MOSIMANN, J. E. 1963. ON THE COMPOUND NEGATIVE MULTINOMIAL DISTRIBUTION AND CORRELATIONS AMONG INVERSELY SAMPLED POLLEN COUNTS. BIOMETRIKA 50(1/2), 47-57.

MOSIMANN, J. E. 1965. STATISTICAL METHODS FOR THE POLLEN ANALYST: MULTINOMIAL AND NEGATIVE MULTINOMIAL TECHNIQUES. PP. 636-673. IN. HANDBOOK OF PALEONTOLOGICAL TECHNIQUES. (B. KUMMEL AND D. RAUP, EDS.), W. H. FREEMAN AND CO., SAN FRANCISCO, CALIFORNIA.

MOSIMANN, J. E. 1970. DISCRETE DISTRIBUTION MODELS ARISING IN POLLEN STUDIES. PP. 1-30. IN, RANDOM COUNTS IN SCIENTIFIC WORK, VOLUME 3, RANDOM COUNTS IN PHYSICAL SCIENCES, GEOSCIENCE, AND BUSINESS. (G. P. PATIL, ED.), PENNSYLVANIA STATE UNIVERSITY PRESS, UNIVERSITY PARK, PENNSYLVANIA. [MULTINOMIAL DISTRIBUTION; NEGATIVE MULTINOMIAL DISTRIBUTION; POISSON DISTRIBUTION; LOG-NORMAL DISTRIBUTION]

MURAI, M. 1966. A STATISTICAL CONSIDERATION ON THE CHANGE OF THE DISTRIBUTION PATTERN BETWEEN TWO SUCCESSIVE STAGES. RES. POPUL. ECOL.-KYOTO 8(1), 8-13. [MODEL; NEGATIVE BINOMIAL DISTRIBUTION; POISSON DISTRIBUTION]

NEYMAN, J. 1939. ON A NEW CLASS OF "CONTAGIOUS" DISTRIBUTIONS APPLICABLE IN ENTOMOLOGY AND BACTERIOLOGY. ANN. MATH. STAT. 10(1), 35-57.

NEYMAN, J. 1950. HYPOTHESES RELATING TO THE HYPERGEOMETRIC RANDOM VARIABLE. FISH TAGGING EXPERIMENTS. PP. 335-338. IN, FIRST COURSE IN PROBABILITY AND STATISTICS, HENRY HOLT AND CO., NEW YORK, NEW YORK.

NEYMAN, J. 1965. CERTAIN CHANCE MECHANISMS INVOLVING DISCRETE DISTRIBUTIONS. PP. 4-14. IN, CLASSICAL AND CONTAGIOUS DISCRETE DISTRIBUTIONS. (G. P. PATIL, ED.). STATISTICAL PUBLISHING SOCIETY, CALCUTTA, INDIA. [GEOMETRIC DISTRIBUTION; CLUSTERING NEGATIVE BINOMIAL DISTRIBUTION]

OTTESTAD, P. 1934. LETTER TO THE EDITOR. J. CONS. PERM. INT. EXPLOR. MER 9(2), 249-253. [CONCERNING BUCHANAN-WOLLASTON'S PAPER; FREQUENCY FUNCTIONS; MEAN; EXPONENTIAL DISTRIBUTION; CURVE FITTING; NORMAL DISTRIBUTION]

PALOHEIMO, J. E. 1972. A SPATIAL BIVARIATE POISSON DISTRIBUTION. BIOMETRIKA 59(2), 489-492. [NEAREST NEIGHBOR; MODEL OF ASSOCIATION]

PATIL, G. P. AND R. W. SHORROCK. 1966. STOCHASTIC AND SAMPLING PROCESSES FOR THE LOGRITHMIC SERIES DISTRIBUTION. U. S. AIR FORCE, OFFICE OF AEROSPACE RESEARCH, WRIGHT-PATTERSON AIR FORCE BASE, OHIO. REPORT ARL 66-0147, IV, 82 PP.

PATIL, G. P. AND S. BILDIKAR. 1966. CERTAIN STUDIES ON THE MULTIVARIATE LOGARITHMIC SERIES DISTRIBUTION. U. S. AIR FORCE, OFFICE OF AEROSPACE RESEARCH, WRIGHT-PATTERSON AIR FORCE BASE, OHIO. REPORT ARL 66-0148, IV, 41 PP.

PATIL, G. P. AND S. BILDIKAR. 1967. MULTIVARIATE LOGARITHMIC SERIES DISTRIBUTION AS A PROBABILITY MODEL IN POPULATION AND COMMUNITY ECOLOGY AND SOME OF ITS STATISTICAL PROPERTIES. J. AMER. STAT. ASSOC. 62(318), 655-674.

PIELOU, E. C. 1960. A SINGLE MECHANISM TO ACCOUNT FOR REGULAR, RANDOM AND AGGREGATED POPULATIONS. J. ECOL. 48(3), 575-584. [CHI-SQUARE; ARTIFICIAL POPULATION; DISTANCE MEASUREMENTS; SAMPLING; MODEL]

RAO, C. R. 1971. SOME COMMENTS ON THE LOGARITHMIC SERIES DISTRIBUTION IN THE ANALYSIS OF INSECT TRAP DATA. PP. 131-141. IN, STATISTICAL ECOLOGY, VOLUME 1, SPATIAL PATTERNS AND STATISTICAL DISTRIBUTIONS. (G. P. PATIL, E. C. PIELOU AND W. E. WATERS, EDS.). PENNSYLVANIA STATE UNIVERSITY PRESS, UNIVERSITY PARK, PENNSYLVANIA.

ROBINSON, P. 1954. THE DISTRIBUTION OF PLANT POPULATIONS. ANN. BOT. (LONDON) NEW SERIES 18(69), 35-45. [POISSON DISTRIBUTION; LOG-NORMAL DISTRIBUTION; NEGATIVE BINOMIAL DISTRIBUTION; NEYMAN'S DISTRIBUTION; THOMAS'S DISTRIBUTION; PEARSON TYPE I DISTRIBUTION]

SERFLING, R. E. 1949. QUANTITATIVE ESTIMATION OF PLANKTON FROM SMALL SAMPLES OF SEDGEWICK-RAFTER-CELL MOUNTS OF CONCENTRATE SAMPLES. TRANS. AMER. MICROSC. SOC. 68(3), 185-199. [POISSON DISTRIBUTION; TRANSFORMATION; ANALYSIS OF VARIANCE; PROBABILITY; SAMPLING; FISHER'S T; U-TEST]

SEVERS, R. K. 1971. USE OF STATISTICAL TECHNIQUES TO ASSESS PROGRESS TOWARD A CLEAN AIR ENVIRONMENT: HOUSTON, TEXAS. ATMOS. ENVIRON. 5(10), 853-861. [COMPUTER; GEOMETRIC MEAN]

SKELLAM, J. G. 1954. APPENDIX. (B. HOPKINS, A NEW METHOD FOR DETERMINING THE TYPE OF DISTRIBUTION OF PLANT INDIVIDUALS). ANN. BOT. (LONDON) NEW SERIES 18(70), 226-227.

SKELLAM, J. G. 1958. ON THE DERIVATION AND APPLICABILITY OF NEYMAN'S TYPE A DISTRIBUTION. BIOMETRIKA 45(1/2), 32-36.

SMITH, J. H. G. AND J. W. KER. 1957. SOME DISTRIBUTIONS ENCOUNTERED IN SAMPLING FOREST STANDS. FOREST SCI. 3(2), 137-144.

TALLIS, G. M. AND A. D. DONALD. 1964. MODELS FOR THE DISTRIBUTION ON PASTURE OF INFECTIVE LARVAE OF THE GASTROINTESTINAL NEMATODE PARASITES OF SHEEP. AUST. J. BIOL. SCI. 17, 504-513. [POISSON DISTRIBUTION; NEGATIVE BINOMIAL DISTRIBUTION; PROBABILITY GENERATING FUNCTION]

TALLIS, G. M. AND A. D. DONALD. 1970. FURTHER MODELS FOR THE DISTRIBUTION ON PASTURE OF INFECTIVE LARVAE OF THE STRONGYLOID NEMATODE PARASITES OF SHEEP. MATH. BIOSCI. 7, 179-190. [PROBABILITY]

TANAKA, S. 1962. A METHOD OF ANALYSING A POLYMODAL FREQUENCY DISTRIBUTION AND ITS APPLICATION TO THE LENGTH DISTRIBUTION OF THE PORGY, TAIUS TUMIFRONS (T. AND S.). J. FISH. RES. BOARD CAN. 19(6), 1143-1159. [SUBSAMPLING]

TAYLOR, B. J. R. 1965. THE ANALYSIS OF POLYMODAL FREQUENCY DISTRIBUTIONS. J. ANIM. ECOL. 34(2), 445-452. [MOVING AVERAGES; GAUSSIAN DISTRIBUTION]

TAYLOR, L. R. 1965. A NATURAL LAW FOR THE SPATIAL DISPOSITION OF INSECTS. PROC. INTERN. CONGR. ENTOMOL. 12, 396-397. [DISTRIBUTION; POWER LAW; TRANSFORMATION FUNCTION; VARIANCE]

TAYLOR, L. R. 1971. AGGREGATION AS A SPECIES CHARACTERISTIC. PP. 357-372. IN. STATISTICAL ECOLOGY, VOLUME 1. SPATIAL PATTERNS AND STATISTICAL DISTRIBUTIONS. (G. P. PATIL, E. C. PIELOU AND W. E. WATERS, EDS.). PENNSYLVANIA STATE UNIVERSITY PRESS, UNIVERSITY PARK, PENNSYLVANIA. [POISSON DISTRIBUTION]

THOMAS, M. 1949. A GENERALIZATION OF POISSON'S BINOMIAL LIMIT FOR USE IN ECOLOGY. BIOMETRIKA 36(1/2), 18-25. [DOUBLE POISSON DISTRIBUTION]

THOMAS, M. 1951. SOME TESTS FOR RANDOMNESS IN PLANT POPULATIONS. BIOMETRIKA 38(1/2), 102-111. [DOUBLE POISSON DISTRIBUTION; INDEX OF DISPERSION; MAXIMUM LIKELIHOOD]

THOMPSON, H. R. 1954. A NOTE ON CONTAGIOUS DISTRIBUTIONS. BIOMETRIKA 41(1/2), 268-271. [POISSON DISTRIBUTION; NEYMAN'S TYPE A AND C DISTRIBUTIONS; THOMAS'S DISTRIBUTION; DARWIN'S MODEL; INDEX OF DISPERSION]

THOMSON, G. W. 1952. MEASURES OF PLANT AGGREGATION BASED ON CONTAGIOUS DISTRIBUTION. CONTRIB. LAB. VERTEBR. BIOL., UNIVERSITY OF MICHIGAN, ANN ARBOR, MICHIGAN, NO. 53. 17 PP. [NEYMAN'S TYPE A DISTRIBUTION; THOMAS'S DISTRIBUTION; POISSON DISTRIBUTION; MAXIMUM LIKELIHOOD; MOMENTS]

TRAJSTMAN, A. C. 1973. THE NECESSITY OF THE POISSON DISTRIBUTION FOR THE EQUIVALENCE OF SOME RANDOM MATING MODELS. MATH. BIOSCI. 17, 1-10.

VAN DER PLANK, J. E. 1946. A METHOD FOR ESTIMATING THE NUMBER OF RANDOM GROUPS OF ADJACENT DISEASED PLANTS IN A HOMOGENEOUS FIELD. TRANS. R. SOC. S. AFR. 31(3), 269-278. [BINOMIAL SERIES TEST; GEOMETRIC SERIES TEST; POISSON DISTRIBUTION; RANDOM]

WADLEY, F. M. 1950. NOTES ON THE FORM OF DISTRIBUTION OF INSECT AND PLANT POPULATIONS. ANN. ENTOMOL. SOC. AMER. 43(4), 581-586. [POISSON SERIES; NEGATIVE BINOMIAL DISTRIBUTION; TRANSFORMATION; NEYMAN'S DISTRIBUTION]

WATERS, W. E. 1959. A QUANTITATIVE MEASURE OF AGGREGATION IN INSECTS. J. ECON. ENTOMOL. 52(6), 1180-1184. [NEGATIVE BINOMIAL DISTRIBUTION]

WATERS, W. E. AND W. R. HENSON. 1959. SOME SAMPLING ATTRIBUTES OF THE NEGATIVE BINOMIAL DISTRIBUTION WITH SPECIAL REFERENCE TO FOREST INSECTS. FOREST SCI. 5(4), 397-412.

WIERZBOWSKA, T. 1970. TRUNCATED DISTRIBUTIONS AND THEIR APPLICATION IN ECOLOGY. EKOLOGIYA POLSKA-SERIA A 18(40), 837-848. [GEOMETRICAL DISTRIBUTION; EXPONENTIAL DISTRIBUTION; TAG-RECAPTURE]

WILLIAMS, C. B. 1947. THE LOGARITHMIC SERIES AND ITS APPLICATION TO BIOLOGICAL PROBLEMS. J. ECOL. 34(2), 253-272.

YOSHIHARA, T. 1957. ON THE TYPE OF FREQUENCY CURVE OF THE CATCH - II. BULL. JAPANESE SOC. SCI. FISH. 22(10), 618-620. [LOGARITHMIC NORMAL CURVE]

ZOHRER, F. 1972. THE BETA-DISTRIBUTION FOR BEST FIT OF STEM-DIAMETER-DISTRIBUTIONS. PP. 91-106. IN. 3RD CONFERENCE ADVISORY GROUP OF FOREST STATISTICIANS. INSTITUT NATIONAL DE LA RECHERCHE AGRONOMIQUE, PARIS, FRANCE. I.N.R.A. PUBL. 72-3. [MODEL; UNIMODAL DISTRIBUTIONS; DECREASING DISTRIBUTIONS]

GROWTH OF INDIVIDUALS

ALLEN, K. R. 1966. A METHOD OF FITTING GROWTH CURVES OF THE VON BERTALANFFY TYPE TO OBSERVED DATA. J. FISH. RES. BOARD CAN. 23(2), 163-179.

ALLEN, K. R. 1969. APPLICATION OF THE BERTALANFFY GROWTH EQUATION TO PROBLEMS OF FISHERIES MANAGEMENT: A REVIEW. J. FISH. RES. BOARD CAN. 26(9), 2267-2281.

ARABINA, I. P. 1971. GROWTH OF BITHYNIA TENTACULATA (L.). HYDROBIOL. J. 7(2), 105-107. (ENGLISH TRANSL.). [GROWTH EQUATIONS]

BAGENAL, T. B. 1957. THE BREEDING AND FECUNDITY OF THE LONG ROUGH DAB HIPPOGLOSSOIDES PLATESSOIDES (FABR.) AND THE ASSOCIATED CYCLE IN CONDITION. J. MAR. BIOL. ASSOC. U. K. 36(2), 339-375. [GROWTH EQUATIONS; REGRESSION]

BAKER, G. A. 1943. LENGTH-GROWTH CURVES FOR THE RAZOR CLAM. GROWTH 7(4), 439-443.

BEHNKEN, D. W. AND D. G. WATTS. 1972. BAYESIAN ESTIMATION AND DESIGN OF EXPERIMENTS FOR GROWTH RATES WHEN SAMPLING FROM THE POISSON DISTRIBUTION. BIOMETRICS 28(4), 999-1009. [ALGAE]

BERNSTEIN, F. 1934. GROWTH AND DECAY. COLD SPRING HARBOR SYMPOSIA ON QUANTITATIVE BIOLOGY 2, 209-216.

BHATTACHARYA, C. G. 1966. FITTING A CLASS OF GROWTH CURVES. SANKHYA 28B(1/2), 1-10. [VON BERTALANFFY CURVE; GOMPERTZ CURVE]

BLACKITH, R. E. 1960. A SYNTHESIS OF MULTIVARIATE TECHNIQUES TO DISTINQUISH PATTERNS OF GROWTH IN GRASSHOPPERS. BIOMETRICS 16(1), 28-40.

BLUMBERG, A. A. 1968. LOGISTIC GROWTH RATE FUNCTIONS. J. THEOR. BIOL. 21(1), 42-44.

BOWDEN, D. C. AND R. K. STEINHORST. 1973. TOLERANCE BANDS FOR GROWTH CURVES. BIOMETRICS 29(2), 361-371. [FISH; MULTIVARIATE NORMAL TOLERANCE ELLIPSOIDS]

BRYUZGIN, V. L. 1970. USE OF EMPERICAL SCALES TO STUDY THE GROWTH OF FISHES. HYDROBIOL. J. 6(1), 85-94. (ENGLISH TRANSL.)

BUCHANAN-WOLLASTON, H. J. 1934. THE THEORY OF VARIATION, CORRELATION AND REGRESSION. ITS RELEVANCE IN RESEARCHES ON PROPORTIONAL GROWTH. RAPP. PROCES-VERB. REUNIONS J. CONS. PERM. INT. EXPLOR. MER 89(3), 33-44.

CAMPBELL, N. A. AND B. F. PHILLIPS. 1972. THE VON BERTALANFFY GROWTH CURVE AND ITS APPLICATION TO CAPTURE-RECAPTURE DATA IN FISHERIES BIOLOGY. J. CONS. PERM. INT. EXPLOR. MER 34(2), 295-299.

CARLANDER, K. D. AND L. L. SMITH, JR. 1944. SOME USES OF NOMOGRAPHS IN FISH GROWTH STUDIES. COPEIA 1944(3), 157-162. [EQUATIONS]

CLUTTER, J. L. 1963. COMPATIBLE GROWTH AND YIELD MODELS FOR LOBLOLLY PINE. FOREST SCI. 9(3), 354-371.

CLYMO, R. S. 1970. THE GROWTH OF SPHAGNUM: METHODS OF MEASUREMENT. J. ECOL. 58(1), 13-49. [EQUATIONS]

COCK, A. G. 1966. GENETICAL ASPECTS OF METRICAL GROWTH AND FORM IN ANIMALS. QUART. REV. BIOL. 41(2), 131-190. [MULTIVARIATE ANALYSIS; ALLOMETRY; MODEL; ANALYSIS OF VARIANCE; ANALYSIS OF COVARIANCE;]

COHEN, D. 1971. MAXIMIZING FINAL YIELD WHEN GROWTH IS LIMITED BY TIME OR BY LIMITING RESOURCES. J. THEOR. BIOL. 33(2), 299-307. [MODEL; PLANTS]

COLLOT, F. 1968. BIOMATHEMATICAL INTERPRETATION OF ORGANISMIC GROWTH. PP. 133-135. IN, QUANTITATIVE BIOLOGY OF METABOLISM. (A. LOCKER, ED.). SPRINGER-VERLAG NEW YORK INC., NEW YORK, NEW YORK.

COOMBE, D. E. 1960. AN ANALYSIS OF THE GROWTH OF TREMA QUINEENSIS. J. ECOL. 48(1), 219-231. [ASSIMILATION RATE]

COOPER, C. F. 1961. EQUATIONS FOR THE DESCRIPTION OF PAST GROWTH IN EVEN-AGED STANDS OF PONDEROSA PINE. FOREST SCI. 7(1), 72-80. [LOG TRANSFORMATION; LOGISTIC]

CURTIS, R. O. 1964. A STEM-ANALYSIS APPROACH TO SITE-INDEX CURVES. FOREST SCI. 10(2), 241-256. [GROWTH CURVES]

CZARNOWSKI, M. S. 1968. NOTES ON GROWTH IN STAND-COMPACTNESS AS A FUNCTION OF GROWTH IN HEIGHT. EKOLOGIYA POLSKA-SERIA A 16(42), 833-841. [EQUATIONS]

DAVIDSON, F. A. 1930. GRAPHICAL AND MATHEMATICAL TREATMENTS IN GROWTH STUDIES. PP. 246-252. IN, CONTRIBUTIONS TO MARINE BIOLOGY. STANFORD UNIVERSITY PRESS, STANFORD UNIVERSITY, CALIFORNIA. [GROWTH CURVES]

DAVIDSON, J. 1944. ON THE RELATIONSHIP BETWEEN TEMPERATURE AND RATE OF DEVELOPMENT OF INSECTS AT CONSTANT TEMPERATURES. J. ANIM. ECOL. 13(1), 26-38. [FORM OF LOGISTIC CURVE]

DEAKIN, M. A. B. 1970. GOMPERTZ CURVES, ALLOMETRY AND EMBRYOGENESIS. BULL. MATH. BIOPHYS. 32, 445-452. [GROWTH CURVES; VON BERTALANFFY CURVE]

DIAZ, E. L. 1963. AN INCREMENT TECHNIQUE FOR ESTIMATING GROWTH PARAMETERS OF TROPICAL TUNAS, AS APPLIED TO YELLOWFIN TUNA (THUNNUS ALBACARES). BULL. INTER-AMER. TROP. TUNA COMM. 8(7), 383-405. [VON BERTALANFFY'S EQUATION]

EBERHARDT, L. L. AND R. E. NAKATANI. 1968. A POSTULATED EFFECT OF GROWTH ON RETENTION TIME OF METABOLITES. J. FISH. RES. BOARD CAN. 25(3), 591-596. [MODEL; LOG TRANSFORMATION]

EBERT, T. A. 1973. ESTIMATING GROWTH AND MORTALITY RATES FROM SIZE DATA. OECOLOGIA 11(3), 281-298. [MODEL; BRODY-BERTALANFFY GROWTH FORMULA]

FABENS, A. J. 1965. PROPERTIES AND FITTING OF THE VON BERTALANFFY GROWTH CURVE. GROWTH 29(3), 265-289. [COMPUTER]

FARRIS, D. A. 1960. THE EFFECT OF THREE DIFFERENT TYPES OF GROWTH CURVES ON ESTIMATES OF LARVAL FISH SURVIVAL. J. CONS. PERM. INT. EXPLOR. MER 25(3), 294-306.

FELLER, W. 1939. ON THE LOGISTIC LAW OF GROWTH AND ITS EMPIRICAL VERIFICATIONS IN BIOLOGY. ACTA BIOTHEOR. 5(2), 51-66.

FIL'ROZE, E. M. AND T. M. SHMEL'KOVA. 1971. DYNAMICS OF TREE GROWTH AND SOME METHODS OF ITS MATHEMATICAL DESCRIPTION. EKOLOGIYA 2(2), 15-26. (ENGLISH TRANSL. PP. 101-109.). [LOGISTIC CURVE; GAUSSIAN CURVE; BACKMAN FUNCTION]

FRETWELL, S. D., D. E. BOWEN AND H. A. HESPENHEIDE. 1974. GROWTH RATES OF YOUNG PASSERINES AND THE FLEXIBILITY OF CLUTCH SIZE. ECOLOGY 55(4), 907-909. [MODEL; RICKLEF'S MODEL]

FURNIVAL, G. M. AND R. W. WILSON, JR. 1971. SYSTEMS OF EQUATIONS FOR PREDICTING FOREST GROWTH AND YIELD. PP. 43-56. IN, STATISTICAL ECOLOGY, VOLUME 3. MANY SPECIES POPULATIONS, ECOSYSTEMS, AND SYSTEMS ANALYSIS. (G. P. PATIL, E. C. PIELOU AND W. E. WATERS, EDS.). PENNSYLVANIA STATE UNIVERSITY PRESS, UNIVERSITY PARK, PENNSYLVANIA.

GARROD, D. J. 1963. THE APPLICATION OF A METHOD FOR THE ESTIMATION OF GROWTH PARAMETERS FROM TAGGING DATA AT UNEQUAL TIME INTERVALS. PP. 258-261. IN, NORTH ATLANTIC FISH MARKING SYMPOSIUM. INTERN. COMM. NORTHWEST ATLANTIC FISH. SPEC. PUBL. NO. 4. [VON BERTALANFFY'S EQUATION]

GLENDAY, A. C. 1955. THE MATHEMATICAL SEPARATION OF PLANT AND WEATHER EFFECTS IN FIELD GROWTH STUDIES. AUST. J. AGRIC. RES. 6(6), 813-822. [MODEL; ORTHOGONAL; NON-ORTHOGONAL; NORMAL EQUATIONS; ANALYSIS OF VARIANCE]

GOLDSTEIN, R. A. AND J. B. MANKIN. 1972. SPACE-TIME CONSIDERATIONS IN MODELING THE DEVELOPMENT OF VEGETATION. PP. 87-97. IN, MODELING THE GROWTH OF TREES. (C. E. MURPHY, JR., J. D. HESKETH AND B. R. STRAIN, EDS.). AEC REPT. EDFB-IBP-72-11. OAK RIDGE NATL. LAB., OAK RIDGE, TENNESSEE. [MODEL]

GOULD, S. J. 1971. GEOMETRIC SIMILARITY IN ALLOMETRIC GROWTH; A CONTRIBUTION TO THE PROBLEM OF SCALING IN THE EVALUATION OF SIZE. AMER. NAT. 105(942), 113-136.

GRAY, J. 1929. THE KINETICS OF GROWTH. BRIT. J. EXPTL. BIOL. 6(3), 248-274. [EXPONENTIAL; LOGARITHMIC CURVE; SIGMOID CURVE]

GROSENBAUGH, L. R. 1965. GENERALIZATION AND REPARAMETERIZATION OF SOME SIGMOID AND OTHER NONLINEAR FUNCTIONS. BIOMETRICS 21(3), 708-714. [COMPUTER]

GULLAND, J. A. AND S. J. HOLT. 1959. ESTIMATION OF GROWTH PARAMETERS FOR DATA AT UNEQUAL TIME INTERVALS. J. CONS. PERM. INT. EXPLOR. MER 25(1), 47-49. [VON BERTALANFFY'S EQUATION; FORD-WALFORD PLOT]

HANCOCK, D. A. 1965. GRAPHICAL ESTIMATION OF GROWTH PARAMETERS. J. CONS. PERM. INT. EXPLOR. MER 29(3), 340-351. [FORD-WALFORD PLOT]

HAYASHI, S. AND K. KONDO. 1962. GROWTH OF THE JAPANESE ANCHOVEY I. SEASONAL FLUCTUATION IN THE CONDITION COEFFICIENT. BULL. TOKAI REG. FISH. RES. LAB. 9, 179-192. [VARIANCE; ANALYSIS OF COVARIANCE; REGRESSION]

HILE, R. 1970. BODY-SCALE RELATION AND CALCULATION OF GROWTH IN FISHES. TRANS. AMER. FISH. SOC. 99(3), 468-474.

HOEL, P. G. 1964. METHODS FOR COMPARING GROWTH TYPE CURVES. BIOMETRICS 20(4), 859-872.

ITO, T. 1953. ON THE STATISTICAL METHOD OF WEIGHT-LENGTH RELATIONSHIP IN FISH POPULATION. BULL. JAPANESE SOC. SCI. FISH. 19(8), 905-911. [ALLOMETRY; COEFFICIENT OF FATNESS]

JOHNSON, L. 1966. EXPERIMENTAL DETERMINATION OF FOOD CONSUMPTION OF PIKE, ESOX LUCIUS, FOR GROWTH AND MAINTENANCE. J. FISH. RES. BOARD CAN. 23(10), 1495-1505. [EQUATIONS; BEVERTON-HOLT EQUATION; KOSTITZIN]

JOLICOEUR, P. 1963. THE MULTIVARIATE GENERALIZATION OF THE ALLOMETRY EQUATION. BIOMETRICS 19(3), 497-499.

JONES, R. 1958. LEE'S PHENOMENON OF "APPARENT CHANGE IN GROWTH-RATE" WITH PARTICULAR REFERENCE TO HADDOCK AND PLAICE. PP. 229-242. IN, SOME PROBLEMS FOR BIOLOGICAL FISHERY SURVEY AND TECHNIQUES FOR THEIR SOLUTION. INTERN. COMM. NORTHWEST ATLANTIC FISH. SPEC. PUBL. NO. 1. [GROWTH EQUATIONS; PROBABILITY OF SURVIVAL; MODEL]

KERR, S. R. 1971. ANALYSIS OF LABORATORY EXPERIMENTS ON GROWTH EFFICIENCY OF FISHES. J. FISH. RES. BOARD CAN. 28(6), 801-808. [MODEL]

KILMER, W. L. 1971. ON GROWING PINE CONES AND OTHER FIBONACCI FRUITS-MCCULLOCH'S LOCALIZED ALGORITHM. MATH. BIOSCI. 11, 53-57. [FIBONACCI NUMBERS]

KITAHARA, T. 1966. VARIATION OF LENGTH COMPOSITION DUE TO CHANGE IN INTENSITY OF EXPLOITATION TO A STOCK-I. BULL. JAPANESE SOC. SCI. FISH. 32(3), 242-247. [VON BERTALANFFY'S GROWTH CURVES; MODEL]

KNIGHT, W. 1968. ASYMPTOTIC GROWTH: AN EXAMPLE OF NONSENSE DISGUISED AS MATHEMATICS. J. FISH. RES. BOARD CAN. 25(6), 1303-1307. [LOGISTIC; VON BERTALANFFY GROWTH EQUATION]

KNIGHT, W. 1969. A FORMULATION OF THE VON BERTALANFFY GROWTH CURVE WHEN THE GROWTH RATE IS ROUGHLY CONSTANT. J. FISH. RES. BOARD CAN. 26(11), 3069-3076.

KODAMA, Y. 1958. STUDIES ON THE SPECIFIC GROWTH OF FRESHWATER FISHES REFERRING TO THE WEIGHT INCREASE OF INDIVIDUAL AND POPULATION - I. VARIATION OF THE SPECIFIC GROWTH AND THE RELATION BETWEEN THE AVERAGE AND THE INDIVIDUAL SPECIFIC GROWTH BY A SMALL POPULATION OF THE BROWN TROUT (SALMO TRUTTA L.) REARED IN POND. BULL. JAPANESE SOC. SCI. FISH. 24(6/7), 435-440. [GROWTH EQUATIONS]

KOLESNIK, YU. A. 1972. SOME QUESTIONS RELATING TO GROWTH AND METABOLIC RATE IN ANIMALS. J. ICHTHYOLOGY 12(5), 860-863. (ENGLISH TRANSL.). [EQUATIONS; TAYLOR AND VINBERG GROWTH CURVES; VON BERTALANFFY'S GROWTH CURVES; 'SHMAL'GAUZE GROWTH CURVES;N]

KOSTITZIN, V. A. 1970. SUR LA LOI LOGISTIQUE ET SES GENERALIZATIONS. ACTA BIOTHEOR. (3), 155-159. [LOGISTIC]

RUGER, F. 1968. CONTRIBUTIONS O THE ENERGETICS OF ANIMAL GROWTH. P. 113-122. IN. QUANTITATIVE IOLOGY OF METABOLISM. (A. LOCKER, D.). SPRINGER-VERLAG NEW YORK NC., NEW YORK, NEW YORK. [VON ERTALANFFY'S EQUATION; UTTER-BERTALANFFY'S EQUATION]

AIRD, A. K., A. D. BARTON AND S. . TYLER. 1968. GROWTH AND IME: AN INTERPRETATION OF LLOMETRY. GROWTH 32, 347-354.

LEA, E. 1938. A MODIFICATION OF THE FORMULA FOR CALCULATION OF THE GROWTH OF HERRING. RAPP. PROCES-VERB. REUNIONS J. CONS. PERM. INT. EXPLOR. MER 108(1), 13-22.

LEMMON, P. E. AND F. X. SCHUMACHER. 1963. THEORETICAL GROWTH AND YIELD OF HYPOTHETICAL PONDEROSA PINE STANDS UNDER DIFFERENT THINNING REGIMES. FOREST SCI. 9(1), 33-43. [MODEL]

LINDNER, M. J. 1953. ESTIMATION OF GROWTH RATE IN ANIMALS BY MARKING EXPERIMENTS. U. S. FISH WILDL. SERV. FISH. BULL. 54, 65-69.

LOVTRUP, S. AND B. VON SYDOW. 1974. D'ARCY THOMPSON'S THEOREMS AND THE SHAPE OF MOLLUSCAN SHELL. BULL. MATH. BIOL. 36(5/6), 567-575. [MODEL]

LUMER, H. 1939. THE DIMENSIONS AND INTERRELATIONSHIP OF THE RELATIVE GROWTH CONSTANTS. AMER. NAT. 73(747), 339-346. [EQUATIONS]

MAKAROVA, N. P. AND V. YE. ZAIKA. 1971. RELATIONSHIP BETWEEN ANIMAL GROWTH AND QUANTITY OF ASSIMILATED FOOD. HYDROBIOL. J. 7(3), 1-8. (ENGLISH TRANSL.)

MURPHY, C. E., JR., J. D. HESKETH AND B. R. STRAIN. (EDS.). 1972. MODELING THE GROWTH OF TREES. PROC. OF WORKSHOP ON TREE GROWTH DYNAMICS AND MODELING, DUKE UNIVERSITY, OCTOBER 11-12, 1971. AEC REPT. EDFB-IBP-72-11. OAK RIDGE NATL. LAB., OAK RIDGE, TENNESSEE. VI, 199 PP.

NAIR, K. R. 1954. THE FITTING OF GROWTH CURVES. PP. 119-132. IN. STATISTICS AND MATHEMATICS IN BIOLOGY. (O. KEMPTHORNE, T. A. BANCROFT, J. W. GOWEN AND J. L. LUSH, EDS.). IOWA STATE COLLEGE PRESS, AMES, IOWA.

NAKAI, Z. AND S. HAYASHI. 1962. GROWTH OF THE JAPANESE SARDINE - II. ON A DOMINANT YEAR-CLASS IN THE PACIFIC WATERS OFF THE NORTHEASTERN HONSHU DURING JULY 1951 THROUGH JANUARY 1953. BULL. TOKAI REG. FISH. RES. LAB. 9, 97-107. [MALTHUSIAN CURVE; GROWTH COEFFICIENT]

NECHVALENKO, S. P. 1973. RELATIVE WEIGHT GAIN OF CHIRONOMID LARVAE. HYDROBIOL. J. 9(2), 82-87. (ENGLISH TRANSL.). [KONSTANTINOV'S FORMULA; GROWTH EQUATIONS]

NELDER, J. A. 1961. A NOTE ON SOME GROWTH PATTERNS IN A SIMPLE THEORETICAL ORGANISM. BIOMETRICS 17(2), 220-228. [EQUATIONS]

OLIVER, F. R. 1964. METHODS OF ESTIMATING THE LOGISTIC GROWTH FUNCTION. APPL. STAT. 13(2), 57-66.

OPIE, J. E. 1968. PREDICTABILITY OF INDIVIDUAL TREE GROWTH USING VARIOUS DEFINITIONS OF COMPETING BASAL AREA. FOREST SCI. 14(3), 314-323. [ZONE COUNT MODEL]

OSBURN, J. W. 1974. ON THE CONTROL OF TOOTH REPLACEMENT IN REPTILES AND ITS RELATIONSHIP TO GROWTH. J. THEOR. BIOL. 46(2), 509-527. [EXPONENTIAL MODEL]

OTTESTAD, P. 1938. NOTES ON STATISTICAL METHODS OF GROWTH STUDIES. RAPP. PROCES-VERB. REUNIONS J. CONS. PERM. INT. EXPLOR. MER 108(1), 9-12. [SIGMOID CURVE; GAUSSIAN FREQUENCY FUNCTION]

OTTESTAD, P. 1938. ON THE RELATION BETWEEN THE GROWTH OF THE FISH AND THE GROWTH OF THE SCALES. RAPP. PROCES-VERB. REUNIONS J. CONS. PERM. INT. EXPLOR. MER 108(1), 23-31.

PALOHEIMO, J. E. AND L. M. DICKIE. 1965. FOOD AND GROWTH OF FISHES. I. A GROWTH CURVE DERIVED FROM EXPERIMENTAL DATA. J. FISH. RES. BOARD CAN. 22(2), 521-542. [PARKIN-LARKIN EQUATION; VON BERTALANFFY'S EQUATION]

PALOHEIMO, J. E. AND L. M. DICKIE. 1966. FOOD AND GROWTH OF FISHES. III. RELATIONS AMONG FOOD, BODY SIZE, AND GROWTH EFFICIENCY. J. FISH. RES. BOARD CAN. 23(8), 1209-1248. [EQUATIONS]

PARKER, R. R. AND P. A. LARKIN. 1959. A CONCEPT OF GROWTH IN FISHES. J. FISH. RES. BOARD CAN. 16(5), 721-745. [GROWTH CURVES; VON BERTALANFFY'S GROWTH CURVES]

PASTERNACK, B. S. AND R. R. GIANUTSOS. 1969. APPLICATION OF THE EXPONENTIAL AND POWER FUNCTIONS TO THE STUDY OF ALLOMETRIC GROWTH, WITH PARTICULAR REFERENCE TO DORYLINE ANTS. AMER. NAT. 103(931), 225-234.

PAULIK, G. J. AND L. E. GALES. 1964. ALLOMETRIC GROWTH AND THE BEVERTON AND HOLT YIELD EQUATION. TRANS. AMER. FISH. SOC. 93(4), 369-381.

PAYANDEH, B. 1974. FORMULATED SITE INDEX CURVES FOR MAJOR TIMBER SPECIES IN ONTARIO. FOREST SCI. 20(2), 143-144. [GROWTH MODEL]

PAYNE, P. R. AND E. F. WHEELER. 1968. MODELS OF THE GROWTH OF ORGANISMS UNDER NUTRIENT LIMITING CONDITIONS. PP. 127-132. IN, QUANTIATIVE BIOLOGY OF METABOLISM. (A. LOCKER, ED.). SPRINGER-VERLAG NEW YORK INC., NEW YORK, NEW YORK.

PIENAAR, L. V. AND J. A. THOMPSON. 1969. ALLOMETRIC WEIGHT-LENGTH REGRESSION MODEL. J. FISH. RES. BOARD CAN. 26(1), 123-131.

PIENAAR, L. V. AND K. J. TURNBULL. 1973. THE CHAPMAN-RICHARDS GENERALIZATION OF VON BERTALANFFY'S GROWTH MODEL FOR BASAL AREA GROWTH AND YIELD IN EVEN-AGED STANDS. FOREST SCI. 19(1), 2-22. [MAXIMUM LIKELIHOOD; ALLOMETRY]

PITCHER, T. J. AND P. D. M. MACDONALD. 1973. TWO MODELS FOR SEASONAL GROWTH IN FISHES. J. APPL. ECOL. 10(2), 599-606. [VON BERTALANFFY MODEL; SWITCHED GROWTH VON BERTALANFFY MODEL; SINE WAVE GROWTH VON BERTALANFFY MODEL; COMPUTER]

RAFAIL, S. Z. 1972. FITTING A PARABOLA TO GROWTH DATA OF FISHES AND SOME APPLICATIONS TO FISHERIES. MARINE BIOL. 15(3), 255-264. [MODEL; VON BERTALANFFY GROWTH EQUATION]

RAFAIL, S. Z. 1973. A SIMPLE AND PRECISE METHOD FOR FITTING A VON BERTALANFFY GROWTH CURVE. MARINE BIOL. 19(4), 354-358.

RAPOPORT, A. 1955. SOME THEORETICAL CONSEQUENCES OF THE ALLOMETRIC GROWTH EQUATIONS. BULL. MATH. BIOPHYS. 17, 155-167.

RASHEVSKY, N. 1944. STUDIES IN THE PHYSICOMATHEMATICAL THEORY OF ORGANIC GROWTH. BULL. MATH. BIOPHYS. 6, 1-59. [PLANT FORM; SNAKE FORM; SNAKE LOCOMOTION; BIRD AND INSECT FLIGHT; QUADRUPED LOCOMOTION]

REED, H. S. 1921. A METHOD FOR OBTAINING CONSTANTS FOR FORMULAS OF ORGANIC GROWTH. PROC. NATL. ACAD. SCI. U.S.A. 7(11), 311-316.

REED, H. S. 1929. QUANTITATIVE ASPECTS OF THE PROBLEM OF GROWTH AND DIFFERENTIATION. PROC. INTERN. CONGR. PLANT SCI. 2, 1095-1106.

REED, H. S. 1932. THE GROWTH OF SCENEDESMUS ACUTUS. PROC. NATL. ACAD. SCI. U.S.A. 18(1), 23-30. [GROWTH FORMULA]

RICHARDS, O. W. AND A. J. KAVANAGH. 943. THE ANALYSIS OF THE ELATIVE GROWTH GRADIENTS AND HANGING FORM OF GROWING ORGANISMS: LLUSTRATED BY THE TOBACCO LEAF. MER. NAT. 77(772), 385-399.

ICKLEFS, R. E. 1967. A RAPHICAL METHOD OF FITTING QUATIONS TO GROWTH CURVES. COLOGY 48(6), 978-983. [GROWTH NDEX; LOGISTIC EQUATION; GOMPERTZ QUATION; VON BERTALANFFY'S QUATION]

ICKLEFS, R. E. 1969. RELIMINARY MODELS FOR GROWTH RATES N ALTRICIAL BIRDS. ECOLOGY (6), 1031-1039. [ENERGY MODEL]

RIFFENBURGH, R. H. 1960. A NEW METHOD FOR ESTIMATING PARAMETERS FOR THE GOMPERTZ GROWTH CURVE. J. CONS. PERM. INT. EXPLOR. MER 25(3), 285-293.

ROSTON, S. 1962. ON BIOLOGICAL GROWTH. BULL. MATH. BIOPHYS. 24, 369-373. [GROWTH EQUATIONS]

SCHROEDER, L. 1969. POPULATION GROWTH EFFICIENCIES OF LABORATORY HYDRA PSEUDOLIGATIS HYMAN POPULATIONS. ECOLOGY 50(1), 81-86. [EQUATIONS]

SHIMO, S. AND S. NAKATANI. 1969. STUDIES ON ARTIFICIAL MASS CULTURE OF PORPHYRA TERRA - I. EFFECT OF LIGHT INTENSITY AND POPULATION DENSITY ON THE GROWTH RATE IN PORPHYRA FRONDS. BULL. JAPANESE SOC. SCI. FISH. 35(6), 524-532. [EQUATIONS]

SIBERT, J. AND R. R. PARKER. 1972. A MODEL OF JUVENILE PINK SALMON GROWTH IN THE ESTUARY. FISH. RES. BOARD CAN. TECH. REPT. NO. 321. 62 PP. [PREDATOR-PREY MODEL; COMPUTER; SIMULATION]

SILLIMAN, R. P. 1969. COMPARISON BETWEEN GOMPERTZ AND VON BERTALANFFY CURVES FOR EXPRESSING GROWTH IN WEIGHT OF FISHES. J. FISH. RES. BOARD CAN. 26(1), 161-165.

SKELLAM, J. G., M. V. BRIAN AND J. R. PROCTOR. 1959. THE SIMULTANEOUS GROWTH OF INTERACTING SYSTEMS. ACTA BIOTHEOR. 13(2/3), 131-144. [MODEL]

SMITH, C. E. AND H. C. TUCKWELL. 1974. SOME STOCHASTIC GROWTH PROCESSES. PP. 211-225. IN, MATHEMATICAL PROBLEMS IN BIOLOGY. LECTURE NOTES IN BIOMATHEMATICS 2. (P. VAN DEN DRIESSCHE, ED.). SPRINGER-VERLAG NEW YORK INC., NEW YORK. [MODEL; MALTHUSIAN GROWTH; STRATONOVICH CALCULUS; GOMPERTZ EQUATION]

SOUTHWARD, G. M. AND D. G. CHAPMAN. 1965. UTILIZATION OF PACIFIC HALIBUT STOCKS: STUDY OF BERTALANFFY'S GROWTH EQUATION. INTERN. PACIFIC HALIBUT COMM. REPT. NO. 39. 33 PP. [BEVERTON-HOLT MODEL; COMPUTER]

STINNER, R. E., A. P. GUTIERREZ AND G. D. BUTLER, JR. 1974. AN ALGORITHM FOR TEMPERATURE-DEPENDENT GROWTH RATE SIMULATION. CAN. ENTOMOL. 106(5), 519-524. [COMPUTER; MODEL; SIGMOID FUNCTION; SIMULATION]

SUDO, R. AND S. AIBA. 1971. GROWTH RATE OF VORTICELLIDAE ISOLATED FROM ACTIVATED SLUDGE. JAPANESE J. ECOL. 21(1/2), 70-76. [MASS BALANCE; EQUATIONS]

SULLIVAN, A. D. AND J. L. CLUTTER. 1972. A SIMULTANEOUS GROWTH AND YIELD MODEL FOR LOBLOLLY PINE. FOREST SCI. 18(1), 76-86.

SUNDBERG, R. 1974. ON THE ESTIMATION OF POLLUTION-CAUSED GROWTH REDUCTION IN FOREST TREES. PP. 167-175. IN, STATISTICAL AND MATHEMATICAL ASPECTS OF POLLUTION PROBLEMS. (J. W. PRATT, ED.). MARCEL DEKKER, INC., NEW YORK, NEW YORK. [MODEL]

TANAKA, S. 1957. RELATION BETWEEN BARANOV'S MATHEMATICAL MODEL AND SIGMOID CURVE. BULL. JAPANESE SOC. SCI. FISH. 23(1), 12-18.

TAYLOR, C. C. 1962. GROWTH EQUATIONS WITH METABOLIC PARAMETERS. J. CONS. PERM. INT. EXPLOR. MER 27(3), 270-286.

TESCH, F. W. 1968. AGE AND GROWTH. PP. 93-123. IN, METHODS FOR ASSESSMENT OF FISH PRODUCTION IN FRESH WATERS. (W. E. RICKER, ED.). BLACKWELL SCIENTIFIC PUBLICATIONS, OXFORD, ENGLAND. [GROWTH EQUATIONS; COEFFICIENT OF CONDITION]

THORNLEY, J. H. M. AND J. D. HESKETH. 1972. GROWTH AND RESPIRATION IN COTTON BOLLS. J. APPL. ECOL. 9(1), 315-317. [EQUATIONS]

TOMLINSON, P. K. AND N. J. ABRAMSON. 1961. FITTING A VON BERTALANFFY GROWTH CURVE BY LEAST SQUARES: INCLUDING TABLES OF POLYNOMIALS. CALIF. DEPT. FISH GAME FISH BULL. NO. 116. 69 PP.

TRUCCO, E. AND G. I. BELL. 1970. A NOTE ON THE DISPERSIONLESS GROWTH LAW FOR SINGLE CELLS. BULL. MATH. BIOPHYS. 32, 475-483.

TURNER, M. E., JR., B. A. BLUMENSTEIN AND J. L. SEBAUGH. 1969. A GENERALIZATION OF THE LOGISTIC LAW OF GROWTH. BIOMETRICS 25(3), 577-580.

URSIN, E. 1967. A MATHEMATICAL MODEL OF SOME ASPECTS OF FISH GROWTH, RESPIRATION, AND MORTALITY. J. FISH. RES. BOARD CAN. 24(11), 2355-2453.

UTZ, W. R. AND P. E. WALTMAN. 1963. PERIODICITY AND BOUNDNESS OF SOLUTIONS OF GENERALIZED DIFFERENTIAL EQUATIONS OF GROWTH. BULL. MATH. BIOPHYS. 25, 75-93. [COMPETITION; VOLTERRA'S EQUATIONS; HUTCHINSON'S EQUATIONS]

VAN DE SANDE-BAKHUYZEN, H. L. 1926. GROWTH AND GROWTH FORMULAS IN PLANTS. SCIENCE 64(1670), 653-654.

VAN DE SANDE-BAKHUYZEN, H. L. AND C. L. ALSBERG. 1927. THE GROWTH CURVE IN ANNUAL PLANTS. PHYSIOL. REV. 7(1), 151-187.

VAN DEN BERGH, J. P. AND W. T. ELBERSE. 1970. YIELDS OF MONOCULTURES AND MIXTURES OF TWO GRASS SPECIES DIFFERING IN GROWTH HABIT. J. APPL. ECOL. 7(2), 311-320.

VAN DER VAART, H. R. 1953. ADULT AGE, AN INVESTIGATION BASED ON CERTAIN ASPECTS OF GROWTH CURVES. ACTA BIOTHEOR. 10(3/4), 139-212. [EQUATIONS]

VON BERTALANFFY, L. 1938. A QUANTITATIVE THEORY OF ORGANIC GROWTH (INQUIRIES ON GROWTH LAWS II). HUMAN BIOL. 10(2), 181-213. [GROWTH FUNCTION]

VON BERTALANFFY, L. 1949. PROBLEMS OF ORGANIC GROWTH. NATURE 163(4135), 156-158.

VON BERTALANFFY, L. 1957. QUANTITATIVE LAWS IN METABOLISM AND GROWTH. QUART. REV. BIOL. 32(2), 217-231.

VON BERTALANFFY, L. 1960. PRINCIPLES AND THEORY OF GROWTH. PP. 137-259. IN, FUNDAMENTAL ASPECTS OF NORMAL AND MALIGNANT GROWTH. (W. W. NOWINSKI, ED.). ELSEVIER PUBLISHING CO., NEW YORK, NEW YORK. [EQUATIONS]

WALFORD, L. A. 1946. A NEW GRAPHIC METHOD OF DESCRIBING THE GROWTH OF ANIMALS. BIOL. BULL. 90(2), 141-147. [GEOMETRIC SERIES]

WANGERSKY, P. J. AND W. J. CUNNINGHAM. 1956. ON TIME LAGS IN EQUATIONS OF GROWTH. PROC. NATL. ACAD. SCI. U.S.A. 42(9), 699-702.

WILSON, E. B. 1942. NORMS OF GROWTH. SCIENCE 95(2457), 112-113. [VARIANCE]

WINSOR, C. P. 1932. THE GOMPERTZ CURVE AS A GROWTH CURVE. PROC. NATL. ACAD. SCI. U. S. A. 18(1), 1-8.

WOHLSCHLAG, D. E. 1962. ANTARCTIC FISH GROWTH AND METABOLIC DIFFERENCES RELATED TO SEX. COLOGY 43(4), 589-597. MULTIPLE REGRESSION; LOG RANSFORMATION; BEVERTON-HOLT ODEL]

YAGI, N. 1926. ANALYSIS OF THE GROWTH CURVES OF THE INSECT LARVAE. MEM. COLL. AGRIC. KYOTO UNIV., NO. 1, 35 PP.

YOSHIHARA, T. 1950. ON A GRAPHICAL METHOD OF DETERMINATION OF PARAMETERS CONTAINED IN THE LOGISTIC TYPE CURVES. BULL. JAPANESE SOC. SCI. FISH. 16(7), 323-328.

YOSHIHARA, T. 1951. ON THE FITTING, THE SUMMATION, AND AN APPLICATION OF THE LOGISTIC CURVE. J. TOKYO UNIV. FISHERIES 38(2), 181-195.

YOSHIHARA, T. 1953. ON THE RELATION BETWEEN THE LAW OF GROWTH AND DEATH. BULL. JAPANESE SOC. SCI. FISH. 19(4), 273-274. [GOMPERTZ LAW; LOGISTIC LAW; MAKEHAM LAW]

ZUCKERMAN, S. (LEADERSHIP OF) 1950. A DISCUSSION ON THE MEASUREMENT OF GROWTH AND FORM. PROC. ROYAL SOC. EDINB. SECT. B. BIOL. SERIES 137, 433-523.

GROWTH OF POPULATIONS

BANNISTER, T. T. 1974. A GENERAL THEORY OF STEADY STATE PHYTOPLANKTON GROWTH IN A NUTRIENT SATURATED MIXED LAYER. LIMNOL. OCEANOGR. 19(1), 13-30. [EQUATIONS]

BECKING, L. G. M. 1946. ON THE ANALYSIS OF SIGMOID CURVES. ACTA BIOTHEOR. 8(1/2), 42-59.

CAPERON, J. 1967. POPULATION GROWTH IN MICRO-ORGANISMS LIMITED BY FOOD SUPPLY. ECOLOGY 48(5), 715-722. [HYPERBOLIC EQUATION; GROWTH MODEL]

CAPERON, J. 1969. TIME LAG POPULATION GROWTH RESPONSE OF ISOCHRYSIS GALBANA TO A VARIABLE NITRATE ENVIRONMENT. ECOLOGY 50(2), 188-192. [GROWTH EQUATIONS]

CAPILDEO, R. AND J. B. S. HALDANE. 1954. THE MATHEMATICS OF BIRD POPULATION GROWTH AND DECLINE. J. ANIM. ECOL. 23(2), 215-223.

CAPOCELLI, R. M. AND L. M. RICCIARDI. 1974. A DIFFUSION MODEL FOR POPULATION GROWTH IN RANDOM ENVIRONMENT. THEOR. POPUL. BIOL. 5(1), 28-41. [DELTA-CORRELATED NORMAL PROCESS; FOKKER-PLANCK TYPE DIFFUSION EQUATION]

CHRISTIAN, J. J., J. A. LLOYD, D. E. GOLDMAN AND D. E. DAVIS. 1971. AN EMPIRICAL FORMULA FOR THE GROWTH OF SOME VERTEBRATE POPULATIONS. CURR. MOD. BIOL. 4(1971), 26-34.

CROW, J. F. AND M. KIMURA. 1970. MODELS OF POPULATION GROWTH. PP. 3-30. IN, AN INTRODUCTION TO POPULATION GENETICS THEORY. (J. F. CROW AND M. KIMURA, EDS.). HARPER AND ROW PUBLISHERS, NEW YORK, NEW YORK.

CUNNINGHAM, W. J. 1955. SIMULTANEOUS NONLINEAR EQUATIONS OF GROWTH. BULL. MATH. BIOPHYS. 17, 101-110. [COMPETITION; VOLTERRA'S EQUATIONS; GAUSE AND WITT EQUATIONS; HUTCHINSON'S EQUATIONS; STABILITY]

DARWIN, J. H. 1953. POPULATION DIFFERENCES BETWEEN SPECIES GROWING ACCORDING TO SIMPLE BIRTH AND DEATH PROCESSES. BIOMETRIKA 40(3/4), 370-382.

DEMETRIUS, L. 1969. THE SENSITIVITY OF POPULATION GROWTH RATE TO PERTURBATIONS IN THE LIFE CYCLE COMPONENTS. MATH. BIOSCI. 4, 129-136. [MODEL; COLONIZING SPECIES; LESLIE'S MODEL; PROBABILITY; MATRIX]

DICKSON, E. M. 1970. MODEL FOR ZERO POPULATION GROWTH. BIOSCIENCE 20(23), 1245-1246.

DROOP, M. R. 1966. VITAMIN B12 AND MARINE ECOLOGY III. AN EXPERIMENT WITH A CHEMOSTAT. J. MAR. BIOL. ASSOC. U. K. 46(3), 659-671. [GROWTH EQUATIONS]

DROOP, M. R. 1968. VITAMIN B12 AND MARINE ECOLOGY. IV. THE KINETICS OF UPTAKE, GROWTH AND INHIBITION IN MONOCHRYSIS LUTHERI. J. MAR. BIOL. ASSOC. U. K. 48(3), 689-733. [EQUATIONS]

DUNKEL, G. 1968. SINGLE SPECIES MODEL FOR POPULATION GROWTH DEPENDING ON PAST HISTORY. PP. 92-99. IN, SEMINAR ON DIFFERENTIAL EQUATIONS AND DYNAMIC SYSTEMS, LECTURE NOTES IN MATHEMATICS, VOL. 60. [VERHULST; PEARL; LOTKA'S EQUATION; VOLTERRA]

FLETCHER, R. I. 1974. THE QUADRIC LAW OF DAMPED EXPONENTIAL GROWTH. BIOMETRICS 30(1), 111-124. [DETERMINISTIC MODEL; CANONICAL FORM; LOGISTIC LAW; VON BERTALANFFY'S LAW; VERHULST-PEARL-REED LOGISTIC LAW; GOMPERTZ LAW]

FORREST, H. 1963. IMMORTALITY AND PEARL'S SURVIVORSHIP CURVE FOR HYDRA. ECOLOGY 44(3), 609-610.

FRANK, P. W. 1960. PREDICTION OF POPULATION GROWTH FORM IN DAPHNIA PULEX CULTURES. AMER. NAT. 94(878), 357-372. [MODEL; LOGISTIC CURVE; LESLIE'S MODEL; RICKER'S MODEL]

FREEDMAN, H. I. 1975. A PERTURBED KOLMOGOROV-TYPE MODEL FOR THE GROWTH PROBLEM. MATH. BIOSCI. 23(1/2), 127-149. [PREDATOR-PREY MODEL]

FUJITA, H. 1951. A NOTE ON PEARL'S COEFFICIENT FOR THE INCREASE OF BACTERIA IN A CULTURE MEDIUM. BULL. JAPANESE SOC. SCI. FISH. 17(1), 24-26.

FUJITA, H. AND S. UTIDA. 1953. THE EFFECT OF POPULATION DENSITY ON THE GROWTH OF AN ANIMAL POPULATION. ECOLOGY 34(3), 488-498.

GAUSE, G. F., O. K. NASTUKOVA AND W. W. ALPATOV. 1934. THE INFLUENCE OF BIOLOGICALLY CONDITIONED MEDIA ON THE GROWTH OF A MIXED POPULATION OF PARAMECIUM CAUDATUM AND P. AURELIA. J. ANIM. ECOL. 3(2), 222-230.

GOLDMAN, J. C. AND E. J. CARPENTER. 1974. A KINETIC APPROACH TO THE EFFECT OF TEMPERATURE ON ALGAL GROWTH. LIMNOL. OCEANOGR. 19(5), 756-766. [EQUATIONS; MONOD'S MODEL]

GOODMAN, L. 1967. ON THE RECONCILIATION OF MATHEMATICAL THEORIES OF POPULATION GROWTH. J. ROYAL STAT. SOC. SER. A 130(4), 541-553.

GOODMAN, L. A. 1953. POPULATION GROWTH OF THE SEXES. BIOMETRICS 9(2), 212-225.

GOODMAN, L. A. 1968. STOCHASTIC MODELS FOR THE POPULATION GROWTH OF THE SEXES. BIOMETRIKA 55(3), 469-487.

GOODMAN, L. A. 1969. THE ANALYSIS OF POPULATION GROWTH WHEN THE BIRTH AND DEATH RATES DEPEND UPON SEVERAL FACTORS. BIOMETRICS 25(4), 659-681.

GRACE, J. AND H. W. WOOLHOUSE. 1974. A PHYSIOLOGICAL AND MATHEMATICAL STUDY OF GROWTH AND PRODUCTIVITY OF A CALLUNA-SPHAGNUM COMMUNITY. IV. A MODEL OF GROWING CALLUNA. J. APPL. ECOL. 11(1), 281-295. [COMPUTER]

GREVILLE, T. N. E. AND N. KEYFITZ. 1974. BACKWARD POPULATION PROJECTION BY A GENERALIZED INVERSE. THEOR. POPUL. BIOL. 6(2), 135-142. [LESLIE MATRIX]

GULLAND, J. A. 1955. ESTIMATION OF GROWTH AND MORTALITY IN COMMERCIAL FISH POPULATIONS. FISH. INVEST. MINIST. AGRIC. FISH. FOOD (G. B.) SERIES II 18(9), 1-46.

HAIRSTON, N. G., D. W. TINKLE AND H. M. WILBUR. 1970. NATURAL SELECTION AND THE PARAMETERS OF POPULATION GROWTH. J. WILDL. MANAGE. 34(4), 681-690.

HARDIN, G. 1945. A MORE MEANINGFUL FORM OF THE "LOGISTIC" EQUATION. AMER. NAT. 79(782), 279-280.

HERON, A. C. 1972. POPULATION ECOLOGY OF A COLONIZING SPECIES: THE PELAGIC TUNICATE THALIA DEMOCRATICA II. POPULATION GROWTH RATE. OECOLOGIA 10, 294-312. [EQUATIONS; SURVIVORSHIP; INTRINSIC RATE; LOTKA'S EQUATION; AGE DISTRIBUTION]

HUGHES, A. P. AND P. R. FREEMAN. 1967. GROWTH ANALYSIS USING FREQUENT SMALL HARVESTS. J. APPL. ECOL. 4(2), 553-560. [PLANTS; EQUATIONS; REGRESSION; COMPUTER]

JANNASCH, H. W. 1974. STEADY STATE AND THE CHEMOSTAT IN ECOLOGY. LIMNOL. OCEANOGR. 19(4), 716-720. [GROWTH FUNCTION]

KENDALL, D. G. 1948. ON SOME MODES OF POPULATION GROWTH LEADING TO R. A. FISHER'S LOGARITHMIC SERIES DISTRIBUTION. BIOMETRIKA 35(1/2), 6-15.

KENDALL, D. G. 1949. STOCHASTIC PROCESSES AND POPULATION GROWTH. J. ROYAL STAT. SOC. SER. B 11(2), 230-264.

KING, C. E. AND W. W. ANDERSON. 1971. AGE-SPECIFIC SELECTION. II. THE INTERACTION BETWEEN R AND K DURING POPULATION GROWTH. AMER. NAT. 105(942), 137-156. [MODEL; LOGISTIC; R AND K REFERRING TO GENETIC FITNESS]

LABEYRIE, V. 1971. THE VARIABILITY OF THE PHYSIOLOGICAL AND ETHOLOGICAL ACTIVITIES AND THE POPULATION GROWTH OF INSECTS. PP. 313-335. IN, STATISTICAL ECOLOGY, VOLUME 2. SAMPLING AND MODELING BIOLOGICAL POPULATIONS AND POPULATION DYNAMICS. (G. P. PATIL, E. C. PIELOU AND W. E. WATERS, EDS.). PENNSYLVANIA STATE UNIVERSITY PRESS, UNIVERSITY PARK, PENNSYLVANIA.

LANDAHL, H. D. 1957. POPULATION GROWTH UNDER THE INFLUENCE OF RANDOM DISPERSAL. BULL. MATH. BIOPHYS. 19, 171-186. [LOGISTIC GROWTH]

LANDAHL, H. D. 1959. A NOTE ON POPULATION GROWTH UNDER RANDOM DISPERSAL. BULL. MATH. BIOPHYS. 21, 153-159.

LEFKOVITCH, L. P. 1964. THE GROWTH OF RESTRICTED POPULATIONS OF LASIODERMA SERRICORNE (F.) (COLEOPTERA, ANOBIIDAE). BULL. ENTOMOL. RES. 55(1), 87-96. [COMPLEX NUMBER; MODEL; DEMOIVRE'S THEOREM; OSCILLATION; MATRIX]

LEFKOVITCH, L. P. 1965. THE STUDY OF POPULATION GROWTH IN ORGANISMS GROUPED BY STAGES. BIOMETRICS 21(1), 1-18. [MATRIX]

LEFKOVITCH, L. P. 1966. A POPULATION GROWTH MODEL INCORPORATING DELAYED RESPONSES. BULL. MATH. BIOPHYS. 28, 219-233. [TIME LAG MODEL]

LEFKOVITCH, L. P. 1971. SOME COMMENTS ON THE INVARIANTS OF POPULATION GROWTH. PP. 337-359. IN, STATISTICAL ECOLOGY, VOLUME 2. SAMPLING AND MODELING BIOLOGICAL POPULATIONS AND POPULATION DYNAMICS. (G. P. PATIL, E. C. PIELOU AND W. E. WATERS, EDS.). PENNSYLVANIA STATE UNIVERSITY PRESS, UNIVERSITY PARK, PENNSYLVANIA. [COMPUTER]

LESLIE, P. H. 1957. AN ANALYSIS OF THE DATA FOR SOME EXPERIMENTS CARRIED OUT BY GAUSE WITH POPULATIONS OF THE PROTOZOA, PARAMECIUM AURELIA AND PARAMECIUM CAUDATUM. BIOMETRIKA 44(3/4), 314-327. [MODEL]

LESLIE, P. H. 1959. THE PROPERTIES OF A CERTAIN LAG TYPE OF POPULATION GROWTH AND THE INFLUENCE OF AN EXTERNAL RANDOM FACTOR ON A NUMBER OF SUCH POPULATIONS. PHYSIOL. ZOOL. 32(3), 151-159.

LEVINS, R. 1969. THE EFFECT OF RANDOM VARIATIONS OF DIFFERENT TYPES ON POPULATION GROWTH. PROC. NATL. ACAD. SCI. U. S. A. 62(4), 1061-1065. [NORMAL DISTRIBUTION; HARMONIC MEAN]

LEWIS, E. G. 1942. ON THE GENERATION AND GROWTH OF A POPULATION. SANKHYA 6, 93-96. [MATRIX]

LEWIS, E. R. 1972. DELAY-LINE MODELS OF POPULATION GROWTH. ECOLOGY 53(5), 797-807. [RELATIONSHIP TO LESLIE MODEL]

LEWONTIN, R. C. AND D. COHEN. 1969. ON POPULATION GROWTH IN A RANDOMLY VARYING ENVIRONMENT. PROC. NATL. ACAD. SCI. U. S. A. 62(4), 1056-1060. [MODEL; NORMAL DISTRIBUTION]

LINDSEY, J. K. 1970. EXACT STATISTICAL INFERENCES ABOUT THE PARAMETER FOR AN EXPONENTIAL GROWTH CURVE FOLLOWING A POISSON DISTRIBUTION. J. FISH. RES. BOARD CAN. 27(1), 172-174.

LONG, G. E., P. H. DURAN, R. O. JEFFORDS AND D. N. WELDON. 1974. AN APPLICATION OF THE LOGISTIC EQUATION TO THE POPULATION DYNAMICS OF SALT-MARSH GASTROPODS. THEOR. POPUL. BIOL. 5(3), 450-459. [MODEL; COMPUTER; SIMULATION]

LOTKA, A. J. 1932. THE GROWTH OF MIXED POPULATIONS: TWO SPECIES COMPETING FOR A COMMON FOOD SUPPLY. J. WASH. ACAD. SCI. 22(16,17), 461-469. [VOLTERRA; VERHULST-PEARL EQUATION; CHARACTERISTIC EQUATION]

MACDONALD, P. D. M. 1973. ON THE STATISTICS OF CELL PROLIFERATION. PP. 303-314, IN, THE MATHEMATICAL THEORY OF THE DYNAMICS OF BIOLOGICAL POPULATIONS. (M. S. BARTLETT AND R. W. HIORNS, EDS.), ACADEMIC PRESS, NEW YORK, NEW YORK. [MODEL; EXPONENTIAL GROWTH; PROBABILITY; GAMMA DISTRIBUTION; MAXIMUM LIKELIHOOD; MATRIX]

MARTINEZ, H. M. 1966. ON THE DERIVATION OF A MEAN GROWTH EQUATION FOR CELL CULTURES. BULL. MATH. BIOPHYS. 28, 411-416. [HARRIS-BELLMAN EQUATION; VON FOERSTER'S EQUATION; HIRSCH AND ENGELBERG EQUATION]

MCCLAMROCH, N. H. 1972. FUNCTIONAL DIFFERENTIAL EQUATIONS AND AGE DEPENDENT POPULATION GROWTH. MATH. BIOSCI. 14, 255-280.

MCLAREN, I. A. 1963. EFFECTS OF TEMPERATURE ON GROWTH OF ZOOPLANKTON, AND THE ADAPTIVE VALUE OF VERTICAL MIGRATION. J. FISH. RES. BOARD CAN. 20(3), 685-727. [TEMPERATURE FUNCTION; VON BERTALANFFY]

MCNEIL, D. R. 1974. PEARL-REED TYPE STOCHASTIC MODELS FOR POPULATION GROWTH. THEOR. POPUL. BIOL. 5(3), 358-365. [TRANSFORMATION]

MEATS, A. 1971. QUANTITATIVE ANALYSIS OF DATA ON POPULATION GROWTH. PROC. ECOL. SOC. AUSTRALIA 6, 76-83.

MEGARD, R. O. AND P. D. SMITH. 1974. MECHANISMS THAT REGULATE GROWTH RATES OF PHYTOPLANKTON IN SHEGAWA LAKE, MINNESOTA. LIMNOL. OCEANOGR. 19(2), 229-296. [PRODUCTION MODEL; PHOTOSYNTHETIC RATE EQUATION]

MENSHUTKIN, V. V. AND YU. YA. KISLYAKOV. 1967. MODELING OF THE POPULATION OF THE WHITE FISH WITH THE ACCOUNT OF VARIABLE GROWTH RATE. ZOOLOGISCHESKII ZHURNAL 46(6), 805-810. [COMPUTER; MODEL]

MERRELL, M. 1931. THE RELATIONSHIP OF INDIVIDUAL GROWTH TO AVERAGE GROWTH. HUMAN BIOL. 3(1), 37-70. [LOGISTIC]

MERTZ, D. B. 1972. THE TRIBOLIUM MODEL AND THE MATHEMATICS OF POPULATION GROWTH. PP. 51-78. IN, ANNUAL REVIEW OF ECOLOGY AND SYSTEMATICS, VOLUME 3. (R. F. JOHNSTON, P. W. FRANK AND C. D. MICHENER, EDS.), ANNUAL REVIEWS INC., PALO ALTO, CALIFORNIA.

MORISITA, M. 1965. THE FITTING OF THE LOGISTIC EQUATION TO THE RATE OF INCREASE OF POPULATION DENSITY. RES. POPUL. ECOL.-KYOTO 7(1), 52-55.

MOSER, H. 1957. STRUCTURE AND DYNAMICS OF BACTERIAL POPULATIONS MAINTAINED IN THE CHEMOSTAT. COLD SPRING HARBOR SYMPOSIA ON QUANTITATIVE BIOLOGY 22, 121-137. [GROWTH RATE]

NAKAMURA, M. 1973. A GENERAL LIMIT THEOREM FOR DYNAMIC SYSTEMS WITH AN APPLICATION TO POPULATION GROWTH. MATH. BIOSCI. 16, 177-187. [MATRIX]

PADGETT, W. J. AND C. P. TSOKOS. 1973. A NEW STOCHASTIC FORMULATION OF A POPULATION GROWTH PROBLEM. MATH. BIOSCI. 17, 105-120. [MARKOV INEQUALITY; VOLTERRA TYPE; MODEL; DETERMINISTIC MODEL; STOCHASTIC MODEL; POISSON DISTRIBUTION; PROBABILITY]

PEARL, R. 1924. THE CURVE OF POPULATION GROWTH. PROC. AMER. PHILOS. SOC. 63(1), 10-17. [LOGISTIC; PROBABLE ERROR]

PEARL, R. AND L. J. REED 1923. ON THE MATHEMATICAL THEORY OF POPULATION GROWTH. METRON 3(1), 1-14. [LOGISTIC]

PEARL, R. AND L. J. REED. 1920. ON THE RATE OF GROWTH OF THE POPULATION OF THE UNITED STATES SINCE 1790 AND ITS MATHEMATICAL REPRESENTATION. PROC. NATL. ACAD. SCI. U. S. A. 6(6), 275-288.

PEARL, R. AND L. J. REED. 1922. A FURTHER NOTE ON THE MATHEMATICAL THEORY OF POPULATION GROWTH. PROC. NATL. ACAD. SCI. U. S. A. 8(12), 365-368.

POLLAK, E. AND O. KEMPTHORNE. 1970. MALTHUSIAN PARAMETERS IN GENETIC POPULATIONS. I. HAPLOID AND SELFING MODELS. THEOR. POPUL. BIOL. 1(3), 315-345. [LESLIE'S THEORY OF POPULATION GROWTH; MATRIX]

POLLAK, E. AND O. KEMPTHORNE. 1971. MALTHUSIAN PARAMETERS IN GENETIC POPULATIONS PART II. RANDOM MATING POPULATIONS IN INFINITE HABITATS. THEOR. POPUL. BIOL. 2(4), 357-390. [MODEL; FISHER'S FUNDAMENTAL THEOREM; LESLIE'S THEORY OF POPULATION GROWTH]

POLLARD, J. H. 1970. ON SIMPLE APPROXIMATE CALCULATIONS APPROPRIATE TO POPULATIONS WITH RANDOM GROWTH RATES. THEOR. POPUL BIOL. 1(2), 208-218.

RABINOVICH, J. E. 1969. THE APPLICABILITY OF SOME POPULATION GROWTH MODELS TO A SINGLE-SPECIES LABORATORY POPULATION. ANN. ENTOMOL. SOC. AMER. 62(2), 437-442. [COMPUTER]

RAFAIL, S. Z. 1971. A NEW GROWTH MODEL FOR FISHES AND THE ESTIMATION OF OPTIMUM AGE OF FISH POPULATIONS. MARINE BIOL. 10(1), 13-21. [VON BERTALANFFY GROWTH EQUATION]

RAMKRISHNA, D., A. G. FREDRICKSON AND H. M. TSUCHIYA. 1968. ON RELATIONSHIPS BETWEEN VARIOUS DISTRIBUTION FUNCTIONS IN BALANCED UNICELLULAR GROWTH. BULL. MATH. BIOPHYS. 30, 319-323. [EXPONENTIAL GROWTH]

REID, A. T. 1953. AN AGE-DEPENDENT STOCHASTIC MODEL OF POPULATION GROWTH. BULL. MATH. BIOPHYS. 15, 361-365. [BELLMAN-HARRIS PROCESS]

ROONEY, D. W. 1971. SKEWED ALGAL DIVISION PATTERNS: EFFECTS OF AUTOSPORE YIELD ON COMPUTED SYNCHRONY INDICES. MATH. BIOSCI. 12, 367-373. [EQUATIONS]

ROONEY, D. W. 1972. SYNCHRONY INDICES BASED ON COMPUTER-CORRECTED ALGAL CELL DATA. MATH. BIOSCI. 14, 59-64. [EQUATIONS]

SHINOZAKI, K. 1953. ON THE GENERALIZATION OF THE LOGISTIC CURVE. VI. VARIOUS POPULATION CURVES. J. OSAKA CITY MED. CENTER 3(1), 21-29.

SHINOZAKI, K. 1953. ON THE GENERALIZATION OF THE LOGISTIC CURVE, I. GENERAL DISCUSSIONS. J. OSAKA CITY MED. CENTER 2(2), 143-147.

SHINOZAKI, K. 1953. ON THE GENERALIZATION OF THE LOGISTIC CURVE, III. POPULATION CURVE FOR JAPAN AND THE SUPPLEMENTARY NOTES ON THE S-N DIAGRAM. J. OSAKA CITY MED. CENTER 2(4), 265-272.

SHINOZAKI, K. 1953. ON THE GENERALIZATION OF THE LOGISTIC CURVE, IV. GENERAL PROPERTIES OF POPULATION CURVE. J. OSAKA CITY MED. CENTER 3(1), 12-14.

SHINOZAKI, K. 1953. ON THE GENERALIZATION OF THE LOGISTIC CURVE, V. PEARL'S POPULATION CURVES. J. OSAKA CITY MED. CENTER 3(1), 15-20.

SLOBODKIN, L. B. 1953. AN ALGEBRA OF POPULATION GROWTH. ECOLOGY 34(3), 513-519.

SMITH, F. E. 1954. QUANTITATIVE ASPECTS OF POPULATION GROWTH. PP. 277-294. IN. DYNAMICS OF GROWTH PROCESSES. (E. J. BOELL, ED.). PRINCETON UNIVERSITY PRESS, PRINCETON, NEW JERSEY. [MODEL]

STANLEY, J. 1932. A MATHEMATICAL THEORY OF THE GROWTH OF POPULATIONS OF THE FLOUR BEETLE, TRIBOLIUM CONFUSUM, DUV. II. THE DISTRIBUTION BY AGES IN THE EARLY STAGES OF POPULATION GROWTH. CAN. J. RES. 7(4), 426-433. [VOLTERRA'S MODEL]

STANLEY, J. 1932. A MATHEMATICAL THEORY OF THE GROWTH OF POPULATIONS OF THE FLOUR BEETLE, TRIBOLIUM CONFUSUM, DUV. CAN. J. RES. 6(6), 632-671. [MODEL]

STANLEY, J. 1941. A MATHEMATICAL THEORY OF THE GROWTH OF POPULATIONS OF THE FLOUR BEETLE TRIBOLIUM CONFUSUM DUV. IV. A MODIFIED THEORY DESCRIPTIVE OF THE RELATION BETWEEN THE LIMITING VALUE OF EGG-POPULATIONS IN THE ABSENCE OF HATCHING, AND THE VOLUME (OR WEIGHT) OF FLOUR USED IN THE CULTURE. ECOLOGY 22(1), 23-37. [MODEL]

STANLEY, J. 1942. A MATHEMATICAL THEORY OF THE GROWTH OF POPULATIONS OF THE FLOUR BEETLE TRIBOLIUM CONFUSUM DUV. V. THE RELATION BETWEEN THE LIMITING VALUE OF EGG- POPULATIONS IN THE ABSENCE OF HATCHING AND THE SEX-RATIO OF THE GROUP OF ADULT BEETLES USED IN A CULTURE. ECOLOGY 23(1), 24-31. [MODEL]

STANLEY, J. 1943. A MATHEMATICAL THEORY OF THE GROWTH OF POPULATIONS OF THE FLOUR BEETLE TRIBOLIUM CONFUSUM DUV. VI. EGG POPULATIONS IN WHICH THE INITIAL NUMBER OF EGGS IS ABOVE THE LIMITING VALUE. ECOLOGY 24(3), 323-328. [MODEL]

STANLEY, J. 1964. A MATHEMATICAL THEORY OF THE GROWTH OF POPULATIONS OF THE FLOUR BEETLE TRIBOLIUM CONFUSUM DUVAL. VIII. A FURTHER STUDY OF THE "RE-TUNNELLING" PROBLEM. CAN. J. ZOOL. 42(2), 201-227.

STANLEY, J. 1966. A MATHEMATICAL THEORY OF THE GROWTH OF POPULATIONS OF THE FLOUR BEETLE TRIBOLIUM CONFUSUM. DUVAL IX. THE PERSISTENCE OF TUNNELS IN THE FLOUR MASS. J. THEOR. BIOL. 13, 379-411. [EQUATIONS]

STEPANOVA, L. A. 1970. RELATIVE RATE OF GROWTH OF ZOOPLANKTON IN HEAVILY EXPLOITED FATTENING PONDS HYDROBIOL. J. 6(2), 110-113. (ENGLISH TRANSL.). [GROWTH EQUATIONS]

STRAND, L. 1972. A MODEL FOR STAND GROWTH. PP. 207-216. IN, 3RD CONFERENCE ADVISORY GROUP OF FOREST STATISTICIANS. INSTITUT NATIONAL DE LA RECHERCHE AGRONOMIQUE, PARIS, FRANCE. I.N.R.A. PUBL. 72-3. [COMPUTER; SIMULATION]

TANNER, J. T. 1966. EFFECTS OF POPULATION DENSITY ON GROWTH RATES OF ANIMAL POPULATIONS. ECOLOGY 47(5), 733-745. [RANDOM NUMBERS; LOG TRANSFORMATION; INTRINSIC RATE OF INCREASE]

TUCKWELL, H. C. 1974. A STUDY OF SOME DIFFUSION MODELS OF POPULATION GROWTH. THEOR. POPUL. BIOL. 5(3), 345-357. [STOCHASTIC TRANSFORMATION; ORNSTEIN-UHLENBECK PROCESS; MALTHUSIAN GROWTH; PEARL-VERHULST AND LEVINS MODELS]

UTZ, W. R. 1961. THE EQUATIONS OF POPULATION GROWTH. BULL. MATH. BIOPHYS. 23, 261-262. [COMPETITION; DIFFERENTIAL EQUATION]

VERHULST, P. F. 1845. RECHERCHES MATHEMATIQUES SUR LA LOI D'ACCROISSEMENT DE LA POPULATION. NOUV. MEM. DE L'ACAD. ROY. DES SCI. ET BELLES-LETT. DE BRUXELLES 18, 1-38.

VOLTERRA, V. 1938. POPULATION GROWTH, EQUILIBRIA, AND EXTINCTION UNDER SPECIFIED BREEDING CONDITIONS: A DEVELOPMENT AND EXTENSION OF THE THEORY OF THE LOGISTIC CURVE. HUMAN BIOL. 10(1), 1-11.

WALTMAN, P. E. 1964. THE EQUATIONS OF GROWTH. BULL. MATH. BIOPHYS. 26, 39-43. [COMPETITION; VOLTERRA'S EQUATIONS; HUTCHINSON'S EQUATIONS]

WILSON, E. B. 1925. THE LOGISTIC OR AUTOCATALYTIC GRID. PROC. NATL. ACAD. SCI. U.S.A. 11(8), 451-456. [GRAPH; COORDINATE PAPER]

WINSOR, C. P. 1934. MATHEMATICAL ANALYSIS OF GROWTH OF MIXED POPULATIONS. COLD SPRING HARBOR SYMPOSIA ON QUANTITATIVE BIOLOGY 2, 181-186.

YOSHIZAWA, S. 1970. POPULATION GROWTH PROCESS DESCRIBED BY A SEMILINEAR PARABOLIC FUNCTION. MATH. BIOSCI. 7, 291-303.

YULE, G. U. 1924. THE GROWTH OF POPULATION AND THE FACTORS WHICH CONTROL IT. J. ROYAL STAT. SOC. SER. A 88(1), 1-62. [LOGISTIC]

ZWANZIG, R. 1973. GENERALIZED VERHULST LAWS FOR POPULATION GROWTH. PROC. NATL. ACAD. SCI. U.S.A, 70(11), 3048-3051. [VERHULST MODEL; VOLTERRA-LOTKA MODEL; MONTROLLS' MODEL; GOMPERTZ LAW]

MODELS (MATHEMATICAL)

ALDERDICE, D. F. 1965. ANALYSIS OF EXPERIMENTAL MULTIVARIABLE ENVIRONMENTS RELATED TO THE PROBLEMS OF AQUATIC POLLUTION. PP. 320-325. IN, BIOLOGICAL PROBLEMS IN WATER POLLUTION, THIRD SEMINAR AUGUST 13-17, 1962. (C. M. TARZWELL, COMP.). R. A. TAFT SANITARY ENGINEERING CENTER, CINCINNATI, OHIO. ENVIRONMNETAL HEALTH SERIES, WATER SUPPLY AND POLLUTION CONTROL PUBLIC HEALTH SERVICE PUBL. NO. 999-WP-25. [POLYNOMIAL REGRESSION; RESPONSE SURFACE; MULTIVARIATE ANALYSIS]

ALEXANDER, R. M. 1969. THE ORIENTATION OF MUSCLE FIBRES IN THE MYOMERES OF FISHES. J. MAR. BIOL. ASSOC. U. K. 49(2), 263-290. [EQUATIONS]

ALLEN, T. F. H. AND J. F. KOONCE. 1973. MULTIVARIATE APPROACHES TO ALGAL STRATAGEMS AND TACTICS IN SYSTEMS ANALYSIS OF PHYTOPLANKTON. ECOLOGY 54(6), 1234-1246. [COMPUTER; PRINCIPAL COMPONENT ANALYSIS; NUMERICAL CLASSIFICATION]

ANDERSON, E. 1956. NATURAL HISTORY, STATISTICS, AND APPLIED MATHEMATICS. AMER. J. BOT. 43(10), 882-889.

ANDERSON, R. M. 1974. MATHEMATICAL MODELS OF HOST-HELMINTH PARASITE INTERACTIONS. PP. 43-69. IN, ECOLOGICAL STABILITY. (M. B. USHER AND M. H. WILLIAMSON, EDS.). HALSTED PRESS, NEW YORK, NEW YORK. [STOCHASTIC MODEL; NEGATIVE BINOMIAL DISTRIBUTION; DETERMINISTIC MODEL]

ANONYMOUS. 1972. DISCUSSION OF THE APPLICATION OF MATHEMATICAL MODELS TO THE ASSESSMENT OF THE CHANGES LIKELY FROM THE CONSTRUCTION OF AN ESTUARINE BARRAGE IN MORECAMBE BAY. PP. 345-354. IN, MATHEMATICAL MODELS IN ECOLOGY. (J. N. R. JEFFERS, ED.). BLACKWELL SCIENTIFIC PUBLICATIONS, OXFORD, ENGLAND.

AR, A., C. V. PAGANELLI, R. B. REEVES, D. G. GREENE AND H. RAHN. 1974. THE AVIAN EGG: WATER VAPOR CONDUCTANCE, SHELL THICKNESS, AND FUNCTIONAL PORE AREA. CONDOR 76(2), 153-158. [MODEL]

ARGIERO, L., A. M. SERRA-GENTILI AND G. ZOLI. 1971. MATHEMATICAL MODEL OF MARINE DIFFUSION OF CONTAMINANTS AND RELEVANT EXPERIMENTAL MEASURES. PP. 957-976. IN, PROC. INTERN. SYMP. RADIOECOLOGY APPLIED TO THE PROTECTION OF MAN AND HIS ENVIRONMENT. COMMISSION OF THE EUROPEAN COMMUNITIES, REPT. EUR-4800 D-F-I-E. [TAG-RECAPTURE; MATRIX; CONDITIONAL LIKELIHOOD; MOMENT EQUATIONS]

ASMUNDSON, V. S., G. A. BAKER AND J. T. EMLEN. 1943. CERTAIN RELATIONS BETWEEN THE PARTS OF BIRDS' EGGS. AUK 60(1), 34-44. [LOG TRANSFORMATION; FUNCTION; SIMPLE LINEAR REGRESSION]

ATTIWILL, P. M. 1966. A METHOD FOR ESTIMATING CROWN WEIGHT IN EUCALYPTUS AND SOME IMPLICATIONS OF RELATIONSHIPS BETWEEN CROWN WEIGHT AND STEM DIAMETER. ECOLOGY 47(5), 795-804. [MODEL; VARIANCE]

BAILEY, R. L. AND J. L. CLUTTER. 1974. BASE-AGE INVARIANT POLYMORPHIC SITE CURVES. FOREST SCI. 20(2), 155-159. [MODEL]

BAKER, J. R. 1936. NOMOGRAMS FOR SATURATION DEFICIENCY. J. ANIM. ECOL. 5(1), 94-96. [ATMOSPHERE]

BAKER, R., C. L. MAURER AND R. A. MAURER. 1967. ECOLOGY OF PLANT PATHOGENS IN SOIL. VIII. MATHEMATICAL MODELS AND INOCULUM DENSITY. PHYTOPATHOLOGY 57(7), 662-666.

BAKUZIS, E. V. AND R. M. BROWN. 1962. ELEMENTS OF MODEL CONSTRUCTION AND THE USE OF TRIANGULAR MODELS IN FORESTRY RESEARCH. FOREST SCI. 8(2), 119-131.

BARBIERI, J. F. AND R. I. POLLACK. 1973. NUMERICAL SOLUTIONS OF ECOSYSTEM MODELS. AEC REPT. UCID-16337. LAWRENCE LIVERMORE LAB., UNIVERSITY OF CALIFORNIA. II, 37 PP.

BARTHOLOMAY, A. F. 1968. THE CASE FOR MATHEMATICAL BIOLOGY OR THE MATHEMATICAL DESTINY OF BIOLOGY. BIOSCIENCE 18(7), 717-726.

BARTOS, D. L. AND D. A. JAMESON. 1974. A DYNAMIC ROOT MODEL AMER. MIDL. NAT. 91(2), 499-504. [LOGISTIC FUNCTION]

BELLMAN, R., H. KAGIWADA AND R. KALABA. 1966. INVERSE PROBLEMS IN ECOLOGY. J. THEOR. BIOL. 11(1), 164-167.

BEN-YAAKOV, S. AND I. R. KAPLAN 1969. DETERMINATION OF CARBONATE SATURATION OF SEAWATER WITH A CARBONATE SATUROMETER. LIMNOL. OCEANOGR. 14(6), 874-882. [MODEL]

BENEDICT, B. A., J. L. ANDERSON AND E. L. YANDELL, JR. 1974. ANALYTICAL MODELING OF THERMAL DISCHARGES: A REVIEW OF THE STATE OF THE ART . AEC REPT. ANL/ES-18. ARGONNE NATL. LAB., ARGONNE, ILLINOIS. 321 PP.

BENNETT, E. B. 1959. SOME OCEANOGRAPHIC FEATURES OF THE NORTHEAST PACIFIC OCEAN DURING AUGUST 1955. J. FISH. RES. BOARD CAN. 16(5), 565-633. [EQUATIONS]

BENNETT, L. 1973. EFFECTIVENESS AND FLIGHT OF SMALL INSECTS. ANN. ENTOMOL. SOC. AMER. 66(6), 1187-1190. [MODEL]

BERGEN, J. D. 1971. VERTICAL PROFILES OF WINDSPEED IN A PINE STAND. FOREST SCI. 17(3), 315-321. [MODEL]

BISCOE, P. V., J. A. CLARK, K. GREGSON, M. MCGOWAN, J. L. MONTEITH AND R. K. SCOTT. 1975. BARLEY AND ITS ENVIRONMENT. I. THEORY AND PRACTICE. J. APPL. ECOL. 12(1), 227-257. [TRANSFER EQUATIONS; BROWN RATIO METHOD; AERODYNAMIC METHOD; COMPUTER]

BLANC, F. AND P. KERAMBRUN. 1972. ESSAI D'APPRECIATION D'UNE PHENOMENE D'ALLOMETRIE CHIMIQUE DANS UN UNIVERS MULTIVARIE. APPLICATION A L'ETUDE DES POPULATIONS NATURALLES DE SPHAEROMA HOOKERI (CRUSTACEA: ISOPODA FLABELLIFERA). MARINE BIOL. 17(2), 158-161. [MULTIVARIATE MODEL; ALLOMETRY; PRINCIPAL AXES]

BLANC, F., M. LEVEAU, M.-C. BONIN AND A. LAUREC. 1972. ECOLOGIE D'UN MILIEU EUTROPHIQUE: TRAITEMENT MATHEMATIQUE DES DONNEES. MARINE BIOL. 14(2), 120-129. [MULTIVARIATE; CLUSTER ANALYSIS]

BLANTON, J. O. 1973. VERTICAL ENTRAINMENT INTO THE EPILIMNIA OF STRATIFIED LAKES. LIMNOL. OCEANOGR. 18(5), 697-704. [MODEL]

BLOOM, S. G., A. A. LEVIN AND G. E. RAINES. 1974. FUTURE USES OF MATHEMATICAL MODELS TO PREDICT ECOLOGICAL EFFECTS FROM THERMAL DISCHARGES. PP. 174-190. IN, ENERGY PRODUCTION AND THERMAL EFFECTS. (B. J. GALLAGHER, ED.). LIMNETICS INC., MILWAUKEE, WISCONSIN.

BOORMAN, S. A. AND P. R. LEVITT. 1973. A FREQUENCY-DEPENDENT NATURAL SELECTION MODEL FOR THE EVOLUTION OF SOCIAL COOPERATION NETWORKS. PROC. NATL. ACAD. SCI. U.S.A. 70(1), 187-189.

BOORMAN, S. A. AND P. R. LEVITT. 1973. GROUP SELECTION ON THE BOUNDARY OF A STABLE POPULATION. THEOR. POPUL. BIOL. 4(1), 85-128. [MODEL; LEVINS' MODEL; LOGISTIC]

BOULDIN, D. R. 1968. MODELS FOR DESCRIBING THE DIFFUSION OF OXYGEN AND OTHER MOBILE CONSTITUENTS ACROSS THE MUD-WATER INTERFACE. J. ECOL. 56(1), 77-87.

BOULDING, K. E. 1972. ECONOMICS AS A NOT VERY BIOLOGICAL SCIENCE. PP. 357-375. IN, CHALLENGING BIOLOGICAL PROBLEMS. (J. A. BEHNKE, ED.). OXFORD UNIVERSITY PRESS, NEW YORK, NEW YORK. [POSSIBILITY BOUNDARY; MODEL; COMPETITION]

BOYCE, F. M. 1974. SOME ASPECTS OF GREAT LAKES PHYSICS OF IMPORTANCE TO BIOLOGICAL AND CHEMICAL PROCESSES. J. FISH. RES. BOARD CAN. 31(5), 689-730. [MODEL]

BRADSHAW, A. 1973. THE EFFECT OF CARBON DIOXIDE ON THE SPECIFIC VOLUME OF SEAWATER. LIMNOL. OCEANOGR. 18(1), 95-105. [EQUATIONS]

BRADSHAW, W. E. 1973. HOMEOSTASIS AND POLYMORPHISM IN VERNAL DEVELOPMENT OF CHAOBORUS AMERICANUS. ECOLOGY 54(6), 1247-1259. [MODEL]

BROCKINGTON, N. R. 1972. SUMMARY AND ASSESSMENT: AN AGRICULTURAL RESEARCH SCIENTIST'S POINT OF VIEW. PP. 361-365. IN, MATHEMATICAL MODELS IN ECOLOGY. (J. N. R. JEFFERS, ED.). BLACKWELL SCIENTIFIC PUBLICATIONS, OXFORD, ENGLAND.

BROWN, C. E. AND B. S. MUIR. 1970. ANALYSIS OF RAM VENTILATION OF FISH GILLS WITH APPLICATION TO SKIPJACK TUNA(KATSUWONUS PELAMIS). J. FISH. RES. BOARD CAN. 27(9), 1637-1652. [EQUATIONS]

BRUCE, D., R. O. CURTIS AND C. VANCOEVERING. 1968. DEVELOPMENT OF A SYSTEM OF TAPER AND VOLUME TABLES FOR RED ALDER. FOREST SCI. 14(3), 339-350. [EQUATIONS]

BYRAM, G. M. AND R. E. MARTIN. 1970. THE MODELING OF FIRE WHIRLWINDS. FOREST SCI. 16(4), 386-399.

CALHOON, R. E. AND D. L. JAMESON. 1970. CANONICAL CORRELATION BETWEEN VARIATION IN WEATHER AND VARIATION IN SIZE IN THE PACIFIC TREE FROG, HYLA REGILLA, IN SOUTHERN CALIFORNIA. COPEIA 1970(1), 124-134. [DISCRIMINANT ANALYSIS; COMPUTER]

CALLAHAN, P. S. 1965. INTERMEDIATE AND FAR INFRARED SENSING OF NOCTURNAL INSECTS. PART I. EVIDENCES FOR A FAR INFRARED (FIR) ELECTROMAGNETIC THEORY OF COMMUNICATION AND SENSING IN MOTHS AND ITS RELATIONSHIP TO THE LIMITING BIOSPHERE OF THE CORN EARWORM. ANN. ENTOMOL. SOC. AMER. 58(5), 727-745. [EQUATIONS]

CHARNOV, E. L. AND J. R. KREBS. 1975. THE EVOLUTION OF ALARM CALLS: ALTRUISM OR MANIPULATION? AMER. NAT. 109(965), 107-112. [MODEL; PROBABILITY]

CHARTERS, A. C., M. NEUSHUL AND D. COON. 1973. THE EFFECT OF WATER MOTION ON ALGAL SPORE ADHESION. LIMNOL. OCEANOGR. 18(6), 884-896. [EQUATIONS]

CHARTIER, PH. 1970. A MODEL OF CO2 ASSIMILATION IN THE LEAF. PP. 307-315. IN, PREDICTION AND MEASUREMENT OF PHOTOSYNTHETIC PRODUCTIVITY. (I. SETLIK, ED.). CENTRE FOR AGRICULTURAL PUBLISHING AND DOCUMENTATION, WAGENINGEN, NETHERLANDS.

CHELSKY, M. AND J. J. ANGULO. 1973. TWO MODELS FOR ESTIMATION OF SOME PARAMETERS OF DISEASE SPREAD. MATH. BIOSCI. 18, 119-131. [EPIDEMIOLOGY]

CHEN, C. W. 1970. CONCEPTS AND UTILITIES OF ECOLOGIC MODEL. AMER. SOC. CIVIL ENG., SAN. ENG. DIV. PROC. SA 5, 1085-1097. [PLANKTON MODELS]

CIONCO, R. M. 1965. A MATHEMATICAL MODEL FOR AIR FLOW IN A VEGETATIVE CANOPY. J. APPL. METEOROL. 4(4), 517-522.

COBBLE, M. H. 1971. THE SHAPE OF PLANT STEMS. AMER. MIDL. NAT. 86(2), 371-378.

CODY, M. L. 1973. CHARACTER CONVERGENCE. PP. 189-211. IN, ANNUAL REVIEW OF ECOLOGY AND SYSTEMATICS, VOLUME 4. (R. F. JOHNSTON, P. W. FRANK AND C. D. MICHENER, EDS.). ANNUAL REVIEWS, INC., PALO ALTO, CALIFORNIA. [VOLTERRA-GAUSE EQUATIONS; TERRITORIALITY; MAHALANOBIS' D2]

CONE, C. D., JR. 1964. A MATHEMATICAL ANALYSIS OF THE DYNAMIC SOARING FLIGHT OF THE ALBATROSS WITH ECOLOGICAL INTERPRETATIONS. VA. INST. MAR. SCI. SPEC. SCI. REPT. 50. 104 PP.

CONNORS, D. N. AND P. K. WEYL. 1968. THE PARTIAL EQUIVALENT CONDUCTANCES OF SALTS IN SEAWATER AND THE DENSITY/CONDUCTANCE RELATIONSHIPS. LIMNOL. OCEANOGR. 13(1), 39-50. [EQUATIONS]

CONWAY, G. R. 1973. EXPERIENCE IN INSECT PEST MODELLING: A REVIEW OF MODELS, USES AND FUTURE DIRECTIONS. PP. 1-29. IN, INSECT STUDIES IN POPULATION MANAGEMENT. (P. W. GEIER, L. R. CLARK, D. J. ANDERSON AND H. A. NIX, EDS.). ECOLOGICAL SOCIETY OF AUSTRALIA MEMOIRS 1.

COPPINGER, R. P. AND L. K. BAXTER. 1969. INFORMATION THEORY: A CRITICISM OF ITS APPLICATION TO BATESIAN MIMICRY. AMER. NAT. 103(933), 551-552. [SHANNON'S FORMULA]

COULSON, C. A. 1973. MATHEMATICS AND BIOLOGY. PP. 1-2. IN, THE MATHEMATICAL THEORY OF THE DYNAMICS OF BIOLOGICAL POPULATIONS. (M. S. BARTLETT AND R. W. HIORNS, EDS.). ACADEMIC PRESS, NEW YORK, NEW YORK. [HISTORY]

COWAN, I. R. 1965. TRANSPORT OF WATER IN THE SOIL-PLANT-ATMOSPHERE SYSTEM. J. APPL. ECOL. 2(1), 221-239. [EQUATIONS]

COWAN, I. R. 1971. LIGHT IN PLANT STANDS WITH HORIZONTAL FOLIAGE. J. APPL. ECOL. 8(2), 579-580.

CRIMINALE, W. O., JR. AND D. F. WINTER. 1974. THE STABILITY OF STEADY-STATE DEPTH DISTRIBUTIONS OF MARINE PHYTOPLANKTON. AMER. NAT. 108(963), 679-687. [MODEL; EIGENVALUE; RILEY'S MODEL]

CRISP, D. J. AND R. WILLIAMS. 1971. DIRECT MEASUREMENT OF PORE-SIZE DISTRIBUTION ON ARTIFICIAL AND NATURAL DEPOSITS AND PREDICTION OF PORE SPACE ACCESSIBLE TO INTERSTITIAL ORGANISMS. MARINE BIOL. 10(3), 214-226. [EQUATIONS]

CROWLEY, P. H. 1973. FILTERING RATE INHIBITION OF DAPHNIA PULEX IN WINTERGREEN LAKE WATER. LIMNOL. OCEANOGR. 18(3), 394-402. [MODEL]

CROZIER, W. J. 1929. THE STUDY OF LIVING ORGANISMS. PP. 45-127. IN, THE FOUNDATIONS OF EXPERIMENTAL PSYCHOLOGY. CLARK UNIVERSITY PRESS, WORCESTER, MASSACHUSETTS.

CROZIER, W. J. 1935. DETERMINISME ET VARIABILITE DANS LE COMPORTEMENT DES ORGANISMES. ACTUAL. SCI. IND. 261. EXPOSES DE BIOMETRIE ET DE STATISTIQUE BIOLOGIQUE. VII. 56 PP.

CZARNOWSKI, M. AND J. SLOMKA. 1959. SOME REMARKS ON THE PERCOLATION OF LIGHT THROUGH THE FOREST CANOPY. ECOLOGY 40(2), 312-315. [EQUATIONS]

DALE, M. B. 1971. VALIDITY AND UTILITY OF INFORMATION THEORY IN ECOLOGICAL RESEARCH. PROC. ECOL. SOC. AUSTRALIA 6, 7-17.

DALY, M. T. 1972. CONCEPTUAL MODELS OF THE CITY. PROC. ECOL. SOC. AUSTRALIA 7, 89-103. [SYSTEMS ANALYSIS; INFORMATION; COST FUNCTION]

DAVIS, R. B. 1974. STRATIGRAPHIC EFFECTS OF TUBIFICIDS IN PROFUNDAL LAKE SEDIMENTS. LIMNOL. OCEANOGR. 19(3), 466-488. [MODEL; COMPUTER]

DAVIS, R. L. 1954. STRUCTURES OF DOMINANCE RELATIONS. BULL. MATH. BIOPHYS. 16, 131-140. [BEHAVIOR]

DAYKIN, P. N., F. E. KELLOGG AND R. H. WRIGHT. 1965. HOST-FINDING AND REPULSION OF AEDES AEGYPTI. CAN. ENTOMOL. 97(3), 239-263. [MODEL]

DEAKIN, M. A. B. 1975. THE STEADY STATES OF ECOSYSTEMS. MATH. BIOSCI. 24(3/4), 319-331. [DIFFERENTIAL EQUATION; MODEL]

DIAMOND, J. M. 1972. BIOGEOGRAPHIC KINETICS: ESTIMATION OF RELAXATION TIMES FOR AVIFAUNAS OF SOUTHWEST PACIFIC ISLANDS. PROC. NATL. ACAD. SCI. U.S.A. 69(11), 3199-3203. [EQUATIONS]

DINGLE, H. 1972. AGGRESSIVE BEHAVIOR IN STOMATOPODS AND THE USE OF INFORMATION THEORY IN THE ANALYSIS OF ANIMAL COMMUNICATION. PP. 126-156. IN. BEHAVIOR OF MARINE ANIMALS, VOLUME 1, INVERTEBRATES. (H. E. WINN AND B. L. OLLA, EDS.). PLENUM PRESS, NEW YORK, NEW YORK.

DOLININ, V. A. 1973. THEORETICAL ASPECTS OF THE EXPERIMENTAL STUDY OF RESPIRATORY METABOLISM IN FISHES. J. ICHTHYOLOGY 13(2), 245-253. [EQUATIONS]

DOURY, A. AND C. BADIE. 1973. PRACTICAL METHOD GIVING A QUANTITATIVE ESTIMATE OF OCEANIC POLLUTION. CEA CENTRE D'ETUDES DE BRUYERES-LE-CHATEL, MONTROUGE, FRANCE. REPORT CEA-R-4512. 23 PP. [MODEL; GAUSSIAN SOLUTION; DIFFUSION EQUATION]

DRURY, W. H. AND I. C. T. NISBET. 1971. INTER-RELATIONS BETWEEN DEVELOPMENTAL MODELS IN GEOMORPHOLOGY, PLANT ECOLOGY, AND ANIMAL ECOLOGY. GENERAL SYSTEMS 16, 57-68.

DUEDALL, I. W. AND P. K. WEYL. 1967. THE PARTIAL EQUIVALENT VOLUMES OF SALTS IN SEAWATER. LIMNOL. OCEANOGR. 12(1), 52-59. [EQUATIONS]

DUTTON, J. A. AND R. A. BRYSON. 1962. HEAT FLUX IN LAKE MENDOTA. LIMNOL. OCEANOGR. 7(1), 80-97. [EQUATIONS]

DYUL'DIN, A. A. 1973. THE COEFFICIENT OF VARIATION AND ALLOMETRY. EKOLOGIYA 4(6), 97-99. (ENGLISH TRANSL. PP. 546-548).

EDWARDS, A. W. F. 1963. NATURAL SELECTION AND THE SEX RATIO: THE APPROACH TO EQUILIBRIUM. AMER. NAT. 97(897), 397-400. [MODEL]

EICKWORT, K. R. 1973. CANNIBALISM AND KIN SELECTION IN LABIDOMERA CLIVICOLLIS (COLEOPTERA: CHRYSOMELIDAE). AMER. NAT. 107(955), 452-453. [MODEL]

EIJKMAN, E. G. 1970. LEARNING AND FLUCTUATING BEHAVIOUR. EXPERIMENTS WITH GOLDFISH. J. THEOR. BIOL. 26(2), 251-264. [DETECTION AND DECISION THEORY; PROBABILITY]

EMLEN, J. M. 1968. BATESIAN MIMICRY: A PRELIMINARY THEORETICAL INVESTIGATION OF QUANTITATIVE ASPECTS. AMER. NAT. 102(925), 235-241. [MODEL]

EMLEN, J. M. 1968. A NOTE ON NATURAL SELECTION AND THE SEX RATIO. AMER. NAT. 102(923), 94-95. [MACARTHUR'S MODEL]

EMLEN, J. M. 1968. SELECTION FOR SEX RATIO. AMER. NAT. 102(928), 589-591. [MODEL; GENETICS]

ENDELMAN, F. J., G. E. P. BOX, J. R. BOYLE, R. R. HUGHES, D. R. KEENEY, M. L. NORTHUP AND P. G. SAFFIGNA. 1974. THE MATHEMATICAL MODELING OF SOIL-WATER-NITROGEN PHENOMENA. AEC REPT. EDFB-IBP-74-8. OAK RIDGE NATL. LAB., OAK RIDGE, TENNESSEE. IX, 66 PP. [NON-LINEAR MODEL; COMPUTER]

ESHEL, I. 1971. ON EVOLUTION IN A POPULATION WITH AN INFINITE NUMBER OF TYPES. THEOR. POPUL. BIOL. 2(2), 209-236. [MODEL]

ESPEY, W. H., JR., D. R. BETTERTON, A. J. MARY AND W. M. MAYER. 1971. MODELING OF THERMAL DISCHARGES IN SHALLOW ESTUARIES. PROC. AMER. POWER CONF. 33, 457-464.

ESSER, M. H. M. 1946. TREE TRUNKS AND BRANCHES AS OPTIMUM MECHANICAL SUPPORTS OF THE CROWN: I. THE TRUNK. BULL. MATH. BIOPHYS. 8, 65-74.

ESSER, M. H. M. 1946. TREE TRUNKS AND BRANCHES AS OPTIMUM MECHANICAL SUPPORTS OF THE CROWN: II. THE BRANCHES. BULL. MATH. BIOPHYS. 8, 95-100.

EVANS, W. G. 1966. PERCEPTION OF INFRARED RADIATION FROM FOREST FIRES MELANOPHILA ACUMINATA DE GEER (BUPRESTIDAE, COLEOPTERA). ECOLOGY 47(6), 1061-1065. [EQUATIONS]

FECHTER, H. 1973. ON THE FUNCTION OF ANAL-CONE ROTATION IN DIADEMATID SEA URCHINS. MARINE BIOL. 22(4), 347-351. [EQUATIONS]

FEDOROV, V. D. AND S. A. SOKOLOVA. 1973. AN ATTEMPT AT EVALUATING THE STABILITY OF AN AQUATIC ECOSYSTEM. HYDROBIOL. J. 9(2), 6-8. (ENGLISH TRANSL). [MEASURE OF STABILITY]

FLEMINGER, A. AND R. I. CLUTTER. 1965. AVOIDANCE OF TOWED NETS BY ZOOPLANKTON. LIMNOL. OCEANOGR. 10(1), 96-104. [EQUATIONS]

FOLIE, G. M., D. T. HOWELL AND R. F. WARNER. 1972. QUANTITATIVE MODELS FOR URBAN SYSTEMS. PROC. ECOL. SOC. AUSTRALIA 7, 105-124. [GRAVITY MODELS; PREDICTIVE MODELS; INPUT-OUTPUT MODELS; COMPUTER; GAMING SIMULATION; PRESCRIPTIVE MODELS]

FOREST H. S. AND H. GREENSTEIN. 1966. BIOLOGISTS AS PHILOSOPHERS. BIOSCIENCE 16(11), 783-788. [MODEL; INFORMATION]

FORSYTHE, W. L. AND O. L. LOUCKS. 1972. A TRANSFORMATION FOR SPECIES RESPONSE TO HABITAT FACTORS. ECOLOGY 53(6), 1112-1119. [PARABOLA TRANSFORMATION; REGRESSION; CURVE FITTING]

FOSBERG, M. A. 1971. CLIMATOLOGICAL INFLUENCES ON MOISTURE CHARACTERISTICS OF DEAD FUEL: THEORETICAL ANALYSIS. FOREST SCI. 17(1), 64-72. [SIMULATION; MODEL]

FOSBERG, M. A. 1972. THEORY OF PRECIPITATION EFFECTS ON DEAD CYLINDRICAL FUELS. FOREST SCI. 18(2), 98-108. [MODEL]

FOSTER, C. AND A. RAPOPORT. 1956. PARASITISM AND SYMBIOSIS IN AN N-PERSON NON-CONSTANT-SUM CONTINUOUS GAME. BULL. MATH. BIOPHYS. 18, 219-231.

FOX, D. L., H. U. SVERDRUP AND J. P. CUNNINGHAM. 1937. THE RATE OF WATER PROPULSION BY THE CALIFORNIA MUSSEL. BIOL. BULL. 72(3), 417-438. [EQUATIONS]

FREESE, M. 1973. THE AUTOMATIC DETECTION OF PARASITES BY THE ULTRASONIC ECHO METHOD. FISH. RES. BOARD CAN. TECH. REPT. NO. 346. VIII, 81 PP. [EQUATIONS]

GABOVICH, YE.YA. AND YE. M. MALKIN. 1974. AN ATTEMPT TO GIVE FORMAL DEFINITION TO THE BIOSTATIC METHOD OF ESTIMATING THE RELATIVE ABUNDANCE OF FOOD FISHES. J. ICHTHYOLOGY 14(1), 20-27. (ENGLISH TRANSL.). [ALGORITHM; COMPUTER]

GADGIL, M. 1971. DISPERSAL: POPULATION CONSEQUENCES AND EVOLUTION. ECOLOGY 52(2), 253-261.

GADGIL, M. AND W. H. BOSSERT. 1970. LIFE HISTORICAL CONSEQUENCES OF A NATURAL SELECTION. AMER. NAT. 104(935), 1-24. [MALTHUSIAN PARAMETER; MODEL; COLE'S MODEL; PROBABILITY]

GALE, J. 1972. ELEVATION AND TRANSPIRATION: SOME THEORETICAL CONSIDERATIONS WITH SPECIAL REFERENCE TO MEDITERRANEAN-TYPE CLIMATE. J. APPL. ECOL. 9(3), 691-702. [EQUATIONS]

GALLAGHER, J. N., P. V. BISCOE AND R. K. SCOTT. 1975. BARLEY AND ITS ENVIRONMENT V. STABILITY OF GRAIN WEIGHT. J. APPL. ECOL. 12(1), 319-336. [MODEL]

GALLUCCI, V. F. 1973. ON THE PRINCIPLES OF THERMODYNAMICS IN ECOLOGY. PP. 329-357. IN, ANNUAL REVIEW OF ECOLOGY AND SYSTEMATICS, VOLUME 4. (R. F.JOHNSTON, P. W. FRANK AND C. D. MICHENER, EDS.). ANNUAL REVIEWS, INC., PALO ALTO, CALIFORNIA. [INFORMATION THEORY; INFORMATION INDEX; LAWS OF THERMODYNAMICS; MODEL; LOTKA-VOLTERRA EQUATION]

GANI, J. 1965. ON A PARTIAL DIFFERENTIAL EQUATION OF EPIDEMIC THEORY. I. BIOMETRIKA 52(3/4), 617-622. [MODEL]

GAUSE, G. F. AND A. A. WITT. 1935. BEHAVIOR OF MIXED POPULATIONS AND THE PROBLEM OF NATURAL SELECTION. AMER. NAT. 69(725), 596-609.

GEMPERLE, M. E. AND F. W. PRESTON. 1955. VARIATION OF SHAPE IN THE EGGS OF THE COMMON TERN IN THE CLUTCH-SEQUENCE. AUK 72(2), 184-198. [EQUATIONS; RANDOM NUMBERS; PROBABILITY OF SELECTING TERMINAL EGG]

GERWITZ, A. AND E. R. PAGE. 1974. AN EMPIRICAL MATHEMATICAL MODEL TO DESCRIBE PLANT ROOT SYSTEMS. J. APPL. ECOL. 11(2), 773-781.

GIESKES, J. M. 1969 EFFECT OF TEMPERATURE ON THE PH OF SEAWATER. LIMNOL. OCEANOGR. 14(5), 679-685. [EQUATIONS]

GILLETT, B. E. 1970. INFORMAL REPORT: CRITICAL STUDY OF APPROXIMATING FUNCTIONS (AND METHODS) AS APPLIED TO OCEAN STATION DATA. PROJECT I. REGRESSION ANALYSIS. NAVAL OCEANOGRAPHIC OFFICE, WASHINGTON, D. C. REPORT, IR NO. 70-36. VI, 51 PP.

GLASS, N. R. 1967. A TECHNIQUE FOR FITTING NONLINEAR MODELS TO BIOLOGICAL DATA. ECOLOGY 48(6), 1010-1013. [ITERATIVE LEAST SQUARES]

GOFFMAN, W. AND V. A. NEWILL. 1964. GENERALIZATION OF EPIDEMIC THEORY: AN APPLICATION TO THE TRANSMISSION OF IDEAS. NATURE 204(4953), 225-228. [MODEL]

GOH, B. S., T. L. VINCENT AND D. J. WILSON. 1974. A METHOD FOR FORMULATING SUBOPTIMAL POLICIES FOR CRUDELY MODELLED ECOSYSTEMS. PP. 405-405. IN, PROC. FIRST INTERN. CONGR. ECOL. CENTRE FOR AGRICULTURAL PUBLISHING AND DOCUMENTATION, WAGENINGEN, NETHERLANDS. [OPTIMAL CONTROL THEORY]

GOODALL, D. W. 1970. STUDYING THE EFFECTS OF ENVIRONMENTAL FACTORS ON ECOSYSTEMS. PP. 19-28. IN, ANALYSIS OF TEMPERATE FOREST ECOSYSTEMS. (D. E. REICHLE, ED.). SPRINGER-VERLAG, NEW YORK, NEW YORK. [MODEL]

GOODALL, D. W. 1972. BUILDING AND TESTING ECOSYSTEM MODELS. PP. 173-194. IN, MATHEMATICAL MODELS IN ECOLOGY. (J. N. R. JEFFERS, ED.). BLACKWELL SCIENTIFIC PUBLICATIONS, OXFORD, ENGLAND.

GOODALL, D. W. 1974. PROBLEMS OF SCALE AND DETAIL IN ECOLOGICAL MODELLING. J. ENVIRON. MANAGE. 2(2), 149-157.

GOODMAN, A. S. AND W. E. DOBBINS. 1966. MATHEMATICAL MODEL FOR WATER POLLUTION CONTROL STUDIES. AMER. SOC. CIVIL ENG., SAN. ENG. DIV. PROC. 92(SA6), 1-19. [COMPUTER]

GORE, A. J. P. AND J. S. OLSON. 1968. PRELIMINARY MODELS FOR THE ACCUMULATION OF ORGANIC MATTER IN A ERIOPHORUM CALLUNA ECOSYSTEM. AQUILO SER. BOTANICA 6, 297-313.

GRIFFIN, D. R. 1952. BIRD NAVIGATION. BIOL. REV. 27(4), 359-393. [MATHEMATICS OF EXPLORATION]

GRIFFITHS, K. J. 1969. THE IMPORTANCE OF COINCIDENCE IN THE FUNCTIONAL AND NUMERICAL RESPONSES OF TWO PARASITES OF THE EUROPEAN PINE SAWFLY, NEODIPRION SERTIFER. CAN. ENTOMOL. 101(7), 673-713. [MODEL; WATT'S MODEL; HOLLING MODEL; DISC EQUATION; INDEX OF SYNCHRONY; DISPERSION COEFFICIENT; HOST-PARASITE MODEL]

HAFLEY, W. L. 1969. CALCULATION AND MISCALCULATION OF THE ALLOMETRIC EQUATION RECONSIDERED. BIOSCIENCE 19(11), 974-975, 983.

HAMMERTON, M. 1964. A CASE OF AN INAPPROPRIATE MODEL. NATURE 203(4940), 63-64.

HANKS, R. J., D. D. AUSTIN AND W. T. ONDRECHEN. 1971. SOIL TEMPERATURE ESTIMATION BY A NUMERICAL METHOD. SOIL SCI. SOC. AMER. PROC. 35(5), 665-667.

HANSELL, R. I. C. AND E. MARCHI. 1974. ASPECTS OF EVOLUTIONARY THEORY AND THE THEORY OF GAMES. PP. 66-72. IN, MATHEMATICAL PROBLEMS IN BIOLOGY. LECTURE NOTES IN BIOMATHEMATICS 2. (P. VAN DEN DRIESSCHE, ED.). SPRINGER-VERLAG NEW YORK INC., NEW YORK, NEW YORK. [GAME THEORY MODELS]

HANSEN, D. V. AND M. RATTRAY, JR. 1966. NEW DIMENSIONS IN ESTUARY CLASSIFICATION. LIMNOL. OCEANOGR. 11(3), 319-326. [EQUATIONS]

HARRIS, J. G. K. 1968. A MATHEMATICAL MODEL DESCRIBING THE POSSIBLE BEHAVIOUR OF A COPEPOD FEEDING CONTINUOUSLY IN A RELATIVELY DENSE RANDOMLY DISTRIBUTED POPULATION OF ALGAL CELLS. J. CONS. PERM. INT. EXPLOR. MER 32(1), 83-92.

HASKELL, E. F. 1940. MATHEMATICAL SYSTEMATIZATION OF "ENVIRONMENT", "ORGANISM" AND "HABITAT". ECOLOGY 21(1), 1-16. [CARTESIAN FRAMEWORK; ENTROPY]

HAUFE, W. O. 1963. ETHOLOGICAL AND STATISTICAL ASPECTS OF A QUANTAL RESPONSE IN MOSQUITOES TO ENVIRONMENTAL STIMULI. BEHAVIOUR 20(1/2), 221-241. [NORMAL DISTRIBUTION; MODEL]

HELLKVIST, J., G. P. RICHARDS AND P. G. JARVIS. 1974. VERTICAL GRADIENTS OF WATER POTENTIAL AND TISSUE WATER RELATIONS IN SITKA SPRUCE TREES MEASURED WITH THE PRESSURE CHAMBER. J. APPL. ECOL. 11(2), 637-667. [MODEL]

HESSLEIN, R. AND P. QUAY. 1973. VERTICAL EDDY DIFFUSION STUDIES IN THE THERMOCLINE OF A SMALL STRATIFIED LAKE. J. FISH. RES. BOARD CAN. 30(10), 1491-1500. [EQUATIONS]

HETHCOTE, H. W. 1974. ASYMPTOTIC BEHAVIOR AND STABILITY IN EPIDEMIC MODELS. PP. 83-92. IN, MATHEMATICAL PROBLEMS IN BIOLOGY. LECTURE NOTES IN BIOMATHEMATICS 2. (P. VAN DEN DRIESSCHE, ED.), SPRINGER-VERLAG NEW YORK INC., NEW YORK, NEW YORK. [DETERMINISTIC MODEL]

HILCHEY, J. D. AND R. D. COOPER. 1960. DOSIMETRY FOR STUDIES ON THE RADIOBIOLOGY OF TRIBOLIUM CASTANEUM USING THE VAN DE GRAAFF ELECTRON ACCELERATOR. J. ECON. ENTOMOL. 53(4), 496-500. [EQUATIONS]

HINCKLEY, T. M. AND G. A. RITCHIE. 1973. A THEORETICAL MODEL FOR CALCULATION OF XYLEM SAP PRESSURE FROM CLIMATOLOGICAL DATA. AMER. MIDL. NAT. 90(1), 56-69.

HOEM, J. M. 1971. ON THE INTERPRETATION OF CERTAIN VITAL RATES AS AVERAGES OF UNDERLYING FORCES OF TRANSITION. THEOR. POPUL. BIOL. 2(4), 454-468.

HOLLAND, D. L. AND P. A. GABBOTT. 1971. A MICRO-ANALYTICAL SCHEME FOR THE DETERMINATION OF PROTEIN, CARBOHYDRATE, LIPID AND RNA LEVELS IN MARINE INVERTEBRATE LARVAE. J. MAR. BIOL. ASSOC. U. K. 51(3), 659-668. [VARIANCE; REGRESSION]

HOLLING, C. S. 1966. THE STRATEGY OF BUILDING MODELS OF COMPLEX ECOLOGICAL SYSTEMS. PP. 195-214. IN, SYSTEMS ANALYSIS IN ECOLOGY. (K. E. F. WATT, ED.). ACADEMIC PRESS, NEW YORK, NEW YORK.

HONDA, H. 1971. DESCRIPTION OF THE FORM OF TREES BY THE PARAMETERS OF TREE-LIKE BODY: EFFECTS OF THE BRANCHING ANGLE AND BRANCH LENGTH ON THE SHAPE OF THE TREE-LIKE BODY. J. THEOR. BIOL. 31(2), 331-338.

HOPKINS, J. W. 1966. SOME CONSIDERATIONS IN MULTIVARIATE ALLOMETRY. BIOMETRICS 22(4), 747-760.

HOPPENSTEADT, F. 1974. THRESHOLDS FOR DETERMINISTIC EPIDEMICS. PP. 96-101. IN, MATHEMATICAL PROBLEMS IN BIOLOGY. LECTURE NOTES IN BIOMATHEMATICS 2. (P. VAN DEN DRIESSCHE, ED.). SPRINGER-VERLAG NEW YORK INC., NEW YORK, NEW YORK. [DETERMINISTIC MODEL]

HOPPENSTEADT, F. AND P. WALTMAN. 1970. A PROBLEM IN THE THEORY OF EPIDEMICS. MATH. BIOSCI. 9, 71-91. [MODEL]

HOPPENSTEADT, F. AND P. WALTMAN. 1971. A PROBLEM IN THE THEORY OF EPIDEMICS, II. MATH. BIOSCI. 12, 133-145. [DETERMINISTIC MODEL]

HORN, H. S. 1968. THE ADAPTIVE SIGNIFICANCE OF COLONIAL NESTING IN THE BREWER'S BLACKBIRD (EUPHAGUS CYANOCEPHALUS). ECOLOGY 49(4), 682-694. [FORAGING MODEL]

HOUSEHOLDER, A. S. 1945. DYNAMICS OF QUADRUPEDAL LOCOMOTION. BULL. MATH. BIOPHYS. 7, 53-57.

HOWLAND, H. C. 1974. OPTIMAL STRATEGIES FOR PREDATOR AVOIDANCE: THE RELATIVE IMPORTANCE OF SPEED AND MANEUVERABILITY. J. THEOR. BIOL. 47(2), 333-350. [PREDATOR-PREY MODEL; EQUATIONS; COMPUTER]

HUANG, H.-W. AND H. J. MOROWITZ. 1972. A METHOD FOR PHENOMENOLOGICAL ANALYSIS OF ECOLOGICAL DATA. J. THEOR. BIOL. 35(3), 489-503. [DIFFERENTIAL EQUATION; VOLTERRA'S EQUATIONS; ECOLOGICAL INDEX; STABILITY]

HUDDLESTON, J. V. 1974. OPTIMALITY IN THE CONTROL OF ENVIRONMENTAL SYSTEMS. BULL. MATH. BIOL. 36(4), 341-345. [DIFFERENTIAL EQUATION; VOLTERRA'S EQUATIONS]

HUDDLESTON, J. V., C. G. DEWALD AND H. N. JAGADEESH. 1974. A DYNAMIC MODEL OF AN ENVIRONMENTAL SYSTEM WITH N INTERACTING COMPONENTS AND P DEGREES OF FREEDOM. BULL. MATH. BIOL. 36(1), 91-96. [VOLTERRA'S EQUATIONS; DIFFERENTIAL EQUATION]

HUTCHINSON, G. E. 1941. LIMNOLOGICAL STUDIES IN CONNECTICUT IV. THE MECHANISMS OF INTERMEDIARY METOBOLISM IN STRATIFIED LAKES. ECOL. MONOGR. 11(1), 21-60. [MODEL]

IBBITT, R. P. 1972. EFFECTS OF RANDOM DATA ERRORS ON THE PARAMETER VALUES FOR A CONCEPTUAL MODEL. WATER RESOUR. RES. 8(1), 70-78. [RUNOFF MODEL]

IDSO, S. B. 1973. ON THE CONCEPT OF LAKE STABILITY. LIMNOL. OCEANOGR. 18(4), 681-683. [EQUATIONS]

IDSO, S. B. AND G. A. COLE. 1973. STUDIES ON A KENTUCKY KNOBS LAKE. V. SOME ASPECTS OF THE VERTICAL TRANSPORT OF HEAT IN THE HYPOLIMNION. J. ECOL. 61(2), 413-420. [EQUATIONS]

IDSO, S. B. AND R. G. GILBERT. 1974. ON THE UNIVERSALITY OF THE POOLE AND ATKINS SECCHI DISK-LIGHT EXTINCTION EQUATION. J. APPL. ECOL. 11(1), 399-401.

INNIS, G. 1974. DYNAMIC ANALYSIS IN "SOFT SCIENCE" STUDIES: IN DEFENSE OF DIFFERENCE EQUATIONS. PP. 102-122. IN, MATHEMATICAL PROBLEMS IN BIOLOGY. LECTURE NOTES IN BIOMATHEMATICS 2. (P. VAN DEN DRIESSCHE, ED.). SPRINGER-VERLAG NEW YORK INC., NEW YORK, NEW YORK. [LOTKA-VOLTERRA SYSTEM]

ISAACS, J. D. 1972. UNSTRUCTURED MARINE FOOD WEBS AND "POLLUTANT ANALOGUES". U. S. FISH WILDL. SERV. FISH. BULL. 70(3), 1053-1059. [MATRIX; EQUATIONS]

ISHIHARA, M., K. HOZUMI AND K. SHINOZAKI. 1972. A MATHEMATICAL MODEL OF THE FOOD-CONSUMER SYSTEM. I. A CASE WITHOUT FOOD REPLENISHMENT. RES. POPUL. ECOL.-KYOTO 13(2), 114-126.

IVAKHNENKO, A. G. AND N. V. GULYAN 1972. A MATHEMATICAL MODEL OF ARTIFICIAL AERATION OF A POND. HYDROBIOL. J. 8(1), 59-64. (ENGLISH TRANSL.)

IVANKOV, V. N. AND V. L. ANDREYEV, 1971. THE SOUTH KURIL' CHUM (ONCORHYNCHUS KETA (WALB.))-ECOLOGY, POPULATION STRUCTURE AND THE MODELLING OF THE POPULATION. J. ICHTHYOLOGY 11(4), 511-524. (ENGLISH TRANSL.). [CYBERNETIC MODEL; COMPUTER]

JACOBS, J. 1974. QUANTITATIVE MEASUREMENT OF FOOD SELECTION. A MODIFICATION OF THE FORAGE RATIO AND IVLEV'S ELECTRIVITY INDEX. OECOLOGIA 14, 413-417. [EQUATIONS]

JEFFERS, J. N. R. 1972. INTRODUCTION-THE CHALLENGE OF MATHEMATICS TO THE ECOLOGIST. PP. 1-11. IN, MATHEMATICAL MODELS IN ECOLOGY. (J. N. R. JEFFERS, ED.). BLACKWELL SCIENTIFIC PUBLICATIONS, OXFORD, ENGLAND.

JEFFERS, J. N. R. 1972. SUMMARY AND ASSESSMENT: A RESEARCH DIRECTOR'S POINT OF VIEW. PP. 355-359. IN, MATHEMATICAL MODELS IN ECOLOGY. (J. N. R. JEFFERS, ED.). BLACKWELL SCIENTIFIC PUBLICATIONS, OXFORD, ENGLAND.

JEFFERS, J. N. R. 1974. A MULTIVARIATE ANALYSIS OF BIOLOGICAL ACTIVITY IN WOODLAND SOILS. (ABSTRACT). BIOMETRICS 30(2), 387-388.

JEFFRIES, C. 1974. QUALITATIVE STABILITY AND DIGRAPHS IN MODEL ECOSYSTEMS. ECOLOGY 55(6), 1415-1419. [MATRIX]

JEFFRIES, C. 1975. STABILITY OF ECOSYSTEMS WITH COMPLEX FOOD WEBS. THEOR. POPUL. BIOL. 7(2), 149-155. [LOOPS; MODEL; MATRIX]

JOWETT, D., J. A. BROWNING AND B. C. HANING. 1974. NON-LINEAR DISEASE PROGRESS CURVES. PP. 115-136. IN, EPIDEMICS OF PLANT DISEASES: MATHEMATICAL ANALYSIS AND MODELING. (J. KRANZ, ED.). SPRINGER-VERLAG NEW YORK INC., NEW YORK, NEW YORK. [MODEL; DIFFERENTIAL EQUATION; LOGISTIC MODEL; COMPUTER SIMULATION]

KALMUS, H. AND S. MAYNARD SMITH. 1966. SOME EVOLUTIONARY CONSEQUENCES OF PEGMATYPIC MATING SYSTEMS (IMPRINTING). AMER. NAT. 100(916), 619-635. [GENETICS]

KATZ, P. L. 1974. A LONG-TERM APPROACH TO FORAGING OPTIMIZATION. AMER. NAT. 108(964), 758-782. [MODEL; DIFFERENCE EQUATION; BIRDS]

KAWAKAMI, T. 1964. MECHANICS OF TOWED NETS. BULL. JAPANESE SOC. SCI. FISH. 30(10), 858-871. [EQUATIONS]

KEMP, C. D. 1971. ON ECOSTRUCTURE AND SECOND-ORDER ENTROPY. (ABSTRACT). BIOMETRICS 27(1), 253-254.

KERNER, E. H. 1957. A STATISTICAL MECHANICS OF INTERACTING BIOLOGICAL SPECIES. BULL. MATH. BIOPHYS. 19, 121-146.

KERNER, E. H. 1959. FURTHER CONSIDERATIONS ON THE STATISTICAL MECHANICS OF BIOLOGICAL ASSOCIATIONS. BULL. MATH. BIOPHYS. 21, 217-255. [VOLTERRA]

KERNER, E. H. 1961. ON THE VOLTERRA-LOTKA PRINCIPLE. BULL. MATH. BIOPHYS. 23, 141-157.

KERNER, E. H. 1962. GIBBS ENSEMBLE AND BIOLOGICAL ENSEMBLE. ANN. N. Y. ACAD. SCI. 96(ART. 4), 975-984. [VOLTERRA-LOTKA MODEL]

KERNER, E. H. 1964. DYNAMICAL ASPECTS OF KINETICS. BULL. MATH. BIOPHYS. 26, 333-349.

KERNER, E. H. 1972. INTRODUCTION. PP. 1-66. IN, GIBBS ENSEMBLES: BIOLOGICAL ENSEMBLE. THE APPLICATION OF STATISTICAL MECHANICS TO ECOLOGICAL, NEURAL, AND BIOLOGICAL NETWORKS. (E. H. KERNER). GORDON AND BREACH, SCIENCE PUBLISHERS, INC., NEW YORK, NEW YORK.

KETCHUM, B. H. 1954. RELATION BETWEEN CIRCULATION AND PLANKTONIC POPULATIONS IN ESTUARIES. ECOLOGY 35(2), 191-200. [MODEL]

KIEFER, D. A. AND R. W. AUSTIN. 1974. THE EFFECT OF VARYING PHYTOPLANKTON CONCENTRATION ON SUBMARINE LIGHT TRANSMISSION IN THE GULF OF CALIFORNIA. LIMNOL. OCEANOGR. 19(1), 55-64. [EQUATIONS]

KIMURA, K. 1937. THICKNESS AND DENSITY OF THE FISH BODY. BULL. JAPANESE SOC. SCI. FISH. 6(2), 69-72. [EQUATIONS]

KINERSON, R. S., JR. 1974. SELECTED ASPECTS OF CO2 AND H2O EXCHANGE IN A DOUGLAS FIR STAND. AMER. MIDL. NAT. 91(1), 170-181. [EQUATIONS]

KITAGAWA, T. 1971. A CONTRIBUTION TO THE METHODOLOGY OF BIOMATHEMATICS: INFORMATION SCIENCE APPROACH TO BIOMATHEMATICS, I. MATH. BIOSCI. 12, 329-345. [INCLUDES ECOLOGY; SYSTEMS ANALYSIS]

KLOTZ, I. M. 1963. VARIATION OF SOLUBILITY WITH DEPTH IN THE OCEAN: A THERMODYNAMIC ANALYSIS. LIMNOL. OCEANOGR. 8(2), 149-151. [EQUATIONS]

KOIVO, A. J. AND G. R. PHILLIPS. 1971. IDENTIFICATION OF MATHEMATICAL MODELS FOR DO AND BOD CONCENTRATIONS IN POLLUTED STREAMS FROM NOISE CORRUPTED MEASUREMENTS. WATER RESOUR. RES. 7(4), 853-862. [COMPUTER]

KONDOH, M. 1969. BIOECONOMIC STUDIES ON THE COLONY OF AN ANT SPECIES, FORMICA JAPONICA MOTSCHULSKY 4. ALLOMETRIC STUDY OF THE BODY WEIGHT OF WORKERS IN RELATION TO THE HEAD, THORAX AND ABDOMEN WEIGHT. JAPANESE J. ECOL. 19(3), 96-102.

KOVESSI, F. 1938. ON EQUATIONS EXPRESSING THE DEVELOPMENT-RHYTHM OF LIVING BEINGS. ACTA BIOTHEOR. 4(2), 97-110.

KOWAL, N. E. 1971. A RATIONALE FOR MODELING DYNAMIC ECOLOGICAL SYSTEMS. PP. 123-194. IN, SYSTEMS ANALYSIS AND SIMULATION IN ECOLOGY, VOLUME 1. (B. C. PATTEN, ED.). ACADEMIC PRESS, NEW YORK, NEW YORK.

KOZAK, A. 1972. A SIMPLE METHOD TO TEST PARALLELISM AND COINCIDENCE FOR CURVILINEAR, MULTIPLE LINEAR AND MULTIPLE CURVILINEAR REGRESSION. PP. 133-144. IN, 3RD CONFERENCE ADVISORY GROUP OF FOREST STATISTICIANS. INSTITUT NATIONAL DE LA RECHERCHE AGRONOMIQUE, PARIS, FRANCE, I.N.R.A. PUBL. 72-3. [SEQUENTIAL ANALYSIS; MODEL; FORESTRY]

KRANZ, J. 1974. THE ROLE AND SCOPE OF MATHEMATICAL ANALYSIS AND MODELING IN EPIDEMIOLOGY. PP. 7-54. IN, EPIDEMICS OF PLANT DISEASES: MATHEMATICAL ANALYSIS AND MODELING. (J. KRANZ, ED.). SPRINGER-VERLAG NEW YORK INC., NEW YORK, NEW YORK.

KRAPIVIN, V. F. 1973. THE VERTICAL STRUCTURE OF AN AQUATIC ECOSYSTEM BASED ON ITS MATHEMATICAL MODEL. HYDROBIOL. J. 9(2), 1-5. (ENGLISH TRANSL.). [LYAPUNOV'S MODEL]

KRAVCHUK, T. S. 1973. METHODS OF ANALYZING MATHEMATICAL MODELS FOR REGULATION OF BIOCENOSES. HYDROBIOL. J. 9(1), 65-70. (ENGLISH TRANSL.). [DIFFERENTIAL EQUATION COEFFICIENT OF NATURAL INCREASE; COMPETITION]

KUZNETSOV, E. V., YU. D. CHUGUNOV AND V. YA. BRODSKII. 1972. DIURNAL RHYTHMS IN THE LOCOMOTOR AND MITOTIC ACTIVITY OF FROGS UNDER NATURAL CONDITIONS. EKOLOGIYA 3(1), 16-23. (ENGLISH TRANSL. PP. 10-15. [MULTIPLE REGRESSION MODEL]

LANDAU, H. G. 1951. ON DOMINANCE RELATIONS AND THE STRUCTURE OF ANIMAL SOCIETIES: I. EFFECT OF INHERENT CHARACTERISTICS. BULL. MATH. BIOPHYS. 13, 1-19.

LANDAU, H. G. 1951. ON DOMINANCE RELATIONS AND THE STRUCTURE OF ANIMAL SOCIETIES: II. SOME EFFECTS OF POSSIBLE SOCIAL FACTORS. BULL. MATH. BIOPHYS. 13, 245-262.

LANDAU, H. G. 1953. ON DOMINANCE RELATIONS AND THE STRUCTURE OF ANIMAL SOCIETIES: III. THE CONDITION FOR A SCORE STRUCTURE. BULL. MATH. BIOPHYS. 15, 143-148.

LANDSBERG, J. J. AND G. B. JAMES. 1971. WIND PROFILES IN PLANT CANOPIES: STUDIES ON AN ANALYTICAL MODEL. J. APPL. ECOL. 8(3), 729-741.

LASIEWSKI, R. C. AND W. R. DAWSON. 1969. CALCULATION AND MISCALCULATION OF THE EQUATIONS RELATING AVIAN STANDARD METABOLISM TO BODY WEIGHT. CONDOR 71(3), 335-336.

LEARY, R. A. AND K. E. SKOG. 1972. A COMPUTATIONAL STRATEGY FOR SYSTEM IDENTIFICATION IN ECOLOGY. ECOLOGY 53(5), 969-973. [RECURRENCE RELATION; MODEL]

LEBEDEVA, L. P. 1971. A METHOD OF STUDY OF THE STRUCTURE OF PHYTOPLANKTON COMMUNITIES WITH MATHEMATICAL MODELS. EKOLOGIYA (4), 5-11. (ENGLISH TRANSL. PP. 291-295).

LEE, K. K. AND J. A. LIGGETT. 1970. COMPUTATION FOR CIRCULATION IN STRATIFIED LAKES. AMER. SOC. CIVIL ENGRS., HYDRAULICS DIV. 96(HY10), 2089-2115.

LEGAY, J. M. 1971. CONTRIBUTION A L'ETUDE DE LA FORME DES PLANTES: DISCUSSION D'UN MODELE DE RAMIFICATION. BULL. MATH. BIOPHYS. 33, 387-401. [MODELS OF PLANT FORM AND BRANCHING]

LEGAY, J. M. AND R. PERNET. 1971. MODELE GEOMETRIQUE DE L'OEUF DE VER A SOIE. ACTA BIOTHEOR. 20(1/2), 18-28. [MODEL; GEOMETRY OF AN EGG]

LETTAU, H. 1974. APPENDIX: EVAPOTRANSPIRATION CLIMATONOMY OF DRAINLESS AREAS. (LETTAU, K., MODELING OF THE ANNUAL CYCLE OF SOIL MOISTURE.). PP. 349-352. IN: PHENOLOGY AND SEASONALITY MODELING. (H. LIETH, ED.). SPRINGER-VERLAG NEW YORK INC., NEW YORK, NEW YORK. [MODEL]

LEVIN, S. A. 1972. A MATHEMATICAL ANALYSIS OF THE GENETIC FEEDBACK MECHANISM. AMER. NAT. 106(948), 145-164. [PIMENTEL'S MODEL; PREDATOR-PREY MODEL]

LEVIN, S. A. AND R. T. PAINE. 1975. THE ROLE OF DISTURBANCE IN MODELS OF COMMUNITY STRUCTURE. PP. 56-63. IN: ECOSYSTEM ANALYSIS AND PREDICTION. (S. A. LEVIN, ED.). PROCEEDINGS OF A SIAM-SIMS CONFERENCE HELD AT ALTA, UTAH, JULY 1-5, 1974.

LEVINE, S. H. 1974. OPTIMAL ALLOCATION OF TIME IN RESOURCE HARVESTING. MATH. BIOSCI. 20, 171-178. [MODEL; VECTOR; NICHE]

LEVINE, S. H. 1975. DISCRETE TIME MODELING OF ECOSYSTEMS WITH APPLICATIONS IN ENVIRONMENTAL ENRICHMENT. MATH. BIOSCI. 24(3/4), 307-317. [MODEL; GRAPHICAL TECHNIQUE; PREDATOR-PREY MODEL]

LEVINS, R. 1963. THEORY OF FITNESS IN A HETEROGENEOUS ENVIRONMENT II. DEVELOPMENTAL FLEXIBILITY AND NICHE SELECTION. AMER. NAT. 97(893), 75-90. [MODEL; PROBABILITY]

LEVINS, R. 1968. EVOLUTIONARY CONSEQUENCES OF FLEXIBILITY. PP. 67-70. IN, POPULATION BIOLOGY AND EVOLUTION. (R. C. LEWONTIN, ED.). SYRACUSE UNIVERSITY PRESS, SYRACUSE, NEW YORK. [MODEL]

LEVINS, R. 1970. EXTINCTION. PP. 75-107. IN, SOME MATHEMATICAL PROBLEMS IN BIOLOGY. (M. GERSTENHABER, ED .). AMER. MATH. SOC., PROVIDENCE, RHODE ISLAND. [MODEL]

LEVINS, R. 1975. COMMENTS ON THE WOODIN-YORKE PAPER. PP. 42. IN, ECOSYSTEM ANALYSIS AND PREDICTION. (S. A. LEVIN, ED.), PROCEEDINGS OF A SIAM-SIMS CONFERENCE HELD AT ALTA, UTAH, JULY 1-5, 1974. [MODEL; COMPETITION]

LEVINS, R. 1975. PROBLEMS OF SIGNED DIGRAPHS IN ECOLOGICAL THEORY. PP. 264-277. IN, ECOSYSTEM ANALYSIS AND PREDICTION. (S. A. LEVIN, ED.). PROCEEDINGS OF A SIAM-SIMS CONFERENCE HELD AT ALTA, UTAH, JULY 1-5, 1974. [GRAPH; DIFFERENTIAL EQUATION; MATRIX]

LEVINS, R. (CHAIRMAN). 1966. PART II. ASPECTS OF THEORETICAL ECOLOGY. PP. 139-174. IN, PROCEEDINGS OF A CONFERENCE ON THEORETICAL BIOLOGY. (G. J. JACOBS, ED.). NATIONAL AERONAUTICS AND SPACE ADMINISTRATION REPORT NASA SP-104. WASHINGTON, D. C.

LEVINS, R. AND R. MACARTHUR. 1969. AN HYPOTHESIS TO EXPLAIN THE INCIDENCE OF MONOPHAGY. ECOLOGY 50(5), 910-911. [LEVINS' FORMULA; MODEL]

LOTKA, A. J. 1921. NOTE ON MOVING EQUILIBRA. PROC NATL. ACAD. SCI. U.S.A. 7(6), 168-172. [DIFFERENTIAL EQUATION]

LOTKA, A. J. 1923. CONTRIBUTION TO A QUANTITATIVE PARASITOLOGY. J. WASH. ACAD. SCI. 13(8), 152-158.

LOTKA, A. J. 1932. CONTRIBUTION TO THE MATHEMATICAL THEORY OF CAPTURE. I. CONDITIONS FOR CAPTURE. PROC. NATL. ACAD. SCI. U. S. A. 18(2), 172-178. [LOTKA-VOLTERRA EQUATION; PREDATOR-PREY MODEL]

LOVGREN, B. 1958. A MATHEMATICAL TREATMENT OF THE DEVELOPMENT OF COLONIES OF DIFFERENT KINDS OF SOCIAL WASPS. BULL. MATH. BIOPHYS. 20, 119-148.

LOWRY, W. P. 1959. ON THE USE OF MATHEMATICAL NOTATION IN ECOLOGICAL LITERATURE. ECOLOGY 40(3), 492.

LOWRY, W. P. 1962. STANDARDIZING FIELD ESTIMATES OF EVAPORATIVE SOIL MOISTURE LOSS RATES. ECOLOGY 43(4), 757-760. [EQUATIONS]

LOWRY, W. P. 1972. ATMOSPHERIC POLLUTION AND GLOBAL CLIMATIC CHANGE. ECOLOGY 53(5), 908-914. [MODEL]

MAGNUSON, J. J. 1970. HYDROSTATIC EQUILIBRIUM OF EUTHYNNUS AFFINIS, A PELAGIC TELEOST WITHOUT A GAS BLADDER. COPEIA 1970(1), 56-85. [MODEL]

MAHONEY, J. J. 1970. THE USE OF BOOLEAN ALGEBRA PARTITION ANALYSIS IN POLLUTION ECOLOGY. J. APPL. ECOL. 7(1), 169-176.

MAJOR, J. 1951. A FUNCTIONAL FACTORIAL APPROACH TO PLANT ECOLOGY. ECOLOGY 32(3), 392-413. [EQUATIONS]

MAJOR, J. 1974. KINDS AND RATES OF CHANGES IN VEGETATION AND CHRONOFUNCTIONS. PP. 7-18. IN, VEGETATION DYNAMICS. (R. KNAPP, ED.). DR. W. JUNK PUBLISHERS, THE HAGUE, NETHERLANDS. [EQUATIONS]

MANLY, B. F. J. 1974. A MODEL FOR CERTAIN TYPES OF SELECTION EXPERIMENTS. BIOMETRICS 30(2), 281-294. [NATURAL SELECTION; STOCHASTIC MODEL; PREDATOR-PREY MODEL; COMPETITION]

MATUDA, K. AND T. KAWAKAMI. 1968. EXPERIMENTAL VERIFICATION FOR SIMILARITY LAW ON DISTORTED MODEL NET. BULL. JAPANESE SOC. SCI. FISH. 34(1), 23-28. [EQUATIONS]

MAYNARD SMITH, J. 1966. SYMPATRIC SPECIATION. AMER. NAT. 100(916), 637-650. [NICHE; GENETICS]

MAYNARD SMITH, J. 1974. THE THEORY OF GAMES AND THE EVOLUTION OF ANIMAL CONFLICTS. J. THEOR. BIOL. 47(1), 209-219. [BEHAVIOR; MODEL]

MAZANOV, A. AND K. P. TOGNETTI. 1974. TAYLOR SERIES EXPANSION OF DELAY DIFFERENTIAL EQUATIONS - A WARNING. J. THEOR. BIOL. 46(1), 271-282. [POPULATION MODEL; MALTHUS EQUATION; COMPUTER; HUTCHINSON TIME LAG]

MCALLISTER, C. D. 1971. SOME ASPECTS OF NOCTURNAL AND CONTINUOUS GRAZING BY PLANKTONIC HERBIVORES IN RELATION TO PRODUCTION STUDIES. FISH. RES. BOARD CAN. TECH. REPT. NO. 248. XIX, 281 PP. [MODEL; COMPUTER; SIMULATION]

MCCORMICK, J. AND P. A. HARCOMBE. 1968. PHYTOGRAPH: USEFUL TOOL OR DECORATIVE DOODLE? ECOLOGY 49(1), 13-20. [EQUATIONS]

MCLAY, C. 1970. A THEORY CONCERNING THE DISTANCE TRAVELLED BY ANIMALS ENTERING THE DRIFT OF A STREAM. J. FISH. RES. BOARD CAN. 27(2), 359-370. [MODEL]

MCNEIL, R. AND F. CADIEUX. 1972. NUMERICAL FORMULAE TO ESTIMATE FLIGHT RANGE OF SOME NORTH AMERICAN SHOREBIRDS FROM FRESH WEIGHT AND WING LENGTH. BIRD-BANDING 43(2), 107-113. [SIMPLE LINEAR REGRESSION; CORRELATION COEFFICIENT]

MEATS, A. 1974. A POPULATION MODEL FOR TWO SPECIES OF TIPULA (DIPTERA, NEMATOCERA) DERIVED FROM DATA ON PHYSIOLOGICAL RELATIONS WITH THEIR ENVIRONMENT. OECOLOGIA 16(2), 119-138.

MENSHUTKIN, V. V. AND A. A. UMNOV. 1970. A MATHEMATICAL MODEL OF A VERY SIMPLE AQUATIC ECOSYSTEM. HYDROBIOL. J. 6(2), 18-23. (ENGLISH TRANSL.)

MENSHUTKIN, V. V. AND A. A. UMNOV. 1970. MATHEMATICAL MODEL OF THE ECOLOGICAL SYSTEM OF DRIVYATY LAKE. EKOLOGIYA 1(4), 3-10. (ENGLISH TRANSL. PP. 275-279). [COMPUTER; BIOMASS; MORTALITY; ENERGY]

MENSHUTKIN, V. V. AND T. I. PRIKHOD'KO. 1971. MATHEMATICAL MODEL OF VERTICAL DISTRIBUTION AND PRODUCTION OF PHYTOPLANKTON. HYDROBIOL. J. 7(2), 1-6. (ENGLISH TRANSL.) [SIMULATION; COMPUTER]

MILLER, P. C. 1969. TESTS OF SOLAR RADIATION MODELS IN THREE FOREST CANOPIES. ECOLOGY 50(5), 878-885.

MINER, J. R. 1922. THE PROBABLE ERROR OF THE VITAL INDEX OF A POPULATION. PROC. NATL. ACAD. SCI. U.S.A. 8(5), 106-108.

MISHIN, A. S. AND F. N. SEMEVSKII. 1971. GENERALIZATION OF ECOLOGICAL INFORMATION BY MEANS OF A MATHEMATICAL MODEL WITH LOCAL CORRECTIONS. EKOLOGIYA 2(5), 5-10. (ENGLISH TRANSL. PP. 391-395).

MITCHELL, R. 1969. A MODEL ACCOUNTING FOR SYMPATRY IN WATER MITES. AMER. NAT. 103(932), 331-346. [HOST-PARASITE MODEL]

MITTELSTAEDT, H. 1962. CONTROL SYSTEMS OF ORIENTATION IN INSECTS. ANNU. REV. ENTOMOL. 7, 177-198. [CONTROL THEORY]

MIYATA, M. 1970. COMPLEX GENERALIZATION OF CANONICAL CORRELATION AND ITS APPLICATION TO A SEA-LEVEL STUDY. J. MARINE RES. 28(2), 202-214. [COMPUTER; TIME SERIES; MULTIPLE COHERENCE]

MONTEITH, J. L. AND T. A. BULL. 1970. A DIFFUSIVE RESISTANCE POROMETER FOR FIELD USE II. THEORY, CALIBRATION AND PERFORMANCE. J. APPL. ECOL. 7(3), 623-638. [EQUATIONS]

MORALES, R. 1975. A PHILOSOPHICAL APPROACH TO MATHEMATICAL APPROACHES TO ECOLOGY. PP. 334-337. IN, ECOSYSTEM ANALYSIS AND PREDICTION. (S. A. LEVIN, ED.). PROCEEDINGS OF A SIAM-SIMS CONFERENCE HELD AT ALTA, UTAH, JULY 1-5, 1974. [MODEL]

MOSIMANN, J. E. 1958. AN ANALYSIS OF ALLOMETRY IN THE CHELONIAN SHELL. REV. CAN. BIOL. 17(2), 137-228. [LOG TRANSFORMATION]

MOSIMANN, J. E. 1958. THE EVOLUTIONARY SIGNIFICANCE OF RARE MATINGS IN ANIMAL POPULATIONS. EVOLUTION 12(2), 246-261. [PROBABILITY; LOGISTIC]

MOSS, R. 1973. THE DIGESTION AND INTAKE OF WINTER FOODS BY WILD PTARMIGAN IN ALASKA. CONDOR 75(3), 293-300. [MODEL]

MUNSON, S. C. AND J. F. YEAGER. 1949. BLOOD VOLUME AND CHLORIDE NORMALITY IN ROACHES (PERIPLANETA AMERICANA (L.)) INJECTED WITH SODIUM CHLORIDE SOLUTIONS. ANN. ENTOMOL. SOC. AMER. 42(2), 165-173. [EQUATIONS]

MURRAY, S. D. 1972. TURBULENT DIFFUSION OF OIL IN THE OCEAN. LIMNOL. OCEANOGR. 17(5), 651-660. [EQUATIONS]

NAIR, K. R. AND H. K. MUKERJI. 1960. CLASSIFICATION OF NATURAL AND PLANTATION TEAK (TECTONA GRANDIS) GROWN AT DIFFERENT LOCALITIES OF INDIA AND BURMA WITH RESPECT TO ITS PHYSICAL AND MECHANICAL PROPERTIES. SANKHYA 22(1/2), 1-20. [MAHALANOBIS' D2]

NAKAMURA, K. 1972. THE INGESTION IN WOLF SPIDERS. II. THE EXPRESSION OF DEGREE OF HUNGER AND AMOUNT OF INGESTION IN RELATION TO SPIDER'S HUNGER. RES. POPUL. ECOL.-KYOTO 14(1), 82-96. [MODEL; LOG TRANSFORMATION; COMPUTER]

NAKASUJI, F. AND K. KIRITANI. 1972. DESCRIPTIVE MODELS FOR THE SYSTEM OF THE NATURAL SPREAD OF INFECTION OF RICE DWARF VIRUS (RDV) BY THE GREEN RICE LEAFHOPPER, NEPHOTETTIX CINCTICEPS UHLER (HEMIPTERA: DELTOCEPHALIDAE). RES. POPUL. ECOL.-KYOTO 14(1), 18-35. [DETERMINISTIC MODEL; LOG TRANSFORMATION]

NAROLL, R. S. AND L. VON BERTALANFFY. 1956. THE PRINCIPLE OF ALLOMETRY IN BIOLOGY AND SOCIAL SCIENCES. GENERAL SYSTEMS 1, 76-89. [ALLOMETRIC EQUATION]

NELDER, J. A. 1972. SUMMARY AND ASSESSMENT: A STATISTICIAN'S POINT OF VIEW. PP. 367-373. IN, MATHEMATICAL MODELS IN ECOLOGY. (J. N. R. JEFFERS, ED.). BLACKWELL SCIENTIFIC PUBLICATIONS, OXFORD, ENGLAND.

NOBLE, V. E. AND J. C. AYERS. 1961. A PORTABLE PHOTOCELL FLUOREMETER FOR DILUTION MEASUREMENTS IN NATURAL WATERS. LIMNOL. OCEANOGR. 6(4), 457-461. [EQUATIONS]

NOONEY, G. C. 1965. MATHEMATICAL MODELS, REALITY AND RESULTS. J. THEOR. BIOL. 9(2), 239-252.

NORRIS, K. S. AND J. L. KAVANAU. 1966. THE BURROWING OF THE WESTERN SHOVEL-NOSED SNAKE, CHIONACTIS OCCIPITALIS HALLOWELL, AND THE UNDERSAND ENVIRONMENT. COPEIA 1966(4), 650-664. [EARTH PRESSURE MODEL]

NOUR, A. A. 1972. A STATISTICAL METHODOLOGY FOR PREDICTING THE POLLUTANTS IN A RIVER. WATER RESOUR. BULL. 8(1), 15-23. [COMPUTER; REGRESSION MODEL; ANALYSIS OF VARIANCE]

O'BRIEN, J. J. AND J. S. WROBLEWSKI. 1973. ON ADVECTION IN PHYTOPLANKTON MODELS. J. THEOR. BIOL. 38(1), 197-202.

O'REGAN, W. G. 1964. LIMITS FO[R] THE SUM OF PREDICTIONS IN REGRESSION ANALYSIS. FOREST SCI. 10(3), 300-301.

OKUBO, A. AND H. C. CHIANG. 1974. AN ANALYSIS OF THE KINEMATICS OF SWARMING OF ANARETE PRITCHARDI KIM (DIPTERA: CECIDOMYIIDAE). RES. POPUL. ECOL.-KYOTO 16(1), 1-42. [MODEL; LAGRANGIAN AUTOCORRELATION COEFFICIENT OF INSECT VELOCITIES; EFFECTIVE ADVECTION VELOCITY; EFFECTIVE DIFFUSIVITY FOR SWARMING; VARIANCE]

OPATOWSKI, I. 1944. ON THE FORM AND STRENGTH OF TREES: PART I. THE TRUNK. BULL. MATH. BIOPHYS. 6, 113-118.

OPATOWSKI, I. 1944. ON THE FORM AND STRENGTH OF TREES: PART II. THE PRIMARY BRANCHES. BULL. MATH. BIOPHYS. 6, 153-156.

OPATOWSKI, I. 1946. ON OBLIQUE GROWTH OF TREES UNDER THE ACTION OF WINDS. BULL. MATH. BIOPHYS. 8, 41-49.

ORITSLAND, N. A. 1974. A WINDCHILL AND SOLAR RADIATION INDEX FOR HOMEOTHERMS. J. THEOR. BIOL. 47(2), 413-420.

OSBORN, T. R. AND P. H. LEBLOND. 1974. STATIC STABILITY IN FRESHWATER LAKES. LIMNOL. OCEANOGR. 19(3), 544-545. [EQUATIONS; REPLY BY A. H. LEE AND G. K. RODGERS 19(3), 546.]

OSBORNE, M. F. M. 1951. AERODYNAMICS OF FLAPPING FLIGHT WITH APPLICATION TO INSECTS. J. EXPTL. BIOL. 28(2), 221-245. [EQUATIONS]

OVERTON, W. S. 1972. TOWARD A GENERAL MODEL STRUCTURE FOR A FOREST ECOSYSTEM. PP. 37-47. IN, PROCEEDINGS-RESEARCH ON CONIFEROUS FOREST ECOSYSTEMS-A SYMPOSIUM. (J. . FRANKLIN, L. J. DEMPSTER AND R. . WARING, EDS.). U. S. FOR. SERV. PACIFIC NORTHWEST FOREST EXPT. STN., PORTLAND, OREGON.

PAGANELLI, C. V., A. OLSZOWKA AND . AR. 1974. THE AVIAN EGG: SURFACE AREA, VOLUME, AND DENSITY. CONDOR 76(3), 319-325. [ALLOMETRIC EQUATION]

PARK, B. C. AND B. B. DAY. 1942. A SIMPLIFIED METHOD FOR DETERMINING THE CONDITION OF WHITE-TAILED DEER HERDS IN RELATION TO AVAILABLE FORAGE. U. S. DEPT. AGRICULTURE, TECH. BULL. NO. 840, 60 PP. [ANALYSIS OF VARIANCE; DISCRIMINANT ANALYSIS; T-TEST]

PARKER, R. A. 1975. STABILITY OF A NON-AUTONOMOUS ECOSYSTEM MODEL. INTERN. J. SYSTEMS SCI. 6(2), 197-200.

PARLANGE, J.-Y. 1974. APPENDIX: ANALYTIC SOLUTION TO MODEL PASSAGES THROUGH PHENOPHASES. (WAGGONER, P. E., MODELING SEASONALITY.). PP. 323-326. IN, PHENOLOGY AND SEASONALITY MODELING. (H. LIETH, ED.). SPRINGER-VERLAG NEW YORK INC., NEW YORK, NEW YORK.

PATLAK, C. S. 1953. A MATHEMATICAL CONTRIBUTION TO THE STUDY OF ORIENTATION OF ORGANISMS. BULL. MATH. BIOPHYS. 15, 431-476.

PATTEN, B. C. 1959. AN INTRODUCTION TO THE CYBERNETICS OF THE ECOSYSTEM: THE TROPHIC-DYNAMIC ASPECT. ECOLOGY 40(2), 221-231. [ENTROPY; INFORMATION THEORY; SHANNON'S FORMULA; WIENNER]

PATTEN, B. C. 1964. LETTER TO THE EDITOR-SALMON MIGRATION. J. CONS. PERM. INT. EXPLOR. MER 28(3), 443-444. [RANDOM WALK; PROBABILITY]

PEARSON, E. S. 1923. NATURAL SELECTION AND THE AGE AND AREA THEORY OF DR. J. C. WILLIS. BIOMETRIKA 15(1/2), 89-108. [DISTRIBUTION OF SPECIES LAW]

PEARSON, K. 1906. MATHEMATICAL CONTRIBUTIONS TO THE THEORY OF EVOLUTION. XV. MATHEMATICAL THEORY OF RANDOM MIGRATION. DRAPERS COMPANY, RES. MEMOIRS. BIOM. SER. 3, 1-54.

PENNYCUICK, C. J. 1969. THE MECHANICS OF BIRD MIGRATION. IBIS 111(4), 525-556. [MODEL]

PETARD, H. 1938. A CONTRIBUTION TO THE MATHEMATICAL THEORY OF BIG GAME HUNTING. AMER. MATH. MONTHLY 45(7), 446-447. [ALGORITHM; HUMOR]

PETIPA, T. S. AND N. P. MAKAROVA. 1969. DEPENDENCE OF PHYTOPLANKTON PRODUCTION ON RHYTHM AND RATE OF ELIMINATION. MARINE BIOL. 3(3), 191-195. [GRAZING MODEL; SINUSOIDAL EQUATION]

PIELOU, E. C. 1974. VEGETATION ZONES: REPETITION OF ZONES ON A MONOTONIC ENVIRONMENTAL GRADIENT J. THEOR. BIOL. 47(2), 485-489. [LOTKA-VOLTERRA EQUATION; MODEL; OSCILLATION; TIME LAG]

PLINSTON, D. T. 1972. PARAMETER SENSITIVITY AND INTERDEPENDENCE IN HYDROLOGICAL MODELS. PP. 237-247. IN, MATHEMATICAL MODELS IN ECOLOGY. (J. N. R. JEFFERS, ED.). BLACKWELL SCIENTIFIC PUBLICATIONS, OXFORD, ENGLAND.

POLICASTRO, A. J. AND J. V. TOKAR. 1972. HEATED-EFFLUENT DISPERSION IN LARGE LAKES: STATE-OF-THE-ART OF ANALYTICAL MODELING. PART 1. CRITIQUE OF MODEL FORMULATIONS. AEC REPT. ANL/ES/11. ARGONNE NATL. LAB., ARGONNE, ILLINOIS. XVIII, 374 PP.

POOLE, R. W. 1974. THE USE OF SIMULTANEOUS LINEAR REGRESSION EQUATIONS AS EMPIRICAL MODELS OF COMMUNITY STRUCTURE. MATH. BIOSCI. 20, 105-116. [MAXIMUM LIKELIHOOD; FACTOR ANALYSIS]

POPE, J. A. 1964. STATISTICAL ANALYSIS OF BODY DIMENSIONS. J. MAR. BIOL. ASSOC. U. K. 44(3), 703-709. [VARIANCE-COVARIANCE MATRIX; REGRESSION]

PORTER, W. P., J. W. MITCHELL, W. A. BECKMAN AND C. B. DE WIT. 1973. BEHAVIORAL IMPLICATIONS OF MECHANISTIC ECOLOGY. THERMAL AND BEHAVIORAL MODELING OF DESERT ECOTHERMS AND THEIR MICROENVIRONMENT. OECOLOGIA 13, 1-54. [ENERGY BALANCE; COMPUTER]

POUGH, F. H., L. P. BROWER, H. R. MECK AND S. R. KESSELL. 1973. THEORETICAL INVESTIGATIONS OF AUTOMIMICRY: MULTIPLE TRIAL LEARNING AND THE PALATABILITY SPECTRUM. PROC. NATL. ACAD. SCI. U.S.A. 70(8), 2261-2265. [MODEL; PREDATOR-PREY MODEL]

POWER, D. M. 1969. EVOLUTIONARY IMPLICATIONS OF WING AND SIZE VARIATION IN THE RED-WINGED BLACKBIRD IN RELATION TO GEOGRAPHIC AND CLIMATIC FACTORS: A MULTIPLE REGRESSION ANALYSIS. SYST. ZOOL. 18(4), 363-373. [TAXONOMY]

POWERS, J. E. AND R. T. LACKEY. 1974. MODELING ECOLOGICAL PROCESSES AS A TOOL IN NATURAL RESOURCE MANAGEMENT. (ABSTRACT). VIRGINIA J. SCI. 25(2), 52. [QUEUEING THEORY; MODEL; COMPETITION AND PREDATION]

PRESTON, F. W. 1953. THE SHAPES OF BIRDS' EGGS. AUK 70(2), 160-182.

PRESTON, F. W. 1966. THE MATHEMATICAL REPRESENTATION OF MIGRATION. ECOLOGY 47(3), 375-392.

PRESTON, F. W. 1968. THE SHAPES OF BIRDS' EGGS: MATHEMATICAL ASPECTS. AUK 85(3), 454-463.

PRESTON, F. W. 1974. ANCIENT ERROR IN A 1955 AUK PAPER. AUK 91(2), 417-418. [REFERS TO GEMPERLE AND PRESTON]

PRIKHOD'KO, T. I. 1971. A MATHEMATICAL MODEL OF THE POPULATION OF NEUTRODIAPTOMUS ANGUSTILOBUS IN DAL'NEE LAKE. ZOOL. ZH. 50(12), 1785-1794. (ENGLISH TRANSL. OAK RIDGE NATL. LAB., OAK RIDGE, TENNNESSEE, REPORT ORNL-TR-2870).

PULLIAM, H. R. 1973. ON THE ADVANTAGES OF FLOCKING. J. THEOR BIOL. 38(2), 419-422.

RAGOTZKIE, R. A. AND R. A. BRYSON. 1953. CORRELATION OF CURRENTS WITH THE DISTRIBUTION OF ADULT DAPHNIA IN LAKE MENDOTA. J. MARINE RES. 12(2), 157-172. [EQUATIONS]

RAHN, H. AND A. AR. 1974. THE AVIAN EGG: INCUBATION TIME AND WATER LOSS. CONDOR 76(2), 147-152. [MODEL]

RAPOPORT, A. 1956. SOME GAME-THEORETICAL ASPECTS OF PARASITISM AND SYMBIOSIS. BULL. MATH. BIOPHYS. 18, 15-30.

RAPPORT, D. J. AND J. E. TURNER 1970. DETERMINATION OF PREDATOR FOOD PREFERENCES. J. THEOR. BIOL. 26(3), 365-372. [PREFERENCE COEFFICIENT; POISSON DISTRIBUTION; MODEL; PROBABILITY]

REDDINGIUS, J. 1963. A MATHEMATICAL NOTE ON A MODEL OF CONSUMER-FOOD RELATION IN WHICH THE FOOD IS CONTINUALLY REPLACED. ACTA BIOTHEOR. 16(3/4), 184-198.

REDDINGIUS, J. 1971. MODELS AS RESEARCH TOOLS. PP. 64-74. IN, DYNAMICS OF POPULATIONS. (P. J. DEN BOER AND G. R. GRADWELL, EDS.). CENTRE FOR AGRICULTURAL PUBLISHING AND DOCUMENTATION, WAGENINGEN, NETHERLANDS.

REDDINGIUS, J. 1971. NOTES ON THE MATHEMATICAL THEORY OF EPIDEMICS. ACTA BIOTHEOR. 20(3/4), 125-157. [BAILEY'S MODEL; KERMACK-MCKENDRICK MODEL]

REDDINGIUS, J. 1974. MODELS IN BIOLOGY: A COMMENT. AMER. NAT. 108(961), 393-394. [REDDINGIUS-DEN BOER MODEL; NUMBER-OF-FACTORS MODEL]

EDETZKE, K. A. 1973. A MATRIX ODEL OF A RANGELAND GRAZING YSTEM. DISSERTATION. COLORADO TATE UNIVERSITY, FT. COLLINS, OLORADO. 143 PP.

EED, K. L. AND W. L. WEBB. 1972. RITERIA FOR SELECTING AN OPTIMAL ODEL: TERRESTRIAL PHOTOSYNTHESIS. P. 227-236. IN, ROCEEDINGS-RESEARCH ON CONIFEROUS OREST ECOSYSTEMS-A SYMPOSIUM. (J. . FRANKLIN, L. J. DEMPSTER AND R. WARING, EDS.). U. S. FOR. SERV. ACIFIC NORTHWEST FOREST EXPT. N., PORTLAND, OREGON.

REIGNER, I. C. 1966. A METHOD OF ESTIMATING STREAMFLOW LOSS OF EVAPOTRANSPIRATION FROM THE RIPARIAN ZONE. FOREST SCI. 12(2), 130-139. [MODEL]

RILEY, G. A. 1967. MATHEMATICAL MODEL OF NUTRIENT CONDITIONS IN COASTAL WATERS. BULL. BINGHAM OCEANOGR. COLLECT. YALE UNIV. 19, 72-80.

ROBINS, C. R. AND R. W. CRAWFORD. 1954. A SHORT ACCURATE METHOD FOR ESTIMATING THE VOLUME OF STREAM FLOW. J. WILDL. MANAGE. 18(3), 366-369.

ROBSON, D. S. 1969. EXPLORING THE RELATIONSHIP BETWEEN DEGREE DAYS AND DATE OF METAMORPHOSIS. BIOMETRICS UNIT, CORNELL UNIVERSITY, BU-330-M. MIMEO. 6 PP.

ROGERS, P. 1969. A GAME THEORY APPROACH TO THE PROBLEMS OF INTERNATIONAL RIVER BASINS. WATER RESOUR. RES. 5(4), 749-760.

ROSENZWEIG, M. L. 1974. ON THE EVOLUTION OF HABITAT SELECTION. PP. 401-404. IN, PROCEEDINGS OF THE FIRST INTERNATIONAL CONGRESS ON ECOLOGICAL CENTRE FOR AGRICULTURAL PUBLISHING AND DOCUMENTATION, WAGENINGEN, NETHERLANDS. [MODEL]

RUSHTON, S. AND A. J. MAUTNER. 1955. THE DETERMINISTIC MODEL OF A SIMPLE EPIDEMIC FOR MORE THAN ONE COMMUNITY. BIOMETRIKA 42(1/2), 126-132.

RUTTER, A. J. 1972. SUMMARY AND ASSESSMENT: AN ECOLOGIST'S POINT OF VIEW. PP. 375-380. IN, MATHEMATICAL MODELS IN ECOLOGY. (J. N. R. JEFFERS, ED.). BLACKWELL SCIENTIFIC PUBLICATIONS, OXFORD, ENGLAND.

RUTTER, A. J., A. J. MORTON AND P. C. ROBINS. 1975. A PREDICTIVE MODEL OF RAINFALL INTERCEPTION IN FORESTS. II. GENERALIZATION OF THE MODEL AND COMPARISON WITH OBSERVATIONS IN SOME CONIFEROUS AND HARDWOOD STANDS. J. APPL. ECOL. 12(1), 367-380. [COMPUTER]

RYLANDER, M. K. AND E. G. BOLEN. 1970. ECOLOGICAL AND ANATOMICAL ADAPTATIONS OF NORTH AMERICAN TREE DUCKS. AUK 87(1), 72-90. [MODEL]

SCHNIEWIND, A. P. 1962. HORIZONTAL SPECIFIC GRAVITY VARIATION IN TREE STEMS IN RELATION TO THEIR SUPPORT. FOREST SCI. 8(2), 111-118. [EQUATIONS]

SCOTT, D. 1966. INTERPRETATION OF ECOLOGICAL DATA BY PATH ANALYSIS. PROC. N. Z. ECOL. SOC. 13, 1-4.

SELIGER, H. H., J. H. CARPENTER, M. LOFTUS, W. H. BIGGLEY AND W. D. MCELROY. 1971. BIOLUMINESCENCE AND PHYTOPLANKTON SUCCESSIONS IN BAHIA FOSFORESCENTE, PUERTO RICO. LIMNOL. OCEANOGR. 16(4), 608-622. [MODEL]

SELIGER, H. H., W. G. FASTIE AND W. D. MCELROY. 1969. TOWABLE PHOTOMETER FOR RAPID AREA MAPPING OF CONCENTRATIONS OF BIOLUMINESCENT MARINE DINOFLAGELLATES. LIMNOL. OCEANOGR. 14(5), 806-813. [EQUATIONS]

SERFLING, R. E. 1952. HISTORICAL REVIEW OF EPIDEMIC THEORY. HUMAN BIOL. 24(3), 145-166.

SHAW, C. F. 1930. POTENT FACTORS IN SOIL FORMATION. ECOLOGY 11(2), 239-245. [EQUATIONS]

SKELLAM, J. G. 1969. MODELS, INFERENCE, AND STRATEGY. BIOMETRICS 25(3), 457-475.

SKELLAM, J. G. 1972. SOME PHILOSOPHICAL ASPECTS OF MATHEMATICAL MODELLING IN EMPIRICAL SCIENCE WITH SPECIAL REFERENCE TO ECOLOGY. PP. 13-28. IN, MATHEMATICAL MODELS IN ECOLOGY. (J. N. R. JEFFERS, ED.). BLACKWELL SCIENTIFIC PUBLICATIONS, OXFORD, ENGLAND.

SLIFER, E. H. 1954. A METHOD FOR CALCULATING THE SURFACE AREA OF THE BODY OF GRASSHOPPERS AND LOCUSTS (ORTHOPTERA, ACRIDIDAE). ANN. ENTOMOL. SOC. AMER. 47(2), 265-271. [EQUATIONS; REGRESSION]

SLOBODKIN, L. B. 1958. FORMAL PROPERTIES OF ANIMAL COMMUNITIES. GENERAL SYSTEMS 3, 73-100.

SLOBODKIN, L. B. 1958. META-MODELS IN THEORETICAL ECOLOGY. ECOLOGY 39(3), 550-551.

SLOBODKIN, L. B. 1961. PRELIMINARY IDEAS FOR A PREDICTIVE THEORY OF ECOLOGY. AMER. NAT. 95(882), 147-153. [MODEL; BOUNDARY CONDITIONS]

SLOBODKIN, L. B. 1965. ON THE PRESENT INCOMPLETENESS OF MATHEMATICAL ECOLOGY. AMER. SCI. 53(3), 347-357.

SMITH, A. P. 1972. BUTTRESSING OF TROPICAL TREES: A DESCRIPTIVE MODEL AND NEW HYPOTHESIS. AMER. NAT. 106(947), 32-46. [EQUATIONS]

SMITH, C. C. 1968. THE ADAPTIVE NATURE OF SOCIAL ORGANIZATION IN THE GENUS OF THREE SQUIRRELS TAMIASCIURUS. ECOL. MONOGR. 38(1), 31-63. [DISTANCE TRAVELED MODELS]

SMITH, R. C. 1968. THE OPTICAL CHARACTERIZATION OF NATURAL WATERS BY MEANS OF AN "EXTINCTION COEFFICIENT". LIMNOL. OCEANOGR. 13(3), 423-429. [EQUATIONS]

SMITH, R. C. G. AND W. A. WILLIAMS 1973. MODEL DEVELOPMENT FOR A DEFERRED-GRAZING SYSTEM. J. RANG MANAGE. 26(6), 454-460. [SIMULATION; COMPUTER]

SNOW, D. R. 1975. SOME APPLICATIONS OF FUNCTIONAL EQUATIONS IN ECOLOGY AND BIOLOGY. PP. 306-313. IN, ECOSYSTEM ANALYSI AND PREDICTION. (S. A. LEVIN, ED.) PROCEEDINGS OF A SIAM-SIMS CONFERENCE HELD AT ALTA, UTAH, JUL 1-5, 1974. [CAUCHY FUNCTIONAL EQUATIONS]

SOKAL, R. R. AND F. J. SONLEITNER. 1968. THE ECOLOGY OF SELECTION IN HYBRID POPULATIONS OF TRIBOLIUM CASTANEUM. ECOL. MONOGR. 38(4), 345-379. [GENETICS; CANNIBALISM MODEL; FECUNDITY MODEL]

SOWER, L. L., R. S. KAAE AND H. H. SHOREY. 1973. SEX PHEROMONES OF LEPIDOPTERA. XLI. FACTORS LIMITING POTENTIAL DISTANCE OF SEX PHEROMONE COMMUNICATION IN TRICHOPLUSIA NI. ANN. ENTOMOL. SOC. AMER. 66(5), 1121-1122. [EQUATIONS]

SPOTILA, J. R., P. W. LOMMEN, G. S. BAKKEN AND D. M. GATES. 1973. A MATHEMATICAL MODEL FOR BODY TEMPERATURES OF LARGE REPTILES, IMPLICATIONS FOR DINOSAUR ECOLOGY. AMER. NAT. 107(955), 391-404.

SPRENT, P. 1972. THE MATHEMATICS OF SIZE AND SHAPE. BIOMETRICS 28(1), 23-37. [ALLOMETRY]

STAHL, W. R. 1961. DIMENSIONAL ANALYSIS IN MATHEMATICAL BIOLOGY. I. GENERAL DISCUSSION. BULL. MATH. BIOPHYS. 23, 355-376. [INFORMATION THEORY]

STAHL, W. R. 1962. DIMENSIONAL ANALYSIS IN MATHEMATICAL BIOLOGY. II. BULL. MATH. BIOPHYS. 24(1), 31-108. [INFORMATION THEORY]

STIVEN, A. E. 1964. EXPERIMENTAL STUDIES ON THE EPIDEMIOLOGY OF THE HOLT-PARASITE SYSTEM, HYDRA AND HYDRAMOEBA HYDROXENA (ENTZ). II. THE COMPONENTS OF A SIMPLE EPIDEMIC. COL. MONOGR. 34(2), 119-142. LOGISTIC CURVE; LOGIT; COMPUTER]

TOUT, B. B., J. M. DESCHENES AND . F. OHMANN. 1975. ULTI-SPECIES MODEL OF A DECIDUOUS OREST. ECOLOGY 56(1), 226-231. REGRESSION; LEAST SQUARES; OMPUTER; ECONOMETRIC APPROACH]

UKHANOV, V. V. AND A. P. SHAPIRO. 971. SELECTIVE FEEDING OF QUATIC ORGANISMS. HYDROBIOL. J. (2), 50-55. (ENGLISH TRANSL.) MODEL]

SVERDRUP, H. U. 1953. ON CONDITIONS FOR THE VERNAL BLOOMING OF PHYTOPLANKTON. J. CONS. PERM. INT. EXPLOR. MER 18(3), 287-295. [PHOTOSYNTHESIS MODEL; DEPTH MODEL]

SWARTZMAN, G. L. 1975. AN INTRINSIC ENVIRONMENTAL FRAMEWORK FOR ADAPTATION. PP. 330-333. IN. ECOSYSTEM ANALYSIS AND PREDICTION. (S. A. LEVIN, ED.). PROCEEDINGS OF A SIAM-SIMS CONFERENCE HELD AT ALTA, UTAH, JULY 1-5, 1974. [VECTOR; EQUATIONS]

SWARTZMAN, G. L. (ED.). 1970. SOME CONCEPTS OF MODELING. U. S. IBP GRASSLAND BIOME, COLORADO STATE UNIVERSITY, FT. COLLINS, COLORADO. TECH. REPT. 32. 142 PP.

SWEERS, H. E. 1968. TWO METHODS OF DESCRIBING THE "AVERAGE" VERTICAL TEMPERATURE DISTRIBUTION OF A LAKE. J. FISH. RES. BOARD CAN. 25(9), 1911-1922. [EQUATIONS]

SWEERS, H. E. 1970. VERTICAL DIFFUSIVITY COEFFICIENT IN A THERMOCLINE. LIMNOL. OCEANOGR. 15(2), 273-280. [EQUATIONS]

TABATA, S. 1958. HEAT BUDGET OF THE WATER IN THE VICINITY OF TRIPLE ISLAND, BRITISH COLUMBIA J. FISH. RES. BOARD CAN. 15(3), 429-451. [EQUATIONS]

TAKENOUTI, Y. 1936. EFFECTS OF OBSTACLES UPON THE SUBMARINE ILLUMINATIONS. I. THE RECTANGULAR MEMBRANE IMMERSED IN WATER. BULL. JAPANESE SOC. SCI. FISH. 5(4), 224-226. [EQUATIONS]

TAKENOUTI, Y. 1941. ON THE DIFFERENTIAL EQUATION OF FISHING NET. BULL. JAPANESE SOC. SCI. FISH. 10(1), 1-6. [VECTOR]

TALBOT, J. W. AND J. L. HENRY. 1968. THE ABSORPTION OF RHODAMINE-B ON TO MATERIALS CARRIED IN SUSPENSION BY INSHORE WATERS. J. CONS. PERM. INT. EXPLOR. MER 32(1), 7-16. [EQUATIONS]

TAYLOR, S. E. AND O. J. SEXTON. 1972. SOME IMPLICATIONS OF LEAF TEARING IN MUSACEAE. ECOLOGY 53(1), 143-149. [MODEL]

TEISSIER, G. 1948. LA RELATION D'ALLOMETRIE: SA SIGNIFICATION STATISTIQUE ET BIOLOGIQUE. BIOMETRICS 4(1), 14-48. [ALLOMETRY]

TEMPLETON, A. R. AND E. D. ROTHMAN. 1974. EVOLUTION IN HETEROGENEOUS ENVIRONMENTS. AMER. NAT. 108(962), 409-428. [FITNESS; DECISION THEORY; LOSS FUNCTION; LEVINS' ADAPTIVE FUNCTION; LEVINS; POPULATION; MAXIMIN POPULATION]

TERHUNE, L. D. B. 1958. THE MARK VI GROUNDWATER STANDPIPE FOR MEASURING SEEPAGE THROUGH SALMON SPAWNING GRAVEL. J. FISH. RES. BOARD CAN. 15(5), 1027-1063. [EQUATIONS]

TEZUKA, Y. 1968. A METHOD FOR ESTIMATING BACTERIAL RESPIRATION IN NATURAL WATER. JAPANESE J. ECOL. 18(2), 60-65. [EQUATIONS]

THEODARIDIS, G. C. AND L. STARK. 1971. INFORMATION AND BIOTECHNOLOGICAL SYSTEMS. MATH. BIOSCI. 12, 375-388.

THOMANN, R. V. 1974. SOME STATISTICAL ANALYSES OF WATER QUALITY IN THE DELAWARE RIVER. PP. 297-311. IN, STATISTICAL AND MATHEMATICAL ASPECTS OF POLLUTION PROBLEMS. (J. W. PRATT, ED.). MARCEL DEKKER, INC., NEW YORK, NEW YORK. [MODEL; AUTOCORRELATION]

THOMPSON, W. R. 1924. LA THEORIE MATHEMATIQUE DE L'ACTION DES PARASITES ENTOMOPHAGES ET LE FACTEUR DU HAZARD. ANN. FAC. SCI. MARSEILLE SER. 2, 69-89.

THORNTON, K. W. AND R. J. MULHOLLAND. 1974. LAGRANGE STABILITY AND ECOLOGICAL SYSTEMS. J. THEOR. BIOL. 45(2), 473-485. [LINEAR AND NON-LINEAR SYSTEMS]

THORPE, M. R. 1974. RADIANT HEATING OF APPLES. J. APPL. ECOL. 11(2), 755-760. [MODEL]

TURNER, B. J. 1974. APPLICATIONS OF CLUSTER ANALYSIS IN NATURAL RESOURCES RESEARCH. FOREST SCI. 20(4), 343-349. [MULTIVARIATE ANALYSIS; MODEL; COMPUTER; ALGORITHM]

TYLER, J. E. 1959. NATURAL WATER AS A MONOCHROMATOR. LIMNOL. OCEANOGR. 4(1), 102-105. [EQUATIONS]

ULTSCH, G. R. 1974. THE ALLOMETRIC RELATIONSHIP BETWEEN METABOLIC RATE AND BODY SIZE: ROLE OF THE SKELETON. AMER. MIDL. NAT. 92(2), 500-504. [EQUATIONS]

UMNOV, A. A. 1972. A MATHEMATIC MODEL OF THE BIOTIC CIRCULATION WITHIN THE LAKE ECOSYSTEM. HYDROBIOL. J. 8(5), 1-8. (ENGLISH TRANSL.).

UPCHURCH, S. B. AND D. C. N. ROBB. 1972. MATHEMATICAL MODELS: PLANNING TOOLS FOR THE GREAT LAKES. WATER RESOUR. BULL. 8(2), 338-348.

USHER, M. B. 1970. AN ALGORITHM FOR ESTIMATING THE LENGTH AND DIRECTION OF SHADOWS WITH REFERENCE TO THE SHADOWS OF SHELTER BELTS. J. APPL. ECOL. 7(1), 141-145.

VAN VALEN, L. 1974. ERRATUM: MULTIVARIATE STRUCTURAL STATISTICS IN NATURAL HISTORY. J. THEOR. BIOL. 48(2), 501. [MAHALANOBIS' D2]

VAN VALEN, L. 1974. MULTIVARIATE STRUCTURAL STATISTICS IN NATURAL HISTORY. J. THEOR. BIOL. 45(1), 235-247. [VARIANCE; CORRELATION; INFORMATION; DISTANCE MODALITY]

VAUGHAN, B. E. 1972. ECOLOGICAL AND ENVIRONMENTAL PROBLEMS IN THE APPLICATION OF BIOMATHEMATICS. PROC. SIXTH BERKELEY SYMP. MATH. STAT. AND PROB. 6, 495-509. [SIMULATION]

VAUX, W. G. 1968. INTRAGRAVEL FLOW AND INTERCHANGE OF WATER IN A STREAMBED. U. S. FISH WILDL. SERV. FISH. BULL. 66(3), 479-489. [EQUATIONS; ANALOG MODEL]

VERDUIN, J. 1952. THE CALCULUS AND THE INOPERABLE EXPRESSION. ECOLOGY 33(1), 116. [COMMENT ON USE OF CALCULUS]

VERHOFF, F. H. AND F. E. SMITH. 1971. THEORETICAL ANALYSIS OF A CONSERVED NUTRIENT ECOSYSTEM. J. THEOR. BIOL. 33(1), 131-147. [MODEL; MATRIX; STABILITY; SIMULATION; COMPUTER; TROPHIC LEVELS]

VINBERG, G. G. 1971. SEMINAR ON MATHEMATICAL MODELS OF AQUATIC ECOSYSTEMS. HYDROBIOL. J. 7(5), 113-116. (ENGLISH TRANSL.)

VINBERG, G. G. AND S. I. ANISIMOV. 1966. MATHEMATICAL MODEL OF AN AQUATIC ECOLOGICAL SYSTEM. PP. 173-181. IN, PHOTOSYNTHESIS OF PRODUCTIVE SYSTEMS. (A. A. NICHIPOROVICH, ED.). ACAD. SCI. U.S.S.R. (ENGLISH TRANSL., 1967. IDST CAT. NO. 2182.).

WAGGONER, P. E. 1970. CONSULTATION ON HOW MODELS ARE MADE, HOW THEY ARE TESTED AND WHAT THEY TELL US OF EXPERIMENTS TO BE DONE. PP. 575-578. IN, PREDICTION AND MEASUREMENT OF PHOTOSYNTHETIC PRODUCTIVITY. (I. SETLIK, ED.). CENTRE FOR AGRICULTURAL PUBLISHING AND DOCUMENTATION, WAGENINGEN, NETHERLANDS.

WAGGONER, P. E. 1974. MODELING SEASONALITY. PP. 301-323, 326-327. IN, PHENOLOGY AND SEASONALITY MODELING. (H. LIETH, ED.). SPRINGER-VERLAG NEW YORK INC., NEW YORK, NEW YORK. [SIMULATION; COMPUTER]

WAGNER, S. S. AND S. A. ALTMANN 1973. WHAT TIME DO THE BABOONS COME DOWN FROM THE TREES? (AN ESTIMATION PROBLEM). BIOMETRICS 29(4), 623-635. [PROBABILITY]

WALKER, K. F. 1974. THE STABILITY OF MEROMICTIC LAKES IN CENTRAL WASHINGTON. LIMNOL. OCEANOGR. 19(2), 209-222. [SCHMIDT EQUATION]

WALTER, G. G. 1973. DELAY-DIFFERENTIAL EQUATION MODELS FOR FISHERIES. J. FISH. RES. BOARD CAN. 30(7), 939-945. [MODEL; TIME LAG]

WALTERS, C. J. 1971. SYSTEMS ECOLOGY: THE SYSTEMS APPROACH AND MATHEMATICAL MODELS IN ECOLOGY. PP. 276-292. IN, FUNDAMENTALS OF ECOLOGY. (3RD. EDITION). (E. P. ODUM). W. B. SAUNDERS CO., PHILADELPHIA, PENNSYLVANIA.

WARBURTON, F. E. 1967. A MODEL OF NATURAL SELECTION BASED ON A THEORY OF GUESSING GAMES. J. THEOR. BIOL. 16(1), 78-96. [INFORMATION; DIGITAL REPRESENTATION]

WARNER, R. E., K. K. PETERSON AND L. BORGMAN 1966. BEHAVIOURAL PATHOLOGY IN FISH: A QUANTITATIVE STUDY OF SUBLETHAL PESTICIDE TOXICATION. J. APPL. ECOL. 3(SUPPL.), 223-247. [COMPUTER; ANALYSIS OF VARIANCE; STATISTICAL MODEL]

WARNER, R. R. 1975. THE ADAPTIVE SIGNIFICANCE OF SEQUENTIAL HERMAPHRODITISM IN ANIMALS. AMER. NAT. 109(965), 61-82. [MODEL]

WARRICK, A. W. 1970. A MATHEMATICAL SOLUTION TO A HILLSIDE SEEPAGE PROBLEM. SOIL SCI. SOC. AMER. PROC. 34(6), 849-853.

WATSON, G. S. 1965. THE DISTRIBUTION OF ORGANISMS. BIOMETRICS 21(3), 543-550.

WATT, K. E. F. 1962. USE OF MATHEMATICS IN POPULATION ECOLOGY. ANNU. REV. ENTOMOL. 7, 243-260. [COMPUTER]

WESTENBERG, J. 1960. AN ANALYSIS OF PERSISTENCE IN POPULATION SYSTEMS. ACTA BIOTHEOR. 13(4), 145-160. [MODEL]

WESTOBY, M. 1974. AN ANALYSIS OF DIET SELECTION BY LARGE GENERALIST HERBIVORES. AMER. NAT. 108(961), 290-304. [OPTIMIZATION MODEL; RELATIVE PREFERENCE INDEX]

WHITE, J. F. AND S. J. GOULD. 1965. INTERPRETATION OF THE COEFFICIENT IN THE ALLOMETRIC EQUATION. AMER. NAT. 99(904), 5-18.

WHITLOCK, J. H., H. D. CROFTON AND D. S. ROBSON. 1969. DATA ANALYSIS OF HYPERBOLAS DERIVED FROM ECOLOGICAL OBSERVATIONS. CORNELL VET. 59(3), 439-452. [MULTIPLE REGRESSION; SIMPLE LINEAR REGRESSION; HYPERBOLA]

WHITNEY, L. V. AND R. L. PIERCE. 1957. FACTORS CONTROLLING THE INPUT OF ELECTRICAL ENERGY INTO A FISH (CYPRINUS CARPIO L.) IN AN ELECTRICAL FIELD. LIMNOL. OCEANOGR. 2(2), 55-61. [EQUATIONS]

WILKINS, J. E., JR. 1945. THE DIFFERENTIAL DIFFERENCE EQUATION FOR EPIDEMICS. BULL. MATH. BIOPHYS. 7, 149-150.

WILKINSON, D. H. 1952. THE RANDOM ELEMENT IN BIRD "NAVIGATION". J. EXPTL. BIOL. 29(4), 532-560. [PROBABILITY; BESSEL FUNCTION]

WILLIAMS, E. J. 1969. THE DEVELOPMENT OF BIOMATHEMATICAL MODELS. BULL. INTERN. STAT. INST. 42(1), 131-140. [POPULATION; SPATIAL DISTRIBUTION; GAME THEORY; STOCHASTIC]

WILSON, A. L. AND A. W. DOUGLAS. 1969. A NOTE ON NON-LINEAR CURVE FITTING. AMER. STATISTICIAN 23(4), 37-38. [FORESTRY]

WILSON, D. S. 1975. A THEORY OF GROUP SELECTION. PROC. NATL. ACAD. SCI. U.S.A. 72(1), 143-146. [MODEL]

WILSON, K. J. 1974. THE RELATIONSHIP OF OXYGEN SUPPLY FOR ACTIVITY TO BODY TEMPERATURE IN FOUR SPECIES OF LIZARDS. COPEIA 1974(4), 920-934. [EQUATIONS]

WILSON, L. O. 1972. AN EPIDEMIC MODEL INVOLVING A THRESHOLD. MATH. BIOSCI. 15, 109-121.

WOODGER, J. H. 1965. THEOREMS ON RANDOM EVOLUTION. BULL. MATH. BIOPHYS. 27(SPECIAL ISSUE), 145-150. [SET THEORY]

WOODING, R. A. 1972. DRAG ON A SURFACE COVERED BY AN ARRAY OF OBSTACLES, AND ITS IMPLICATIONS FOR HEAT AND MASS TRANSFER. PP. 49053. IN, MODELING THE GROWTH OF TREES. (C. E. MURPHY, JR., J. D. HESKETH AND B. R. STRAIN, EDS.). AEC REPT. EDFB-IBP-72-11. OAK RIDGE NATL. LAB., OAK RIDGE, TENNESSEE.

WU, L. S.-Y. 1975. ON THE STABILITY OF ECOSYSTEMS. PP. 155-165. IN, ECOSYSTEM ANALYSIS AND PREDICTION. (S. A. LEVIN, ED.). PROCEEDINGS OF A SIAM-SIMS CONFERENCE HELD AT ALTA, UTAH, JULY 1-5, 1974. [MODEL; EQUATIONS]

WYATT, T. 1974. RED TIDES AND ALGAL STRATEGIES. PP. 35-40. IN, ECOLOGICAL STABILITY. (M. B. USHER AND M. H. WILLIAMSON, EDS.). HALSTED PRESS, NEW YORK, NEW YORK. [MODEL; TIME LAG]

WYATT, T. AND J. HORWOOD. 1973. MODEL WHICH GENERATES RED TIDES. NATURE 244(5413), 238-240.

YAMAMURA, Y. 1960. STUDIES ON THE MECHANISM OF CONSUMPTION AND ACCUMULATION OF VITAMIN A IN FISH. BULL. JAPANESE SOC. SCI. FISH. 26(5), 490-499. [EQUATIONS]

YARRANTON, G. A. 1969. PLANT ECOLOGY: A UNIFYING MODEL. J. ECOL. 57(1), 245-250.

YARRANTON, G. A. 1971. MATHEMATICAL REPRESENTATIONS AND MODELS IN PLANT ECOLOGY: RESPONSE TO A NOTE BY R. MEAD. J. ECOL. 59(1), 221-224.

YEAGER, J. F. AND S. C. MUNSON. 1948. A RATIO HYPOTHESIS PERTAINING TO THE BIOLOGICAL ACTIO OF POISONS AND DRUGS ANN. ENTOMOL. SOC. AMER. 41(3), 377-383 [EQUATIONS]

YIM, Y.-J., H. OGAWA AND T. KIRA. 1969. LIGHT INTERCEPTION BY STEM IN PLANT COMMUNITIES. JAPANESE ECOL. 19(6), 233-238. [EQUATIONS

YU, P. 1972. SOME HOST PARASI GENETIC-INTERACTION MODELS. THEOR. POPUL. BIOL. 3(3), 347-357.

ZAIKA, V. E. AND N. A. OSTROVSKAYA. 1973. INDICATORS OF THE AVAILABILITY OF FOOD TO FISH LARVAE 2. THE PERCENTAGE OF FEEDING CONSUMERS IN AN UNSTEADY FEEDING REGIME. THE AVERAGE TIME TAKEN TO SEEK OUT FOOD. J. ICHTHYOLOGY 13(1), 120-128. [EQUATIONS]

ZAR, J. H. 1968. CALCULATION AND MISCALCULATION OF THE ALLOMETRIC EQUATION AS A MODEL IN BIOLOGICAL DATA. BIOSCIENCE 18(12), 1118-1120.

ZAR, J. H. 1968. THE EFFECT OF THE CHOICE OF TEMPERATURE SCALE ON SIMPLE LINEAR REGRESSION EQUATIONS. ECOLOGY 49(6), 1161. [FORMULA]

MODELS (POPULATION)

ABBOTT, M. B. AND I. R. WARREN. 1974. A DYNAMIC POPULATION MODEL. J. ENVIRON. MANAGE. 2(4), 281-297. [SIMULATION; COMPUTER]

ABDELKADER, M. A. 1974. EXACT SOLUTIONS OF LOTKA-VOLTERRA EQUATIONS. MATH. BIOSCI. 20, 293-297. [COMPETITION]

ADKE, S. R. 1964. THE MAXIMUM POPULATION SIZE IN THE FIRST N GENERATIONS OF A BRANCHING PROCESS. BIOMETRICS 20(3), 649-651.

ANDERSEN, F. S. 1965. SIMPLE POPULATION MODELS AND THEIR APPLICATION TO THE FORMULATION OF COMPLEX MODELS. PROC. INTERN. CONGR. ENTOMOL. 12, 620-622. [WATT'S MODEL; EXPONENTIAL MODEL]

ANDREWARTHA, H. G. 1957. THE USE OF CONCEPTUAL MODELS IN POPULATION ECOLOGY. COLD SPRING HARBOR SYMPOSIA ON QUANTITATIVE BIOLOGY 22, 219-232. [DETERMINISTIC MODEL]

AUER, C. 1971. A SIMPLE MATHEMATICAL MODEL FOR "KEY-FACTOR" ANALYSIS AND COMPARISON IN POPULATION RESEARCH WORK. PP. 33-46, IN, STATISTICAL ECOLOGY, VOLUME 2. SAMPLING AND MODELING BIOLOGICAL POPULATIONS AND POPULATION DYNAMICS. (G. P. PATIL, E. C. PIELOU AND W. E. WATERS, EDS.). PENNSYLVANIA STATE UNIVERSITY PRESS, UNIVERSITY PARK, PENNSYLVANIA.

AUER, C. 1972. REFLECTIONS ON MATHEMATIC STATISTICAL POPULATION MODELS. ZEITSCHRIFT FUR ANGEWANDTE ENTOMOLOGIE 70, 1-7.

BAZIN, M. J., V. RAPA AND P. T. SAUNDERS. 1974. THE INTEGRATION OF THEORY AND EXPERIMENT IN THE STUDY OF PREDATOR-PREY DYNAMICS. PP. 159-164. IN, ECOLOGICAL STABILITY. (M. B. USHER AND M. H. WILLIAMSON, EDS.). HALSTED PRESS, NEW YORK, NEW YORK. [LOTKA-VOLTERRA EQUATION; MODEL]

BECKMANN, M. J. 1957. ON THE EQUILIBRIUM DISTRIBUTION OF POPULATION IN SPACE. BULL. MATH. BIOPHYS. 19, 81-90.

BEDDINGTON, J. R. 1974. AGE DISTRIBUTIONS AND STABILITY OF SIMPLE DISCRETE TIME POPULATION MODELS. J. THEOR. BIOL. 47(1), 65-74. [LESLIE MATRIX; SIMULATION; COMPUTER; DIFFERENCE EQUATION]

BELL, G. I. 1973. PREDATOR-PRE EQUATIONS SIMULATING AN IMMUNE RESPONSE. MATH. BIOSCI. 16, 291-314. [MODEL]

BEYER, W. A. 1970. SOLUTION TO A MATHEMATICAL MODEL OF CELL GROWTH, DIVISION, AND DEATH. MATH. BIOSCI. 6, 431-436.

BOLING, R. H., JR. 1973. TOWAR A STATE-SPACE MODEL FOR BIOLOGICAL POPULATIONS. J. THEOR. BIOL. 40(3), 485-506. [LOTKA-VOLTERRA; VECTOR; MATRIX; SIMULATION; COMPUTER]

BOORMAN, S. A. AND P. R. LEVITT. 1972. GROUP SELECTION ON THE BOUNDARY OF A STABLE POPULATION. PROC. NATL. ACAD. SCI. U. S. A. 69(9), 2711-2713. [CARRYING CAPACITY; MODEL; EXTINCTION; DIFFUSION EQUATION]

BOSCH, C. A. 1971. REDWOODS: A POPULATION MODEL. SCIENCE 172(3981), 345-349. [MATRIX]

BRANNEN, J. P. 1975. A POPULATION MODEL FOR DERMESTID BEETLE SURVIVAL UNDER STARVATION AND CANNIBALISM. J. THEOR. BIOL. 49(1), 179-189. [PROBABILITY; COMPUTER]

BRUSSARD, P. F., S. A. LEVIN, L. N MILLER AND R. H. WHITTAKER. 1971 REDWOODS: A POPULATION MODEL DEBUNKED. SCIENCE 174(4007), 435-436. [REFERS TO C. A. BOSCH, SCIENCE 172.]

BUTT, D. J. AND D. J. ROYLE. 1974. MULTIPLE REGRESSION ANALYSIS IN THE EPIDEMIOLOGY OF PLANT DISEASES. PP. 78-114. IN, EPIDEMICS OF PLANT DISEASES: MATHEMATICAL ANALYSIS AND MODELING (J. KRANZ, ED.). SPRINGER-VERLAG NEW YORK INC., NEW YORK, NEW YORK.

BUZAS, M. A. 1971. ANALYSES OF SPECIES DENSITIES BY THE MULTIVARIATE GENERAL LINEAR MODEL. LIMNOL. OCEANOGR. 16(4), 667-670. [COMPUTER; MULTIPLE REGRESSION; U-TEST; CHI-SQUARE; LOG TRANSFORMATION]

CALOW, P. 1973. THE RELATIONSHIP BETWEEN FECUNDITY, PHENOLOGY, AND LONGEVITY: A SYSTEMS APPROACH. AMER. NAT. 107(956), 559-574. [MODEL]

CAMPBELL, R. W. 1967. THE ANALYSIS OF NUMERICAL CHANGE IN GYPSY MOTH POPULATIONS. FOREST SCI. MONOGR. 15. 33 PP. [MODEL; LIFE TABLE]

CAUGHLEY, G. 1970. ERUPTION OF UNGULATE POPULATIONS, WITH EMPHASIS ON HIMALAYAN THAR IN NEW ZEALAND. ECOLOGY 51(1), 53-72. [PROBIT; LIFE TABLE; MODEL]

CHARNOV, E. L. AND W. M. SCHAFFER. 1973. LIFE-HISTORY CONSEQUENCES OF NATURAL SELECTION: COLE'S RESULT REVISITED. AMER. NAT. 107(958), 791-793. [MODEL]

CLARK, J. P. 1971. THE SECOND DERIVATIVE AND POPULATION MODELING. ECOLOGY 52(4), 606-613. [SEE ALSO INNIS, ECOL. 53(4).]

COHEN, D. 1967. OPTIMIZATION OF SEASONAL MIGRATORY BEHAVIOR. AMER. NAT. 101(917), 5-17. [MODEL; PROBABILITY; HARMONIC MEAN; VARIANCE]

COHEN, D. 1970. A THEORETICAL MODEL FOR THE OPTIMAL TIMING OF DIAPAUSE. AMER. NAT. 104(938), 389-400.

COLGAN, P., F. COOKE AND J. T. SMITH. 1974. AN ANALYSIS OF GROUP COMPOSITION IN ASSORTATIVELY MATING POPULATIONS. BIOMETRICS 30(4), 693-696. [LESSER SNOW GEESE FLOCKS; PROBABILITY]

COMINS, H. N. AND D. W. E. BLATT. 1974. PREY-PREDATOR MODELS IN SPATIALLY HETEROGENEOUS ENVIRONMENTS. J. THEOR. BIOL. 48(1), 75-83. [STABILITY ANALYSIS]

COOKE, K. L. 1975. A DISCRETE-TIME EPIDEMIC MODEL WITH CLASSES OF INFECTIVES AND SUSCEPTIBLES. THEOR. POPUL. BIOL. 7(2), 175-196.

COUTINHO, A. B. AND F. A. B. COUTINHO. 1973. SNAIL POPULATION IN RUNNING WATER. BULL. MATH. BIOL. 35(4), 449-458. [MODEL]

CROFTON, H. D. 1971. A MODEL OF HOST-PARASITE RELATIONSHIPS. PARASITOLOGY 63(3), 343-364. [DETERMINISTIC MODEL; NEGATIVE BINOMIAL DISTRIBUTION]

CULL, P. AND A. VOGT. 1973. MATHEMATICAL ANALYSIS OF THE ASYMPTOTIC BEHAVIOR OF THE LESLIE POPULATION MATRIX MODEL. BULL. MATH. BIOL. 35(5/6), 645-661.

DAS GUPTA, P. D. 1972. ON TWO-SEX MODELS LEADING TO STABLE POPULATIONS. THEOR. POPUL. BIOL. 3(3), 358-375. [METHOD ANALOGOUS TO LOTKA'S]

DAVIDSON, R. A. AND R. A. DUNN. 1967. A CORRELATION APPROACH TO CERTAIN PROBLEMS OF POPULATION-ENVIRONMENT RELATIONS. AMER. J. BOT. 54(5), 529-538. [MODEL; SET; MATRIX; COMPUTER; CORRELATION]

DE WIT, C. T. 1970. DYNAMIC CONCEPTS IN BIOLOGY. PP. 17-23. IN, PREDICTION AND MEASUREMENT OF PHOTOSYNTHETIC PRODUCTIVITY. (I. SETLIK, ED.). CENTRE FOR AGRICULTURAL PUBLISHING AND DOCUMENTATION, WAGENINGEN, NETHERLANDS. [MODEL; COMPUTER; SIMULATION]

DEAKIN, M. A. B. 1968. CONTINUOUS CLINES AND POPULATION STRUCTURE. AMER. NAT. 102(925), 295-296.

DIAMOND, P. 1973. THE EFFECT OF MULTIPLE PARASITOID INTRODUCTIONS UPON THE EQUILIBRIUM VALUE OF HOST DENSITY. OECOLOGIA 13, 279-290. [WATT'S MODEL; DISC MODEL; NICHOLSON'S MODEL; HASSELL-VARLEY MODEL]

DILL, L. M. 1973. AN AVOIDANCE LEARNING SUBMODEL FOR A GENERAL PREDATION MODEL. OECOLOGIA 13, 291-312. [PREDATOR-PREY MODEL; HOLLING MODEL; COMPUTER; SIMULATION]

DIXON, K. R. AND G. W. CORNWELL. 1970. A MATHEMATICAL MODEL FOR PREDATOR AND PREY POPULATIONS. RES. POPUL. ECOL.-KYOTO 12(2), 127-136. [COMPUTER; SIMULATION]

ESTABROOK, G. F. AND H. H. ROBINSON. 1974. A COMMENT ON VINE'S PREDATOR-PREY VISUAL DETECTION MODEL. J. THEOR. BIOL. 47(1), 245-247.

EYLAND, E. A. 1971. MORAN'S ISLAND MIGRATION MODEL. GENETICS 69(3), 399-403.

FLEMING, R. H. 1939. THE CONTROL OF DIATOM POPULATIONS BY GRAZING. J. CONS. PERM. INT. EXPLOR. MER 14(2), 3-20. [MODEL]

FREDIN, R. A. 1964. OCEAN MORTALITY AND MATURITY SCHEDULES OF KARLUK RIVER SOCKEYE SALMON AND SOME COMPARISONS OF MARINE GROWTH AND MORTALITY RATES. U. S. FISH WILDL. SERV. FISH. BULL. 63(3), 551-574. [POPULATION MODELS]

GANI, J. 1971. SOME ATTACHMENT MODELS ARISING IN VIRUS POPULATIONS. PP. 49-86. IN, STATISTICAL ECOLOGY, VOLUME 2. SAMPLING AND MODELING BIOLOGICAL POPULATIONS AND POPULATION DYNAMICS. (G. P. PATIL, E. C. PIELOU AND W. E. WATERS, EDS.). PENNSYLVANIA STATE UNIVERSITY PRESS, UNIVERSITY PARK, PENNSYLVANIA.

GARFINKEL, D. A. 1967. EFFECT ON STABILITY OF LOTKA-VOLTERRA ECOLOGICAL SYSTEMS OF IMPOSING STRICT TERRITORIAL LIMITS ON POPULATIONS. J. THEOR. BIOL. 14(3), 325-327. [SIMULATION; COMPUTER]

GAUSE, G. F. 1934. UBER EINIGE QUANTITATIVE BEZIEHUNGEN IN DER INSEKTEN-EPIDEMIOLOGIE. ZEITSCHRIFT FUR ANGEWANDTE ENTOMOLOGIE 20(4), 478-670.

GETZ, W. M. 1975. OPTIMAL CONTROL OF A BIRTH-AND-DEATH PROCESS POPULATION MODEL. MATH. BIOSCI. 23(1/2), 87-111. [STOCHASTIC POPULATION; COST INDEX; SYSTEMS THEORY]

GILBERT, E. E. 1971. TIME AND MOTION STUDIES OF TRIBOLIUM. PP. 285-311. IN, STATISTICAL ECOLOGY, VOLUME 2. SAMPLING AND MODELING BIOLOGICAL POPULATIONS AND POPULATION DYNAMICS. (G. P. PATIL, E. C. PIELOU AND W. E. WATERS, EDS.). PENNSYLVANIA STATE UNIVERSITY PRESS, UNIVERSITY PARK, PENNSYLVANIA.

GILBERT, N. 1973. ANIMAL POPULATIONS. PP. 97-102. IN, BIOMETRICAL INTERPRETATION. OXFORD UNIVERSITY PRESS, OXFORD, ENGLAND. [ESTIMATION; LOG TRANSFORMATION; REGRESSION; STOCHASTIC THEORY]

GILLESPIE, D. M. 1969. POPULATION STUDIES OF FOUR SPECIES OF MOLLUSCS IN THE MADISON RIVER, YELLOWSTONE NATIONAL PARK. LIMNOL. OCEANOGR. 14(1), 101-114. [POPULATION MODEL; PRODUCTION MODEL]

GORDON, G., M. O'CALLAGHAN AND G. M. TALLIS. 1970. A DETERMINISTIC MODEL FOR THE LIFE CYCLE OF A CLASS OF INTERNAL PARASITES OF SHEEP. MATH. BIOSCI. 8, 209-226.

GULLAND, J. A. 1962. THE APPLICATION OF MATHEMATICAL MODELS TO FISH POPULATIONS. PP. 204-217. IN, THE EXPLOITATION OF NATURAL ANIMAL POPULATIONS. (E. D. LE CREN AND M. W. HOLGATE, EDS.). BLACKWELL SCIENTIFIC PUBLICATIONS, OXFORD, ENGLAND.

HAIGH, J. 1974. THE EXISTENCE OF EVOLUTIONARY STABLE STRATEGIES. J. THEOR. BIOL. 47(1), 219-221. [APPENDIX TO J. MAYNARD SMITH; GAME THEORY]

HASSELL, M. P. 1969. A POPULATION MODEL FOR THE INTERACTION BETWEEN CYZENIS ALBICANS (FALL.) (TACHINIDAE) AND OPEROPHTERA BRUMATA (L.) (GEOMETRIDAE) AT WYTHAM, BERKSHIRE. J. ANIM. ECOL. 38(3), 567-576.

HASSELL, M. P. AND C. B. HUFFAKER. 1969. REGULATORY PROCESSES AND POPULATION CYCLICITY IN LABORATORY POPULATION OF ANAGASTA KUHNIELLA (ZELLER) (LEPIDOPTERA: PHYCITIDAL) III. THE DEVELOPMENT OF POPULATION MODELS. RES. POPUL. ECOL.-KYOTO 11(2), 186-210.

HASSELL, M. P. AND D. J. ROGERS. 1972. INSECT PARASITE RESPONSES IN THE DEVELOPMENT OF POPULATION MODELS. J. ANIM. ECOL. 41(3), 661-676.

HASSELL, M. P. AND G. C. VARLEY. 1969. NEW INDUCTIVE POPULATION MODEL FOR INSECT PARASITES AND ITS BEARING ON BIOLOGICAL CONTROL. NATURE 223(5211), 1133-1137.

HENNY, C. J., W. S. OVERTON AND H. M. WIGHT. 1970. DETERMINING PARAMETERS FOR POPULATIONS BY USING STRUCTURAL MODELS. J. WILDL. MANAGE. 34(4), 690-703.

HETHCOTE, H. W. 1970. NOTE ON DETERMINING THE LIMITING SUSCEPTIBLE POPULATION IN AN EPIDEMIC MODEL. MATH. BIOSCI. 9, 161-163.

HETHCOTE, H. W. 1973. ASYMPTOTIC BEHAVIOR IN A DETERMINISTIC EPIDEMIC MODEL. BULL. MATH. BIOL. 35(5/6), 607-614. [BAILEY'S MODEL]

HIRSCH, H. R. AND J. ENGELBERG. 1966. DECAY OF CELL SYNCHRONIZATION: SOLUTIONS OF THE CELL-GROWTH EQUATION. BULL. MATH. BIOPHYS. 28, 391-409. [GAUSSIAN DISTRIBUTION; POISSON DISTRIBUTION]

HOLLING, C. S. 1963. AN EXPERIMENTAL COMPONENT ANALYSIS OF POPULATION PROCESSES. MEM. ENTOMOL. SOC. CAN. 32, 22-32. [MODEL; PREDATOR-PREY MODEL]

HUGHES, R. D. AND N. GILBERT. 1968. A MODEL OF AN APHID POPULATION - A GENERAL STATEMENT. J. ANIM. ECOL. 37(3), 553-563. [COMPUTER]

INNIS, G. 1972. THE SECOND DERIVATIVE AND POPULATION MODELING: ANOTHER VIEW. ECOLOGY 53(4), 720-723.

JENKS, R. D. 1969. QUADRATIC DIFFERENTIAL SYSTEMS FOR INTERACTIVE POPULATION MODELS. J. DIFFL. EQUANS. 5(3), 497-514. [VOLTERRA]

JUNGE, C. O. 1966. DEPENSATORY PROCESS BASED ON THE CONCEPT OF HUNGER. J. FISH. RES. BOARD CAN. 23(5), 689-699. [REPRODUCTION CURVE OF RICKER; DEPENSATORY CURVE OF LARKIN ET AL.; PREDATOR-PREY MODEL]

KEEN, R. 1973. A PROBABILISTIC APPROACH TO THE DYNAMICS OF NATURAL POPULATIONS OF THE CHYDORIDAE (CLADOCERA, CRUSTACEA). ECOLOGY 54(3), 524-534. [MODEL; STOCHASTIC]

KESTEVEN, G. L. 1966. QUANTITATIVE METHODS IN THE STUDY OF PROCESSES IN MARINE POPULATIONS. PROC. ECOL. SOC. AUSTRALIA 1, 65-74. [MODEL]

KIERSTEAD, H. AND L. B. SLOBODKIN. 1953. THE SIZE OF WATER MASSES CONTAINING PLANKTON BLOOMS. J. MARINE RES. 12(1), 141-147. [POPULATION MODEL]

LANDAHL, H. D. 1955. A MATHEMATICAL MODEL FOR THE TEMPORAL PATTERN OF A POPULATION STRUCTURE, WITH PARTICULAR REFERENCE TO THE FLOUR BEETLE. BULL. MATH. BIOPHYS. 17, 63-77.

LANDAHL, H. D. 1955. A MATHEMATICAL MODEL FOR THE TEMPORAL PATTERN OF A POPULATION STRUCTURE, WITH PARTICULAR REFERENCE TO THE FLOUR BEETLE: II. COMPETITION BETWEEN SPECIES. BULL. MATH. BIOPHYS. 17, 131-140.

LEVANDOWSKY, M. 1974. A FURTHER COMMENT ON THE MODEL OF REDDINGIUS AND DEN BOER. AMER. NAT. 108(961), 396-397. [NUMBER-OF-FACTORS MODEL]

LEVINS, R. 1966. THE STRATEGY OF MODEL BUILDING IN POPULATION BIOLOGY. AMER. SCI. 54(4), 421-431.

LEVINS, R. AND H. HEATWOLE. 1973. BIOGEOGRAPHY OF THE PUERTO RICAN BANK: INTRODUCTION OF SPECIES ONTO PALOMINITOS ISLAND. ECOLOGY 54(5), 1056-1064. [POPULATION MODEL; PROBABILITY; GEOMETRIC DISTRIBUTION]

MACARTHUR, R. H. AND E. O. WILSON. 1963. AN EQUILIBRIUM THEORY OF INSULAR ZOOGEOGRAPHY. EVOLUTION 17(4), 373-387. [MODEL; MEAN DISPERSAL DISTANCE]

MATESSI, C. AND R. CORI. 1972. MODELS OF POPULATION GENETICS OF BATESIAN MIMICRY. THEOR. POPUL. BIOL. 3(1), 41-68.

MAY, R. M. 1973. TIME-DELAY VERSUS STABILITY IN POPULATION MODELS WITH TWO AND THREE TROPHIC LEVELS. ECOLOGY 54(2), 315-325. [LOTKA-VOLTERRA; MODEL; TIME LAG; LOGISTIC; GEOMETRIC MEAN; PREDATOR-PREY MODEL]

MAZANOV, A. 1973. A MODEL FOR A CLASS OF BIOLOGICAL SYSTEMS. J. ENVIRON. MANAGE. 1(3), 229-238. [SIMULATION; POPULATION MODEL; METABOLISM MODEL]

MAZANOV, A. 1973. A MULTI-STAGE POPULATION MODEL. J. THEOR. BIOL. 39(3), 581-587. [SIMULATION; COMPUTER]

MCFADDEN, J. T. 1963. AN EXAMPLE OF INACCURACIES INHERENT IN INTERPRETATION OF ECOLOGICAL FIELD DATA. AMER. NAT. 97(893), 99-116. [REGRESSION]

MOTT, D. G. 1966. THE ANALYSIS OF DETERMINATION IN POPULATION. PP. 179-194. IN. SYSTEMS ANALYSIS IN ECOLOGY, (K. E. F. WATT, ED.). ACADEMIC PRESS, NEW YORK, NEW YORK.

MOTT, D. G. 1969. DYNAMIC MODELS FOR POPULATION SYSTEMS. PP. 53-72. IN. FOREST INSECT POPULATION DYNAMICS. (ANONYMOUS, ED.) U. S. FOR. SERV. NORTHEASTERN FOREST EXPT. STN. RES. PAPER NE-125. [COMPUTER]

NIVEN, B. S. 1970. MATHEMATICS OF POPULATIONS OF THE QUOKKA, SETONIX BRACHYURUS (MARSUPIALIA). I. A SIMPLE DETERMINISTIC MODEL FOR QUOKKA POPULATIONS. AUST. J. ZOOL. 18(2), 209-214. [COMPUTER]

PARK, T., D. B. MERTZ, W. GRODZINSKI AND T. PRUS. 1965. CANNIBALISTIC PREDATION IN POPULATIONS OF FLOUR BEETLES. PHYSIOL. ZOOL. 38(3), 289-321. [MODEL]

PARLETT, B. 1970. ERGODIC PROPERTIES OF POPULATIONS I: THE ONE SEX MODEL. THEOR. POPUL. BIOL. 1(2), 191-207.

POOLE, R. W. 1972. AN AUTOREGRESSIVE MODEL OF POPULATION DENSITY CHANGE IN AN EXPERIMENTAL POPULATION OF DAPHNIA MAGNA. OECOLOGIA 10(3), 205-221. [MONTE CARLO; COMPUTER]

RIGLER, F. H. AND J. M. COOLEY. 1974. THE USE OF FIELD DATA TO DERIVE POPULATION STATISTICS OF MULTIVOLTINE COPEPODS. LIMNOL. OCEANOGR. 19(4), 636-655. [POPULATION EQUATIONS; GROWTH CURVES]

RILEY, G. A. 1946. FACTORS CONTROLLING PHYTOPLANKTON POPULATIONS ON GEORGES BANK. J. MARINE RES. 6(1), 54-73. [MODEL]

RILEY, G. A. 1947. A THEORETICAL ANALYSIS OF THE ZOOPLANKTON POPULATION OF GEORGES BANK. J. MARINE RES. 6(2), 104-113. [MODEL]

RILEY, G. A. 1965. A MATHEMATICAL MODEL OF REGIONAL VARIATIONS IN PLANKTON. LIMNOL. OCEANOGR. 10(SUPPLEMENT), R202-R215.

ROFF, D. A. 1974. THE ANALYSIS OF A POPULATION MODEL DEMONSTRATING THE IMPORTANCE OF DISPERSAL IN A HETEROGENEOUS ENVIRONMENT. OECOLOGIA 15(3), 259-275. [SIMULATION; CARRYING CAPACITY]

ROFF, D. A. 1974. A COMMENT ON THE NUMBER-OF-FACTORS MODEL OF REDDINGIUS AND DEN BOER. AMER. NAT. 108(961), 391-393.

ROGERS, D. 1972. RANDOM SEARCH AND INSECT POPULATION MODELS. J. ANIM. ECOL. 41(2), 369-383.

ROGERS, D. 1975. A MODEL FOR AVOIDANCE OF SUPERPARASITISM BY SOLITARY INSECT PARASITOIDS. J. ANIM. ECOL. 44(2), 623-638. [COMPUTER; BINOMIAL DISTRIBUTION; POISSON DISTRIBUTION]

ROTENBERG, M. 1972. THEORY OF POPULATION TRANSPORT. J. THEOR. BIOL. 37(2), 291-305. [MALTHUS EQUATION; VERHULST AND VOLTERRA EQUATIONS; PROBABILITY; TRANSPORT EQUATION; BEHAVIOR]

ROYAMA, T. 1970. FACTORS GOVERNING THE HUNTING BEHAVIOUR AND SELECTION OF FOOD BY THE GREAT TIT (PARUS MAJOR L.). J. ANIM. ECOL. 39(3), 619-668. [MODEL]

SAIDEL, G. M. 1968. BACTERIAL CELL POPULATIONS IN A CONTINUOUSLY CHANGING ENVIRONMENT. J. THEOR. BIOL. 19(3), 287-296. [MODEL]

SHIYOMI, M. 1974. AN INTERPRETATION OF APHID POPULATION DEVELOPMENT BY STATISTICAL MODEL. RES. POPUL. ECOL.-KYOTO 16(1), 69-79. [PROBABILITY; NEGATIVE BINOMIAL DISTRIBUTION; COMPUTER]

SIMPSON, H. R. 1958. THE EFFECT OF STERILIZED MALES ON A NATURAL TSETSE FLY POPULATION. BIOMETRICS 14(2), 159-173. [MODEL; POISSON DISTRIBUTION; MATRIX]

SINKO, J. W. AND W. STREIFER. 1971. A MODEL FOR POPULATIONS REPRODUCING BY FISSION. ECOLOGY 52(2), 330-335.

SMITH, R. H. 1973. THE ANALYSIS OF INTRA-GENERATION CHANGE IN ANIMAL POPULATIONS. J. ANIM. ECOL. 42(3), 611-622. [MODEL; KEY FACTOR ANALYSIS; MULTIVARIATE; PRINCIPAL COMPONENTS ANALYSIS; MORRIS'S REGRESSION MODEL; VARIANCE; COVARIANCE; MAXIMUM LIKELIHOOD; MATRIX]

SOKAL, R. R. AND E. H. BRYANT. 1967. COMPUTING POPULATION BUDGET FROM SEQUENTIALLY SACRIFICED, REPLICATED CULTURES. RES. POPUL. ECOL.-KYOTO 9(1), 10-18. [TRANSITION PROBABILITIES; RANDOMIZED CENSUS MATRICES; TRINOMIAL PROBABILITIES; COMPUTER; CONFIDENCE REGIONS]

STALEY, D. H., W. N. CANNON, JR. AND W. R. HOLT. 1971. A MATHEMATICAL MODEL OF AN INSECT POPULATION WITH OVERLAPPING GENERATIONS WHERE THE FEMALES ARE POLYANDROUS AND THE MALES ARE SUBJECT TO AUTO-STERILIZATION. ANN. ENTOMOL. SOC. AMER. 64(2), 325-330.

STALEY, D. H., W. R. HOLT AND W. N. CANNON, JR. 1974. NECESSARY AND SUFFICIENT CONDITION FOR A MODEL INSECT POPULATION TO GO TO EXTINCTION. BULL. MATH. BIOL. 36(5/6), 527-533.

STEELE, J. 1974. STABILITY OF PLANKTON ECOSYSTEMS. PP. 179-191. IN, ECOLOGICAL STABILITY. (M. B. USHER AND M. H. WILLIAMSON, EDS.). HALSTED PRESS, NEW YORK, NEW YORK. [MODEL; LINEAR THEORY; STABILITY ANALYSIS; LOTKA-VOLTERRA EQUATION; NON-LINEAR THEORY]

STIRZAKER, D. 1975. ON A POPULATION MODEL. MATH. BIOSCI. 23(3/4), 329-336. [VOLE POPULATION; PERTURBATION ANALYSIS; NON-LINEAR DELAY DIFFERENTIAL EQUATION]

STREIFER, W. 1974. REALISTIC MODELS IN POPULATION ECOLOGY. ADV. ECOL. RES. 8, 199-266. [AGE-SPECIFIC; SINGLE SPECIES; SPECIES INTERACTIONS; CRITICAL VARIABLE FORMULATION; APPLICATIONS]

SUBRAMANIAN, G., D. RANKRISHNA, A. G. FREDRICKSON AND H. M. TSUCHIYA. 1970. ON THE MASS DISTRIBUTION MODEL FOR MICROBIAL CELL POPULATIONS. BULL. MATH. BIOPHYS. 32, 521-537. [GROWTH MODEL; TIME LAG]

TAYLOR, N. W. 1967. A MATHEMATICAL MODEL FOR TRIBOLIUM CONFUSUM POPULATIONS. ECOLOGY 48(2), 290-294. [COMPUTER]

TOGNETTI, K. 1975. THE TWO STAGE INTEGRAL POPULATION MODEL. MATH. BIOSCI. 24(1/2), 61-70. [VOLTERRA INTEGRAL EQUATIONS]

TOGNETTI, K. P. AND A. MAZANOV. 1970. A TWO-STAGE POPULATION MODEL. MATH. BIOSCI. 8, 371-378. [LIKE MALTHUSIAN MODEL]

VAN DEN DRIESSCHE, P. 1974. STABILITY OF LINEAR POPULATION MODELS. J. THEOR. BIOL. 48(2), 473-476.

VOLTERRA, V. 1926. FLUCTUATIONS IN THE ABUNDANCE OF SPECIES CONSIDERED MATHEMATICALLY. NATURE 118(2972), 558-560. [COMPETITION]

WANGERSKY, P. J. AND W. J. CUNNINGHAM. 1957. TIME LAG IN POPULATION MODELS. COLD SPRING HARBOR SYMPOSIA ON QUANTITATIVE BIOLOGY 22, 329-337.

WANGERSKY, P. J. AND W. J. CUNNINGHAM. 1957. TIME LAG IN PREY-PREDATOR POPULATION MODELS. ECOLOGY 38(1), 136-139.

WATT, K. E. F. 1963. MATHEMATICAL POPULATION MODELS FOR FIVE AGRICULTURAL CROP PESTS. MEM. ENTOMOL. SOC. CAN. 32, 83-91.

WHITLOCK, J. H. AND D. S. ROBSON. 1969. THE ANALYSIS OF POPULATION AND ENVIRONMENTAL DISEASE MECHANISMS. CORNELL VET. 59(3), 453-465. [MODEL; POISSON DISTRIBUTION; LOG TRANSFORMATION; REGRESSION]

WILLIAMS, E. J. 1961. THE GROWTH AND AGE-DISTRIBUTION OF A POPULATION OF INSECTS UNDER UNIFORM CONDITIONS. BIOMETRICS 17(3), 349-358. [MODEL]

WILSON, E. O. 1973. GROUP SELECTION AND ITS SIGNIFICANCE FOR ECOLOGY. BIOSCIENCE 23(11), 631-638. [R AND K EXTINCTION; LEVINS' MODEL; BOORMAN-LEVITT MODEL]

MODELS (STOCHASTIC)

AHMED, M. S. 1963. A STOCHASTIC MODEL FOR THE TUNNELLING AND RETUNNELLING OF THE FLOUR BEETLE. BIOMETRICS 19(2), 341-351. [MARKOV CHAIN]

ANDRZEJEWSKI, R. AND T. WIERZBOWSKA. 1970. ESTIMATE OF THE NUMBER OF TRAPS VISITED BY SMALL MAMMALS BASED ON A PROBABILISTIC MODEL. ACTA THERIOL. 15(1), 1-14.

AOYAMA, I. AND Y. INOUE. 1974. A STOCHASTIC STUDY ON THE CONCENTRATION PROCESS OF RADIOACTIVE SUBSTANCES TO AQUATIC ORGANISMS. HEALTH PHYSICS 26(2), 191-198. [TIME SERIES; PROBABILITY; MODEL; POISSON DISTRIBUTION]

ASHFORD, J. R., K. L. Q. READ AND G. G. VICKERS. 1970. A SYSTEM OF STOCHASTIC MODELS APPLICABLE TO STUDIES OF ANIMAL POPULATION DYNAMICS. J. ANIM. ECOL. 39(1), 29-50.

BAILEY, N. T. J. 1950. A SIMPLE STOCHASTIC EPIDEMIC. BIOMETRIKA 37(3/4), 193-202. [MODEL]

BAILEY, N. T. J. 1953. THE TOTAL SIZE OF A GENERAL STOCHASTIC EPIDEMIC. BIOMETRIKA 40(1/2), 177-185. [MODEL]

BAILEY, N. T. J. 1963. THE SIMPLE STOCHASTIC EPIDEMIC: A COMPLETE SOLUTION IN TERMS OF KNOWN FUNCTIONS. BIOMETRIKA 50(3/4), 235-240.

BAILEY, N. T. J. 1968. A PERTURBATION APPROXIMATION TO THE SIMPLE STOCHASTIC EPIDEMIC IN A LARGE POPULATION. BIOMETRIKA 55(1), 199-209. [MODEL]

BAILEY, N. T. J. 1968. STOCHASTIC BIRTH, DEATH AND MIGRATION PROCESSES FOR SPATIALLY DISTRIBUTED POPULATIONS. BIOMETRIKA 55(1), 189-198.

BAILEY, N. T. J. AND A. S. THOMAS. 1971. THE ESTIMATION OF PARAMETERS FROM POPULATION DATA ON THE GENERAL STOCHASTIC EPIDEMIC. THEOR. POPUL. BIOL. 2(3), 253-270. [SMALL POX]

BAKER, M. C. 1973. STOCHASTIC PROPERTIES OF THE FORAGING BEHAVIOR OF SIX SPECIES OF MIGRATORY SHOREBIRDS. BEHAVIOUR 45(3/4) 242-270. [COMPUTER; GAMMA DISTRIBUTION; PROBABILITY; POISSON PROCESS; INFORMATION; MARKOV CHAIN; MODEL; MATRIX]

BAKKER, K., H. J. P. EIJSACKERS, J. C. VAN LENTEREN AND E. MEELIS. 1972. SOME MODELS DESCRIBING THE DISTRIBUTION OF EGGS OF THE PARASITE PSEUDEUCOILA BOCHEI (HYM., CYNIP.) OVER ITS HOSTS, LARVAE OF DROSOPHILA MELANOGASTER. OECOLOGIA 10, 29-57. [POISSON DISTRIBUTION; PROBABILITY]

BARAKAT, R. 1959. A NOTE ON THE TRANSIENT STAGE OF THE RANDOM DISPERSAL OF LOGISTIC POPULATIONS. BULL. MATH. BIOPHYS. 21, 141-151. [STURM-LIOUVILLE SERIES]

BARTHOLOMAY, A. F. 1958. ON THE LINEAR BIRTH AND DEATH PROCESSES OF BIOLOGY AS MARKOFF CHAINS. BULL. MATH. BIOPHYS. 20, 97-118 [STOCHASTIC MODEL; Q-MATRIX METHOD OF DOOB]

BARTLETT, M. S. 1949. SOME EVOLUTIONARY STOCHASTIC PROCESSES. J. ROYAL STAT. SOC. SER. B 11(2), 221-229.

BARTLETT, M. S. 1956. DETERMINISTIC AND STOCHASTIC MODELS FOR RECURRENT EPIDEMICS. PROC. THIRD BERKELEY SYMP. MATH. STAT. AND PROB. 4, 81-109.

BARTLETT, M. S. 1960. MONTE CARLO STUDIES IN ECOLOGY AND EPIDEMIOLOGY. PROC. FOURTH BERKELEY SYMP. MATH. STAT. AND PROB. 4, 39-55. [COMPUTER]

BARTLETT, M. S. 1973. EQUATIONS AND MODELS OF POPULATION CHANGE. PP. 5-21. IN, THE MATHEMATICAL THEORY OF THE DYNAMICS OF BIOLOGICAL POPULATIONS. (M. S. BARTLETT AND R. W. HIORNS, EDS.). ACADEMIC PRESS, NEW YORK, NEW YORK. [STOCHASTIC; DETERMINISTIC MODEL; PROBABILITY; MATRIX; VARIANCE; AGE DISTRIBUTION; EPIDEMIC MODEL]

BARTLETT, M. S., J. C. GOWER AND P. H. LESLIE. 1960. A COMPARISON OF THEORETICAL AND EMPIRICAL RESULTS FOR SOME STOCHASTIC POPULATION MODELS. BIOMETRIKA 47(1/2), 1-11.

BARTLETT, M. S., J. M. BRENNAN AND J. N. POLLOCK. 1971. STOCHASTIC ANALYSIS OF SOME EXPERIMENTS ON THE MATING OF BLOWFLIES. BIOMETRICS 27(3), 725-730. [STOCHASTIC MODEL; DETERMINISTIC MODEL; LIKELIHOOD FUNCTION]

BECKER, N. G. 1970. A STOCHASTIC MODEL FOR TWO INTERACTING POPULATIONS. J. APPL. PROBAB. 7(3), 544-564. [DETERMINISTIC MODEL; LINEAR BIRTH AND DEATH PROCESS; NON-LINEAR INTERACTION]

BECKER, N. G. 1973. INTERACTIONS BETWEEN SPECIES: SOME COMPARISONS BETWEEN DETERMINISTIC AND STOCHASTIC MODELS. ROCKY MOUNTAIN. J. MATH. 3(1), 53-68. [LOTKA-VOLTERRA EQUATION; COMPUTER; MONTE CARLO; PREDATOR-PREY MODEL; PROBABILITY; POISSON PROCESS]

BEYER, J. E. AND N. KEIDING. 1975. REMARKS ON THE PAPER, A STOCHASTIC BIVARIATE ECOLOGY MODEL FOR COMPETING SPECIES, BY CHRIS P. TSOKOS AND SIDNEY W. HINKLEY. MATH. BIOSCI. 24(3/4), 351-352.

BEYER, W. A., D. R. HARRIS AND R. J. RYAN. 1971. A STOCHASTIC MODEL OF THE ISLE ROYALE BIOME. AEC REPT. LA-DC-12961. LOS ALAMOS SCI. LAB., LOS ALAMOS, NEW MEXICO. 33 PP. [POISSON DISTRIBUTION; PROBABILITY; SIMULATION; COMPUTER]

BHARGAVA, T. N. AND M. C. WILSON. 1973. ON SOME PROBLEMS IN STATISTICAL ANALYSIS OF AN AQUATIC ECOSYSTEM. BULL. INTERN. STAT. INST. 45(1), 169-186 [COMPUTER; MODEL; TRANSFER MATRIX]

BHARUCHA-REID, A. T. 1958. COMPARISON OF POPULATIONS WHOSE GROWTH CAN BE DESCRIBED BY A BRANCHING STOCHASTIC PROCESS-WITH SPECIAL REFERENCE TO A PROBLEM IN EPIDEMIOLOGY. SANKHYA 19(1/2), 1-14.

BROADBENT, S. R. AND D. G. KENDALL. 1953. THE RANDOM WALK OF TRICHOSTRONGYLUS RETORTAEFORMIS. BIOMETRICS 9(4), 460-466.

BRYANT, E. H. 1969. A MODEL OF PLANT-TO-PLANT MOVEMENT OF APHIDS: A NEW APPROACH. RES. POPUL. ECOL.-KYOTO 11(1), 34-44. [BINOMIAL DISTRIBUTION; POISSON DISTRIBUTION; STOCHASTIC MODEL]

BRYANT, E. H. 1969. A MODEL OF POPULATION FITNESS AND THE MAINTENANCE OF GENETIC POLYMORPHISM WHEN TWO ECOLOGICAL PATCHES ARE AVAILABLE. BULL. MATH. BIOPHYS. 31, 669-680. [MOVEMENT; MARKOV CHAIN]

CHAPMAN, D. G. 1967. STOCHASTIC MODELS IN ANIMAL POPULATION ECOLOGY. PP. 147-162. IN. BIOLOGY AND PROBLEMS OF HEALTH. PROC. FIFTH BERKELEY SYMP. MATH. STAT. PROBAB. VOL. 4. UNIVERSITY OF CALIFORNIA PRESS, BERKELEY, CALIFORNIA. [LOTKA-VOLTERRA MODEL; MONTE CARLO; SIMULATION]

CHAPMAN, R. C. 1969. MODELING FOREST INSECT POPULATIONS-THE STOCHASTIC APPROACH. PP. 73-88. IN. FOREST INSECT POPULATION DYNAMICS. (ANONYMOUS, ED.). U. S. FOR. SERV. NORTHEASTERN FOREST EXPT. STN. RES. PAPER NE-125.

CHIANG, C. L. 1957. AN APPLICATION OF STOCHASTIC PROCESSES TO EXPERIMENTAL STUDIES ON FLOUR BEETLES. BIOMETRICS 13(1), 79-97.

CHOI, K. AND D. S. ROBSON. NO DATE A STOCHASTIC MODEL OF A "PUT-AND-TAKE" FISHERY. BIOMETRICS UNIT, CORNELL UNIVERSITY, BU-284-M. 18 PP.

COHEN, J. E. 1969. NATURAL PRIMATE TROOPS AND A STOCHASTIC POPULATION MODEL. AMER. NAT. 103(933), 455-477. [BIRTH AND DEATH RATE; BINOMIAL DISTRIBUTION; POISSON DISTRIBUTION; GOODNESS OF FIT; MAXIMUM LIKELIHOOD; JACKNIFE ESTIMATE; NEGATIVE BINOMIAL DISTRIBUTION]

COHEN, J. E. 1970. A MARKOV CONTINGENCY-TABLE MODEL FOR REPLICATED LOTKA-VOLTERRA SYSTEMS NEAR EQUILIBRIUM. AMER. NAT. 104(940), 547-560.

COHEN, J. E. 1972. MARKOV POPULATION PROCESSES AS MODELS OF PRIMATE SOCIAL AND POPULATION DYNAMICS. THEOR. POPUL. BIOL. 3(2), 119-134.

CRUMPS, K. S. 1970. MIGRATORY POPULATIONS IN BRANCHING PROCESSES. J. APPL. PROBAB. 7(3), 565-572. [AGE DEPENDENT; STOCHASTIC MODEL]

CUSTER, S. W. AND R. G. KRUTCHKOFF. 1969. STOCHASTIC MODEL FOR BOD AND DO IN ESTUARIES. AMER. SOC. CIVIL ENG., SAN. ENG. DIV. PROC. 95(SA 5), 865-885. [PROBABILITY; POISSON DISTRIBUTION; COMPUTER]

DACEY, M. F. 1971. REGULARITY IN SPATIAL DISTRIBUTIONS: A STOCHASTIC MODEL OF THE IMPERFECT CENTRAL PLACE PLANE PP. 287-309, IN, STATISTICAL ECOLOGY, VOLUME 1. SPATIAL PATTERNS AND STATISTICAL DISTRIBUTIONS. (G. P. PATIL, E. C. PIELOU AND W. E. WATER, EDS.). PENNSYLVANIA STATE UNIVERSITY PRESS, UNIVERSITY PARK, PENNSYLVANIA.

DANIELS, H. E. 1972. AN EXACT RELATION IN THE THEORY OF CARRIER-BORNE EPIDEMICS. BIOMETRIKA 59(1), 211-213. [RANDOM WALK; MODEL]

DARWIN, J. H. 1956. THE BEHAVIOUR OF AN ESTIMATOR FOR A SIMPLE BIRTH AND DEATH PROCESS. BIOMETRIKA 43(1/2), 23-31.

DAWSON, D. A. 1972. STOCHASTIC EVOLUTION EQUATIONS. MATH. BIOSCI. 15, 287-316. [POPULATION MODEL; BAILEY'S MODEL]

DEISTLER, M. AND G. FEICHTINGER. 1973. ON THE LINEAR STOCHASTIC MODEL IN POPULATION DYNAMICS. BULL. INTERN. STAT. INST. 45(1), 301-308. [DETERMINISTIC MATRIX MODEL; LESLIE MATRIX]

DENTON, G. M. 1972. ON DOWNTON'S CARRIER-BORNE EPIDEMIC. BIOMETRIKA 59(2), 455-461. [STOCHASTIC MODEL; MATRIX; TOTAL SIZE]

DOUBLEDAY, W. G. 1973. ON LINEAR BIRTH-DEATH PROCESSES WITH MULTIPLE BIRTHS. MATH. BIOSCI. 17, 43-56. [MODEL; PROBABILITY GENERATING FUNCTION; MARKOVIAN; MARKOV CHAIN; MATRIX; STOCHASTIC PATH INTEGRALS; MAXIMUM LIKELIHOOD ESTIMATION]

DOWNTON, F. 1967. A NOTE ON THE ULTIMATE SIZE OF A GENERAL STOCHASTIC EPIDEMIC. BIOMETRIKA 54(1/2), 314-316. [RANDOM WALK]

DOWNTON, F. 1968. THE ULTIMATE SIZE OF CARRIER-BORNE EPIDEMICS. BIOMETRIKA 55(2), 277-289. [MODEL]

ETTER, D. O., JR. 1971. SOME NUMERICAL COMPARISONS OF DETERMINISTIC AND STOCHASTIC PREDATION MODELS. PP. 267-283. IN, STATISTICAL ECOLOGY, VOLUME 2. SAMPLING AND MODELING BIOLOGICAL POPULATIONS AND POPULATION DYNAMICS. (G. P. PATIL, E. C. PIELOU AND W. E. WATERS, EDS.). PENNSYLVANIA STATE UNIVERSITY PRESS, UNIVERSITY PARK, PENNSYLVANNIA.

FOSTER, F. G. 1955. A NOTE ON BAILEY'S AND WHITTLE'S TREATMENT OF A GENERAL STOCHASTIC EPIDEMIC. BIOMETRIKA 42(1/2), 123-125.

GANI, J. 1973. STOCHASTIC FORMULATIONS FOR LIFE TABLES, AGE DISTRIBUTIONS AND MORTALITY CURVES. PP. 291-302. IN, THE MATHEMATICAL THEORY OF THE DYNAMICS OF BIOLOGICAL POPULATIONS. (M. S. BARTLETT AND R. W. HIORNS, EDS.). ACADEMIC PRESS, NEW YORK, NEW YORK. [SURVIVAL RATE; MARKOV CHAIN FORMULATION; PROBABILITY DISTRIBUTION]

GART, J. J. 1972. THE STATISTICAL ANALYSIS OF CHAIN-BINOMIAL EPIDEMIC MODELS WITH SEVERAL KINDS OF SUSCEPTIBLES. BIOMETRICS 28(4), 921-930.

GOODMAN, L. A. 1967. THE PROBABILITIES OF EXTINCTION FOR BIRTH-AND-DEATH PROCESSES THAT ARE AGE-DEPENDENT OR PHASE-DEPENDENT. BIOMETRIKA 54(3/4), 579-596.

GRINGORTEN, I. I. 1971. MODELING CONDITIONAL PROBABILITY. J. APPL. METEOROL. 10(4), 646-657.

HADIDI, N. 1973. A NOTE ON GENERALIZED BIRTH-DEATH PROCESSES. BULL. INTERN. STAT. INST. 45(1), 487-491. [MARKOVIAN BIRTH-DEATH PROCESS]

HAILMAN, J. P. 1974. A STOCHASTIC MODEL OF LEAF-SEARCHING BOUTS IN TWO EMBERIZINE SPECIES. WILSON BULL. 86(3), 296-298.

HAMANN, J. R. AND L. M. BRANCH. 1970. STOCHASTIC POPULATION MECHANICS IN THE RELATIONAL SYSTEMS FORMALISM: VOLTERRA-LOTKA ECOLOGICAL DYNAMICS. J. THEOR. BIOL. 28(2), 175-184. [GIBBS STATISTICAL MECHANICS]

HASKEY, H. W. 1954. A GENERAL EXPRESSION FOR THE MEAN IN A SIMPLE STOCHASTIC EPIDEMIC. BIOMETRIKA 41(1/2), 272-275.

HILL, R. T. AND N. C. SEVERO. 1969. THE SIMPLE STOCHASTIC EPIDEMIC FOR SMALL POPULATIONS WITH ONE OR MORE INITIAL INFECTIVES. BIOMETRIKA 56(1), 183-196. [MODEL]

HOEL, D. G. AND T. J. MITCHELL. 1971. THE SIMULATION, FITTING AND TESTING OF A STOCHASTIC CELLULAR PROLIFERATION MODEL. BIOMETRICS 27(1), 191-199.

HOLGATE, P. 1971. RANDOM WALK MODELS FOR ANIMAL BEHAVIOR. PP. 1-12. IN, STATISTICAL ECOLOGY, VOLUME 2. SAMPLING AND MODELING BIOLOGICICAL POPULATIONS AND POPULATION DYNAMICS. (G. P. PATIL, E. C. PIELOU AND W. E. WATERS, EDS.). PENNSYLVANIA STATE UNIVERSITY PRESS, UNIVERSITY PARK, PENNSYLVANIA.

JAQUETTE, D. L. 1970. A STOCHASTIC MODEL FOR THE OPTIMAL CONTROL OF EPIDEMICS AND PEST POPULATIONS. MATH. BIOSCI. 8, 343-354. [COST FUNCTION]

KATTI, S. K. AND L. E. SLY 1965. ANALYSIS OF CONTAGIOUS DATA THROUGH BEHAVIORISTIC MODELS. PP. 303-319. IN, CLASSICAL AND CONTAGIOUS DISCRETE DISTRIBUTIONS. (G. P. PATIL, ED.). STATISTICAL PUBLISHING SOCIETY, CALCUTTA, INDIA. [POISSON DISTRIBUTION; PROBABILITY]

KENDALL, D. G. 1948. ON THE GENERALIZED "BIRTH-AND-DEATH" PROCESS. ANN. MATH. STAT. 19(1), 1-15.

KENDALL, D. G. 1948. ON THE ROLE OF VARIABLE GENERATION TIME IN THE DEVELOPMENT OF A STOCHASTIC BIRTH PROCESS. BIOMETRIKA 35(3/4), 316-330.

KERRIDGE, D. 1964. PROBABILISTIC SOLUTION OF THE SIMPLE BIRTH PROCESS. BIOMETRIKA 51(1/2), 258-259.

KIESTER, A. R. AND R. BARAKAT. 1974. EXACT SOLUTIONS TO CERTAIN STOCHASTIC DIFFERENTIAL EQUATION MODELS OF POPULATION GROWTH. THEOR. POPUL. BIOL. 6(2), 199-216.

KOROSTYSHEVSKY, M. A., M. R. SCHTABNOY AND V. A. RATNER. 1974. ON SOME PRINCIPLES OF EVOLUTION VIEWED AS A STOCHASTIC PROCESS. J. THEOR. BIOL. 48(1), 85-103. [MODEL]

KRYSCIO, R. J. 1972. ON ESTIMATING THE INFECTION RATE OF THE SIMPLE STOCHASTIC EPIDEMIC. BIOMETRIKA 59(1), 213-214. [MAXIMUM LIKELIHOOD; MODEL]

KRYSCIO, R. J. 1974. ON THE EXTENDED SIMPLE STOCHASTIC EPIDEMIC MODEL. BIOMETRIKA 61(1), 200-202.

KRYSCIO, R. J. AND N. C. SEVERO. 1969. SOME PROPERTIES OF AN EXTENDED SIMPLE STOCHASTIC EPIDEMIC MODEL INVOLVING TWO ADDITIONAL PARAMETERS. MATH. BIOSCI. 5, 1-8.

LEAK, W. B. 1970. SAPLING STAND DEVELOPMENT: A COMPOUND EXPONENTIAL PROCESS. FOREST SCI. 16(2), 177-180. [STOCHASTIC MODEL]

LESLIE, P. H. 1958. A STOCHASTIC MODEL FOR STUDYING THE PROPERTIES OF CERTAIN BIOLOGICAL SYSTEMS BY NUMERICAL METHODS. BIOMETRIKA 45(1/2), 16-31. [COMPUTER]

LESLIE, P. H. 1960. A NOTE ON SOME APPROXIMATIONS TO THE VARIANCE IN DISCRETE-TIME STOCHASTIC MODELS FOR BIOLOGICAL SYSTEMS. BIOMETRIKA 47(1/2), 196-197.

LESLIE, P. H. 1962. A STOCHASTIC MODEL FOR TWO COMPETING SPECIES OF TRIBOLIUM AND ITS APPLICATION TO SOME EXPERIMENTAL DATA. BIOMETRIKA 49(1/2), 1-25.

LESLIE, P. H. AND J. C. GOWER. 1958. THE PROPERTIES OF A STOCHASTIC MODEL FOR TWO COMPETING SPECIES. BIOMETRIKA 45(3/4), 316-330. [COMPUTER]

LESLIE, P. H. AND J. C. GOWER. 1960. THE PROPERTIES OF A STOCHASTIC MODEL FOR THE PREDATOR-PREY TYPE OF INTERACTION BETWEEN TWO SPECIES. BIOMETRIKA 47(3/4), 219-234. [COMPUTER]

LESLIE, P. H., T. PARK AND D. B. MERTZ. 1968. THE EFFECT OF VARYING THE INITIAL NUMBERS ON THE OUTCOME OF COMPETITION BETWEEN TWO TRIBOLIUM SPECIES. J. ANIM. ECOL. 37(1), 9-23. [DISCRETE-TIME STOCHASTIC MODEL OF LESLIE AND GOWER]

LEYTON, M. K. 1968. STOCHASTIC MODELS IN POPULATIONS OF HELMINTHIC PARASITES IN THE DEFINITIVE HOST. II: SEXUAL MATING FUNCTIONS. MATH. BIOSCI. 3, 413-419.

LLOYD, M. 1962. PROBABILITY AND STOCHASTIC PROCESSES IN ECOLOGY. PP. 238-245. IN: THE CULLOWHEE CONFERENCE ON TRAINING IN BIOMATHEMATICS. (H. L. LUCAS, ED.). TYPING SERVICE,, RALEIGH, NORTH CAROLINA.

LUDWIG, D. 1974. QUALITATIVE BEHAVIOR OF STOCHASTIC EPIDEMICS. PP. 152-154. IN: MATHEMATICAL PROBLEMS IN BIOLOGY. LECTURE NOTES IN BIOMATHEMATICS 2. (P. VAN DEN DRIESSCHE, ED.). SPRINGER-VERLAG NEW YORK INC., NEW YORK, NEW YORK. [MODEL]

LUDWIG, D. 1975. FINAL SIZE DISTRIBUTION FOR EPIDEMICS. MATH. BIOSCI. 23(1/2), 33-46. [MARKOV CHAIN]

MCMICHAEL, F. C. AND J. S. HUNTER. 1972. STOCHASTIC MODELING OF TEMPERATURE AND FLOW IN RIVERS. WATER RESOUR. RES. 8(1), 87-98.

MCNEIL, D. R. 1972. ON THE SIMPLE STOCHASTIC EPIDEMIC. BIOMETRIKA 59(2), 494-497. [MODEL]

MODE, C. J. 1964. A STOCHASTIC MODEL OF THE DYNAMICS OF HOST-PATHOGEN SYSTEMS WITH MUTATION. BULL. MATH. BIOPHYS. 26, 205-233.

NAMKOONG, G. 1972. PERSISTENCE OF VARIANCES FOR STOCHASTIC, DISCRETE-TIME, POPULATION GROWTH MODELS. THEOR. POPUL. BIOL. 3(4), 507-518. [EQUATIONS; LESLIE MATRIX]

NEYMAN, J. AND E. L. SCOTT. 1957. ON A MATHEMATICAL THEORY OF POPULATIONS CONCEIVED AS CONGLOMERATIONS OF CLUSTERS. COLD SPRING HARBOR SYMPOSIA ON QUANTITATIVE BIOLOGY 22, 109-120. [STOCHASTIC MODEL; PROBABILITY; POISSON-WISE DISTRIBUTION]

NEYMAN, J. AND E. L. SCOTT. 1959. STOCHASTIC MODELS OF POPULATION DYNAMICS. SCIENCE 130(3371), 303-308.

NIVEN, B. S. 1971. MATHEMATICS OF POPULATIONS OF THE QUOKKA, SETONIX BRACHYURUS (MARSUPIALIA). II. A STOCHASTIC MODEL FOR QUOKKA POPULATIONS. AUST. J. ZOOL. 19, 393-399. [COMPUTER]

PAKES, A. G. 1972. LIMIT THEOREMS FOR AN AGE-DEPENDENT BRANCHING PROCESS WITH IMMIGRATION. MATH. BIOSCI. 14, 221-234.

PALOHEIMO, J. E. 1971. ON A THEORY OF SEARCH. BIOMETRIKA 58(1), 61-75. [CONTAGIOUS DISTRIBUTION; STOCHASTIC MODEL; PREDATOR-PREY MODEL]

PALOHEIMO, J. E. 1971. A STOCHASTIC THEORY OF SEARCH: IMPLICATIONS FOR PREDATOR-PREY SITUATIONS. MATH. BIOSCI. 12, 105-132.

PARKER, R. A. 1974. SOME CONSEQUENCES OF STOCHASTICIZING AN ECOLOGICAL SYSTEM MODEL. PP. 174-183. IN, MATHEMATICAL PROBLEMS IN BIOLOGY. LECTURE NOTES IN BIOMATHEMATICS 2. (P. VAN DEN DRIESSCHE, ED.), SPRINGER VERLAG NEW YORK INC., NEW YORK, NEW YORK. [MODEL]

PATTEN, B. C. 1964. THE RATIONAL DECISION PROCESS IN SALMON MIGRATION. J. CONS. PERM. INT. EXPLOR. MER 28(3), 410-417. [PROBABILITY; MODEL; RANDOM WALK; MONTE CARLO]

PEARSON, K. ET AL. 1919. ON THE NEST AND EGGS OF THE COMMON TERN (S. FLUVIATILIS). A COOPERATIVE STUDY. BIOMETRIKA 12(3/4), 308-354. [PROBABILITY; MULTIVARIATE ANALYSIS]

PEDEN, L. M., J. S. WILLIAMS AND W. E. FRAYER. 1973. A MARKOV MODEL FOR STAND PROJECTION. FOREST SCI. 19(4), 303-314. [STOCHASTIC MODEL; DIFFERENCE EQUATION]

PELLA, J. J. 1969. A STOCHASTIC MODEL FOR PURSE SEINING IN A TWO-SPECIES FISHERY. J. THEOR. BIOL. 22(2), 209-226. [SEMI-MARKOV PROCESS; POISSON PROCESS]

POLLARD, J. H. 1966. ON THE USE OF THE DIRECT MATRIX PRODUCT IN ANALYSING CERTAIN STOCHASTIC POPULATION MODELS. BIOMETRIKA 53(3/4), 397-415. [COMPUTER]

POOLE, R. W. 1974. A DISCRETE TIME STOCHASTIC MODEL OF TWO PREY, ONE PREDATOR SPECIES INTERACTION. THEOR. POPUL. BIOL. 5(2), 208-228. [MONTE CARLO SIMULATION; TIME LAG; LOGISTIC MODEL]

PULLIAM, H. R. 1974. ON THE THEORY OF OPTIMAL DIETS. AMER. NAT. 108(959), 59-74. [STOCHASTIC MODEL; PREDATOR-PREY MODEL]

RAND, W. M. 1973. A STOCHASTIC MODEL OF THE TEMPORAL ASPECT OF BREEDING STRATEGIES. J. THEOR. BIOL. 40(2), 337-351. [NORMAL DISTRIBUTION]

RAPOPORT, A. 1949. OUTLINE OF A PROBABILISTIC APPROACH TO ANIMAL SOCIOLOGY: I. BULL. MATH. BIOPHYS. 11, 183-196. [PECK ORDER]

RAPOPORT, A. 1949. OUTLINE OF A PROBABILISTIC APPROACH TO ANIMAL SOCIOLOGY: II. BULL. MATH. BIOPHYS. 11, 273-281. [PECK ORDER]

RAPOPORT, A. 1950. OUTLINE OF A PROBABILISTIC APPROACH TO ANIMAL SOCIOLOGY: III. BULL. MATH. BIOPHYS. 12, 7-17. [PECK ORDER]

RAUP, D. M. AND S. J. GOULD. 1974. STOCHASTIC SIMULATION AND EVOLUTION OF MORPHOLOGY - TOWARDS A NOMOTHETIC PALEONTOLOGY. SYST. ZOOL. 23(3), 305-322. [COMPUTER]

REDDY, V. T. N. 1974. ON THE EXISTENCE OF THE STEADY STATE IN THE STOCHASTIC VOLTERRA MODEL. AEC REPT. ORO-3992-159. CENTER FOR PARTICLE THEORY, UNIVERSITY OF TEXAS, AUSTIN, TEXAS. 6 PP.

REVELLE, C., E. JOERES AND W. KIRBY. 1969. THE LINEAR DECISION RULE IN RESERVOIR MANAGEMENT AND DESIGN. 1. DEVELOPMENT OF THE STOCHASTIC MODEL. WATER RESOUR. RES. 5(4), 767-777.

RICHTER-DYN, N. AND N. S. GOEL. 1972. ON THE EXTINCTION OF A COLONIZING SPECIES. THEOR. POPUL. BIOL. 3(4), 406-433. [STOCHASTIC MODEL]

RIFFENBURGH, R. H. 1969. A STOCHASTIC MODEL OF INTERPOPULATION DYNAMICS IN MARINE ECOLOGY. J. FISH. RES. BOARD CAN. 26(11), 2843-2880. [MARKOV CHAIN; COMPUTER]

ROSS, G. J. S. 1972. STOCHASTIC MODEL FITTING BY EVOLUTIONARY OPERATION. PP. 297-308. IN, MATHEMATICAL MODELS IN ECOLOGY. (J. N. R. JEFFERS, ED.). BLACKWELL SCIENTIFIC PUBLICATIONS, OXFORD, ENGLAND.

SAILA, S. B. AND R. A. SHAPPY. 1964. LETTER TO THE EDITOR-SALMON MIGRATION. J. CONS. PERM. INT. EXPLOR. MER 28(3), 440-443. [MODEL; RANDOM WALK; MONTE CARLO]

SAILA, S. B. AND R. A. SHAPPY. 1964. RANDOM MOVEMENT AND ORIENTATION IN SALMON MIGRATION. J. CONS. PERM. INT. EXPLOR. MER 28(1), 153-166. [MODEL; PROBABILITY; RANDOM WALK]

SCHWARTZ, S. 1971. PROBABILISTIC MODELS FOR CALCULATING AIR POLLUTION DAMAGE. J. ENVIRON. SYSTEMS 1(2), 111-131.

SCOTT, E. L. 1965. SUBCLUSTERING. PP. 33-44. IN, CLASSICAL AND CONTAGIOUS DISCRETE DISTRIBUTIONS. (G. P. PATIL, ED.). STATISTICAL PUBLISHING SOCIETY, CALCUTTA, INDIA. [TAG-RECAPTURE ESTIMATION; CLUSTERING; EUCLIDEAN SPACE]

SEVERO, N. C. 1969. GENERALIZATIONS OF SOME STOCHASTIC EPIDEMIC MODELS. MATH. BIOSCI. 4, 395-402.

SEVERO, N. C. 1969. THE PROBABILITIES OF SOME EPIDEMIC MODELS. BIOMETRIKA 56(1), 197-201.

SISKIND, V. 1965. A SOLUTION OF THE GENERAL STOCHASTIC EPIDEMIC. BIOMETRIKA 52(3/4), 613-616. [MODEL]

STOCHASTICS INCORPORATED. 1971. STOCHASTIC MODELING FOR WATER QUALITY MANAGEMENT. EPA, WATER POLLUTION CONTROL RESEARCH SERIES 16090 DUH 02/71. XII, 395 PP.

TALLIS, G. M. 1970. SOME STOCHASTIC EXTENSIONS TO A DETERMINISTIC TREATMENT OF SHEEP PARASITE CYCLES. MATH. BIOSCI. 8, 131-135. [MODEL]

TALLIS, G. M. AND M. K. LEYTON. 1969. STOCHASTIC MODELS OF POPULATIONS OF HELMINTHIC PARASITES IN THE DEFINITIVE HOST, I. MATH. BIOSCI. 4, 39-48.

TALLIS, G. M. AND M. LEYTON. 1966. A STOCHASTIC APPROACH TO THE STUDY OF PARASITE POPULATIONS. J. THEOR. BIOL. 13, 251-260. [MODEL; PROBABILITY]

THAYER, R. P. AND R. G. KRUTCHKOFF. 1967. STOCHASTIC MODEL FOR BOD AND DO IN STREAMS. AMER. SOC. CIVIL ENG., SAN. ENG. DIV. PROC. 93(SA3), 59-72. [PROBABILITY; VARIANCE; COMPUTER]

TSOKOS, C. P. 1969. ON A STOCHASTIC INTEGRAL EQUATION OF THE VOLTERRA TYPE. MATH. SYSTEMS THEORY 3, 222-231.

TSOKOS, C. P. 1975. EXCERPTS FROM A LETTER BY C. TSOKOS TO D. PERKE. MATH. BIOSCI. 24(3/4), 353-354. [STOCHASTIC BIVARIATE ECOLOGY MODEL]

TSOKOS, C. P. AND S. W. HINKLEY 1973. A STOCHASTIC BIVARIATE ECOLOGY MODEL FOR COMPETING SPECIES. MATH. BIOSCI. 16(3/4), 191-208. [DETERMINISTIC MODEL; LOTKA-VOLTERRA MODEL]

UPPULURI, V. R. R. 1974. THEORY AND APPLICATIONS OF PERIODIC MARKOV CHAINS. PP. 761-765. IN, CHRONOBIOLOGY. (L. E. SCHEVING, F. HALBERG AND J. E. PAULY, EDS.). IGAKU SHOIN LTD., TOKYO, JAPAN. [WEATHER ANALYSIS]

WAUGH, W. A. O'N. 1972. THE APPARENT 'LAG PHASE' IN A STOCHASTIC POPULATION MODEL IN WHICH THERE IS NO VARIATION IN THE CONDITIONS OF GROWTH. BIOMETRICS 28(2), 329-336. [MARKOVIAN BRANCHING PROCESS; CELLULAR GROWTH]

WEISS, G. H. AND M. DISHON. 1971. ON THE ASYMPTOTIC BEHAVIOR OF THE STOCHASTIC AND DETERMINISTIC MODELS OF AN EPIDEMIC. MATH. BIOSCI. 11, 261-265. [BAILEY'S MODEL]

WHITTLE, P. 1955. THE OUTCOME OF A STOCHASTIC EPIDEMIC-A NOTE ON BAILEY'S PAPER. BIOMETRIKA 42(1/2), 116-122.

WHITTLE, P. 1964. A BRANCHING PROCESS IN WHICH INDIVIDUALS HAVE VARIABLE LIFETIMES. BIOMETRIKA 51(1/2), 262-264.

WILLIAMS, T. 1965. THE DISTRIBUTION OF RESPONSE TIMES IN A BIRTH-DEATH PROCESS. BIOMETRIKA 52(3/4), 581-585. [HOST-PARASITE MODEL]

WILLIAMS, T. 1965. THE SIMPLE STOCHASTIC EPIDEMIC CURVE FOR LARGE POPULATIONS OF SUSCEPTIBLES. BIOMETRIKA 52(3/4), 571-579.

YANG, G. L. 1972. ON THE PROBABILITY DISTRIBUTIONS OF SOME STOCHASTIC EPIDEMIC MODELS. THEOR. POPUL. BIOL. 3(4), 448-459. [MARKOV MODELS]

OPERATIONS RESEARCH AND PROGRAMMING

ABOUEL NOUR, A.-R. A. 1972. A STATISTICAL METHODOLOGY FOR PREDICTING THE POLLUTANTS IN A RIVER. WATER RESOUR. BULL. 8(1), 15-23. [STEPWISE MULTIPLE REGRESSION]

AMIDON, E. L. AND G. S. AKIN. 1968. DYNAMIC PROGRAMMING TO DETERMINE OPTIMUM LEVELS OF GROWING STOCK. FOREST SCI. 14(3), 287-291. [COMPUTER]

ARIMIZU, T. 1958. REGULATION OF THE CUT BY DYNAMIC PROGRAMMING. JAPANESE J. OPERATIONS RES. 1, 175-182. [FORESTRY]

ARROYO, G. J. 1962. FISH POND DEVELOPMENT PLANNING WITH THE HELP OF LINEAR PROGRAMMING. FAO FISHERIES TECHNICAL PAPER NO. 21. 22 PP.

BARE, B. B. 1971. APPLICATIONS OF OPERATIONS RESEARCH IN FOREST MANAGEMENT: A SURVEY. QUANTITATIVE SCIENCE PAPER NO. 26, CENTER FOR QUANTITATIVE SCIENCE IN FORESTRY, FISHERIES AND WILDLIFE, UNIVERSITY OF WASHINGTON, SEATTLE, WASHINGTON. MIMEO. 51 PP.

BARE, B. B. 1972. APPLICATIONS OF OPERATIONS RESEARCH IN FOREST MANAGEMENT: A SURVEY. (ABSTRACT). BIOMETRICS 28(1), 258.

BAREA, D. J. 1963. ANALYSIS OF BIOLOGICAL ECOSYSTEMS BY MEANS OF LINEAR PROGRAMMING. ARCHIVOS DE ZOOTECNIA 12(47), 252-263.

BARGUR, J. S., H. C. DAVIS AND E. M. LOFTING. 1971. ON INTERSECTORAL PROGRAMING MODEL FOR THE MANAGEMENT OF THE WASTE WATER ECONOMY OF THE SAN FRANCISCO BAY REGION. WATER RESOUR. RES. 7(6), 1393-1409. [COMPUTER]

BELLMAN, R. AND R. KALABA. 1960. SOME MATHEMATICAL ASPECTS OF OPTIMAL PREDATION IN ECOLOGY AND BOVICULTURE. PROC. NATL. ACAD. SCI. U. S. A. 46(5), 718-720. [DYNAMIC PROGRAMMING]

BROIDO, A., R. J. MCCONNEN AND W. G. O'REGAN. 1965. SOME OPERATIONS RESEARCH APPLICATIONS IN THE CONSERVATION OF WILDLAND RESOURCES. MANAGE. SCI. 11(9), 802-814. [LINEAR PROGRAMMING; COMPUTER; SAMPLING; SIMULATION; MODEL]

BUFFINGTON, C. D. 1972. AN ANALYSIS OF THE DECISION-MAKING SYSTEMS WITHIN THE NATIONAL WILDLIFE REFUGE SYSTEM. DISSERTATION. VIRGINIA POLYTECHNIC INSTITUTE, BLACKSBURG, VIRGINIA. 179 PP. [DECISION THEORY; ALGORITHM]

CHILD, G. I. AND H. H. SHUGART, JR. 1972. FREQUENCY RESPONSE ANALYSIS OF MAGNESIUM CYCLING IN A TROPICAL FOREST ECOSYSTEM. PP. 103-135. IN. SYSTEMS ANALYSIS AND SIMULATION IN ECOLOGY, VOLUME II. (B. C. PATTEN, ED.). ACADEMIC PRESS, NEW YORK, NEW YORK. [MODEL; SYSTEMS]

D'AQUINO, S. A. 1974. A CASE STUDY FOR OPTIMAL ALLOCATION OF RANGE RESOURCES. J. RANGE MANAGE. 27(3), 228-233. [LINEAR PROGRAMMING]

DALE, M. B. 1970. SYSTEMS ANALYSIS AND ECOLOGY. ECOLOGY 51(1), 2-16. [COMPUTER]

DANE, C. W. 1965. STATISTICAL DECISION THEORY AND ITS APPLICATION TO FOREST ENGINEERING. J. FORESTRY 63(4), 276-279. [BAYESIAN STATISTICS]

DAVIS, L. S. 1967. DYNAMIC PROGRAMMING FOR DEER MANAGEMENT PLANNING. J. WILDL. MANAGE. 31(4), 667-679. [COMPUTER; LINEAR PROGRAMMING]

EISNER, E. 1971. EXPERIMENTS IN ECOLOGY: A PROBLEM IN SIGNAL EXTRACTION. PP. 237-249. IN. STATISTICAL ECOLOGY, VOLUME 2. SAMPLING AND MODELING BIOLOGICAL POPULATIONS AND POPULATION DYNAMICS. (G. P. PATIL, E. C. PIELOU AND W. E. WATERS, EDS.). PENNSYLVANIA STATE UNIVERSITY PRESS, UNIVERSITY PARK, PENNSYLVANIA.

FIELD, D. B. 1973. GOAL PROGRAMMING FOR FOREST MANAGEMENT. FOREST SCI. 19(2), 125-135. [LINEAR PROGRAMMING; MODEL; COMPUTER]

FIERING, M. B. 1965. AN OPTIMIZATION SCHEME FOR GAGING. WATER RESOUR. RES. 1(4), 463-470. [NON-LINEAR PROGRAMMING; PRINCIPAL COMPONENT ANALYSIS; REGRESSION]

HAZARD, J. W. AND L. C. PROMNITZ. 1974. DESIGN OF SUCCESSIVE FOREST INVENTORIES: OPTIMIZATION BY CONVEX MATHEMATICAL PROGRAMMING. FOREST SCI. 20(2), 117-127. [SAMPLING; COST FUNCTION; COMPUTER; MODEL]

HOOL, J. N. 1966. A DYNAMIC PROGRAMMING - MARKOV CHAIN APPROACH TO FOREST PRODUCTION CONTROL. FOREST SCI. MONOGR. 12. 26 PP. [MODEL; PROBABILITY MATRIX; LINEAR PROGRAMMING]

HOOL, J. N. 1968. AN UNIVARIATE ALLOCATION ALGORITHM FOR USE IN FORESTRY PROBLEMS. J. FORESTRY 66(6), 492-493.

INNIS, G. 1971. AN EXPERIMENTAL UNDERGRADUATE COURSE IN SYSTEMS ECOLOGY. BIOSCIENCE 21(6), 283-284.

KANE, J. 1971. ON THE CONCEPT OF ECOLOGICAL SPECTRA. PP. 231-236. IN, STATISTICAL ECOLOGY, VOLUME 2. SAMPLING AND MODELING BIOLOGICAL POPULATIONS AND POPULATION DYNAMICS. (G. P. PATIL, . C. PIELOU AND W. E. WATERS, EDS.), PENNSYLVANIA STATE UNIVERSITY PRESS, UNIVERSITY PARK, PENNSYLVANIA.

ATZ, P. L. AND M. W. BARTNICK. 1974. INSTANTANEOUS (STATIC) VS ONG-TERM (DYNAMIC) OPTIMIZATION IN COSYSTEMS. PP. 395-400. IN, ROC. FIRST INTERN. CONGR. ECOL. ENTRE FOR AGRICULTURAL PUBLISHING ND DOCUMENTATION, WAGENINGEN, ETHERLANDS. [MODEL]

DD, W. E. 1965. A LINEAR OGRAMMING APPROACH TO EVALUATING REST MANAGEMENT ALTERNATIVES. S. THESIS, VIRGINIA POLYTECHNIC STITUTE, BLACKSBURG, VIRGINIA. 6 PP.

LEAK, W. B. 1964. ESTIMATING MAXIMUM ALLOWABLE TIMBER YIELDS BY LINEAR PROGRAMMING. U. S. FOR. SERV. NORTHEASTERN FOREST EXPT. STN. RES. PAPER NE-17. 9 PP.

LITTSCHWAGER, J. M. AND T. H. TCHENG. 1967. SOLUTION OF A LARGE-SCALE FOREST SCHEDULING PROBLEM BY LINEAR PROGRAMMING DECOMPOSITION. J. FORESTRY 65(9), 644-646. [COMPUTER]

LOMNICKI, A. 1972. PLANNING OF DEER POPULATION MANAGEMENT BY NON-LINEAR PROGRAMMING. ACTA THERIOL. 17(12), 137-150. [COMPUTER; MODEL]

MACARTHUR, R. 1970. GRAPHICAL ANALYSIS OF ECOLOGICAL SYSTEMS. PP. 59-73. IN, SOME MATHEMATICAL QUESTIONS IN BIOLOGY (M. GERSTENHABER, ED.), AMER. MATH. SOC., PROVIDENCE, RHODE ISLAND. [MODEL]

MAY, P. F., A. R. TILL AND M. J. CUMMING. 1973. SYSTEMS ANALYSIS OF THE EFFECT OF APPLICATION METHODS ON THE ENTRY OF SULPHUR INTO PASTURES GRAZED BY SHEEP. J. APPL. ECOL. 10(2), 607-626. [COMPUTER; MODEL]

MILLIER, C., M. POISSONNET AND J. SERRA. 1972. MORPHOLOGIE MATHEMATIQUE ET SYLVICULTURE. PP. 287-306. IN, 3RD CONFERENCE ADVISORY GROUP OF FOREST STATISTICIANS. INSTITUT NATIONAL DE LA RECHERCHE AGRONOMIQUE, PARIS, FRANCE. I.N.R.A. PUBL. 72-3. [VARIOGRAM; POISSON MODEL; OPTIMIZED INVENTORY; REGIONALIZED VARIABLE THEORY; FORESTRY]

MILSUM, J. H. 1968. MATHEMATICAL INTRODUCTION TO GENERAL SYSTEM DYNAMICS. PP. 23-65. IN, POSITIVE FEEDBACK. (J. H. MILSUM, ED.). PERGAMMON PRESS, NEW YORK, NEW YORK.

NAUTIYAL, J. C. AND P. H. PEARSE. 1967. OPTIMIZING THE CONVERSION TO SUSTAINED YIELD - A PROGRAMMING SOLUTION. FOREST SCI. 13(2), 131-139. [ECONOMIC MODEL; COMPUTER]

NORMAN, E. L. AND J. W. CURLIN. 1968. A LINEAR PROGRAMMING MODEL FOR FOREST PRODUCTION CONTROL AT THE AEC OAK RIDGE RESERVATION. OAK RIDGE NATIONAL LAB. AEC REPT. ORNL-4349. OAK RIDGE NATL. LAB., OAK RIDGE, TENNESSEE. 48 PP.

PATTEN, B. C. 1969. ECOLOGICAL SYSTEMS ANALYSIS AND FISHERIES SCIENCE. TRANS. AMER. FISH. SOC. 98(3), 570-581.

REAM, R. R. AND L. F. OHMANN. 1971. ANALYSIS OF WILDERNESS ECOSYSTEMS. PP. 123-131. IN, STATISTICAL ECOLOGY, VOLUME 3. MANY SPECIES POPULATIONS, ECOSYSTEMS, AND SYSTEMS ANALYSIS. (G. P. PATIL, E. C. PIELOU AND W. E. WATERS, EDS.). PENNSYLVANIA STATE UNIVERSITY PRESS, UNIVERSITY PARK, PENNSYLVANIA. [COMPUTER; MULTIVARIATE ANALYSIS; MODEL; COLE'S INDEX OF ASSOCIATION; SYSTEMS ANALYSIS]

ROMESBURG, H. C. 1974. SCHEDULING MODELS FOR WILDERNESS RECREATION. J. ENVIRON. MANAGE. 2(2), 159-177. [COMPUTER; LINEAR PROGRAMMING]

ROSS, P. J., E. F. HENZELL AND D. R. ROSS. 1972. EFFECTS OF NITROGEN AND LIGHT IN GRASS-LEGUME PASTURES- A SYSTEMS ANALYSIS APPROACH. J. APPL. ECOL. 9(2), 535-556.

ROTHSCHILD, B. J. 1971. PREREQUISITES FOR THE APPLICATION OF A SYSTEMS ANALYSIS APPROACH TO RESEARCH AND MANAGEMENT OF AQUATIC RESOURCES. QUANTITATIVE SCIENCE PAPER NO. 19. CENTER FOR QUANTITATIVE SCIENCE IN FORESTRY, FISHERIES AND WILDLIFE, UNIVERSITY OF WASHINGTON, SEATTLE, WASHINGTON. MIMEO, 19 PP.

SAILA, S. B. 1972. SYSTEMS ANALYSIS APPLIED TO SOME FISHERIES PROBLEMS. PP. 331-372. IN, SYSTEMS ANALYSIS AND SIMULATION IN ECOLOGY, VOLUME II. (B. C. PATTEN, ED.). ACADEMIC PRESS, NEW YORK, NEW YORK. [BEVERTON-HOLT MODEL; MODEL; BAYES' THEOREM; FORTRAN; SIMULATION; COMPUTER]

SANCHO, N. G. F. 1973. OPTIMAL POLICIES IN ECOLOGY AND RESOURCE MANAGEMENT. MATH. BIOSCI. 17, 35-41. [DYNAMIC PROGRAMMING; COST FUNCTION; STOCHASTIC MODEL]

SCOTT, D. 1973. PATH ANALYSIS: A STATISTICAL METHOD SUITED TO ECOLOGICAL DATA. PROC. N. Z. ECOL. SOC. 20, 79-95.

SMITH, E. T. AND A. R. MORRIS. 1969. SYSTEMS ANALYSIS FOR OPTIMAL WATER QUALITY MANAGEMENT. J. WATER POLLUT. CONTROL FED. 41(9), 1635-1646. [MODEL; COMPUTER]

SMITH, F. E. 1970. ANALYSIS OF ECOSYSTEMS. PP. 7-18. IN, ANALYSIS OF TEMPERATE FOREST ECOSYSTEMS. (D. E. REICHLE, ED.) SPRINGER-VERLAG, NEW YORK, NEW YORK. [SYSTEMS ANALYSIS]

SWARTZMAN, G. L. (ED.). 1972. OPTIMIZATION TECHNIQUES IN ECOSYSTEM AND LAND USE PLANNING. U. S. IBP GRASSLAND BIOME, COLORADO STATE UNIVERSITY, FT. COLLINS, COLORADO. TECH. REPT. 143. 164 PP. [MODEL; COMPUTER]

SWARTZMAN, G. L. AND J. S. SINGH. 1973. A DYNAMIC PROGRAMMING APPROACH TO OPTIMAL GRAZING STRATEGIES USING A SUCCESSION MODEL FOR A TROPICAL GRASSLAND. J. APPL. ECOL. 11(2), 537-548. [MATRIX; COMPUTER; SIMULATION]

THOMPSON, E. F. AND R. W. HAYNES. 1971. A LINEAR PROGRAMMING-PROBABILISTIC APPROACH TO DECISION MAKING UNDER UNCERTAINTY. FOREST SCI. 17(2), 224-229. [SIMULATION; STOCHASTIC MODEL; COMPUTER]

VAN DYNE, G. M. 1965. APPLICATION OF SOME OPERATIONS RESEARCH TECHNIQUES TO FOOD CHAIN ANALYSIS PROBLEMS. HEALTH PHYSIC 11(12), 1511-1519. [COMPUTER]

VAN DYNE, G. M. 1966. APPLICATION AND INTEGRATION OF MULTIPLE LINEAR REGRESSION AND LINEAR PROGRAMMING IN RENEWABLE RESOURCE ANALYSES. J. RANGE MANAGE. 19(6), 356-362.

VAN DYNE, G. M. 1968. PREDICTION BY NONLINEAR PROGRAMMING OF RELATIVE CHEMICAL COMPOSITION OF DIETARY BOTANICAL COMPONENTS. J. RANGE MANAGE. 21(1), 37-46,60.

VAN DYNE, G. M. 1970. A SYSTEMS APPROACH TO GRASSLANDS. PROC. 11TH INTERN. GRASSLAND CONGRESS. PP. A131-A143. [MODEL; MODELING]

VAN DYNE, G. M. 1972. ORGANIZATION AND MANAGEMENT OF AN INTEGRATED ECOLOGICAL RESEARCH PROGRAM. PP. 111-172. IN, MATHEMATICAL MODELS IN ECOLOGY. (J. N. R. JEFFERS, ED.). BLACKWELL SCIENTIFIC PUBLICATIONS, OXFORD, ENGLAND.

VAN DYNE, G. M., G. S. INNIS AND G. L. SWARTZMAN. 1971. SOME ANALYTICAL AND OPERATIONAL APPROACHES TO DEVELOPING DYNAMIC MODELS OF ECOLOGICAL SYSTEMS. PP. 19-26. IN, PROCEEDINGS ENVIRONMENTAL AWARENESS. SECOND ANNUAL SESSION INSTITUTE ENVIRONMENTAL SCIENCE. COLORADO CHAPTER. (M. LILLYWHITE AND C. MARTIN, EDS.).

VAN DYNE, G. M., W. E. FRAYER AND L. J. BLEDSOE. 1970. SOME OPTIMIZATION TECHNIQUES AND PROBLEMS IN THE NATURAL RESOURCE SCIENCES. PP. 95-124. IN, STUDIES IN OPTIMIZATION I. SOCIETY FOR INDUSTRIAL AND APPLIED MATHEMATICS, PHILADELPHIA, PENNSYLVANIA.

VINCENT, T. L. 1975. PEST MANAGEMENT PROGRAMS VIA OPTIMAL CONTROL THEORY. BIOMETRICS 31(1), -10. [LOTKA-VOLTERRA MODEL; PREDATOR-PREY MODEL; COST FUNCTION]

VINCENT, T. L., R. H. PULLIAM AND G. EVERTT. 1974. MODELING AND MANAGEMENT OF ECOSYSTEMS VIA OPTIMAL CONTROL THEORY. PP. 8-394. IN, PROC. FIRST INTERNATIONAL CONGRESS ON ECOLOGICAL CENTRE FOR AGRICULTURAL PUBLISHING AND DOCUMENTATION, WAGENINGEN, NETHERLANDS. [MODEL]

WALTERS, C. J. AND I. E. EFFORD 72. SYSTEMS ANALYSIS IN THE MARION LAKE IBP PROJECT. OECOLOGIA 11(1), 33-44.

WARE, G. O. AND J. L. CLUTTER. 1971. A MATHEMATICAL PROGRAMMING SYSTEM FOR THE MANAGEMENT OF INDUSTRIAL FORESTS. FOREST SCI. 17(4), 428-445.

WATT, K. E. F. 1963. DYNAMIC PROGRAMMING, "LOOK AHEAD PROGRAMMING", AND THE STRATEGY OF INSECT PEST CONTROL. CAN. ENTOMOL. 95(5), 525-536. [MARKOV PROCESS; COMPUTER; STEPWISE MULTIPLE REGRESSION]

WATT, K. E. F. 1966. ECOLOGY IN THE FUTURE. PP. 253-267. IN, SYSTEMS ANALYSIS IN ECOLOGY. (K. E. F. WATT, ED.). ACADEMIC PRESS, NEW YORK, NEW YORK.

WATT, K. E. F. 1966. THE NATURE OF SYSTEMS ANALYSIS. PP. 1-14. IN, SYSTEMS ANALYSIS IN ECOLOGY. (K. E. F. WATT, ED.). ACADEMIC PRESS, NEW YORK, NEW YORK.

WILSON, E. O. 1968. THE ERGONOMICS OF CASTE IN THE SOCIAL INSECTS. AMER. NAT. 102(923), 41-66. [LINEAR PROGRAMMING]

WOODWORTH, B. M. 1973. OPTIMIZING THE CALF MIX ON RANGE LANDS WITH LINEAR PROGRAMMING. J. RANGE MANAGE. 26(3), 175-178. [COST FUNCTION; MODEL; COMPUTER]

WRIGHT, R. G. AND G. M. VAN DYNE. 1971. COMPARATIVE ANALYTICAL STUDIES OF SITE FACTOR EQUATIONS. PP. 59-195. IN, STATISTICAL ECOLOGY, VOLUME 3, MANY SPECIES POPULATIONS, ECOSYSTEMS, AND SYSTEMS ANALYSIS. (G. P. PATIL, E. C. PIELOU AND W. E. WATERS, EDS.). PENNSYLVANIA STATE UNIVERSITY PRESS, UNIVERSITY PARK, PENNSYLVANIA.

ORDINATION

ALDERDICE, D. F. 1972. FACTOR COMBINATIONS: RESPONSES OF MARINE POIKILOTHERMS TO ENVIRONMENTAL FACTORS ACTING IN CONCERT. PP. 1659-1722. IN, MARINE ECOLOGY, VOLUME I, PART 3. (O. KINNE, ED.), JOHN WILEY AND SONS, NEW YORK, NEW YORK. [RESPONSE SURFACE; MULTIVARIATE ANALYSIS; POLYNOMIAL REGRESSION]

ANDERSON, A. J. B. 1971. ORDINATION METHODS IN ECOLOGY. J. ECOL. 59(3), 713-726. [PRINCIPAL COMPONENT ANALYSIS; NON-METRIC MULTIDIMENSIONAL SCALING]

AUSTIN, M. P. 1968. AN ORDINATION STUDY OF A CHALK GRASSLAND COMMUNITY. J. ECOL. 56(3), 739-757. [MULTIVARIATE; PRINCIPAL COMPONENT ANALYSIS; COMPUTER]

AUSTIN, M. P. AND I. NOY-MEIR. 1971. THE PROBLEM OF NON-LINEARITY IN ORDINATION: EXPERIMENTS WITH TWO-GRADIENT MODELS. J. ECOL. 59(3), 763-773.

AUSTIN, M. P. AND L. ORLOCI. 1966. GEOMETRIC MODELS IN ECOLOGY. II. AN EVALUATION OF SOME ORDINATION TECHNIQUES. J. ECOL. 54(1), 217-227. [BRAY-CURTIS SIMILARITY MODEL; PRINCIPAL COMPONENT ANALYSIS; PERPENDICULAR AXIS ORDINATION]

AUSTIN, M. P. AND P. GREIG-SMITH. 1968. THE APPLICATION OF QUANTITATIVE METHODS TO VEGETATION SURVEY II. SOME METHODOLOGICAL PROBLEMS OF DATA FROM RAIN FOREST. J. ECOL. 56(3), 827-844. [MULTIVARIATE; ORDINATION; ASSOCIATION ANALYSIS; COMPUTER]

BANNISTER, P. 1966. THE USE OF SUBJECTIVE ESTIMATES OF COVER-ABUNDANCE AS THE BASIS FOR ORDINATION. J. ECOL. 54(3), 665-674.

BANNISTER, P. 1968. AN EVALUATION OF SOME PROCEDURES USED IN SIMPLE ORDINATIONS. J. ECOL. 56(1), 27-34. [SIMILARITY COEFFICIENT OF BRAY AND CURTIS; EUCLIDEAN MEASURE OF DISTANCE]

BEALS, E. W. 1973. ORDINATION: MATHEMATICAL ELEGANCE AND ECOLOGICAL NAIVETE. J. ECOL. 61(1), 23-35. [PYTHAGOREAN DISTANCE; FACTOR ANALYSIS; EUCLIDEAN DISTANCE; PRINCIPAL COMPONENTS ANALYSIS; MULTIVARIATE ANALYSIS; MODEL]

BLACKITH, R. E. 1962. THE HANDLING OF MULTIPLE MEASUREMENTS. PP. 37-42. IN, PROGRESS IN SOIL ZOOLOGY. (P. W. MURPHY, ED.), BUTTERWORTHS, LONDON, ENGLAND. [MULTIVARIATE ANALYSIS; COVARIANCE; CANONICAL ANALYSIS; FACTOR AND PRINCIPAL-COMPONENT METHODS; DISCRIMINANT FUNCTION]

COLE, L. C. 1949. THE MEASUREMENT OF INTERSPECIFIC ASSOCIATION. ECOLOGY 30(4), 411-424.

COLE, L. C. 1957. THE MEASUREMENT OF PARTIAL INTERSPECIFIC ASSOCIATION. ECOLOGY 38(2), 226-233.

COTTAM, G., F. G. GOFF AND R. H. WHITTAKER. 1974. WISCONSIN COMPARATIVE ORDINATION. PP. 193-221. IN, ORDINATION AND CLASSIFICATION OF COMMUNITIES. (R. H. WHITTAKER, ED.), DR. JUNK PUBLISHERS, THE HAGUE, NETHERLANDS. [GRADIENT ANALYSIS; POLAR ORDINATION]

CRAWFORD, R. M. M. AND D. WISHART. 1968. A RAPID CLASSIFICATION AND ORDINATION METHOD AND ITS APPLICATION TO VEGETATION MAPPING. J. ECOL. 56(2), 385-404. [ORDINATION; MATRIX; MULTIVARIATE ANALYSIS; VECTOR; QUADRAT; COMPUTER]

DAGNELIE, P. 1965. L'ETUDE DES COMMUNAUTES VEGETALES PAR L'ANALYSE STATISTIQUE DES LIAISONS ENTRE LES ESPECES ET LES VARIABLES ECOLOGIQUES: UN EXEMPLE. BIOMETRICS 21(4), 890-907. [FACTOR ANALYSIS]

DAGNELIE, P. 1974. L'ANALYSE FACTORIELLE. PP. 223-248. IN, ORDINATION AND CLASSIFICATION OF COMMUNITIES. (R. H. WHITTAKER, ED.). DR. JUNK PUBLISHERS, THE HAGUE, NETHERLANDS. [MODEL; FACTOR ANALYSIS]

DICE, L. R. 1945. MEASURES OF THE AMOUNT OF ECOLOGIC ASSOCIATION BETWEEN SPECIES. ECOLOGY 26(3), 297-302. [COEFFICIENT OF ASSOCIATION; COINCIDENCE INDEX]

ERMAN, D. C. 1973. ORDINATION OF SOME LITTORAL BENTHIC COMMUNITIES IN BEAR LAKE, UTAH-IDAHO. OECOLOGIA 13, 211-226. [ORLOCI'S SIMILARITY COEFFICIENT; COMPUTER]

FAGER, E. W. 1957. DETERMINATION AND ANALYSIS OF RECURRENT GROUPS. ECOLOGY 38(4), 586-595.

FRESCO, L. F. M. 1969. FACTOR ANALYSIS AS A METHOD IN SYNECOLOGICAL RESEARCH. ACTA BOTANICA NEERLANDICA 18(3), 477-482. [MATRIX; COMPUTER]

GARN, S. M. 1955. APPLICATIONS OF PATTERN ANALYSIS TO ANTHROPOMETRIC DATA. ANN. N. Y. ACAD. SCI. 63(4), 537-552. [FACTOR ANALYSIS]

AUCH, H. G., JR. 1973. A UANTITATIVE EVALUATION OF THE RAY-CURTIS ORDINATION. ECOLOGY 4(4), 829-836.

AUCH, H. G., JR. 1973. THE ELATIONSHIP BETWEEN SAMPLE IMILARITY AND ECOLOGICAL DISTANCE. COLOGY 54(3), 618-622. SIMILARITY MEASURE; BRAY-CURTIS RDINATION; COEFFICIENT OF OMMUNITY; EUCLIDEAN DISTANCE]

AUCH, H. G., JR. AND R. H. ITTAKER. 1972. COENCLINE IMULATION. ECOLOGY 53(3), 6-451. [PRINCIPAL COMPONENTS ALYSIS; RANDOM NUMBERS; COMPUTER; RMAL CURVE; LOG-NORMAL STRIBUTION; MATRIX]

GAUCH, H. G., JR. AND R. H. WHITTAKER. 1972. COMPARISON OF ORDINATION TECHNIQUES. ECOLOGY 53(5), 868-875. [SIMULATION; COMPUTER]

GAUCH, H. G., JR., G. B. CHASE AND R. H. WHITTAKER. 1974. ORDINATION OF VEGETATION SAMPLES BY GAUSSIAN SPECIES DISTRIBUTION. ECOLOGY 55(6), 1382-1390. [COMPUTER; MODEL; CATENATION]

GITTINS, R. 1965 MULTIVARIATE APPROACHES TO A LIMESTONE GRASSLAND COMMUNITY. II. A DIRECT SPECIES ORDINATION. J. ECOL. 53(2), 403-409. [FRACTURE ANALYSIS; CZEKANOWSKI'S COEFFICIENT; MATRIX]

GITTINS, R. 1965. MULTIVARIATE APPROACHES TO A LIMESTONE GRASSLAND COMMUNITY I. A STAND ORDINATION. J. ECOL. 53(2), 385-401. [MULTIVARIATE; CZEKANOWSKI'S COEFFICIENT; BRAY-CURTIS TECHNIQUE]

GITTINS, R. 1965. MULTIVARIATE APPROACHES TO A LIMESTONE GRASSLAND COMMUNITY. III. A COMPARATIVE STUDY OF ORDINATION AND ASSOCIATION-ANALYSIS. J. ECOL. 53(2), 411-425. [INVERSE ASSOCIATION ANALYSIS; COMPUTER]

GOFF, F. G. AND P. H. ZEDLER. 1972. DERIVATION OF SPECIES SUCCESSION VECTORS. AMER. MIDL. NAT. 87(2), 397-412. [COLE'S INDEX OF ASSOCIATION; MODEL; COMPUTER; PRINCIPAL COMPONENT ANALYSIS; MATRIX]

GOLDSTEIN, R. A. AND D. F. GRIGAL. 1972. DEFINITION OF VEGETATION STRUCTURE BY CANONICAL ANALYSIS. J. ECOL. 60(2), 277-284. [PRINCIPAL COMPONENTS ANALYSIS; COMPUTER]

GOODALL, D. W. 1974. SAMPLE SIMILARITY AND SPECIES CORRELATION. PP. 105-156. IN, ORDINATION AND CLASSIFICATION OF COMMUNITIES. (R. H. WHITTAKER, ED.). DR. JUNK PUBLISHERS, THE HAGUE, NETHERLANDS.

GREIG-SMITH, P., M. P. AUSTIN AND T. C. WHITMORE. 1967. THE APPLICATION OF QUANTITATIVE METHODS TO VEGETATION SURVEY I. ASSOCIATION - ANALYSIS AND PRINCIPAL COMPONENT ORDINATION OF RAIN FOREST. J. ECOL. 55(2), 483-503. [MULTIVARIATE; COMPUTER; ORLOCI'S COEFFICIENT]

GRIGAL, D. F. AND R. A. GOLDSTEIN. 1971. AN INTEGRATED ORDINATION - CLASSIFICATION ANALYSIS OF AN INTENSIVELY SAMPLED OAK-HICKORY FOREST. J. ECOL. 59(2), 481-492. [MODEL; COMPUTER; MUTUAL INFORMATION TECHNIQUE; MEASURE OF SIMILARITY; PRINCIPAL COMPONENTS ANALYSIS; CANONICAL ANALYSIS; MATRIX]

HARBERD, D. J. 1960. ASSOCIATION-ANALYSIS IN PLANT COMMUNITIES. NATURE 185(4705), 53-54.

HARBERD, D. J. 1962. APPLICATION OF A MULTIVARIATE TECHNIQUE TO ECOLOGICAL SURVEY. J. ECOL. 50(1), 1-17. [MAHALANOBIS' D2]

HARRISON, C. M. 1970. THE PHYTOSOCIOLOGY OF CERTAIN ENGLISH HEATHLAND COMMUNITIES. J. ECOL. 58(3), 573-589. [ASSOCIATION ANALYSIS; NODAL ANALYSIS; COMPUTER; NORMAL ANALYSIS]

HILL, M. O. 1973. RECIPROCAL AVERAGING: AN EIGENVECTOR METHOD OF ORDINATION. J. ECOL. 61(1), 237-249. [PRINCIPAL COMPONENTS ANALYSIS; CORRELATION MATRIX; MULTIVARIATE ANALYSIS]

HURLBERT, S. H. 1969. A COEFFICIENT OF INTERSPECIFIC ASSOCIATION. ECOLOGY 50(1), 1-9.

IVIMEY-COOK, R. B. AND M. C. F. PROCTOR. 1966. THE APPLICATION OF ASSOCIATION-ANALYSIS TO PHYTOSOCIOLOGY. J. ECOL. 54(1), 179-192. [MULTIVARIATE; COMPUTER]

IVIMEY-COOK, R. B. AND M. C. F. PROCTOR. 1967. FACTOR ANALYSIS OF DATA FROM AN EAST DEVON HEATH: A COMPARISON OF PRINCIPAL COMPONENT AND ROTATED SOLUTIONS. J. ECOL. 55(2), 405-419.

KERSHAW, K. A. 1960. THE DETECTION OF PATTERN AND ASSOCIATION. J. ECOL. 48(1), 233-242. [ARTIFICIAL POPULATION]

KERSHAW, K. A. 1961. ASSOCIATION AND CO-VARIANCE ANALYSIS OF PLANT COMMUNITIES. J. ECOL. 49(3), 643-654.

KERSHAW, K. A. 1968. CLASSIFICATION AND ORDINATION OF NIGERIAN SAVANNA VEGETATION. J. ECOL. 56(2), 467-482. [MULTIVARIATE; ORLOCI'S COEFFICIENT; SIMILARITY COEFFICIENTS; CHI-SQUARE COEFFICIENTS]

KNIGHT, W. AND A. V. TYLER. 1973. ARE SPECIES ASSOCIATION COEFFICIENTS REALLY NECESSARY? FISH. RES. BOARD CAN. TECH. REPT. NO. 397. 8 PP.

KNIGHT, W. AND A. V. TYLER. 1973. A METHOD FOR COMPRESSION OF SPECIES ASSOCIATION DATA BY USING HABITAT PREFERENCES, INCLUDING AN ANALYSIS OF FISH ASSEMBLAGES ON THE SOUTHWEST SCOTIAN SHELF. FISH. RES. BOARD CAN. TECH. REPT. NO. 402. 18 PP. [CONTINGENCY TABLE; COEFFICIENT OF ASSOCIATION]

LAFRANCE, C. R. 1972. SAMPLING AND ORDINATION CHARACTERISTICS OF COMPUTER-SIMULATED INDIVIDUALISTIC COMMUNITIES. ECOLOGY 53(3), 387-397.

LAMBERT, J. M. AND W. T. WILLIAMS. 1966. MULTIVARIATE METHODS IN PLANT ECOLOGY. VI. COMPARISON OF INFORMATION-ANALYSIS AND ASSOCIATION-ANALYSIS. J. ECOL. 54(3), 635-664.

LANCE, G. N. AND W. T. WILLIAMS. 1968. MIXED-DATA CLASSIFICATORY PROGRAMS II. AUST. COMPUT. J. 1(2), 82-85.

LEVANDOWSKY, M. 1972. AN ORDINATION OF PHYTOPLANKTON POPULATIONS IN PONDS OF VARYING SALINITY AND TEMPERATURE. ECOLOG 53(3), 398-407. [JACCARD'S INDEX INDEX OF SIMILARITY]

LIE, U, AND J. C. KELLY. 1970. BENTHIC INFAUNA COMMUNITIES OFF THE COAST OF WASHINGTON AND IN PUGET SOUND:IDENTIFICATION AND DISTRIBUTION OF THE COMMUNITIES. J. FISH. RES. BOARD CAN. 27(4), 621-651. [FACTOR ANALYSIS; FAGER'S RECURRENT GROUP ANALYSIS; INDEX OF AFFINITY; KENDALL'S RANK CORRELATION COEFFICIENT]

LOUCKS, O. L. 1962. ORDINATING FOREST COMMUNITIES BY MEANS OF ENVIRONMENTAL SCALARS AND PHYTOSOCIOLOGICAL INDICES. ECOL. MONOGR. 32(2), 137-166. [TRANSFORMATION; INDEX OF SIMILARITY; PROBABILITY]

MADGWICK, H. A. I. AND P. A. DESROCHERS. 1972. ASSOCIATION-ANALYSIS AND THE CLASSIFICATION OF FOREST VEGETATION OF THE JEFFERSON NATIONAL FOREST. J. ECOL. 60(2), 285-292. [NORMAL ANALYSIS]

MCINTOSH, R. P. 1974. MATRIX AND PLEXUS TECHNIQUES. PP. 157-191. IN, ORDINATION AND CLASSIFICATION OF COMMUNITIES. (R. H. WHITTAKER, ED.). DR. JUNK PUBLISHERS, THE HAGUE, NETHERLANDS. [INDEX OF SIMILARITY; ORDINATION]

MORISITA, M. 1959. MEASURING OF INTERSPECIFIC ASSOCIATION AND SIMILARITY BETWEEN COMMUNITIES. MEM. FAC. SCI. KYUSHU UNIV. SER. E 3, 65-80.

NASH, C. B. 1950. ASSOCIATIONS BETWEEN FISH SPECIES IN TRIBUTARIES AND SHORE WATERS OF WESTERN LAKE ERIE. ECOLOGY 31(4), 561-566.

ORRIS, J. M. 1971. THE PPLICATION OF MULTIVARIATE NALYSIS TO SOIL STUDIES. I. ROUPING OF SOILS USING DIFFERENT ROPERTIES. J. SOIL SCI. 22(1), 9-80. [PRINCIPAL COMPONENT NALYSIS; MATRIX]

OY-MEIR, I. 1973. DATA RANSFORMATIONS IN ECOLOGICAL RDINATION I. SOME ADVANTAGES OF ON-CENTERING. J. ECOL. 61(2), 9-341. [PRINCIPAL COMPONENTS ALYSIS; MATRIX; EIGENROOTS AND GENVECTORS OF A DISJOINT MATRIX]

NOY-MEIR, I. 1973. DIVISIVE POLYTHETIC CLASSIFICATION OF VEGETATION DATA BY OPTIMIZED DIVISION ON ORDINATION COMPONENTS. J. ECOL. 61(3), 753-760. [PRINCIPAL COMPONENTS; COMPUTER]

NOY-MEIR, I. AND D. J. ANDERSON. 1971. MULTIPLE PATTERN ANALYSIS, OR MULTISCALE ORDINATION: TOWARDS A VEGETATION HOLOGRAM? PP. 207-225. IN, STATISTICAL ECOLOGY, VOLUME 3. MANY SPECIES POPULATIONS, ECOSYSTEMS, AND SYSTEMS ANALYSIS. (G. P. PATIL, E. C. PIELOU AND W. E. WATERS, EDS.). PENNSYLVANIA STATE UNIVERSITY PRESS, UNIVERSITY PARK, PENNSYLVANIA.

NOY-MEIR, I. AND M. P. AUSTIN. 1970. PRINCIPAL COMPONENT ORDINATION AND SIMULATED VEGETATIONAL DATA. ECOLOGY 51(3), 551-552.

NOY-MEIR, I., N. H. TADMOR AND G. ORSHAN. 1970. MULTIVARIATE ANALYSIS OF DESERT VEGETATION. I. ASSOCIATION ANALYSIS AT VARIOUS QUADRAT SIZES. ISRAEL J. BOT. 19, 561-591. [INVERSE ASSOCIATION ANALYSIS; NORMAL ASSOCIATION ANALYSIS; COMPUTER]

ORLOCI, L. 1966. GEOMETRIC MODELS IN ECOLOGY. I. THE THEORY AND APPLICATION OF SOME ORDINATION METHODS. J. ECOL. 54(1), 193-215.

ORLOCI, L. 1967. AN AGGLOMERATIVE METHOD FOR CLASSIFICATION OF PLANT COMMUNITIES. J. ECOL. 55(1), 193-206. [ORDINATION; DISTANCE COEFFICIENT; COMPUTER]

ORLOCI, L. 1972. ON OBJECTIVE FUNCTIONS OF PHYTOSOCIOLOGICAL RESEMBLANCE. AMER. MIDL. NAT. 88(1), 28-55. [SAMPLE SPACE; STOCHASTIC AND DETERMINISTIC FUNCTIONS; DISTANCE FUNCTION; SCALAR PRODUCTS; PROBABILITY-TYPE COEFFICIENT; INFORMATION]

ORLOCI, L. 1974. ORDINATION BY RESEMBLANCE MATRICES. PP. 249-286. IN, ORDINATION AND CLASSIFICATION OF COMMUNITIES. (R. H. WHITTAKER, ED.). DR. JUNK PUBLISHERS, THE HAGUE, NETHERLANDS. [MODEL; ALGORITHM; STOCHASTIC]

POOLE, R. W. 1971. THE USE OF FACTOR ANALYSIS IN MODELING NATURAL COMMUNITIES OF PLANTS AND ANIMALS. ILLINOIS NATURAL HISTORY SURVEY, BIOL. NOTES NO. 72. 14 PP.

RANDAL, J. M. 1969. AN INTRODUCTION TO ASSOCIATION ANALYSIS. PROC. N. Z. ECOL. SOC. 16, 48-52. [DENDOGRAM; CENTROID STRATEGY; SIMILARITY MATRIX]

REYMENT, R. A. 1963. MULTIVARIATE ANALYTICAL TREATMENT OF QUANTITATIVE SPECIES ASSOCIATIONS, AN EXAMPLE FROM PALAEOECOLOGY. J. ANIM. ECOL. 32(3), 535-547. [FACTOR ANALYSIS; PRINCIPAL COMPONENT ANALYSIS; CORRELATION; COMPUTER; MATRIX; EIGENVALUE; CANONICAL ANALYSIS]

REYMENT, R. A. 1969. A MULTIVARIATE PALEONTOLOGICAL GROWTH PROBLEM. BIOMETRICS 25(1), 1-8. [PRINCIPAL COMPONENTS ANALYSIS]

ROCHOW, J. J. 1972. VEGETATIONAL DESCRIPTION OF A MID-MISSOURI FOREST USING GRADIENT ANALYSIS TECHNIQUES. AMER. MIDL. NAT. 87(2), 377-396. [TETRACHORIC CORRELATION; MATRIX; PRINCIPAL COMPONENT ANALYSIS; MODEL]

ROYCE, W. F. 1957. STATISTICAL COMPARISON OF MORPHOLOGICAL DATA. PP. 7-28. IN, CONTRIBUTIONS TO THE STUDY OF SUBPOPULATIONS OF FISHES. (J. C. MARR, COORDINATOR). U. S. FISH AND WILDLIFE SERVICE, SPECIAL SCIENTIFIC REPORT-FISHERIES NO. 208. [COEFFICIENT OF RACIAL LIKENESS; FACTOR ANALYSIS; DISCRIMINANT FUNCTION; D2 STATISTIC OF MAHALANOBIS]

SCOTT, D. 1974. DESCRIPTION OF RELATIONSHIPS BETWEEN PLANTS AND ENVIRONMENT. PP. 47-69. IN, VEGETATION AND ENVIRONMENT. (B. R. STRAIN AND W. D. BILLINGS, EDS.). DR. W. JUNK PUBLISHERS, THE HAGUE, NETHERLANDS. [ORDINATION; FITTING FUNCTION; REGRESSION]

SOBOLEV, L. N. AND V. D. UTEKHIN. 1974. RUSSIAN (RAMENSKY) APPROACHES TO COMMUNITY SYSTEMATIZATION. PP. 75-103. IN, ORDINATION AND CLASSIFICATION OF COMMUNITIES. (R. H. WHITTAKER, ED.). DR. JUNK PUBLISHERS, THE HAGUE, NETHERLANDS. [ORDINATION]

SWAN, J. M. A. 1970. AN EXAMINATION OF SOME ORDINATION PROBLEMS BY USE OF SIMULATED VEGETATIONAL DATA. ECOLOGY 51(1), 89-102. [MODEL]

SWAN, J. M. A., R. L. DIX AND C. F. WEHRHAHN. 1969. AN ORDINATION TECHNIQUE BASED ON THE BEST POSSIBLE STAND-DEFINED AXES AND ITS APPLICATION TO VEGETATIONAL ANALYSIS. ECOLOGY 50(2), 206-212.

TOMASSONE, R. 1972. L'ANALYSE FACTORIELLE DES CORRESPONDANCES. PP. 161-172. IN, 3RD CONFERENCE ADVISORY GROUP OF FOREST STATISTICIANS, INSTITUT NATIONAL DE LA RECHERCHE AGRONOMIQUE, PARIS, FRANCE, I.N.R.A. PUBL. 72-3. [FACTOR ANALYSIS; HIERARCHICAL CLASSIFICATION; CORRESPONDENCES; FORESTRY]

VAN DER MAAREL, E. 1969. ON THE USE OF ORDINATION MODELS IN PHYTOSOCIOLOGY. VEGETATIO 19, 21-46. [INDEX OF SIMILARITY; PRINCIPAL COMPONENT ANALYSIS]

VILKS, G., E. H. ANTHONY AND W. T. WILLIAMS. 1970. APPLICATION OF ASSOCIATION-ANALYSIS TO DISTRIBUTION STUDIES OF RECENT FORAMINIFERA. CAN. J. EARTH SCI. 7(6), 1462-1469.

WALLIS, J. R. 1968. FACTOR ANALYSIS IN HYDROLOGY - AN AGNOSTI VIEW. WATER RESOUR. RES. 4(3), 521-527. [REGRESSION; MODEL; MATRIX; PRINCIPAL COMPONENT ANALYSIS]

WHITTAKER, R. H. 1967. GRADIENT ANALYSIS OF VEGETATION. BIOL. REV. 42(2), 207-264. [ORDINATION; FACTOR ANALYSIS; INDEX OF SIMILARITY]

WHITTAKER, R. H. 1974. DIRECT GRADIENT ANALYSIS: TECHNIQUES. PP. 7-31. IN, ORDINATION AND CLASSIFICATION OF COMMUNITIES. (R. H. WHITTAKER, ED.). DR. JUNK PUBLISHERS, THE HAGUE, NETHERLANDS. [SAMPLE SIZE; ORDINATION; NOMOGRAM]

WHITTAKER, R. H. AND H. G. GAUCH, JR. 1974. EVALUATION OF ORDINATION TECHNIQUES. PP. 287-321. IN, ORDINATION AND CLASSIFICATION OF COMMUNITIES. (R. H. WHITTAKER, ED.). DR. JUNK PUBLISHERS, THE HAGUE, NETHERLANDS.

WILLIAMS, W. T. AND J. M. LAMBERT. 1959. MULTIVARIATE METHODS IN PLANT ECOLOGY. I. ASSOCIATION-ANALYSIS IN PLANT COMMUNITIES. J. ECOL. 47(1), 83-101.

WILLIAMS, W. T. AND J. M. LAMBERT. 1960. MULTIVARIATE METHODS IN PLANT ECOLOGY. II. THE USE OF AN ELECTRONIC DIGITAL COMPUTER FOR ASSOCIATION-ANALYSIS. J. ECOL. 48(3), 689-710.

WILLIAMS, W. T. AND J. M. LAMBERT. 1961. MULTIVARIATE METHODS IN PLANT ECOLOGY. III. INVERSE ASSOCIATION-ANALYSIS. J. ECOL. 49(3), 717-729.

WILLIAMS, W. T. AND J. M. LAMBERT. 1961. NODAL ANALYSIS OF ASSOCIATED POPULATIONS. NATURE 191(4784), 202. [FACTOR ANALYSIS]

WILLIAMS, W. T. AND P. GILLARD. 1971. PATTERN ANALYSIS OF A GRAZING EXPERIMENT. AUST. J. AGRIC. RES. 22(2), 245-260. [COMPUTER; CANONICAL ANALYSIS; ORDINATION; MATRIX; INFORMATION; VECTOR; CORRELATION]

WILLIAMS, W. T., G. N. LANCE, L. J. WEBB, J. G. TRACEY AND M. B. DALE. 1969. STUDIES IN THE NUMERICAL ANALYSIS OF COMPLEX RAIN-FOREST COMMUNITIES. III. THE ANALYSIS OF SUCCESSIONAL DATA. J. ECOL. 57(2), 515-535. [BRILLOUIN'S DIVERSITY; MATRIX; DISCRIMINANT INFORMATION STATISTIC; TRANSITION MATRICES; ORDINATION]

WILLIAMS, W. T., M. B. DALE AND G. N. LANCE. 1971. TWO OUTSTANDING ORDINATION PROBLEMS. AUST. J. BOT. 19, 251-258. [AFTER NUMERICAL CLASSIFICATION; ASYMMETRIC SYSTEM; COMPUTER]

YARRANTON, G. A. 1967. ORGANISMAL AND INDIVIDUALISTIC CONCEPTS AND THE CHOICE OF METHODS OF VEGETATION ANALYSIS. VEGETATIO 15(2), 113-116. [CORRELATION; ORDINATION; PRINCIPAL COMPONENT ANALYSIS; MULTIPLE REGRESSION; CANONICAL ANALYSIS]

ZEDLER, P. H. AND F. G. GOFF. 1973. SIZE-ASSOCIATION ANALYSIS OF FOREST SUCCESSIONAL TRENDS IN WISCONSIN. ECOL. MONOGR. 43(1), 79-94. [MATRIX; INTERSPECIFIC ASSOCIATION; VECTOR; DIMENSIONAL ANALYSIS]

PATTERNS

ALMQUIST, S. 1973. SPIDER ASSOCIATIONS IN COASTAL SAND DUNES. OIKOS 24(3), 444-457. [ASSOCIATION INDEX OF DICE; DISTANCE FUNCTION; COMPUTER]

AMEN, R. D. 1966. A BIOLOGICAL SYSTEMS CONCEPT. BIOSCIENCE 16(6), 396-401.

ANDERSON, D. J. 1971. SPATIAL PATTERNS IN SOME AUSTRALIAN DRYLAND PLANT COMMUNITIES. PP. 271-283. IN, STATISTICAL ECOLOGY, VOLUME 1. SPATIAL PATTERNS AND STATISTICAL DISTRIBUTIONS. (G. P. PATIL, E. C. PIELOU AND W. E. WATERS, EDS.). PENNSYLVANIA STATE UNIVERSITY PRESS, UNIVERSITY PARK, PENNSYLVANIA.

ANDO, S. AND A. OGASAWARA. 1971. TRACKING TELEMETRY OF A PHEASANT (PHASIANUS COLCHICUS). JAPANESE J. ECOL. 21(1/2), 32-36. [CENTER OF GRAVITY OF A TRIANGLE]

AUSTIN, M. P. 1971. ROLE OF REGRESSION ANALYSIS IN PLANT ECOLOGY. PROC. ECOL. SOC. AUSTRALIA 6, 63-75.

AUSTIN, M. P. 1972. MODELS AND ANALYSIS OF DESCRIPTIVE VEGETATION DATA. PP. 61-86. IN, MATHEMATICAL MODELS IN ECOLOGY. (J. N. R. JEFFERS, ED.). BLACKWELL SCIENTIFIC PUBLICATIONS, OXFORD, ENGLAND.

BACHI, R. 1973. GEOSTATISTICAL ANALYSIS OF TERRITORIES. BULL. INT. STATIST. INST. 45(1), 121-133. [MODEL]

BARTLETT, M. S. 1964. A NOTE ON SPATIAL PATTERN. BIOMETRICS 20(4), 891-892.

BERGEN, J. D. 1974. VERTICAL AIR TEMPERATURE PROFILES IN A PINE STAND: SPATIAL VARIATION AND SCALING PROBLEMS. FOREST SCI. 20(1), 64-73. [MODEL]

BERRYMAN, A. A., R. W. STARK AND C. O. DUDLEY. 1970. DATA PREPARATIONS AND ANALYSIS OF "WITHIN-TREE" SAMPLES OF POPULATIONS OF THE WESTERN PINE BEETLE. PP. 37-41. IN, STUDIES ON THE POPULATION DYNAMICS OF THE WESTERN PINE BEETLE DENDROCTOMUS BREVICONIS LECONTE (COLEOPTERA: SCOLYTIDAE). (R. W. STARK AND D. L. DAHLSTEN, EDS.). UNIVERSITY OF CALIFORNIA, DIVISION OF AGRICULTURAL SCIENCES, BERKELEY, CALIFORNIA. [COMPUTER]

BLISS, C. I. 1971. THE AGGREGATION OF SPECIES WITHIN SPATIAL UNITS. PP. 311-333. IN, STATISTICAL ECOLOGY, VOLUME 1. SPATIAL PATTERNS AND STATISTICAL DISTRIBUTIONS. (G. P. PATIL, E. C. PIELOU AND W. E. WATERS, ED.). PENNSYLVANIA STATE UNIVERSITY PRESS, UNIVERSITY PARK, PENNSYLVANIA.

BRADFORD, E. AND J. R. PHILIP. 1970. NOTE ON ASOCIAL POPULATIONS DISPERSING IN TWO DIMENSIONS. J. THEOR. BIOL. 29(1), 27-33. [STABILITY ANALYSIS]

BRADFORD, E. AND J. R. PHILIP. 1970. STABILITY OF STEADY DISTRIBUTIONS OF ASOCIAL POPULATIONS DISPERSING IN ONE DIMENSION. J. THEOR. BIOL. 29(1), 13-26. [STABILITY ANALYSIS]

BRAY, J. R. 1962. USE OF NON-AREA ANALYTIC DATA TO DETERMINE SPECIES DISPERSION. ECOLOGY 43(2), 328-333. [PROBABILITY]

BREDER, C. M., JR. 1954. EQUATIONS DESCRIPTIVE OF FISH SCHOOLS AND OTHER ANIMAL AGGREGATIONS. ECOLOGY 35(3), 361-370.

BROWN, S. AND P. HOLGATE. 1974. THE THINNED PLANTATION. BIOMETRIKA 61(2), 253-261. [NEAREST NEIGHBOR; FORESTRY; GEOMETRICAL PROBABILITY; SKELLAM-MOORE TEST; MONTE CARLO; HOPKINS' TEST]

BRUCE, D. 1972. SOME TRANSFORMATIONS OF THE BEHRE EQUATION OF TREE FORM. FOREST SCI. 18(2), 164-166.

BUCHANAN-WOLLASTON, H. J. 1938. ON THE APPLICATION OF THE STATISTICAL THEORY OF SPACE DISTRIBUTION TO HYDROGRAPHIC AND FISHERY PROBLEMS. J. CONS. PERM. INT. EXPLOR. MER 13(2), 173-186. [BINOMIAL DISTRIBUTION; VECTOR; REGRESSION EQUATION; NORMAL DISTRIBUTION; RANDOM SPREAD]

BUCHANAN-WOLLASTON, H. J. 1958. STATISTICAL TESTS FOR SIGNIFICANCE APPLICABLE TO DISTRIBUTION IN SPACE. J. CONS. PERM. INT. EXPLOR. MER 23(2), 161-172. [CHI-SQUARE; VECTOR; MULTINOMIAL DISTRIBUTION]

BURGE, J. R. AND C. D. JORGENSEN. 1973. HOME RANGE OF SMALL MAMMALS: A RELIABLE ESTIMATE. J. MAMMAL. 54(2), 483-488. [PROBABILITY; TRANSFORMATION; VECTOR]

BUZAS, M. A. 1970. SPATIAL HOMOGENEITY: STATISTICAL ANALYSES OF UNISPECIES AND MULTISPECIES POPULATIONS OF FORAMINIFERA. ECOLOGY 51(5), 874-879. [ANALYSIS OF VARIANCE; CANONICAL ANALYSIS; MULTIVARIATE ANALYSIS]

CASSIE, R. M. 1961. THE CORRELATION COEFFICIENT AS AN INDEX OF ECOLOGICAL AFFINITIES IN PLANKTON POPULATIONS. MEM. INST. ITAL. IDROBIOL. 13, 151-177.

CASSIE, R. M. 1963. MULTIVARIATE ANALYSIS IN THE INTERPRETATION OF NUMERICAL PLANKTON DATA. N. Z. J. SCI. 6(1), 36-59.

CASSIE, R. M. 1969. MULTIVARIATE ANALYSIS IN ECOLOGY. PROC. N. Z. ECOL. SOC. 16, 53-57. [REGRESSION; COMPUTER; VECTOR; EIGENVALUE; CANONICAL CORRELATION; MODEL]

CHACKO, V. J. AND G. S. NEGI. 1965. A STATISTICAL STUDY OF THE SPATIAL DISTRIBUTION OF DEAD TREES IN A CASUARINA PLANTATION. SANKHYA 27B(3/4), 211-224. [FREEMAN'S METHOD; PIELOU'S METHOD]

CLAPHAM, A. R. 1936. OVER-DISPERSION IN GRASSLAND COMMUNITIES AND THE USE OF STATISTICAL METHODS IN PLANT ECOLOGY. J. ECOL. 24(1), 232-251. [RELATIVE VARIANCE; POISSON SERIES; LATIN SQUARE; FREQUENCY INDEX]

CLARK, P. J. 1956. GROUPING IN SPATIAL DISTRIBUTIONS. SCIENCE 123(3192), 373-374.

COOK, C. W. 1960. THE USE OF MULTIPLE REGRESSION AND CORRELATION IN BIOLOGICAL INVESTIGATIONS ECOLOGY 41(3), 556-560.

COOK, C. W. AND R. HURST. 1962. A QUANTITATIVE MEASURE OF PLANT ASSOCIATION ON RANGES IN GOOD AND POOR CONDITION. J. RANGE MANAGE. 15(5), 266-273. [COLE'S MEASURE OF INTERSPECIFIC ASSOCIATION]

COX, D. R. AND W. L. SMITH. 1957. ON THE DISTRIBUTION OF TRIBOLIUM CONFUSUM IN A CONTAINER. BIOMETRIKA 44(3/4), 328-335. [MODEL]

DAHL, E. 1960. SOME MEASURES OF UNIFORMITY IN VEGETATION ANALYSIS. ECOLOGY 41(4), 805-808. [JACCARD'S COEFFICIENT; SORENSEN'S INDEX]

DALE, M. B. 1971. INFORMATION ANALYSIS OF QUANTITATIVE DATA. PP. 133-148. IN, STATISTICAL ECOLOGY, VOLUME 3. MANY SPECIES POPULATIONS, ECOSYSTEMS, AND SYSTEMS ANALYSIS. (G. P. PATIL, E. C. PIELOU AND W. E. WATERS, EDS.). PENNSYLVANIA STATE UNIVERSITY PRESS, UNIVERSITY PARK, PENNSYLVANIA.

DAY, B. AND R. A. FISHER. 1937. THE COMPARISON OF VARIABILITY IN POPULATIONS HAVING UNEQUAL MEANS. AN EXAMPLE OF THE ANALYSIS OF COVARIANCE WITH MULTIPLE DEPENDENT AND INDEPENDENT VARIATES. ANN. EUGEN. 7(4), 333-348. [PLANTAGO MARITIMA]

DAY, J. H., J. G. FIELD AND M. P. MONTGOMERY. 1971. THE USE OF NUMERICAL METHODS TO DETERMINE THE DISTRIBUTION OF THE BENTHIC FAUNA ACROSS THE CONTINENTAL SHELF OF NORTH CAROLINA. J. ANIM. ECOL. 40(1), 93-125. [COMPUTER; JACCARD'S COEFFICIENT; CZEKANOWSKI'S COEFFICIENT; DENDOGRAM]

DE WIT, C. T. 1961. SPACE RELATIONSHIPS WITHIN POPULATIONS OF ONE OR MORE SPECIES. PP. 314-329. IN, MECHANISMS IN BIOLOGICAL COMPETITION. (F. L. MILTHORPE, ED.). CAMBRIDGE UNIVERSITY PRESS, LONDON, ENGLAND.

DICE, L. R. 1949. THE SELECTION INDEX AND ITS TEST OF SIGNIFICANCE. EVOLUTION 3(3), 262-265.

DUNCAN, K. W. 1972. A MODIFICATION TO LEFKOVITCH'S INDEX OF SPATIAL DISTRIBUTION. PROC. N. Z. ECOL. SOC. 19, 158-162. [NEGATIVE BINOMIAL DISTRIBUTION; NEYMAN'S TYPE A DISTRIBUTION; POISSON DISTRIBUTION; TAYLOR'S POWER LAW; VARIANCE; CONFIDENCE INTERVAL; INDEX FOR CONTAGION]

EMLEN, J. M. 1968. OPTIMAL CHOICE IN ANIMALS. AMER. NAT. 102(926), 385-389. [PREDATOR-PREY MODEL; DISPERSAL; MODEL]

ERRINGTON, J. C. 1973. THE EFFECT OF REGULAR AND RANDOM DISTRIBUTIONS ON THE ANALYSIS OF PATTERN. J. ECOL. 61(1), 99-105. [ANALYSIS OF VARIANCE; COMPUTER; ARTIFICIAL POPULATION; MODEL]

EVANS, F. C., P. J. CLARK AND R. H. BRAND. 1955. ESTIMATION OF THE NUMBER OF SPECIES PRESENT ON A GIVEN AREA. ECOLOGY 36(2), 342-343. [SPECIES AREA CURVE]

FAGER, E. W. AND J. A. MCGOWAN. 1963. ZOOPLANKTON SPECIES GROUPS IN THE NORTH PACIFIC. SCIENCE 140(3566), 453-460. [INDEX OF AFFINITY; COMPUTER]

FIENBERG, S. E. 1970. THE ANALYSIS OF MULTIDIMENSIONAL CONTINGENCY TABLES. ECOLOGY 51(3), 419-433.

FISHER, R. A. AND R. E. MILES. 1973. THE ROLE OF SPATIAL PATTERN IN THE COMPETITION BETWEEN CROP PLANTS AND WEEDS. A THEORETICAL ANALYSIS. MATH. BIOSCI. 18, 335-350. [STOCHASTIC MODEL; POISSON DISTRIBUTION; DE WIT'S MODEL]

FRETWELL, S. D. AND H. L. LUCAS, JR. 1969. ON TERRITORIAL BEHAVIOR AND OTHER FACTORS INFLUENCING HABITAT DISTRIBUTION IN BIRDS I. THEORETICAL DEVELOPMENT. ACTA BIOTHEOR. 19(1), 16-36. [MODEL]

FURRER, R. K. 1973. HOMING OF PEROMYSCUS MANICULATUS IN THE CHANNELLED SCABLANDS OF EAST-CENTRAL WASHINGTON. J. MAMMAL. 54(2), 466-482. [TAG-RECAPTURE; PROBABILITY]

GAGE, J. AND A. D. GEEKIE. 1973. COMMUNITY STRUCTURE OF THE BENTHOS IN SCOTTISH SEA-LOCHS. III. FURTHER STUDIES ON PATCHINESS. MARINE BIOL. 20(2), 89-100. [DIVERSITY MEASURE; SIMPSON'S INDEX; MORISITA'S INDEX; SORENSEN'S INDEX OF SIMILARITY; COMPUTER; POISSON DISTRIBUTION]

GALLAGHER, B. S. AND J. E. BURDICK. 1970. MEAN SEPARATION OF ORGANISMS IN THREE DIMENSIONS. ECOLOGY 51(3), 538-540.

GAUSE, G. F. 1930. DIE VARIABILITAT DER ZEICHNUNG BEI DEN BLATTKAFERN DER GATTUNG PHYTODECTA. BIOLOGISCHEN ZENTRALBLATT 50(4), 235-248.

GHENT, A. W. 1972. A GRAPHIC COMPUTATION PROCEDURE FOR KENDALL'S TAU SUITED TO EXTENSIVE SPECIES-DENSITY COMPARISONS. AMER. MIDL. NAT. 87(2), 459-471. [MATRIX]

GIESEL, J. T. 1972. SEX RATIO, RATE OF EVOLUTION, AND ENVIRONMENTAL HETEROGENEITY. AMER. NAT. 106(949), 380-387. [MODEL]

GILLESPIE, J. 1974. POLYMORPHISM IN PATCHY ENVIRONMENTS. AMER. NAT. 108(960), 145-151. [LEVINS-MACARTHUR MODEL; MATRIX; GENETICS]

GITTINS, R. 1968. TREND-SURFACE ANALYSIS OF ECOLOGICAL DATA. J. ECOL. 56(3), 845-869. [MULTIVARIATE ANALYSIS; PRINCIPAL COMPONENT ANALYSIS; HYPERSURFACE]

GODRON, M., PH. DAGET, J. POISSONET AND P. POISSONET. 1971. SOME ASPECTS OF HETEROGENEITY IN GRASSLANDS OF CANTAL (FRANCE). PP. 397-415. IN, STATISTICAL ECOLOGY, VOLUME 3. MANY SPECIES POPULATIONS, ECOSYSTEMS, AND SYSTEMS ANALYSIS. (G. P. PATIL, E. C. PIELOU AND W. E. WATERS, EDS.). PENNSYLVANIA STATE UNIVERSITY PRESS, UNIVERSITY PARK, PENNSYLVANIA.

GOODALL, D. W. 1953. OBJECTIVE METHODS FOR THE CLASSIFICATION OF VEGETATION. I. THE USE OF POSITIVE INTERSPECIFIC CORRELATION. AUST. J. BOT. 1(1), 39-63. [CHI-SQUARE; BINOMIAL DISTRIBUTION]

GOODALL, D. W. 1953. OBJECTIVE METHODS FOR THE CLASSIFICATION OF VEGETATION. II. FIDELITY AND INDICATOR VALUE. AUST. J. BOT. 1(3), 434-456. [DICE AND COLE INDEX OF FIDELITY; DISCRIMINANT FUNCTION]

GOODALL, D. W. 1954. MINIMAL AREA: A NEW APPROACH. INTERN. BOT. CONGR. 8(SEC. 7 AND 8), 19-21. [ANALYSIS OF VARIANCE; SPLIT PLOT]

GOODALL, D. W. 1954. OBJECTIVE METHODS FOR THE CLASSIFICATION OF VEGETATION. III. AN ESSAY IN THE USE OF FACTOR ANALYSIS. AUST. J. BOT. 2(3), 304-324.

GOODALL, D. W. 1963. PATTERN ANALYSIS AND MINIMAL AREA-SOME FURTHER COMMENTS. J. ECOL. 51(3), 705-710. [PROBABILITY; VARIANCE; ANALYSIS OF VARIANCE]

GOODALL, D. W. 1964. A PROBABILISTIC SIMILARITY INDEX. NATURE 203(4949), 1098.

GOODALL, D. W. 1966. CLASSIFICATION, PROBABILITY AND UTILITY. NATURE 211(5044), 53-54.

GOODALL, D. W. 1970. CLUSTER ANALYSIS USING SIMILARITY AND DISSIMILARITY. BIOMETRIE-PRAXIMETRIE 11(1), 34-41.

GOULD, L. L. AND F. HEPPNER. 1974. THE VEE FORMATION OF CANADA GEESE. AUK 91(3), 494-506. [PROJECTIVE GEOMETRY]

GOVIND, C. K. AND A. J. BURTON. 1970. FLIGHT ORIENTATION IN A COREID SQUASH BUG (HETEROPTERA). CAN. ENTOMOL. 102(8), 1002-1007. [WING POSITION EQUATION]

GOWER, J. C. 1971. A GENERAL COEFFICIENT OF SIMILARITY AND SOME OF ITS PROPERTIES. BIOMETRICS 27(4), 857-871.

GREEN, R. H. 1966. MEASUREMENT OF NON-RANDOMNESS IN SPATIAL DISTRIBUTIONS. RES. POPUL. ECOL.-KYOTO 8(1), 1-7. [LITERATURE REVIEW AND DISCUSSION]

GREIG-SMITH, P. 1961. DATA ON PATTERN WITHIN PLANT COMMUNITIES. I. THE ANALYSIS OF PATTERN. J. ECOL. 49(3), 695-702. [BINOMIAL DISTRIBUTION; COVARIANCE; DISCUSSION]

GREIG-SMITH, P. 1961. THE USE OF PATTERN ANALYSIS IN ECOLOGICAL INVESTIGATIONS. PP. 1354-1358. IN, RECENT ADVANCES IN BOTANY, UNIVERSITY OF TORONTO PRESS, TORONTO, CANADA. [MEAN SQUARE VERSUS BLOCK SIZE]

GREIG-SMITH, P. 1971. ANALYSIS OF VEGETATION DATA, THE USER VIEWPOINT. PP. 149-162. IN, STATISTICAL ECOLOGY, VOLUME 3. MANY SPECIES POPULATIONS, ECOSYSTEMS, AND SYSTEMS ANALYSIS. (G. P. PATIL, E. C. PIELOU AND W. E. WATERS, EDS.). PENNSYLVANIA STATE UNIVERSITY PRESS, UNIVERSITY PARK, PENNSYLVANIA.

GREIG-SMITH, P. 1971. APPLICATION OF NUMERICAL METHODS TO TROPICAL FORESTS. PP. 195-204. IN, STATISTICAL ECOLOGY, VOLUME 3. MANY SPECIES POPULATIONS, ECOSYSTEMS, AND SYSTEMS ANALYSIS. (G. P. PATIL, E. C. PIELOU AND W. E. WATERS, EDS.). PENNSYLVANIA STATE UNIVERSITY PRESS, UNIVERSITY PARK, PENNSYLVANIA.

GREIG-SMITH, P., K. A. KERSHAW AND D. J. ANDERSON. 1963. THE ANALYSIS OF PATTERN IN VEGETATION: A COMMENT ON A PAPER BY D. W. GOODALL. J. ECOL. 51(1), 223-229.

HADELER, K. P., U. AN DER HEIDEN AND F. ROTHE. 1974. NONHOMOGENEOUS SPATIAL DISTRIBUTIONS OF POPULATIONS. J. MATH. BIOL. 1(2), 165-176. [MODEL; VOLTERRA'S MODEL]

HAGMEIER, E. M. 1966. A NUMERICAL ANALYSIS OF THE DISTRIBUTIONAL PATTERNS OF NORTH AMERICAN MAMMALS. II. RE-EVALUATION OF THE PROVINCES. SYST. ZOOL. 15(4), 279-299.

HAIRSTON, N. G., R. W. HILL AND U. RITTE. 1971. THE INTERPRETATION OF AGGREGATION PATTERNS. PP. 337-352. IN, STATISTICAL ECOLOGY, VOLUME 1. SPATIAL PATTERNS AND STATISTICAL DISTRIBUTIONS. (G. P. PATIL, E. C. PIELOU AND W. E. WATERS, EDS.). PENNSYLVANIA STATE UNIVERSITY PRESS, UNIVERSITY PARK, PENNSYLVANIA.

HALL, J. W., J. C. ARNOLD, W. T. WALLER AND J. CAIRNS, JR. 1975. A PROCEDURE FOR THE DETECTION OF POLLUTION BY FISH MOVEMENTS. BIOMETRICS 31(1), 11-18. [NEGATIVE BINOMIAL DISTRIBUTION; STOCHASTIC MODEL; TRANSFORMATION]

HAMILTON, W. D. 1971. GEOMETRY FOR THE SELFISH HERD. J. THEOR. BIOL. 31(2), 295-311. [MODEL; NEAREST NEIGHBOR; BEHAVIOR]

HARRISON, J. L. 1958. RANGE OF MOVEMENT OF SOME MALAYAN RATS. J. MAMMAL. 39(2), 190-206. [STANDARD DIAMETER; CENTER OF ACTIVITY; PROBABILITY ZONES]

HATHEWAY, W. H. 1971. CONTINGENCY-TABLE ANALYSIS OF RAIN FOREST VEGETATION. PP. 271-307. IN, STATISTICAL ECOLOGY, VOLUME 3. MANY SPECIES POPULATIONS, ECOSYSTEMS, AND SYSTEMS ANALYSIS. (G. P. PATIL, E. C. PIELOU AND W. E. WATERS, EDS.). PENNSYLVANIA STATE UNIVERSITY PRESS, UNIVERSITY PARK, PENNSYLVANIA.

HAYNE, D. W. 1949. CALCULATION OF SIZE OF HOME RANGE. J. MAMMAL. 30(1), 1-18. [PROBABILITY]

HAZEN, W. E. 1966. ANALYSIS OF SPATIAL PATTERN IN EPIPHYTES. ECOLOGY 47(4), 634-635. [MACARTHUR'S MODEL; POISSON DISTRIBUTION]

HESPENHEIDE, H. A. 1973. ECOLOGICAL INFERENCES FROM MORPHOLOGICAL DATA. PP. 213-229. IN, ANNUAL REVIEW OF ECOLOGY AND SYSTEMATICS, VOLUME 4. (R. F. JOHNSTON, P. W. FRANK AND C. D. MICHENER, EDS.). ANNUAL REVIEWS, INC., PALO ALTO, CALIFORNIA. [R AND K SELECTION; NICHE PARAMETERS; MACARTHUR-LEVINS; LOG-NORMAL DISTRIBUTION]

HILL, M. O. 1973. THE INTENSITY OF SPATIAL PATTERN IN PLANT COMMUNITIES. J. ECOL. 61(1), 225-235. [SAMPLING; PATTERN ANALYSIS; SPECTRAL ANALYSIS; BLOCK-SIZE ANALYSIS; MATRIX]

HOLGATE, P. 1971. NOTES ON THE MARCZEWSKI-STEINHAUS COEFFICIENT OF SIMILARITY. PP. 181-190. IN, STATISTICAL ECOLOGY, VOLUME 3. MANY SPECIES POPULATIONS, ECOSYSTEMS, AND SYSTEMS ANALYSIS. (G. P. PATIL, E. C. PIELOU AND W. E. WATERS, EDS.). PENNSYLVANIA STATE UNIVERSITY PRESS, UNIVERSITY PARK, PENNSYLVANIA.

HOLT, S. J. 1955. ON THE FORAGING ACTIVITY OF THE WOOD RAT. J. ANIM. ECOL. 24(1), 1-34. [MODEL; EXPONENTIAL DISTRIBUTION; HARMONIC MEAN]

HUGHES, R. E. 1961. THE APPLICATION OF CERTAIN ASPECTS OF MULTIVARIATE ANALYSIS TO PLANT ECOLOGY. REC. ADV. BOT. (TORONTO) SECTION 12, 1350-1354. [CANONICAL ANALYSIS; D2 STATISTIC OF MAHALANOBIS; MULTIPLE REGRESSION; DISCRIMINANT ANALYSIS; DISCRIMINANT FUNCTION]

IVIMEY-COOK, R. B., M. C. F. PROCTOR AND D. L. WIGSTON. 1969. ON THE PROBLEM OF THE 'R/Q' TERMINOLOGY IN MULTIVARIATE ANALYSES OF BIOLOGICAL DATA. J. ECOL. 57(3), 673-675. [MATRIX]

IWAO, S. 1968. A NEW REGRESSION METHOD FOR ANALYZING THE AGGREGATION PATTERN OF ANIMAL POPULATIONS. RES. POPUL. ECOL.-KYOTO 10(1), 1-20.

IWAO, S. 1970. PROBLEMS OF SPATIAL DISTRIBUTION IN ANIMAL POPULATION ECOLOGY. PP. 117-149. IN, RANDOM COUNTS IN SCIENTIFIC WORK VOLUME 2. RANDOM COUNTS IN BIOMEDICAL AND SOCIAL SCIENCES. (G. P. PATIL, ED.) THE PENNSYLVANIA STATE UNIVERSITY PRESS, UNIVERSITY PARK, PENNSYLVANIA. [TRUNCATED DISTRIBUTION; REGRESSION; MODEL; POISSON DISTRIBUTION; NEGATIVE BINOMIAL DISTRIBUTION; MONTE CARLO; NEYMAN'S TYPE A, B AND C DISTRIBUTIONS; MORISITA-I; LLOYD'S MEAN CROWDING]

IWAO, S. AND E. KUNO. 1971. AN APPROACH TO THE ANALYSIS OF AGGREGATION PATTERN IN BIOLOGICAL POPULATIONS. PP. 461-512. IN, STATISTICAL ECOLOGY, VOLUME 1. SPATIAL PATTERNS AND STATISTICAL DISTRIBUTIONS. (G. P. PATIL, E. C. PIELOU AND W. E. WATERS, EDS.). PENNSYLVANIA STATE UNIVERSITY PRESS, UNIVERSITY PARK, PENNSYLVANIA.

JACKSON, M. T. AND R. O. PETTY. 1971. AN ASSESSMENT OF VARIOUS SYNTHETIC INDICES IN A TRANSITIONAL OLD-GROWTH FOREST. AMER. MIDL. NAT. 86(1), 13-27.

JENNRICH, R. I. AND F. B. TURNER. 1969. MEASUREMENT OF NON-CIRCULAR HOME RANGE. J. THEOR. BIOL. 22(2), 227-237.

JOHNSON, C. G. 1957. THE DISTRIBUTION OF INSECTS IN THE AIR AND THE EMPIRICAL RELATION OF DENSITY TO HEIGHT. J. ANIM. ECOL. 26(2), 479-494. [LOG TRANSFORMATION]

JORGENSEN, C. D. AND W. W. TANNER. 1963. THE APPLICATION OF THE DENSITY PROBABILITY FUNCTION TO DETERMINE THE HOME RANGES OF UTA STANSBURIANA STANSBURIANA AND CNEMIDOPHORUS TIGRIS TIGRIS. HERPETOLOGICA 19(2), 105-115.

KATZ, J. O. AND F. J. ROHLF. 1973. FUNCTION-POINT CLUSTER ANALYSIS. SYST. ZOOL. 22(3), 295-301. [COMPUTER; ALGORITHM]

KERSHAW, K. A. 1957. THE USE OF COVER AND FREQUENCY IN THE DETECTION OF PATTERN IN PLANT COMMUNITIES. ECOLOGY 38(2), 291-299. [MEAN SQUARE VERSUS BLOCK SIZE]

KIESTER, A. R. AND M. SLATKIN. 1974. A STRATEGY OF MOVEMENT AND RESOURCE UTILIZATION. THEOR. POPUL. BIOL. 6(1), 1-20. [MODEL; DIFFERENCE EQUATION; DIFFERENTIAL EQUATION; COMPUTER SIMULATION; LIZARDS]

KILBURN, P. D. 1966. ANALYSIS OF THE SPECIES-AREA RELATION. ECOLOGY 47(5), 831-843. [EQUATIONS]

KING, C. E. 1971. RESOURCE SPECIALIZATION AND EQUILIBRIUM POPULATION SIZE IN PATCHY ENVIRONMENTS. PROC. NATL. ACAD. SCI. U.S.A. 68(11), 2634-2637. [RESOURCE-EXPLOITATION MODEL]

KOCH, L. F. 1957. INDEX OF BIOTAL DISPERSITY. ECOLOGY 38(1), 145-148.

KOEPPL, J. W., N. A. SLADE AND R. S. HOFFMAN. 1975. A BIVARIATE HOME RANGE MODEL WITH POSSIBLE APPLICATION TO ETHOLOGICAL DATA ANALYSIS. J. MAMMAL. 56(1), 81-90. [COMPUTER; CONFIDENCE ELLIPSES; ACTIVITY CENTER; DISPERSION]

LAMBERT, J. M. AND M. B. DALE. 1964. THE USE OF STATISTICS IN PHYTOSOCIOLOGY. PP. 59-99. IN, ADVANCES IN ECOLOGICAL RESEARCH, VOLUME 2. (J. B. CRAGG, ED.). ACADEMIC PRESS, NEW YORK, NEW YORK.

LAMBERT, J. M. AND W. T. WILLIAMS. 1962. MULTIVARIATE METHODS IN PLANT ECOLOGY. IV. NODAL ANALYSIS. J. ECOL. 50(3), 775-802.

LANDAHL, H. D. 1953. ON THE SPREAD OF INFORMATION WITH TIME AND DISTANCE. BULL. MATH. BIOPHYS. 15, 367-381. [POPULATION]

LANDAU, H. G. AND A. RAPOPORT. 1953. CONTRIBUTION TO THE MATHEMATICAL THEORY OF CONTAGION AND SPREAD OF INFORMATION: I. SPREAD THROUGH A THOROUGHLY MIXED POPULATION. BULL. MATH. BIOPHYS. 15, 173-183.

LEBEDEVA, L. P. 1972. A MODEL OF THE LATITUDINAL DISTRIBUTION OF THE NUMBER OF SPECIES OF PHYTOPLANKTON IN THE SEA. J. CONS. PERM. INT. EXPLOR. MER 34(3), 341-350.

LEFKOVITCH, L. P. 1966. AN INDEX OF SPATIAL DISTRIBUTION. RES. POPUL. ECOL.-KYOTO 8(2), 89-92. [TEST OF SIGNIFICANCE]

LEVIN, D. A. AND H. W. KERSTER. 1971. NEIGHBORHOOD STRUCTURE IN PLANTS UNDER DIVERSE REPRODUCTIVE METHODS. AMER. NAT. 105(944), 345-354. [MODEL]

LEVIN, S. A. AND R. T. PAINE. 1974. DISTURBANCE, PATCH FORMATION, AND COMMUNITY STRUCTURE. PROC. NATL. ACAD. SCI. U.S.A. 71(7), 2744-2747. [MODEL]

LIVSHITS, P. Z. AND N. A. BORISOVA. 1970. THE POSSIBLE QUANTITATIVE EVALUATION OF THE RELATIONSHIPS OF PLANTS TO THE ENVIRONMENT. EKOLOGIYA 1(3), 90-92. (ENGLISH TRANSL. PP. 262-264.) [INDEX OF THE ENVIRONMENTAL FACTOR]

LOOMAN, J. AND J. B. CAMPBELL. 1960. ADAPTATION OF SORENSEN'S K (1948) FOR ESTIMATING UNIT AFFINITIES IN PRAIRIE VEGETATION. ECOLOGY 41(3), 409-416. [SORENSEN'S QUOTIENT OF SIMILARITY; DIVERSITY INDEX; KENDALL'S RANK CORRELATION COEFFICIENT]

LYNCH, D. W. AND F. X. SCHUMACHER. 1941. CONCERNING THE DISPERSION OF NATURAL REGENERATION. J. FORESTRY 39(1), 49-51. [PROBIT; ANALYSIS OF VARIANCE; REGRESSION; POISSON DISTRIBUTION;]

MACARTHUR, R. H. 1971. PATTERNS OF TERRESTRIAL BIRD COMMUNITIES. PP. 189-221. IN, AVIAN BIOLOGY, VOLUME 1. (D. S. FARNER AND J. R. KING, EDS.). ACADEMIC PRESS, NEW YORK, NEW YORK.

MACARTHUR, R. H. AND H. S. HORN. 1969. FOLIAGE PROFILE BY VERTICAL MEASUREMENTS. ECOLOGY 50(5), 802-804. [ESTIMATION; PROBABILITY]

MATERN, B. 1971. DOUBLY STOCHASTIC POISSON PROCESSES IN THE PLANE. PP. 195-212 IN, STATISTICAL ECOLOGY, VOLUME 1. SPATIAL PATTERNS AND STATISTICAL DISTRIBUTIONS. (G. P. PATIL, E. C. PIELOU AND W. E. WATERS, EDS.). PENNSYLVANIA STATE UNIVERSITY PRESS, UNIVERSITY PARK, PENNSYLVANIA.

MAURUS, M. AND H. PRUSCHA. 1973. CLASSIFICATION OF SOCIAL SIGNALS IN SQUIRREL MONKEYS BY MEANS OF CLUSTER ANALYSIS. BEHAVIOUR 47(1/2), 106-128. [TRANSFORMATION VALUES; MATRIX; DENDOGRAM; PAIR-GROUP METHOD; AGGLOMERATIVE CLUSTERING METHOD; ENTROPY; INFORMATION THEORY]

MAZURKIEWICZ, M. 1969. ELLIPTICAL MODIFICATION OF THE HOME RANGE PATTERN. BULL. ACAD. POL. SCI. CL. II 17, 427-431.

MCINTIRE, C. D. 1968. STRUCTURAL CHARACTERISTICS OF BENTHIC ALGAL COMMUNITIES IN LABORATORY STREAMS. ECOLOGY 49(3), 520-537. [REGRESSION MODEL; NORMIT; COMPUTER; PROBIT; POISSON SERIES]

MCINTIRE, C. D. AND W. S. OVERTON. 1971. DISTRIBUTIONAL PATTERNS IN ASSEMBLAGES OF ATTACHED DIATOMS FROM YAQUINA ESTUARY, OREGON. ECOLOGY 52(5), 758-777. [INFORMATION; COMPUTER; DIVERSITY INDEX; SHANNON-WEAVER INDEX; SIMPSON'S INDEX; MCINTOSH'S INDEX; MARGALEF'S INDEX]

MCINTOSH, R. P. 1962. RAUNKIAER'S "LAW OF FREQUENCY," ECOLOGY 43(3), 533-535.

MEAD, R. 1971. A NOTE ON THE USE AND MISUSE OF REGRESSION MODELS IN ECOLOGY. J. ECOL. 59(1), 215-219. [MULTIPLE REGRESSION; POLYNOMIAL EQUATION]

METZGAR, L. H. AND A. L. SHELDON. 1974. AN INDEX OF HOME RANGE SIZE. J. WILDL. MANAGE. 38(3), 546-551.

MOHR, C. 1965. CALCULATION OF AREA OF ANIMAL ACTIVITY BY USE OF MEDIAN AXES AND CENTERS IN SCATTER DIAGRAMS. RES. POPUL. ECOL.-KYOTO 7(2), 73-86. [HOME RANGE]

MOHR, C. O. AND W. A. STUMPF. 1966. COMPARISON OF METHODS FOR CALCULATING AREAS OF ANIMAL ACTIVITY. J. WILDL. MANAGE. 30(2), 293-304. [HOME RANGE]

MORGANS, J. F. C. 1956. NOTES ON THE ANALYSIS OF SHALLOW-WATER SOFT SUBSTRATA. J. ANIM. ECOL. 25(2), 367-387. [QUARTILE; PHI VALUES]

MORISITA, M. 1959. MEASURING OF THE DISPERSION OF INDIVIDUALS AND ANALYSIS OF THE DISTRIBUTIONAL PATTERNS. MEM. FAC. SCI. KYUSHU UNIV. SER. E 2, 215-235.

MORISITA, M. 1962. I-INDEX: A MEASURE OF DISPERSION OF INDIVIDUALS. RES. POPUL. ECOL.-KYOTO 4, 1-7.

MORISITA, M. 1971. COMPOSITION OF THE I-INDEX. RES. POPUL. ECOL.-KYOTO 13(1), 1-27.

MORISITA, M. 1971. MEASURING OF HABITAT VALUE BY THE "ENVIRONMENTAL DENSITY" METHOD. PP. 379-400. IN, STATISTICAL ECOLOGY, VOLUME 1, SPATIAL PATTERNS AND STATISTICAL DISTRIBUTIONS. (G. P. PATIL, E. C. PIELOU AND W. E. WATERS, EDS.). PENNSYLVANIA STATE UNIVERSITY PRESS, UNIVERSITY PARK, PENNSYLVANNIA.

MOUNTFORD, M. D. 1962. AN INDEX OF SIMILARITY AND ITS APPLICATION TO CLASSIFICATORY PROBLEMS. PP. 43-50. IN, PROGRESS IN SOIL ZOOLOGY. (P. W. MURPHY, ED.). BUTTERWORTHS, LONDON, ENGLAND. [JACCARD'S COEFFICIENT; SORENSEN'S INDEX; KULCZYNSKI; LOGARITHMIC SERIES]

MOUNTFORD, M. D. 1971. A TEST OF THE DIFFERENCE BETWEEN CLUSTERS. PP. 237-257. IN, STATISTICAL ECOLOGY, VOLUME 3. MANY SPECIES POPULATIONS, ECOSYSTEMS, AND SYSTEMS ANALYSIS. (G. P. PATIL, E. C. PIELOU AND W. E. WATERS, EDS.). PENNSYLVANIA STATE UNIVERSITY PRESS, UNIVERSITY PARK, PENNSYLVANNIA.

NIELSEN, J. S., R. R. BROOKS, C. R. BOSWELL AND N. J. MARSHALL. 1973. STATISTICAL EVALUATION OF GEOBOTANICAL AND BIOGEOCHEMICAL DATA BY DISCRIMINANT ANALYSIS. J. APPL. ECOL. 10(1), 251-258. [MULTIVARIATE ANALYSIS; D2 STATISTIC OF MAHALANOBIS]

NORRIS, J. M. AND J. P. BARKHAM. 1970. A COMPARISON OF SOME COTSWOLD BEECHWOODS USING MULTIPLE-DISCRIMINANT ANALYSIS. J. ECOL. 58(3), 603-619. [CANONICAL ANALYSIS; COMPUTER]

ORLOCI, L. 1971. AN INFORMATION THEORY MODEL FOR PATTERN ANALYSIS. J. ECOL. 59(2), 343-349.

ORLOCI, L. 1971. INFORMATION THEORY TECHNIQUES FOR CLASSIFYING PLANT COMMUNITIES. PP. 259-270. IN, STATISTICAL ECOLOGY, VOLUME 3. MANY SPECIES POPULATIONS, ECOSYSTEMS, AND SYSTEMS ANALYSIS. (G. P. PATIL, E. C. PIELOU AND W. E. WATERS, EDS.). PENNSYLVANIA STATE UNIVERSITY PRESS, UNIVERSITY PARK, PENNSYLVANIA.

ORLOCI, L. AND M. M. MUKKATTU, 1973. THE EFFECT OF SPECIES NUMBER AND TYPE OF DATA ON THE RESEMBLANCE STRUCTURE OF A PHYTOSOCIOLOGICAL COLLECTION. J. ECOL. 61(1), 37-46. [CORRELATION; MATRIX; SAMPLING MATRIX; EUCLIDEAN DISTANCE; COVARIANCE; STRESS PROFILE]

PALTRIDGE, G. W. 1973. ON THE SHAPE OF TREES. J. THEOR. BIOL. 38(1), 111-137.

PATIL, G. P. AND W. M. STITELER, 1974. CONCEPTS OF AGGREGATION AND THEIR QUANTIFICATION: A CRITICAL REVIEW WITH SOME NEW RESULTS AND APPLICATIONS. RES. POPUL. ECOL.-KYOTO 15(2), 238-254. [SPATIAL PATTERN; MODEL; BINOMIAL DISTRIBUTION; NEGATIVE BINOMIAL DISTRIBUTION; CLUMPING MODEL; HETEROGENEITY MODEL; PURE BIRTH PROCESSES; POISSON DISTRIBUTION; VARIANCE-TO-MEAN RATIO; MORISITA'S INDEX; LLOYD'S MEAN CROWDING; IWAO'S INDEX; VECTOR APPROACH; NEYMAN'S TYPE A DISTRIBUTION]

PAYANDEH, B. 1970. COMPARISON OF METHODS FOR ASSESSING SPATIAL DISTRIBUTION OF TREES. FOREST SCI. 16(3), 312-317. [COMPUTER]

PIDGEON, I. M. AND E. ASHBY, 1942. A NEW QUANTITATIVE METHOD OF ANALYSIS OF PLANT COMMUNITIES. AUST. J. SCI. 5, 19-21.

PIELOU, D. P. AND E. C. PIELOU, 1968. ASSOCIATION AMONG SPECIES OF INFREQUENT OCCURRENCE: THE INSECT AND SPIDER FAUNA OF POLYPORUS BETULINUS (BULLIARD) FRIES. J. THEOR. BIOL. 21(2), 202-216. [TEST FOR ASSOCIATION]

PIELOU, E. C. 1962. RUNS OF ONE SPECIES WITH RESPECT TO ANOTHER IN TRANSECTS THROUGH PLANT POPULATIONS. BIOMETRICS 18(4), 579-593.

PIELOU, E. C. 1963. THE DISTRIBUTION OF DISEASED TREES WITH RESPECT TO HEALTHY ONES IN A PATCHILY INFECTED FOREST. BIOMETRICS 19(3), 450-459.

PIELOU, E. C. 1963. RUNS OF HEALTHY AND DISEASED TREES IN TRANSECTS THROUGH AN INFECTED FOREST. BIOMETRICS 19(4), 603-614.

PIELOU, E. C. 1964. THE SPATIAL PATTERN OF TWO-PHASE PATCHWORKS OF VEGETATION. BIOMETRICS 20(1), 156-167.

PIELOU, E. C. 1965. THE CONCEPT OF RANDOMNESS IN THE PATTERNS OF MOSAICS. BIOMETRICS 21(4), 908-920. [N-PHASE MOSAIC; MARKOV CHAIN; SHANNON AND WEAVER]

PIELOU, E. C. 1965. THE SPREAD OF DISEASE IN PATCHILY-INFECTED FOREST STANDS. FOREST SCI. 11(1), 18-26. [RUNS; PROBABILITY]

PIELOU, E. C. (APPENDIX BY D. S. ROBSON). 1972. 2K CONTINGENCY TABLES IN ECOLOGY. J. THEOR. BIOL. 34(2), 337-352.

PLATT, T., L. M. DICKIE AND R. W. TRITES. 1970. SPATIAL HETEROGENEITY OF PHYTOPLANKTON IN A NEAR-SHORE ENVIRONMENT. J. FISH. RES. BOARD CAN. 27(8), 1453-1473. [ANALYSIS OF VARIANCE; AUTOCOVARIANCE; SERIAL AUTOCORRELATION]

POOLE, R. W. 1974. MEASURING THE STRUCTURAL SIMILARITY OF TWO COMMUNITIES COMPOSED OF THE SAME SPECIES. RES. POPUL. ECOL.-KYOTO 16(1), 138-151. [MODEL; COVARIANCE MATRIX; FACTOR ANALYSIS; PROCRUSTES; DIVERSITY MEASURE]

PRITCHARD, N. M. AND A. J. B. ANDERSON. 1971. OBSERVATIONS ON THE USE OF CLUSTER ANALYSIS IN BOTANY WITH AN ECOLOGICAL EXAMPLE. J. ECOL. 59(3), 727-747. [COMPUTER; NEAREST NEIGHBOR; CLUSTERING]

RAPOPORT, A. 1954. SPREAD OF INFORMATION THROUGH A POPULATION WITH SOCIO-STRUCTURAL BIAS: III. SUGGESTED EXPERIMENTAL PROCEDURES. BULL. MATH. BIOPHYS. 16, 75-81.

READER, R., J. S. RADFORD AND H. LIETH. 1974. MODELING IMPORTANT PHYTOPHENOLOGICAL EVENTS IN EASTERN NORTH AMERICA. PP. 329-342. IN, PHENOLOGY AND SEASONALITY MODELING. (H. LIETH, ED.). SPRINGER-VERLAG NEW YORK INC., NEW YORK, NEW YORK. [COMPUTER MAPPED]

RICHARDSON, R. H. 1970. MODELS, AND ANALYSES OF DISPERSAL PATTERNS. PP. 79-183. IN, MATHEMATICAL TOPICS IN POPULATION GENETICS. (K. KOJIMA, ED.). SPRINGER-VERLAG, NEW YORK, NEW YORK. [PROBABILITY]

ROFF, D. A. 1974. SPATIAL HETEROGENEITY AND THE PERSISTENCE OF POPULATIONS. OECOLOGIA 15(3), 245-258. [POPULATION MODEL; SIMULATION]

ROHLF, F. J. 1969. THE EFFECT OF CLUMPED DISTRIBUTION IN SPARSE POPULATIONS. ECOLOGY 50(4), 716-721.

ROMELL, L. G. 1930. COMMENTS ON RAUNKIAER'S AND SIMILAR METHODS OF VEGETATION ANALYSIS AND THE "LAW OF FREQUENCY". ECOLOGY 11(3), 589-596.

ROTH, L. M. AND S. COHEN. 1973. AGGREGATION IN BLATTARIA. ANN. ENTOMOL. SOC. AMER. 66(6), 1315-1323. [INDEX OF AGGREGATION; SEGREGATION INDEX]

ROZENBERG, G., B. M. MIRKIN AND S. YU. RUDERMAN. 1972. IDENTIFICATION OF PLANTS AS A METHOD OF ASSESSING SOIL SALINITY. EKOLOGIYA 3(6), 31-34. (ENGLISH TRANSL. PP. 509-511). [GEOMETRICAL MODEL; ALGORITHM]

SCHOENER, T. W. 1968. SIZES OF FEEDING TERRITORIES AMONG BIRDS. ECOLOGY 49(1), 123-141. [EQUATIONS]

SCOTT, D. 1969. RELATIONSHIP BETWEEN SOME STATISTICAL METHODS. PROC. N. Z. ECOL. SOC. 16, 58-64. [FACTOR ANALYSIS; PRINCIPAL COMPONENT ANALYSIS; CANONICAL ANALYSIS; MODEL; RANDOMNESS; MULTIVARIATE; ANALYSIS OF VARIANCE; REGRESSION; PATH ANALYSIS; COVARIANCE ANALYSIS; DISCRIMINANT ANALYSIS; CORRELATION COEFFICIENT]

SCOTTER, D. R., K. P. LAMB AND E. HASSAN. 1971. AN INSECT DISPERSAL PARAMETER. ECOLOGY 52(1), 174-177. [RANDOM WALK; DIFFUSION EQUATION]

SEAL, H. L. 1966. TESTING FOR CONTAGION IN ANIMAL POPULATIONS. TRANS. AMER. FISH. SOC. 95(4), 436-437.

SEPKOSKI, J. J., JR. AND M. A. REX. 1974. DISTRIBUTION OF FRESHWATER MUSSELS: COASTAL RIVERS AS BIOGEOGRAPHIC ISLANDS. SYST. ZOOL. 23(2), 165-188. [MULTIPLE REGRESSION MODEL; STOCHASTIC MODEL; IMMIGRATION AND EXTINCTION PROBABILITIES]

SHELDON, A. L. 1972. A QUANTITATIVE APPROACH TO THE CLASSIFICATION OF INLAND WATERS. PP. 205-261. IN, NATURAL ENVIRONMENTS: STUDIES IN THEORETICAL AND APPLIED ANALYSIS. (J. V. KRUTILLA, ED.). JOHNS HOPKINS UNIVERSITY PRESS, BALTIMORE, MARYLAND. [MULTIVARIATE ANALYSIS]

SHIYOMI, M. 1967. A MODEL OF THE PLANT-TO-PLANT MOVEMENT OF APHIDS. I. DESCRIPTION OF THE MODEL. RES. POPUL. ECOL.-KYOTO 9(1), 53-61. [POISSON DISTRIBUTION; BINOMIAL DISTRIBUTION; NEGATIVE BINOMIAL DISTRIBUTION; MOMENTS; VARIANCE]

SHIYOMI, M. 1968. A MODEL OF THE PLANT-TO-PLANT MOVEMENT OF APHIDS. II. ESTIMATION OF PARAMETERS AND THE APPLICATION. RES. POPUL. ECOL.-KYOTO 10(1), 105-114. [PROBABILITY FUNCTIONS; MOMENTS; MAXIMUM LIKELIHOOD; COMPUTER]

SINHA, R. N., H. A. H. WALLACE AND F. S. CHEBIB. 1969. CANONICAL CORRELATION BETWEEN GROUPS OF ACARINE, FUNGAL AND ENVIRONMENTAL VARIABLES IN BULK GRAIN ECOSYSTEMS. RES. POPUL. ECOL.-KYOTO 11(1), 92-104.

SKELLAM, J. G. 1951. RANDOM DISPERSAL IN THEORETICAL POPULATIONS. BIOMETRIKA 38(1/2), 196-218.

SKELLAM, J. G. 1952. STUDIES IN STATISTICAL ECOLOGY: I. SPATIAL PATTERN. BIOMETRIKA 39(3/4), 346-362. [POISSON DISTRIBUTION; NEYMAN'S TYPE A DISTRIBUTION; THOMAS'S DISTRIBUTION; BINOMIAL DISTRIBUTION; POLYA-AEPPLI GENERATING FUNCTION]

SMITH, D. B. 1972. A CHI-SQUARED NOMOGRAM. ECOLOGY 53(3), 529-530.

SMITH, J. H. G. 1959. COMPREHENSIVE AND ECONOMICAL DESIGNS FOR STUDIES OF SPACING AND THINNING. FOREST SCI. 5(3), 237-245.

SORENSEN, T. 1948. A METHOD OF ESTABLISHING GROUPS OF EQUAL AMPLITUDE IN PLANT SOCIOLOGY BASED ON SIMILARITY OF SPECIES CONTENT AND ITS APPLICATION TO ANALYSES OF THE VEGETATION ON DANISH COMMONS. BIOL. SKR. 5(4), 1-34. [SORENSEN'S INDEX OF SIMILARITY]

STEWART, G. AND W. KELLER. 1936. A CORRELATION METHOD FOR ECOLOGY AS EXEMPLIFIED BY STUDIES OF NATIVE DESERT VEGETATION. ECOLOGY 17(3), 500-514. [CORRELATION COEFFICIENT; PARTIAL CORRELATION; MULTIPLE CORRELATION]

STITELER, W. M. AND G. P. PATIL. 1971. VARIANCE-TO-MEAN RATIO AND MORISITA'S INDEX AS MEASURES OF SPATIAL PATTERNS IN ECOLOGICAL POPULATIONS. PP. 423-450. IN: STATISTICAL ECOLOGY, VOLUME 1, SPATIAL PATTERNS AND STATISTICAL DISTRIBUTIONS. (G. P. PATIL, E. C. PIELOU AND W. E. WATERS, EDS.). PENNSYLVANIA STATE UNIVERSITY PRESS, UNIVERSITY PARK, PENNSYLVANIA.

STITELER, W. M., III. 1970. MEASUREMENT OF SPATIAL PATTERNS IN ECOLOGY. DISSERTATION. PENNSYLVANIA STATE UNIVERSITY, UNIVERSITY PARK, PENNSYLVANIA. 139 PP.

STRATHMANN, R. 1974. THE SPREAD OF SIBLING LARVAE OF SEDENTARY MARINE INVERTEBRATES. AMER. NAT. 108(959), 29-44. [LIFE TABLE; MODEL; DISPERSAL]

SWITZER, P. 1971. MAPPING A GEOGRAPHICALLY CORRELATED ENVIRONMENT. PP. 235-267. IN: STATISTICAL ECOLOGY VOLUME 1. SPATIAL PATTERNS AND STATISTICAL DISTRIBUTIONS. (G. P. PATIL, E. C. PIELOU AND W. E. WATERS, EDS.). PENNSYLVANIA STATE UNIVERSITY PRESS, UNIVERSITY PARK, PENNSYLVANIA.

THARU, J. AND W. T. WILLIAMS. 1966. CONCENTRATION OF ENTRIES IN BINARY ARRAYS. NATURE 210(5035), 549.

USHER, M. B. 1969. THE RELATION BETWEEN MEAN SQUARE AND BLOCK SIZE IN THE ANALYSIS OF SIMILAR PATTERNS. J. ECOL. 57(2), 505-514. [MODEL; ANALYSIS OF VARIANCE]

VAN DER AART, P. J. M. 1973. DISTRIBUTION ANALYSIS OF WOLFSPIDERS (ARANEAE, LYCOSIDAE) IN A DUNE AREA BY MEANS OF PRINCIPAL COMPONENT ANALYSIS. NETHERLANDS J. ZOOL. 23(3), 266-329. [MATRIX; COMPUTER; MULTIVARIATE ANALYSIS; VECTOR; MODEL; SIMULATION; ORDINATION]

VAN WINKLE, W. 1975. COMPARISON OF SEVERAL PROBABILISTIC HOME-RANGE MODELS. J. WILDL. MANAGE. 39(1), 118-123. [UNIVARIATE MODELS; BIVARIATE MODEL]

VAN WINKLE, W., JR., D. C. MARTIN AND M. J. SEBETICH. 1973. A HOME-RANGE MODEL FOR ANIMALS INHABITING AN ECOTONE. ECOLOGY 54(1), 205-209. [BIVARIATE MODEL; QUARTILE DEVIATIONS; NON-PARAMETRIC TESTS]

VANDERMEER, J. H. 1972. ON THE COVARIANCE OF THE COMMUNITY MATRIX. ECOLOGY 53(1), 187-189.

VAYNSHTEYN, B. A. 1969. EVALUATION OF THE NUMBER AND THE DISTRIBUTION OF FRESH-WATER INVERTEBRATES IN A WATER RESERVOIR. HYDROBIOL. J. 5(2), 38-39. (ENGLISH TRANSL.). [SPACING COEFFICIENT]

VENRICK, E. L. 1972. SMALL-SCALE DISTRIBUTIONS OF OCEANIC DIATOMS. U. S. NATL. MAR. FISH. SERV. FISH. BULL. 70(2), 363-372. [POISSON DISTRIBUTION; VARIANCE; SAMPLING]

VERHAGEN, A. M. W. 1971. CLUSTERING OF ATTRIBUTES ON REGULAR POINT LATTICES. J. APPL. ECOL. 8(3), 665-682.

VESTAL, A. G. 1949. MINIMUM AREAS FOR DIFFERENT VEGETATIONS, THEIR DETERMINATION FROM SPECIES-AREA CURVES. ILLINOIS BIOL. MONOGR. 20(3), VI, 129 PP.

WALKER, D. 1971. QUANTIFICATION IN HISTORICAL PLANT ECOLOGY. PROC. ECOL. SOC. AUSTRALIA 6, 91-104. [HISTORY; POLLEN ANALYSIS]

WEBB, L. J., J. G. TRACEY, W. T. WILLIAMS AND G. N. LANCE. 1967. STUDIES IN THE NUMERICAL ANALYSIS OF COMPLEX RAIN-FOREST COMMUNITIES. I. A COMPARISON OF METHODS APPLICABLE TO SITE/SPECIES DATA. J. ECOL. 55(1), 171-191.

WELLS, G. P. AND R. P. DALES. 1951. SPONTANEOUS ACTIVITY PATTERNS IN ANIMAL BEHAVIOUR: THE IRRIGATION OF THE BURROW IN THE POLYCHAETES CHAETOPTERUS VARIOPEDATUS RENIER AND NEREIS DIVERSICOLOR O. F. MULLER. J. MAR. BIOL. ASSOC. U. K. 29(3), 661-680. [EQUATIONS]

WHITE, J. E. 1964. AN INDEX OF THE RANGE OF ACTIVITY. AMER. MIDL. NAT. 71(2), 369-373. [HARRISON'S STANDARD DIAMETER CONCEPT]

WIENS, J. A. 1973. PATTERN AND PROCESS IN GRASSLAND BIRD COMMUNITIES. ECOL. MONOGR. 43(2), 237-270. [PRODUCTION MODEL; HETEROGENEITY INDEX; COLE'S INDEX OF ASSOCIATION; DENDOGRAM; INTERSPECIFIC OVERLAP INDEX]

IENS, J. A. 1974. HABITAT ETEROGENEITY AND AVIAN COMMUNITY TRUCTURE IN NORTH AMERICAN RASSLANDS. AMER. MIDL. NAT. 1(1), 195-213. [HETEROGENEITY NDEX; SPATIAL OVERLAP INDEX]

WIERZBOWSKA, T. 1972. STATISTICAL ESTIMATION OF HOME RANGE SIZE OF SMALL RODENTS. EKOLOGIYA POLSKA-SERIA A 20(49), 781-831. [PROBABILISTIC MODEL; MAXIMUM LIKELIHOOD; VARIANCE; CONFIDENCE INTERVAL]

WIERZBOWSKA, T. AND H. CHELKOWSKA. 1970. ESTIMATING SIZE OF HOME RANGE OF APODEMUS AGRARIUS (PALL.). EKOLOGIYA POLSKA-SERIA A 18(1), 1-12. [CALHOUN-CASBY METHOD; MODEL; PROBABILITY]

WILLIAMS, W. T. AND J. M. LAMBERT. 1966. MULTIVARIATE METHODS IN PLANT ECOLOGY. V. SIMILARITY ANALYSIS AND INFORMATION-ANALYSIS. J. ECOL. 54(2), 427-445.

WILLIAMS, W. T. AND M. B. DALE. 1962. PARTITION CORRELATION MATRICES FOR HETEROGENEOUS QUANTITATIVE DATA. NATURE 196(4854), 602.

WILLIAMS, W. T., G. N. LANCE, L. J. WEBB, J. G. TRACEY AND J. H. CONNELL. 1969. STUDIES IN THE NUMERICAL ANALYSIS OF COMPLEX RAIN FOREST COMMUNITIES. IV. A METHOD FOR THE ELUCIDATION OF SMALL-SCALE FOREST PATTERN. J. ECOL. 57(3), 635-654. [COMPUTER; MULTIPLE-NEAREST NEIGHBOR CONCEPT; POINT-CLUMP CONCEPT; INFORMATION ANALYSIS]

WILSON, D. E. AND J. S. FINDLEY. 1972. RANDOMNESS IN BAT HOMING. AMER. NAT. 106(949), 418-424. [PROBABILITY MODEL]

WOLFENBARGER, D. O. 1946. DISPERSION OF SMALL ORGANISMS. DISTANCE DISPERSION RATES OF BACTERIA, SPORES, SEEDS, POLLEN, AND INSECTS; INCIDENCE RATES OF DISEASES AND INJURIES. AMER. MIDL. NAT. 35(1), 1-152. [REGRESSION FORMULAE]

YARRANTON, G. A. 1969. PATTERN ANALYSIS BY REGRESSION. ECOLOGY 50(3), 390-395. [COMPUTER]

PLOTLESS SAMPLING

ANDERSON, D. R. AND R. S. POSPAHALA. 1970. CORRECTION OF BIAS IN BELT TRANSECT STUDIES OF IMMOTILE OBJECTS. J. WILDL. MANAGE. 34(1), 141-146.

ARNEY, J. D. AND D. P. PAINE. 1972. TREE AND STAND VOLUMES USING HEIGHT-ACCUMULATION AND THE TELESCOPIC SPIEGEL-RELASKOP. FOREST SCI. 18(2), 159-163. [COMPUTER; EQUATIONS]

ASHBY, W. C. 1972. DISTANCE MEASUREMENTS IN VEGETATION STUDY. ECOLOGY 53(5), 980-981. [SYSTEMATIC BIAS]

BARTLETT, M. S. 1964. THE SPECTRAL ANALYSIS OF TWO-DIMENSIONAL POINT PROCESSES. BIOMETRIKA 51(3/4), 299-311.

BARTLETT, M. S. 1971. TWO-DIMENSIONAL NEAREST-NEIGHBOR SYSTEMS AND THEIR ECOLOGICAL APPLICATIONS. PP. 179-191. IN, STATISTICAL ECOLOGY, VOLUME 1. SPATIAL PATTERNS AND STATISTICAL DISTRIBUTIONS. (G. P. PATIL, E. C. PIELOU AND W. E. WATERS, EDS.). PENNSYLVANIA STATE UNIVERSITY PRESS, UNIVERSITY PARK, PENNSYLVANIA.

BATCHELER, C. L. 1971. ESTIMATION OF DENSITY FROM A SAMPLE OF JOINT POINT AND NEAREST-NEIGHBOR DISTANCES. ECOLOGY 52(4), 703-709.

BATCHELER, C. L. 1973. ESTIMATING DENSITY AND DISPERSION FROM TRUNCATED OR UNRESTRICTED JOINT POINT-DISTANCE NEAREST-NEIGHBOUR DISTANCES. N. Z. ECOL. SOC. PROC. 20, 131-147. [COMPUTER; SIMULATION; SAMPLING]

BATCHELER, C. L. AND D. J. BELL. 1970. EXPERIMENTS IN ESTIMATING DENSITY FROM JOINT POINT-AND NEAREST-NEIGHBOUR DISTANCE SAMPLES. PROC. N. Z. ECOL. SOC. 17, 111-117. [MODEL; PROBABILITY; CONTAGION]

BAUER, H. L. 1943. THE STATISTICAL ANALYSIS OF CHAPARRAL AND OTHER PLANT COMMUNITIES BY MEANS OF TRANSECT SAMPLES. ECOLOGY 24(1), 45-60. [EXPECTED VALUES; PROBABILITY]

BESAG, J. E. 1973. ON THE DETECTION OF SPATIAL PATTERN IN PLANT COMMUNITIES. BULL. INTERN. STAT. INST. 45(1), 153-165. [T-SQUARE METHOD]

BLACKITH, R. E. 1958. NEAREST-NEIGHBOR DISTANCE MEASUREMENTS FOR THE ESTIMATION OF ANIMAL POPULATIONS. ECOLOGY 39(1), 147-150. [CRAIG'S FORMULA; CLARK-EVANS FORMULA; MODEL]

BROWN, J. K. 1971. A PLANAR INTERSECT METHOD FOR SAMPLING FUEL VOLUME AND SURFACE AREA. FOREST SCI. 17(1), 96-102. [MODEL]

BROWN, J. K. AND P. J. ROUSSOPOULOS. 1974. ELIMINATING BIASES IN THE PLANAR INTERSECT METHOD FOR ESTIMATING VOLUMES OF SMALL FUELS. FOREST SCI. 20(4), 350-356. [EQUATIONS]

CATANA, A. J., JR. 1963. THE WANDERING QUARTER METHOD OF ESTIMATING POPULATION DENSITY. ECOLOGY 44(2), 349-360. [INDEX OF AGGREGATION]

CATANA, A. J., JR. 1964. A DISTRIBUTION-FREE METHOD FOR THE DETERMINATION OF HOMOGENEITY IN DISTANCE DATA. ECOLOGY 45(3), 640-641. [RUN TEST]

CLARK, P. J. AND F. C. EVANS. 1954. DISTANCE TO NEAREST NEIGHBOR AS A MEASURE OF SPATIAL RELATIONSHIPS IN POPULATIONS. ECOLOGY 35(4), 445-453.

CLARK, P. J. AND F. C. EVANS. 1955. ON SOME ASPECTS OF SPATIAL PATTERN IN BIOLOGICAL POPULATIONS. SCIENCE 121(3142), 397-398.

COOPER, C. F. 1957. THE VARIABLE PLOT METHOD FOR ESTIMATING SHRUB DENSITY. J. RANGE MANAGE. 10(3), 111-115.

COOPER, C. F. 1963. AN EVALUATION OF VARIABLE PLOT SAMPLING IN SHRUB AND HERBACEOUS VEGETATION. ECOLOGY 44(3), 565-569. [BIAS; GAGE GEOMETRY]

COTTAM, G. AND J. T. CURTIS. 1949. A METHOD FOR MAKING RAPID SURVEYS OF WOODLANDS BY MEANS OF PAIRS OF RANDOMLY SELECTED TREES. ECOLOGY 30(1), 101-104. [SAMPLING]

COTTAM, G. AND J. T. CURTIS. 1955. CORRECTION FOR VARIOUS EXCLUSION ANGLES IN THE RANDOM PAIRS METHODS. ECOLOGY 36(4), 767.

COTTAM, G. AND J. T. CURTIS. 1956. THE USE OF DISTANCE MEASURES IN PHYTOSOCIOLOGICAL SAMPLING. ECOLOGY 37(3), 451-460. [RANDOM NUMBERS]

COX, F. 1972. ARE THERE ANY CONSISTENT PARAMETERS FOR DISTANCE METHODS IF THE SPATIAL DISTRIBUTION DEVIATE REMARKABLY FROM A BIDIMENSIONAL POISSON PROCESS? PP. 247-259. IN, 3RD CONFERENCE ADVISORY GROUP OF FOREST STATISTICIANS. INSTITUT NATIONAL DE LA RECHERCHE AGRONOMIQUE, PARIS, FRANCE. I.N.R.A. PUBL. 72-3. [SIMULATION; MODEL; POISSON PROCESS; COMPUTER]

DE VRIES, P. G. 1974. MULTI-STAGE LINE INTERSECT SAMPLING. FOREST SCI. 20(2), 129-133.

DICE, L. R. 1952. MEASURE OF THE SPACING BETWEEN INDIVIDUALS WITHIN A POPULATION. CONTRIB. LAB. VERTEBR. BIOL., UNIVERSITY OF MICHIGAN, ANN ARBOR, MICHIGAN, NO. 55, 23 PP.

DUDZINSKI, M. L., P. J. PAHL AND G. W. ARNOLD. 1969. QUANTITATIVE ASSESSMENT OF GRAZING BEHAVIOUR OF SHEEP IN ARID AREAS. J. RANGE MANAGE. 22(4), 230-235. [NEAREST NEIGHBOR DISTANCE; MONTE CARLO; COMPUTER]

EBERHARDT, L. L. 1967. SOME DEVELOPMENTS IN 'DISTANCE SAMPLING'. BIOMETRICS 23(2), 207-216.

EBERHARDT, L. L. 1968. A PRELIMINARY APPRAISAL OF LINE TRANSECTS. J. WILDL. MANAGE. 32(1), 82-88.

ELLIS, J. A., R. L. WESTEMEIR, K. P. THOMAS AND H. W. NORTON. 1969. SPATIAL RELATIONSHIPS AMONG QUAIL COVEYS. J. WILDL. MANAGE. 33(2), 249-254. [NEAREST NEIGHBOR]

FRASER, A. R. AND P. VAN DEN DRIESSCHE. 1972. TRIANGLES, DENSITY AND PATTERN IN POINT POPULATIONS. PP. 277-285. IN, 3RD CONFERENCE ADVISORY GROUP OF FOREST STATISTICIANS. INSTITUT NATIONAL DE LA RECHERCHE AGRONOMIQUE, PARIS, FRANCE. I.N.R.A. PUBL. 72-3. [SPATIAL DISTRIBUTION; LEAST DIAGONAL NEIGHBORS; COMPUTER]

GATES, C. E. 1969. SIMULATION STUDY OF ESTIMATORS FOR THE LINE TRANSECT SAMPLING METHOD. BIOMETRICS 25(2), 317-328.

GATES, C. E., W. H. MARSHALL AND D. P. OLSON. 1968. LINE TRANSECT METHOD OF ESTIMATING GROUSE POPULATION DENSITIES. BIOMETRICS 24(1), 135-145.

GOODALL, D. W. 1952. SOME CONSIDERATIONS IN THE USE OF POINT QUADRATS FOR THE ANALYSIS OF VEGETATION. AUST. J. SCI. RES. SER. B 5, 1-41.

GOODALL, D. W. 1965. PLOT-LESS TESTS OF INTERSPECIFIC ASSOCIATION. J. ECOL. 53(1), 197-210. [RUN TEST; NEAREST NEIGHBOR]

GOODALL, D. W. 1966. A NEW SIMILARITY INDEX BASED ON PROBABILITY. BIOMETRICS 22(4), 882-907. [TAXONOMY; JACCARD'S COEFFICIENT; SIMPLE MATCHING COEFFICIENT; DISTANCE MEASUREMENTS; SORENSEN'S INDEX; COMPUTER]

GROSENBAUGH, L. R. 1952. PLOTLESS TIMBER ESTIMATES-NEW, FAST, EASY. J. FORESTRY 50(1), 32-37. [POINT-SAMPLING]

GROSENBAUGH, L. R. 1958. POINT-SAMPLING AND LINE-SAMPLING. PROBABILITY THEORY, GEOMETRIC IMPLICATIONS, SYNTHESIS. U. S. FOR. SERV. SOUTHERN FOREST EXPT. STN. OCCAS. PAPER 160. 34 PP.

GROSENBAUGH, L. R. 1963. OPTICAL DENDROMETERS FOR OUT-OF-REACH DIAMETERS: A CONSPECTUS AND SOME NEW THEORY. FOREST SCI. MONOGR. 4, 47 PP. [EQUATIONS]

GROSENBAUGH, L. R. AND W. S. STOVER. 1957. POINT-SAMPLING COMPARED WITH PLOT-SAMPLING IN SOUTHEAST TEXAS. FOREST SCI. 3(1), 2-14.

GRUM, L. 1971. SPATIAL DIFFERENTIATION OF THE CARABUS L. (CARABIDAE, COLEOPTERA) MOBILITY. EKOLOGIYA POLSKA-SERIA A 19(1), 1-34. [MODEL; ESTIMATION; TAG-RECAPTURE]

HAGA, T AND K. MAEZAWA. 1959. BIAS DUE TO EDGE EFFECT IN USING THE BITTERLICH METHOD. FOREST SCI. 5(4), 370-376. [SAMPLING]

HASEL, A. A. 1941. ESTIMATION OF VEGETATION-TYPE AREAS BY LINEAR MEASUREMENT. J. FORESTRY 39(1), 34-40.

HAYNE, D. W. 1949. AN EXAMINATION OF THE STRIP CENSUS METHOD FOR ESTIMATING ANIMAL POPULATIONS. J. WILDL. MANAGE. 13(2), 145-157.

HEYTING, A. 1968. DISCUSSION AND DEVELOPMENT OF THE POINT-CENTERED QUARTER METHOD OF SAMPLING GRASSLAND VEGETATION. J. RANGE MANAGE. 21(6), 370-380. [SYSTEMATIC SAMPLING; VARIANCE; RANDOM LINE SAMPLING]

HOLGATE, P. 1964. THE EFFICIENCY OF NEAREST NEIGHBOR ESTIMATORS. BIOMETRICS 20(3), 647-649.

HOLGATE, P. 1965. THE DISTANCE FROM A RANDOM POINT TO THE NEAREST POINT OF A CLOSELY PACKED LATTICE. BIOMETRIKA 52(1/2), 261-263.

HOLGATE, P. 1965. SOME NEW TESTS OF RANDOMNESS. J. ECOL. 53(2), 261-266.

HOLGATE, P. 1965. TESTS OF RANDOMNESS BASED ON DISTANCE METHODS. BIOMETRIKA 52(3/4), 345-353.

HOLGATE, P. 1967. THE ANGLE-COUNT METHOD. BIOMETRIKA 54(3/4), 615-623. [BITTERLICH'S METHOD; FORESTRY]

HUTCHINGS, S. S. AND R. C. HOLMGREN. 1959. INTERPRETATION OF LOOP-FREQUENCY DATA AS A MEASURE OF PLANT COVER. ECOLOGY 40(4), 668-667. [BIAS]

KEMP, C. D. AND A. W. KEMP. 1956. THE ANALYSIS OF POINT QUADRAT DATA. AUST. J. BOT. 4(2), 167-174. [MOMENTS; MAXIMUM LIKELIHOOD; BETA DISTRIBUTION; BINOMIAL DISTRIBUTION]

KOVNER, J. L. AND S. A. PATIL. 1974. PROPERTIES OF ESTIMATORS OF WILDLIFE POPULATION DENSITY FOR THE LINE TRANSECT METHOD. BIOMETRICS 30(2), 225-230. [VARIANCE; EXPONENTIAL SIGHTING PROBABILITY; EFFICIENCY; CONSISTENT ESTIMATOR; MINIMUM VARIANCE; UNBIASED ESTIMATOR; SAMPLING; KING'S METHOD; GATES' METHOD]

MATERN, B. 1972. THE PRECISION OF BASAL AREA ESTIMATES. FOREST SCI. 18(2), 123-125. [POINT SAMPLING]

MAWSON, J. C. 1968. A MONTE CARLO STUDY OF DISTANCE MEASURES IN SAMPLING FOR SPATIAL DISTRIBUTION IN FOREST STANDS. FOREST SCI. 14(2), 127-139. [MODEL; COMPUTER]

MCINTYRE, G. A. 1953. ESTIMATION OF PLANT DENSITY USING LINE TRANSECTS. J. ECOL. 41(2), 319-330.

MCLAREN, A. D. 1967. STATISTICAL ANALYSIS OF THE SPACING OF THE SETTLED APHIDS OVER THE LEAF AND OVER THE GLASS. J. ANIM. ECOL. 36(1), 163-170. [NEAREST NEIGHBOR DISTANCE]

MOORE, P. 1955. THE STRIP INTERSECT CENSUS. TRANS. N. AMER. WILDL. CONF. 20, 390-405.

MOORE, P. G. 1953. A TEST FOR NON-RANDOMNESS IN PLANT POPULATIONS. ANN. BOT. (LONDON) NEW SERIES 17(65), 57-62.

MOORE, P. G. 1954. SPACING IN PLANT POPULATIONS. ECOLOGY 35(2), 222-227. [POISSON DISTRIBUTION; BINOMIAL DISTRIBUTION; NEYMAN'S DISTRIBUTION; THOMAS'S DISTRIBUTION; DISTRIBUTION]

MORISITA, M. 1954. ESTIMATION OF POPULATION DENSITY BY SPACING METHOD. MEM. FAC. SCI. KYUSHU UNIV. SER. E 1, 187-197.

MORISITA, M. 1957. A NEW METHOD FOR THE ESTIMATION OF DENSITY BY THE SPACING METHOD APPLICABLE TO NON-RANDOMLY DISTRIBUTED POPULATIONS. PHYSIOL. ECOL.-KYOTO 7, 134-144. (ENGLISH TRANSL. (1960) DIVISION OF RANGE MANAGEMENT AND WILDLIFE HABITAT RESEARCH, U. S. FOR. SERV., M-5123).

MOUNTFORD, M. D. 1961. ON E. C. PIELOU'S INDEX OF NON-RANDOMNESS. J. ECOL. 49(2), 271-275. [VARIANCE TEST]

MYERS, C. C. AND T. W. BEERS. 1968. POINT SAMPLING FOR FOREST GROWTH ESTIMATION. J. FORESTRY 66(12), 927-929. [PLOT]

NEWNHAM, R. M. 1968. THE GENERATION OF ARTIFICIAL POPULATIONS OF POINTS (SPATIAL PATTERNS) ON A PLANE. FOREST MANAGE. INST. OTTAWA, INFORM. REPT. FMR-X-10. 28 PP. [COMPUTER; NEAREST NEIGHBOR]

ORR, T. J. 1959. TIMBER STAND MAPS, PLOTLESS CRUISING, AND BUSINESS MACHINE COMPUTATION AS ELEMENTS OF A TIMBER SURVEY METHOD. J. FORESTRY 57(8), 567-572. [SAMPLING]

PALLEY, M. N. AND L. G. HORWITZ. 1961. PROPERTIES OF SOME RANDOM AND SYSTEMATIC POINT SAMPLING ESTIMATORS. FOREST SCI. 7(1), 52-65. [VARIANCE; BITTERLICH'S G; PROBABILITY]

PALLEY, M. N. AND W. G. O'REGAN. 1965. A COMPUTER TECHNIQUE FOR THE STUDY OF FOREST SAMPLING METHODS. 1. POINT SAMPLING COMPARED WITH LINE SAMPLING. FOREST SCI. (3), 282-294. [SIMULATION]

PERSSON, O. 1964. DISTANCE METHODS: THE USE OF DISTANCE MEASUREMENTS IN THE ESTIMATION OF SEEDLING DENSITY AND OPEN SPACE FREQUENCY. STUDIA FORESTALIA SUECICA NO. 15. 68 PP. [MODEL; POISSON DISTRIBUTION; SAMPLING; SQUARE LATTICE; DISTRIBUTION FUNCTION; TRUNCATED DISTRIBUTION]

PERSSON, O. 1965. DISTANCE MEASURES II: DISTRIBUTIONS OF DISTANCES FROM A RANDOMLY LOCATED SAMPLE POINT TO PLANTS IN AN EQUILATERAL TRIANGULAR LATTICE. INSTITUTIONEN FOR SKOGLIG MATEMATISK STATISTIK, ROYAL COLLEGE OF FORESTRY, STOCKHOLM, SWEDEN, RESEARCH NOTES NO. 6. 24 PP. [DISTRIBUTION FUNCTION; COMPUTER; SAMPLING]

PERSSON, O. 1971. THE ROBUSTNESS OF ESTIMATING DENSITY BY DISTANCE MEASUREMENTS. PP. 175-187. IN, STATISTICAL ECOLOGY, VOLUME 2, SAMPLING AND MODELING BIOLOGICAL POPULATIONS AND POPULATION DYNAMICS. (G. P. PATIL, E. C. PIELOU AND W. E. WATERS, EDS.), PENNSYLVANIA STATE UNIVERSITY PRESS, UNIVERSITY PARK, PENNSYLVANIA.

PERSSON, O. 1972. THE BORDER EFFECT ON THE DISTANCE BETWEEN SAMPLE POINT AND CLOSEST INDIVIDUAL IN A SQUARE. PP. 241-246. IN, 3RD CONFERENCE ADVISORY GROUP OF FOREST STATISTICIANS. INSTITUT NATIONAL DE LA RECHERCHE AGRONOMIQUE, PARIS, FRANCE. I.N.R.A. PUBL. 72-3. [DISTANCE METHODS; SPATIAL DISTRIBUTION; POISSON DISTRIBUTION; RANDOM NUMBERS; NEAREST NEIGHBOR; STOCHASTIC PROCESS; BINOMIAL PROCESS]

PHILIP, J. R. 1966. SOME INTEGRAL EQUATIONS IN GEOMETRICAL PROBABILITY. BIOMETRIKA 53(3/4), 365-374.

PIELOU, E. C. 1959. THE USE OF POINT-TO-PLANT DISTANCES IN THE STUDY OF THE PATTERN OF PLANT POPULATIONS. J. ECOL. 47(3), 607-613. [INDEX OF NON-RANDOMNESS; CLARK-EVANS FORMULA]

PIELOU, E. C. 1961. SEGREGATION AND SYMMETRY IN TWO-SPECIES POPULATIONS AS STUDIED BY NEAREST NEIGHBOUR RELATIONS. J. ECOL. 49(2), 255-269. [PROBABILITY; COEFFICIENT OF SEGREGATION]

PIELOU, E. C. 1962. THE USE OF PLANT-TO-NEIGHBOR DISTANCES FOR THE DETECTION OF COMPETITION. J. ECOL. 50(2), 357-367.

PIELOU, E. C. 1965. THE CONCEPT OF SEGREGATION PATTERN IN ECOLOGY: SOME DISCRETE DISTRIBUTIONS APPLICABLE TO THE RUN LENGTHS OF PLANTS IN NARROW TRANSECTS. PP. 410-418. IN, CLASSICAL AND CONTAGIOUS DISCRETE DISTRIBUTIONS. (G. P. PATIL, ED.). STATISTICAL PUBLISHING SOCIETY, CALCUTTA, INDIA. [POISSON DISTRIBUTION; GEOMETRIC POISSON DISTRIBUTION; CLUMPING MODEL; GEOMETRIC DISTRIBUTION; MODEL; MARKOV CHAIN; PROBABILITY; MATRIX]

POLLARD, J. H. 1971. ON DISTANCE ESTIMATORS OF DENSITY IN RANDOMLY DISTRIBUTED FORESTS. BIOMETRICS 27(4), 991-1002.

PROUDFOOT, M. J. 1942. SAMPLING WITH TRANSVERSE TRAVERSE LINES. J. AMER. STAT. ASSOC. 37(218), 265-270. [CORRELATION COEFFICIENT]

ROBINETTE, W. L., C. M. LOVELESS AND D. A. JONES. 1974. FIELD TESTS OF STRIP CENSUS METHODS. J. WILDL. MANAGE. 38(1), 81-96. [FORMULAE FOR TEN METHODS TABULATED]

ROBINSON, P. 1955. THE ESTIMATION OF GROUND COVER BY THE POINT QUADRAT METHOD. ANN. BOT. (LONDON) NEW SERIES 19(73), 59-66. [ARCSIN TRANSFORMATION; BINOMIAL DISTRIBUTION; PROBABILITY]

ROTHERY, P. 1974. THE NUMBER OF PINS IN A POINT QUADRAT FRAME. J. APPL. ECOL. 11(2), 745-754. [EXPECTATION; CORRELATION; MARKOV CHAIN; MAXIMUM LIKELIHOOD ESTIMATION]

ROYAMA, T. 1960. THE THEORY AND PRACTICE OF LINE TRANSECTS IN ANIMAL ECOLOGY BY MEANS OF VISUAL AND AUDITORY RECOGNITION. MISC. REPT. YAMASHINA'S INST. ORNITHOL. ZOOL. 2(14), 1-17. [PROBABILITY]

SCHREUDER, H. T. 1970. POINT SAMPLING THEORY IN THE FRAMEWORK OF EQUAL-PROBABILITY CLUSTER SAMPLING. FOREST SCI. 16(2), 240-246.

SEN, A. R. AND J. TOURIGNY. 1972. ON THE LINE TRANSECT SAMPLING METHOD. (ABSTRACT). BIOMETRICS 28(1), 271-272.

SEN, A. R., J. TOURIGNY AND G. E. J. SMITH. 1974. ON THE LINE TRANSECT SAMPLING METHOD. BIOMETRICS 30(2), 329-340. [ESTIMATION OF WILDLIFE DENSITY; FITTING COMPOUND PROBABILITY DISTRIBUTIONS; GATES' METHOD]

SKELLAM, J. G. 1958. THE MATHEMATICAL FOUNDATIONS UNDERLYING THE USE OF LINE TRANSECTS IN ANIMAL ECOLOGY. BIOMETRICS 14(3), 385-400.

SPURR, S. H. 1962. A MEASURE OF POINT DENSITY. FOREST SCI. 8(1), 85-96. [SAMPLE SIZE]

STRONG, C. W. 1966. AN IMPROVED METHOD OF OBTAINING DENSITY FROM LINE-TRANSECT DATA. ECOLOGY 47(2), 311-313. [SAMPLING; STRATIFIED SAMPLING]

THOMPSON, H. R. 1955. SPATIAL POINT PROCESSES, WITH APPLICATIONS TO ECOLOGY. BIOMETRIKA 42(1/2), 102-115.

THOMPSON, H. R. 1956. DISTRIBUTION OF DISTANCE TO NTH NEIGHBOR IN A POPULATION OF RANDOMLY DISTRIBUTED INDIVIDUALS. ECOLOGY 37(2), 391-394.

VAN DYNE, G. M. 1960. A PROCEDURE FOR RAPID COLLECTION, PROCESSING, AND ANALYSIS OF LINE INTERCEPT DATA. J. RANGE MANAGE. 13(5), 247-251. [COMPUTER]

VAN WAGNER, C. E. 1968. THE LINE INTERSECT METHOD IN FOREST FUEL SAMPLING. FOREST SCI. 14(1), 20-26. [EQUATIONS]

VITHAYASAI, C. 1971. RUNS OF HEALTHY AND DISEASED TREES IN TRANSECTS THROUGH AN INFECTED FOREST: STATISTICAL AND COMPUTATIONAL METHODS. BIOMETRICS 27(4), 1003-1016.

VITHAYASAI, C. 1972. STATISTICAL ANALYSIS OF THE ORDERING OF DISEASED AND HEALTHY TREES IN A TRANSECT. FOREST SCI. 18(1), 87-92. [STOCHASTIC MODEL; PROBABILITY; VARIANCE; MATRIX]

WARREN, W. G. 1972. AN EXTENSION OF THE BELT TRANSECT METHOD FOR ASSESSMENT OF FOREST DISEASE. PP. 261-275. IN, 3RD CONFERENCE ADVISORY GROUP OF FOREST STATISTICIANS. INSTITUT NATIONAL DE LA RECHERCHE AGRONOMIQUE, PARIS, FRANCE, I.N.R.A. PUBL. 72-3. [SAMPLING; MARKOV CHAIN; RUNS; COMBINATORIAL ANALYSIS; MODEL; COMPUTER]

WARREN, W. G. AND P. F. OLSEN. 1964. A LINE TRANSECT TECHNIQUE FOR ASSESSING LOGGING WASTE. FOREST SCI. 10(3), 267-276. [EQUATIONS]

YAPP, W. B. 1955. THE THEORY OF LINE TRANSECTS. BIRD STUDY 3(2), 93-104.

YARRANTON, G. A. 1966. A PLOTLESS METHOD OF SAMPLING VEGETATION. J. ECOL. 54(1), 229-237. [CONTINGENCY TABLE; CORRELATION COEFFICIENT; INTERSPECIFIC ASSOCIATION; COMPUTER]

PLOTS AND QUADRATS

ABERDEEN, J. E. C. 1958. THE EFFECT OF QUADRAT SIZE, PLANT SIZE AND PLANT DISTRIBUTION ON FREQUENCY ESTIMATES IN PLANT ECOLOGY. AUST. J. BOT. 6(1), 47-58.

ARCHIBALD, E. E. A. 1949. THE SPECIFIC CHARACTER OF PLANT COMMUNITIES II. A QUANTITATIVE APPROACH. J. ECOL. 37(2), 274-288. [SPECIES AREA CURVE]

BARRETT, J. P. AND W. A. GUTHRIE. 1969. OPTIMUM PLOT SAMPLING IN ESTIMATING BROWSE. J. WILDL. MANAGE. 33(2), 399-403.

BORMANN, F. H. 1953. THE STATISTICAL EFFICIENCY OF SAMPLE PLOT SIZE AND SHAPE IN FOREST ECOLOGY. ECOLOGY 34(3), 474-487. [VARIANCE; RANDOM SAMPLING]

EVANS, F. C. 1952. THE INFLUENCE OF SIZE OF QUADRAT ON THE DISTRIBUTIONAL PATTERNS OF PLANT POPULATIONS. CONTRIB. LAB. VERTEBR. BIOL., UNIVERSITY OF MICHIGAN, ANN ARBOR, MICHIGAN, NO. 54, 15 PP. [POISSON DISTRIBUTION]

GILBERT, N. AND T. C. E. WELLS. 1966. ANALYSIS OF QUADRAT DATA. J. ECOL. 54(3), 675-685.

GRANT, J. A. C. 1951. THE RELATIONSHIP BETWEEN STOCKING AND SIZE OF QUADRAT. UNIV. TORONTO FAC. FORESTRY REPT. BULL. NO. 5, V, 35 PP.

GREIG-SMITH, P. 1952. THE USE OF RANDOM AND CONTIGUOUS QUADRATS IN THE STUDY OF THE STRUCTURE OF PLANT COMMUNITIES. ANN. BOT. (LONDON) NEW SERIES 16(62), 293-316. [POISSON DISTRIBUTION; ARTIFICIAL POPULATION; THOMAS'S DISTRIBUTION; NEYMAN'S DISTRIBUTION; CHI-SQUARE; MEAN SQUARE VERSUS BLOCK SIZE]

HYDER, D. N., C. E. CONRAD, P. T. TUELLER, L. D. CALVIN, C. E. POULTON AND F. A. SNEVA. 1963. FREQUENCY SAMPLING IN SAGEBRUSH-BUNCHGRASS VEGETATION. ECOLOGY 44(4), 740-746. [COMPONENTS OF VARIANCE; QUADRAT; TRANSECTS; COST FUNCTION]

IWAO, S. 1972. APPLICATION OF THE M-M METHOD TO THE ANALYSIS OF SPATIAL PATTERNS BY CHANGING THE QUADRAT SIZE. RES. POPUL. ECOL.-KYOTO 14(1), 97-128.

MATERN, B. 1947. METHODS OF ESTIMATING THE ACCURACY OF LINE AND SAMPLE PLOT SURVEYS. MEDDEL. FRAN STATENS SKOGSFORSKNINGINSTITUT 36(1), 1-138.

MATERN, B. 1972. PERFORMANCE OF VARIOUS DESIGNS OF FIELD EXPERIMENTS WHEN APPLIED IN RANDOM FIELDS. PP. 119-128, IN, 3RD CONFERENCE ADVISORY GROUP OF FOREST STATISTICIANS. INSTITUT NATIONAL DE LA RECHERCHE AGRONOMIQUE, PARIS, FRANCE. I.N.R.A. PUBL. 72-3. [PLOT SIZE; DISTANCE FUNCTION; STOCHASTIC PROCESS; EFFICIENCY OF DESIGNS; MODEL]

MCGINNIES, W. G. 1934. THE RELATION BETWEEN FREQUENCY INDEX AND ABUNDANCE AS APPLIED TO PLANT POPULATIONS IN A SEMIARID REGION. ECOLOGY 15(3), 263-282. [SAMPLING; ALIENATION INDEX; CORRELATION INDEX]

MEAD, R. 1974. A TEST FOR SPATIAL PATTERN AT SEVERAL SCALES USING DATA FROM A GRID OF CONTIGUOUS QUADRATS. BIOMETRICS 30(2), 295-307. [RANDOMIZATION TESTS; SMALL SAMPLE DISTRIBUTIONS; PLANT COMMUNITIES]

NEAL, R. L., JR. 1973. REMEASURING TREE HEIGHTS ON PERMANENT PLOTS USING RECTANGULAR COORDINATES AND ONE ANGLE PER TREE. FOREST SCI. 19(3), 233-236.

PIELOU, E. C. 1957. THE EFFECT OF QUADRAT SIZE ON THE ESTIMATION OF THE PARAMETERS OF NEYMAN'S AND THOMAS'S DISTRIBUTIONS. J. ECOL. 45(1), 31-47.

PRENTICE, R. L. 1973. A DESIGN FOR STUDYING THE CLUSTERING OF PLANT OR ANIMAL SPECIES USING QUADRAT SIZES IN GEOMETRIC PROGRESSION. J. THEOR. BIOL. 39(3), 601-608. [POISSON DISTRIBUTION; NEGATIVE BINOMIAL DISTRIBUTION; POISSON-POISSON DISTRIBUTION; POISSON WITH ZEROS]

SINGH, B. N. AND K. DAS. 1939. PERCENTAGE FREQUENCY AND QUADRAT SIZE IN ANALYTICAL STUDIES OF WEED FLORA. J. ECOL. 27(1), 66-77. [PROBABILITY]

SKELLAM, J. G. 1955. QUADRAT SAMPLING FROM THE MATHEMATICAL STANDPOINT. PROC. LINN. SOC. LOND. SESSION 165, (PT. 2), 1952-53, PP. 95-102. [DENSITY; POISSON DISTRIBUTION; NEYMAN'S TYPE A DISTRIBUTION; NEGATIVE BINOMIAL DISTRIBUTION]

SMITH, R. H. 1968. A COMPARISON OF SEVERAL SIZES OF CIRCULAR PLOTS FOR ESTIMATING DEER PELLET-GROUP DENSITY. J. WILDL. MANAGE. 32(3), 585-591. [SAMPLING; RANDOM SAMPLING; STRATIFIED SAMPLING; TRANSECT]

STEVENS, W. L. 1935. THE RELATION BETWEEN PLANT DENSITY AND NUMBER OF EMPTY QUADRATS. (APPENDIX TO E. ASHBY, THE QUANTITATIVE ANALYSIS OF VEGETATION). ANN. BOT. (LONDON) 49(196), 798-801. [RANDOM; NON-RANDOM DISTRIBUTION; PROBABILITY]

THOMPSON, H. R. 1958. THE STATISTICAL STUDY OF PLANT DISTRIBUTION PATTERNS USING A GRID OF QUADRATS. AUST. J. BOT. 6(4), 322-342. [MODEL; POISSON DISTRIBUTION; NEGATIVE BINOMIAL DISTRIBUTION; NEYMAN'S TYPE A DISTRIBUTION]

WIEGERT, R. G. 1962. THE SELECTION OF AN OPTIMUM QUADRAT SIZE FOR SAMPLING THE STANDING CROP OF GRASSES AND FORBS. ECOLOGY 43(1), 125-129. [MODEL; VARIANCE; RANDOM SAMPLING; COST FUNCTION; CORRELATION]

POPULATION DYNAMICS

ADLER, J. AND M. M. DAHL. 1967. A METHOD FOR MEASURING THE MOTILITY OF BACTERIA AND FOR COMPARING RANDOM AND NON-RANDOM MOTILITY. J. GEN. MICROBIOL. 46(2), 161-173.

ALCALA, A. C. AND W. C. BROWN. 1967. POPULATION ECOLOGY OF THE TROPICAL SCINCOID LIZARD, EMOIA ATROCOSTATA, IN THE PHILIPPINES. COPEIA 1967(3), 596-604. [LESLIE-CHITTY SURVIVAL RATE EQUATION; VARIANCE; TAG-RECAPTURE]

ANDERSON, F. S. 1971. SIMPLE ELEMENTARY MODELS IN POPULATION DYNAMICS. PP. 358-363. IN, DYNAMICS OF POPULATIONS. (P. J. DEN BOER AND G. R. GRADWELL, EDS.). CENTRE FOR AGRICULTURAL PUBLISHING AND DOCUMENTATION, WAGENINGEN,NETHERLANDS.

ANDERSON, R. M. 1974. AN ANALYSIS OF THE INFLUENCE OF HOST MORPHOMETRIC FEATURES ON THE POPULATION DYNAMICS OF DIPLOZOON PARADOXUM (NORDMANN, 1832). J. ANIM. ECOL. 43(3), 873-887. [PRINCIPAL COMPONENTS IN REGRESSION ANALYSIS MODEL]

ANDERSON, R. M. 1974. POPULATION DYNAMICS OF THE CESTODE CARYOPHYLLAEUS LATICEPS (PALLAS, 1781) IN THE BREAM (ABRAMIS BRAMA L.). J. ANIM. ECOL. 43(2), 305-321. [DETERMINISTIC MODEL; CYCLIC FUNCTION]

AUER, C. 1971. SOME ANALYSES OF THE QUANTITATIVE STRUCTURE IN POPULATIONS AND DYNAMICS OF LARCH BUD MOTH 1949-1968. PP. 151-170. IN, STATISTICAL ECOLOGY, VOLUME 2. SAMPLING AND MODELING BIOLOGICAL POPULATIONS AND POPULATION DYNAMICS. (G. P. PATIL, E. C. PIELOU AND W. E. WATERS, EDS.). PENNSYLVANIA STATE UNIVERSITY PRESS, UNIVERSITY PARK, PENNSYLVANIA.

AYERS, J. C. 1956. POPULATION DYNAMICS OF THE MARINE CLAM, MYA ARENARIA. LIMNOL. OCEANOGR. 1(1), 26-34. [LOG TRANSFORMATION; POPULATION MATHEMATICS]

BARKALOW, F., JR., R. B. HAMILTON AND R. F. SOOTS, JR. 1970. THE VITAL STATISTICS OF AN UNEXPLOITED GRAY SQUIRREL POPULATION. J. WILDL. MANAGE. 34(3), 489-500. [LIFE TABLE FORMULAE]

BARTLETT, M. S. 1973. A NOTE ON DAS GUPTA'S TWO-SEX POPULATION MODEL. THEOR. POPUL. BIOL. 3(4), 418-424. [MODEL]

BAWEJA, K. D. 1937. THE CALCULATION OF SOIL POPULATION FIGURES. J. ANIM. ECOL. 6(2), 366-367.

BENSON, W. W. AND T. C. EMMEL. 1973. DEMOGRAPHY OF GREGARIOUSLY ROOSTING POPULATIONS OF THE NYMPHALINE BUTTERFLY MARPESIA BERANIA IN COSTA RICA. ECOLOGY 54(2), 326-335. [RICKER'S RECRUITMENT MODEL; MORTALITY RATE]

BERRYMAN, A. A. 1973. POPULATION DYNAMICS OF THE FIR ENGRAVER, SCOLYTUS VENTRALIS (COLEOPTERA: SCOLYTIDAE) I. ANALYSIS OF POPULATION BEHAVIOR AND SURVIVAL FROM 1964 TO 1971. CAN. ENTOMOL. 105(11), 1465-1488. [POPULATION INDEX]

BIRCH, L. C. 1948. THE INTRINSIC RATE OF NATURAL INCREASE OF AN INSECT POPULATION. J. ANIM. ECOL. 17(1), 15-26.

BIRCH, L. C. 1953. EXPERIMENTAL BACKGROUND TO THE STUDY OF THE DISTRIBUTION AND ABUNDANCE OF INSECTS. I. THE INFLUENCE OF TEMPERATURE, MOISTURE AND FOOD ON THE INNATE CAPACITY FOR INCREASE OF THREE GRAIN BEETLES. ECOLOGY 34(4), 698-711. [INTRINSIC RATE OF INCREASE; LIFE TABLE; STABLE AGE DISTRIBUTION]

BIRCH, L. C. 1953. EXPERIMENTAL BACKGROUND TO THE STUDY OF THE DISTRIBUTION AND ABUNDANCE OF INSECTS. II. THE RELATION BETWEEN INNATE CAPACITY FOR INCREASE IN NUMBERS AND THE ABUNDANCE OF THREE GRAIN BEETLES IN EXPERIMENTAL POPULATIONS. ECOLOGY 34(4), 712-726. [LOGISTIC]

BIRCH, L. C. 1953. EXPERIMENTAL BACKGROUND TO THE STUDY OF THE DISTRIBUTION AND ABUNDANCE OF INSECTS. III. THE RELATION BETWEEN INNATE CAPACITY OF DIFFERENT SPECIES OF BEETLES LIVING TOGETHER ON THE SAME FOOD. EVOLUTION 7(2), 136-144.

BOBEK, B. 1969. SURVIVAL, TURNOVER AND PRODUCTION OF SMALL RODENTS IN A BEECH FOREST. ACTA THERIOL. 14(15), 191-210. [MODEL; LIFE TABLE]

CALHOUN, J. B. 1957. SOCIAL WELFARE AS A VARIABLE IN POPULATION DYNAMICS. COLD SPRING HARBOR SYMPOSIA ON QUANTITATIVE BIOLOGY 22, 339-356. [LOG TRANSFORMATION; PROBABILITY]

CHAPMAN, D. G. 1961. POPULATION DYNAMICS OF THE ALASKA FUR SEAL HERD. TRANS. N. AMER. WILDL. NAT. RESOUR. CONF. 26, 356-369.

CHAPMAN, D. G. 1961. STATISTICAL PROBLEMS IN DYNAMICS OF EXPLOITED FISHERIES POPULATIONS. PROC. FOURTH BERKELEY SYMP. MATH. STAT. AND PROB. 4, 153-168. [MORTALITY RATE; GROWTH RATE; SAMPLING]

CHAPMAN, D. G. 1974. ESTIMATION OF POPULATION PARAMETERS OF ANTARCTIC BALEEN WHALES. PP. 336-351. IN, THE WHALE PROBLEM: A STATUS REPORT. (W. E. SCHEVILL, ED.), HARVARD UNIVERSITY PRESS, CAMBRIDGE, MASSACHUSETTS. [MODEL; AGE AND CATCH/EFFORT DATA ANALYSIS; LEAST SQUARES; LESLIE-DELURY ETHODS; TAG-RECAPTURE METHODS]

HAPMAN, R. N. 1928. THE UANTITATIVE ANALYSIS OF NVIRONMENTAL FACTORS. ECOLOGY (2), 111-122. [TRIBOLIUM STUDY]

LARK, C. W. 1974. POSSIBLE FFECTS OF SCHOOLING ON THE YNAMICS OF EXPLOITED FISH OPULATIONS. J. CONS. PERM. INT. XPLOR. MER 36(1), 7-14. [MODEL; TOCK-RECRUITMENT]

OLE, L. C. 1954. THE OPULATION CONSEQUENCES OF LIFE ISTORY PHENOMENA. QUART. REV. IOL. 29(2), 103-137.

COUTINHO, F. A. B. AND A. B. COUTINHO. 1974. DYNAMICS OF POPULATIONS OF BIOMPHALARIA GLABRATA AND THE VON FOERESTER EQUATION. BULL. MATH. BIOL. 36(1), 29-37. [SNAIL; MODEL]

CRAIGHEAD, J. J., J. R. VARNEY AND F. C. CRAIGHEAD, JR. 1974. A POPULATION ANALYSIS OF THE YELLOWSTONE GRIZZLY BEARS. MONTANA FOR. CONSERV. EXPT. STN. BULL. 40. 20 PP. [MODEL; COMPUTER; SIMULATION]

CUSHING D. H. 1959. THE SEASONAL VARIATION IN OCEANIC PRODUCTION AS A PROBLEM IN POPULATION DYNAMICS. J. CONS. PERM. INT. EXPLOR. MER 24(3), 455-464. [PLANKTON PRODUCTION MODEL]

CUSHING, D. H. 1971. THE DEPENDENCE OF RECRUITMENT ON PARENT STOCK IN DIFFERENT GROUPS OF FISHES. J. CONS. PERM. INT. EXPLOR. MER 33(3), 340-362. [BEVERTON-HOLT EQUATION; RICKER'S CURVE; INDEX OF DENSITY DEPENDENCE; LOG TRANSFORMATION]

CUSHING, D. H. 1973. DEPENDENCE OF RECRUITMENT ON PARENT STOCK. J. FISH. RES. BOARD CAN. 30(12, PART 2), 1965-1976. [MODEL; RICKER'S MODEL; BEVERTON-HOLT MODEL; CUSHING'S MODEL]

CUSHING, D. H. 1974. THE NATURAL REGULATION OF FISH POPULATIONS. PP. 399-412. IN, SEA FISHERIES RESEARCH. (F. R. HARDEN JONES, ED.). JOHN WILEY AND SONS, NEW YORK, NEW YORK. [MODEL; STOCK-RECRUITMENT]

DAVIS, A. W. 1964. ON THE PROBABILITY GENERATING FUNCTION FOR THE CUMULATIVE POPULATION IN A SIMPLE BIRTH-AND-DEATH PROCESS. BIOMETRIKA 51(1/2), 245-249.

DEEVEY, E. S., JR. 1947. LIFE TABLES FOR NATURAL POPULATIONS OF ANIMALS. QUART. REV. BIOL. 22(4), 283-314.

DEMETRIUS, L. 1971. MULTIPLICATIVE PROCESSES. MATH. BIOSCI. 12, 261-272. [MODEL; POPULATION DYNAMICS; MATRIX; QUADRAT]

DEMETRIUS, L. 1972. ON AN INFINITE POPULATION MATRIX. MATH. BIOSCI. 13, 133-137. [LESLIE MATRIX]

DEMETRIUS, L. 1974. DEMOGRAPHIC PARAMETERS AND NATURAL SELECTION. PROC. MATH. ACAD. SCI. U.S.A. 71(12), 4645-4647. [MALTHUSIAN PARAMETER; ENTROPY; FISHER'S FUNDAMENTAL THEOREM; MODEL]

DEMETRIUS, L. 1974. MULTIPLICATIVE PROCESSES-II. MATH. BIOSCI. 20, 345-357. [POPULATION MODEL]

DEMPSTER, J. P. 1957. THE POPULATION DYNAMICS OF THE MOROCCAN LOCUST (DOCIOSTAURUS MAROCCANUS THUNBERG) IN CYPRUS. ANTI-LOCUST BULLETIN 27, ANTI-LOCUST RESEARCH CENTRE, LONDON, ENGLAND. 60 PP. [DISPERSION]

DOI, T. 1962. ON THE PROBLEMS OF POPULATION DYNAMICS RELEVANT TO IMMATURE SALMONS CAUGHT OFFSHORE IN THE NORTH PACIFIC. BULL. TOKAI REG. FISH. RES. LAB. 35, 1-8. [MODEL]

EBERHARDT, L. L. 1969. POPULATION ANALYSIS. PP. 457-495. IN, WILDLIFE MANAGEMENT TECHNIQUES. (3RD REVISED EDITION.). (R. H. GILES, JR., ED.). THE WILDLIFE SOCIETY, WASHINGTON, D. C.

EBERHARDT, L., T. J. PETERLE AND R. SCHOFIELD. 1963. PROBLEMS IN A RABBIT POPULATION STUDY. WILDL. MONOGR. NO. 10. 51 PP.

EMBREE, D. G. 1965. THE POPULATION DYNAMICS OF THE WINTER MOTH IN NOVA SCOTIA, 1954-1962. MEM. ENTOMOL. SOC. CAN. 46. 57 PP. [KEY FACTOR ANALYSIS; SURVIVAL RATE MODEL OF MORRIS; LIFE TABLE; PRECISION; NEGATIVE BINOMIAL DISTRIBUTION]

EVANS, F. C. AND F. E. SMITH. 1952. THE INTRINSIC RATE OF NATURAL INCREASE FOR THE HUMAN LOUSE, PEDICULUS HUMANUS L. AMER. NAT. 86(830), 299-310.

FARNER, D. S. 1955. BIRDBANDING IN THE STUDY OF POPULATION DYNAMICS. PP. 397-449. IN, RECENT ADVANCES IN AVIAN BIOLOGY. (A. WOLFSON, ED.). UNIVERSITY OF ILLINOIS PRESS, URBANA, ILLINOIS.

FENCHEL, T. 1974. INTRINSIC RATE OF NATURAL INCREASE: THE RELATIONSHIP WITH BODY SIZE. OECOLOGIA 14, 317-326. [EQUATIONS]

FINN, R. K. AND R. E. WILSON. 1954. POPULATION DYNAMICS OF A CONTINUOUS PROPAGATOR FOR MICROORGANISMS. J. AGRIC. FOOD CHEM. 2(2), 66-69. [LOGARITHMIC GROWTH MODEL]

FISHER, J. AND L. S. V. VENABLES. 1938. GANNETS (SULA BASSANA) ON NOSS, SHETLAND, WITH AN ANALYSIS OF THE RATE OF INCREASE OF THIS SPECIES. J. ANIM. ECOL. 7(2), 305-313. [GEOMETRIC PROGRESSION]

FOWLER, C. W. AND T. SMITH. 1973. CHARACTERIZING STABLE POPULATIONS: AN APPLICATION TO THE AFRICAN ELEPHANT POPULATION. J. WILDL. MANAGE. 37(4), 513-523. [LESLIE MATRIX; COMPUTER; EIGENVALUE; MODEL]

FRAZER, B. D. 1972. LIFE TABLES AND INTRINSIC RATES OF INCREASE OF APTEROUS BLACK BEAN APHIDS AND PEA APHIDS, ON BROAD BEAN (HOMOPTERA: APHIDIDAE). CAN. ENTOMOL. 104(11), 1717-1722. [COMPUTER]

GARROD, D. J. 1967. POPULATION DYNAMICS OF THE ARCTO-NORWEGIAN COD. J. FISH. RES. BOARD CAN. 24(1), 145-190. [CATCH EQUATION; MORTALITY RATE; DELURY'S METHOD; REGRESSION; RECRUITMENT; GULLAND'S VIRTUAL POPULATION ANALYSIS]

GAUSE, G. F. 1931. THE INFLUENCE OF ECOLOGICAL FACTORS ON THE SIZE OF POPULATION. AMER. NAT. 65(696), 70-76. [LOGISTIC; NORMAL DISTRIBUTION]

GEIER, P. W. AND T. J. HILLMAN. 1971. AN ANALYSIS OF THE LIFE SYSTEM OF THE CODLING MOTH IN APPLE ORCHARDS OF SOUTH-EASTERN AUSTRALIA. PROC. ECOL. SOC. AUSTRALIA 6, 203-243. [COMPUTER]

GOODMAN, L. A. 1971. ON THE SENSITIVITY OF THE INTRINSIC GROWTH RATE TO CHANGES IN THE AGE-SPECIFIC, BIRTH AND DEATH RATES. THEOR. POPUL. BIOL. 2(3), 339-354.

GREENBANK, D. O. 1970. CLIMATE AND THE ECOLOGY OF THE BALSAM WOOLLY APHID. CAN. ENTOMOL. 102(5), 546-578. [POPULATION MODEL]

GULLAND, J. A. 1968. REPORT ON THE POPULATION DYNAMICS OF THE PERUVIAN ANCHOVETA. FAO FISHERIES TECHNICAL PAPER NO. 72, V, 29 PP. [MODEL; SCHAEFER'S MODEL; RECRUITMENT MODEL]

GURTIN, M. E. 1973. A SYSTEM OF EQUATIONS FOR AGE-DEPENDENT POPULATION DIFFUSION. J. THEOR. BIOL. 40(2), 389-392.

HAJNAL, J. 1957. MATHEMATICAL MODELS IN DEMOGRAPHY. COLD SPRING HARBOR SYMPOSIA ON QUANTITATIVE BIOLOGY 22, 97-102.

HALDANE, J. B. S. 1953. SOME ANIMAL LIFE TABLES. J. INST. ACTUARIES 79(1), 83-89. [MAXIMUM LIKELIHOOD]

HALL, D. J. 1964. AN EXPERIMENTAL APPROACH TO THE DYNAMICS OF A NATURAL POPULATION OF DAPHNIA GALEATA MENDOTAE. ECOLOGY 45(1), 94-112. [LIFE TABLE; POPULATION GROWTH]

HAMILTON, W. D. 1966. THE MOULDING OF SENESCENCE BY NATURAL SELECTION. J. THEOR. BIOL. 12(1), 12-45. [MALTHUSIAN PARAMETER; LIFE TABLE; MODEL]

HARCOURT, D. G. 1969. THE DEVELOPMENT AND USE OF LIFE TABLES IN THE STUDY OF NATURAL INSECT POPULATIONS. ANNU. REV. ENTOMOL. 14, 175-196.

HELGESEN, R. G. AND D. L. HAYNES. 1972. POPULATION DYNAMICS OF THE CEREAL LEAF BEETLE, OULEMA MELANOPUS (COLEOPTERA: CHRYSOMELIDAE): A MODEL FOR AGE SPECIFIC MORTALITY. CAN. ENTOMOL. 104(6), 797-814.

HENNY, C. J. 1972. AN ANALYSIS OF THE POPULATION DYNAMICS OF SELECTED AVIAN SPECIES WITH SPECIAL REFERENCE TO CHANGES DURING THE MODERN PESTICIDE ERA. U. S. FISH WILDL. SERV. BUR. SPORT. FISH. WILDL. RES. REPT. 1. 99 PP. [LIFE TABLE; LIFE EQUATION]

HERON, A. C. 1972. POPULATION ECOLOGY OF A COLONIZING SPECIES: THE PELAGIC TUNICATE THALIA DEMOCRATICA I. INDIVIDUAL GROWTH RATE AND GENERATION TIME. OECOLOGIA 10, 269-293. [EQUATIONS]

HOEM, J. M. 1970. PROBABILISTIC FERTILITY MODELS OF THE LIFE TABLE-TYPE. THEOR. POPUL. BIOL. 1(1), 12-38.

HOLGATE, P. 1967. POPULATION SURVIVAL AND LIFE HISTORY PHENOMENA. J. THEOR. BIOL. 14(1), 1-10. [MODEL; GEOMETRIC DISTRIBUTION; PROBABILITY; POISSON DISTRIBUTION; MATRIX]

HOLGATE, P. 1967. THE SIZE OF ELEPHANT HERDS. MATH. GAZ. 51(378), 302-304. [PROBABILITY; MARKOV BIRTH AND DEATH PHENOMENA]

HOLT, S. J., J. A. GULLAND, C. TAYLOR AND S. KURITA. 1959. A STANDARD TERMINOLOGY AND NOTATION FOR FISHERY DYNAMICS. J. CONS. PERM. INT. EXPLOR. MER 24, 239-242.

HYRENIUS, H. AND H. DIAMAND. 1974. DEMOMETRIC TECHNIQUES APPLIED TO ANIMAL POPULATIONS. DEMOGRAPHIC RESEARCH INSTITUTE, UNIVERSITY OF GOTHENBURG, GOTHENBURG, SWEDEN, REPORT 14, 52 PP. [LIFE TABLE; MODEL]

IVLEV, V. S. 1959. MATHEMATICAL ANALYSIS OF POPULATION DYNAMICS OF FISHES. LENINGRAD, VESTNIK LENINGRADSKOGO UNIVERSITETA 14(9), (SERIYA BIOLOGII, NO.2), 119-127.

IVLEV, V. S. 1961. THE PRINCIPLES OF MATHEMATICAL REPRESENTATION OF THE DYNAMICS OF COMMERCIAL POPULATIONS OF FISH. AKADEMIYA NAUK SSSR, TRUDY SOVESHCHANII IKHTIOLOGICHESKOI KOMISSII, NO. 13 (TRUDY SOVESHCHANIYA PO DINAMIKE CHISLENNOSTI RYB, 1960), 185-193.

JOSEPH, J. AND T. P. CALKINS. 1969. POPULATION DYNAMICS OF THE SKIPJACK TUNA (KATSUWONUS PELAMIS) OF THE EASTERN PACIFIC OCEAN. BULL. INTER-AMER. TROP. TUNA COMM. 13(1), 7-139. [TAG-RECAPTURE; LINEAR MODEL; MAXIMUM LIKELIHOOD; VON BERTALANFFY'S EQUATION; CONFIDENCE REGIONS; MORTALITY RATE]

KELKER, G. H. 1945. MEASUREMENT AND INTERPRETATION OF FORCES, THAT DETERMINE POPULATIONS OF MANAGED DEER. PH.D. THESIS, UNIVERSITY OF MICHIGAN, ANN ARBOR, MICHIGAN. 422 PP. [ESTIMATION]

KELKER, G. H. 1947. COMPUTING THE RATE OF INCREASE FOR DEER. J. WILDL. MANAGE. 11(2), 177-183.

KENDALL, D. G. 1950. RANDOM FLUCTUATIONS IN THE AGE-DISTRIBUTION OF A POPULATION WHOSE DEVELOPMENT IS CONTROLLED BY THE SIMPLE "BIRTH-AND-DEATH" PROCESS. J. ROYAL STAT. SOC. SER. B 12(2), 278-285.

KEYFITZ, N. AND E. M. MURPHY. 1967. MATRIX AND MULTIPLE DECREMENT IN POPULATION ANALYSIS. BIOMETRICS 23(3), 485-503. [LESLIE MATRIX; EXTINCTION]

KLOMP, H. 1966. THE DYNAMICS OF A FIELD POPULATION OF THE PINE LOOPER, BUPALUS PINIARIUS L. (LEP., GEOM.). ADV. ECOL. RES. 3, 207-305. [TREE SHOOT MODEL; POISSON DISTRIBUTION; NEGATIVE BINOMIAL DISTRIBUTION]

KUROGANE, K., U. SRIRUANGCHEEP, C. TANTISAWETRAT, S. CHULASORN, S. SUPONGPAN AND U. BOONPRAKOB. 1971. ON THE POPULATION DYNAMICS OF THE INDO-PACIFIC MACKEREL (RASTRELLIGER NEGLECTUS VAN KAMPEN) OF THE GULF OF THAILAND. BULL. TOKAI REG. FISH. RES. LAB. 67, 1-33. [MODEL; MORTALITY COEFFICIENT; BEVERTON-HOLT MODEL]

LACK, D. 1951. POPULATION ECOLOGY IN BIRDS. PROC. INTERN. ORNITHOL. CONGR. 10, 409-448. [MORTALITY RATE]

LAUGHLIN, R. 1965. CAPACITY FOR INCREASE OF THE GARDEN CHAFER, PHYLLOPERTHA HORTICOLA (L.) (COLEOPTERA). PROC. INTERN. CONGR. ENTOMOL. 12, 394. [EQUATIONS]

LAUGHLIN, R. 1965. CAPACITY FOR INCREASE: A USEFUL POPULATION STATISTIC. J. ANIM. ECOL. 34(1), 77-91.

LE CREN, E. D. 1973. SOME EXAMPLES OF THE MECHANISMS THAT CONTROL THE POPULATION DYNAMICS OF SALMONID FISH. PP. 125-135. IN, THE MATHEMATICAL THEORY OF THE DYNAMICS OF BIOLOGICAL POPULATIONS. (M. S. BARTLETT AND R. W. HIORNS, EDS.). ACADEMIC PRESS, NEW YORK, NEW YORK. [MODEL; GROWTH RATE]

LEARY, R. A. 1970. SYSTEM IDENTIFICATION PRINCIPLES IN STUDIES OF FOREST DYNAMICS. U. S FOR. SERV. NORTH CENTRAL FOREST EXPT. STN. RES. PAPER NC-45. 38 PP

LEBOWITZ, J. L. AND S. I. RUBINOW. 1974. A THEORY FOR THE AGE AND GENERATION TIME DISTRIBUTION OF A MICROBIAL POPULATION. J. MATH. BIOL. 1(1), 17-36. [MODEL]

LEFKOVITCH, L. P. 1966. THE EFFECTS OF ADULT EMIGRATION ON POPULATIONS OF LASIODERMA SERRICORNE (F.). OIKOS 15(2), 200-210. [MATRIX]

LEON, J. A. 1975. LIMIT CYCLES IN POPULATIONS WITH SEPARATE GENERATIONS. J. THEOR. BIOL. 49(1), 241-244. [DIFFERENCE EQUATION; STABILITY ANALYSIS]

LESLIE, P. H. 1945. ON THE USE OF MATRICES IN CERTAIN POPULATION MATHEMATICS. BIOMETRIKA 33(3), 183-212.

LESLIE, P. H. 1948. SOME FURTHER NOTES ON THE USE OF MATRICES IN POPULATION MATHEMATICS. BIOMETRIKA 35(3/4), 213-245.

LESLIE, P. H. 1966. THE INTRINSIC RATE OF INCREASE AND THE OVERLAP OF SUCCESSIVE GENERATIONS IN A POPULATION OF GUILLEMOTS (URIA AALGE PONT.). J. ANIM. ECOL. 35(2), 291-301.

LESLIE, P. H. AND T. PARK. 1949. THE INTRINSIC RATE OF NATURAL INCREASE OF TRIBOLIUM CASTANEUM HERBST. ECOLOGY 30(4), 469-477. [LIFE TABLE]

LESLIE, P. H., J. S. TENER, M. VIZOSO AND H. CHITTY. 1955. THE LONGEVITY AND FERTILITY OF THE ORKNEY VOLE, MICROTUS ORCADENSIS AS OBSERVED IN THE LABORATORY. PROC. ZOOL. SOC. (LONDON) 125(1), 115-125. [LIFE TABLE]

LESLIE, P. H., U. M. VENABLES AND L. S. V. VENABLES. 1952. THE FERTILITY AND POPULATION STRUCTURE OF THE BROWN RAT (RATTUS NORVEGICUS) IN CORN-RICKS AND SOME OTHER HABITATS. PROC. ZOOL. SOC. (LONDON) 122(1), 187-238. [LIFE TABLE; PEARSON TYPE I DISTRIBUTION; LOGISTIC]

LOTKA, A. J. 1939. ON AN INTEGRAL EQUATION IN POPULATION ANALYSIS. ANN. MATH. STAT. 10(2), 144-161.

LOTKA, A. J. 1940. THE PLACE OF THE INTRINSIC RATE OF NATURAL INCREASE IN POPULATION ANALYSIS. PROC. AMER. SCI. CONGR. 8TH. PP. 297-313.

LOTKA, A. J. 1945. POPULATION ANALYSIS AS A CHAPTER IN THE MATHEMATICAL THEORY OF EVOLUTION. PP. 355-385. IN, ESSAYS ON GROWTH AND FORM PRESENTED TO D'ARCY WENTWORTH THOMPSON. (CLARK, W. E. LE. AND P. B. MEDAWAR, EDS.). CLARENDAN PRESS, OXFORD, ENGLAND.

LUCK, R. F. 1971. AN APPRAISAL OF TWO METHODS OF ANALYZING INSECT LIFE TABLES. CAN. ENTOMOL. 103(9), 1261-1271.

MACARTHUR, R. 1962. SOME GENERALIZED THEOREMS OF NATURAL SELECTION. PROC. NATL. ACAD. SCI. U.S.A. 48(11), 1893-1897. [FISHER'S FUNDAMENTAL THEOREM; HALDANE]

MACARTHUR, R. 1968. SELECTION FOR LIFE TABLES IN PERIODIC ENVIRONMENTS. AMER. NAT. 102(926), 381-383. [INTRINSIC RATE OF INCREASE; MATRIX; EIGENVALUE]

MACLULICH, D. A. 1957. THE PLACE OF CHANCE IN POPULATION PROCESSES. J. WILDL. MANAGE. 21(3), 293-299. [RANDOM NUMBERS]

MEATS, A. 1971. THE RELATIVE IMPORTANCE TO POPULATION INCREASE OF FLUCTUATIONS IN MORTALITY, FECUNDITY AND THE TIME VARIABLES OF THE REPRODUCTIVE SCHEDULE. OECOLOGIA 6(3), 223-237. [MODEL; LEWONTIN'S MODEL; LOTKA'S EQUATION]

MERTZ, D. B. 1969. AGE-DISTRIBUTION AND ABUNDANCE IN POPULATIONS OF FLOUR BEETLES. I. EXPERIMENTAL STUDIES. ECOL. MONOGR. 39(1), 1-31. [CANNIBALISM MODEL; POPULATION MODEL; INTRINSIC RATE OF INCREASE; LESLIE-BIRTH METHOD; LIFE TABLE]

MERTZ, D. B. 1971. LIFE HISTORY PHENOMENA IN INCREASING AND DECREASING POPULATIONS. PP. 361-392. IN, STATISTICAL ECOLOGY, VOLUME 2, SAMPLING AND MODELING BIOLOGICAL POPULATIONS AND POPULATION DYNAMICS. (G. P. PATIL, E. C. PIELOU AND W. E. WATERS, EDS.). PENNSYLVANIA STATE UNIVERSITY PRESS, UNIVERSITY PARK, PENNSYLVANNIA.

MERTZ, D. B. 1971. THE MATHEMATICAL DEMOGRAPHY OF THE CALIFORNIA CONDOR POPULATION. AMER. NAT. 105(945), 437-453. [LIFE TABLE; EQUATION OF POPULATION GROWTH; DEMOGRAPHIC FUNCTIONS]

METZGAR, L. H. 1972. THE MEASUREMENT OF HOME RANGE SHAPE. J. WILDL. MANAGE. 36(2), 643-645.

MODE, C. J. 1966. A MULTIDIMENSIONAL BIRTH PROCESS AND ITS APPLICATION TO SOME PROBLEMS IN THE DYNAMICS OF BIOLOGICAL POPULATIONS. BULL. MATH. BIOPHYS. 28, 333-345.

MODE, C. J. 1966. RESTRICTED TRANSITION PROBABILITIES AND THEIR APPLICATIONS TO SOME PROBLEMS IN THE DYNAMICS OF BIOLOGICAL POPULATIONS. BULL. MATH. BIOPHYS. 28, 315-331.

MODE, C. J. 1974. DISCRETE TIME AGE-DEPENDENT BRANCHING PROCESSES IN RELATION TO STABLE POPULATION THEORY IN DEMOGRAPHY. MATH. BIOSCI. 19, 73-100. [PROBABILITY]

MORAN, P. A. P. 1950. SOME REMARKS ON ANIMAL POPULATION DYNAMICS. BIOMETRICS 6(3), 250-258. [MODEL; OSCILLATION]

MORRIS, R. F. 1957. THE INTERPRETATION OF MORTALITY DATA IN STUDIES ON POPULATION DYNAMICS. CAN. ENTOMOL. 89(2), 49-69.

MORRIS, R. F. 1959. SINGLE-FACTOR ANALYSIS IN POPULATION DYNAMICS. ECOLOGY 40(4), 580-588.

MORRIS, R. F. 1963. THE DEVELOPMENT OF PREDICTIVE EQUATIONS FOR THE SPRUCE BUDWORM BASED ON KEY-FACTOR ANALYSIS. PP. 116-119. IN, THE DYNAMICS OF EPIDEMIC SPRUCE BUDWORM POPULATIONS. (R. F. MORRIS, ED.). MEM. ENTOMOL. SOC. CAN. NO. 31. [MODEL; REGRESSION]

MORRIS, R. F. 1963. PREDICTIVE POPULATION EQUATIONS BASED ON KEY FACTORS. MEM. ENTOMOL. SOC. CAN. 32, 16-21. [MODEL]

MORRIS, R. F. 1969. APPROACHES TO THE STUDY OF POPULATION DYNAMICS. PP. 9-28. IN, FOREST INSECT POPULATION DYNAMICS. (ANONYMOUS, ED.) U. S. FOR. SERV. NORTHEASTERN FOREST EXPT. STN. RES. PAPER NE-125.

MORRIS, R. F. AND T. ROYAMA. 1969. LOGARITHMIC REGRESSION AS AN INDEX OF RESPONSES TO POPULATION DENSITY. COMMENT ON A PAPER BY M. P. HASSELL AND C. B. HUFFAKER. CAN. ENTOMOL. 101(4), 361-364. [NICHOLSON-BAILEY PARASITE MODEL; CORRELATION COEFFICIENT]

MOTT, D. G. 1963. THE POPULATION MODEL FOR THE UNSPRAYED AREA. PP. 99-109. IN, THE DYNAMICS OF EPIDEMIC SPRUCE BUDWORM POPULATIONS. (R. F. MORRIS, ED.). MEM. ENTOMOL. SOC. CAN. NO. 31. [LINEAR REGRESSION MODEL]

MURPHY, G. I. 1968. PATTERN IN LIFE HISTORY AND THE ENVIRONMENT. AMER. NAT. 102(927), 391-403. [INTRINSIC RATE OF INCREASE; LOTKA'S EQUATION; RICKER'S MODEL; COMPETITION EQUATIONS; PREDATOR-PREY MODEL; MODEL; RANDOM NUMBERS; COMPUTER; SIMULATION]

NEYMAN, J., T. PARK AND E. L. SCOTT. 1955. STRUGGLE FOR EXISTENCE-THE TRIBOLIUM MODEL: BIOLOGICAL AND STATISTICAL ASPECTS. PROC. THIRD BERKELEY SYMP. MATH. STAT. AND PROB. 4, 41-79.

NICHOLSON, A. J. 1954. AN OUTLINE OF THE DYNAMICS OF ANIMAL POPULATIONS. AUST. J. ZOOL. 2(1), 9-65. [LOGISTIC]

OSTER, G. 1974. THE ROLE OF AGE STRUCTURE IN THE DYNAMICS OF INTERACTING POPULATIONS. PP. 166-173. IN, MATHEMATICAL PROBLEMS IN BIOLOGY. LECTURE NOTES IN BIOMATHEMATICS 2. (P. VAN DEN DRIESSCHE, ED.). SPRINGER-VERLAG NEW YORK INC., NEW YORK, NEW YORK. [BALANCE MODEL; HOST-PARASITE MODEL]

PALOHEIMO, J. E. 1974. CALCULATION OF INSTANTANEOUS BIRTH RATE. LIMNOL. OCEANOGR. 19(4), 692-694.

PARK, T. 1939. ANALYTICAL POPULATION STUDIES IN RELATION TO GENERAL ECOLOGY. AMER. MIDL. NAT. 21(1), 235-253. [LOGISTIC CURVE]

PEARL, R. AND L. J. REED. 1925. A LIFE TABLE NOMOGRAM. AMER. J. HYGIENE 5(3), 330-334.

PEARSON, O. P. 1960. A MECHANICAL MODEL FOR THE STUDY OF POPULATION DYNAMICS. ECOLOGY 41(3), 494-508. [MODEL]

PETRUSEWICZ, K. 1970. DYNAMICS AND PRODUCTION OF THE HARE POPULATION IN POLAND. ACTA THERIOL. 15(26), 413-445. [POPULATION MODEL; MORTALITY; BIRTHS; SURVIVAL; LIFE TABLE]

PHILIP, J. R. 1955. NOTE ON THE MATHEMATICAL THEORY OF ANIMAL POPULATION DYNAMICS AND A RECENT FALLACY. AUST. J. ZOOL. 3(3), 287-294.

PRESTON, F. W. AND R. T. NORRIS. 1947. NESTING HEIGHTS OF BREEDING BIRDS. ECOLOGY 28(3), 241-273. [MODEL]

PRESTON, S. H. 1970. THE BIRTH TRAJECTORY CORRESPONDING TO PARTICULAR POPULATION SEQUENCES. THEOR. POPUL. BIOL. 1(4), 346-351. [MODEL]

RADCLIFFE, J. AND P. J. STAFF. 1970. IMMIGRATION-MIGRATION-DEATH PROCESSES WITH MULTIPLE LATENT ROOTS. MATH. BIOSCI. 8, 279-290. [MODEL]

RENSHAW, E. 1972. BIRTH, DEATH AND MIGRATION PROCESSES. BIOMETRIKA 59(1), 49-60. [MODEL; MATRIX]

RESCIGNO, A. AND I. W. RICHARDSON. 1973. THE DETERMINISTIC THEORY OF POPULATION DYNAMICS. PP. 283-360. IN, FOUNDATIONS OF MATHEMATICAL BIOLOGY, VOLUME III. (R. ROSEN, ED.). ACADEMIC PRESS, NEW YORK, NEW YORK. [MALTHUS EQUATION; PEARL-VERHULST EQUATION; LOGISTIC; PREDATOR-PREY MODEL; VOLTERRA; DETERMINANT; COMPETITION]

ROSS, G. G. 1972. A DIFFERENCE-DIFFERENTIAL MODEL IN POPULATION DYNAMICS. J. THEOR. BIOL. 37(3), 477-492. [LOTKA'S MODEL; VOLTERRA'S MODEL]

ROUGHGARDEN, J. 1974. POPULATION DYNAMICS IN A SPATIALLY VARYING ENVIRONMENT; HOW POPULATION SIZE "TRACKS" SPATIAL VARIATION IN CARRYING CAPACITY. AMER. NAT. 108(963), 649-664. [LOTKA-VOLTERRA EQUATION; COMPETITION; DISPERSAL; SIMULATION; COMPUTER; STABILITY ANALYSIS]

SAMAME, L. M. AND K. OKADA. 1973. ON THE AGE DETERMINATION AND POPULATION DYNAMICS OF THE CACHEMA, CYNOSCION ANALIS JENYNS, IN THE NORTHERN COASTAL WATERS OF PERU. BULL. TOKAI REG. FISH. RES. LAB. 73, 23-68. [POPULATION MODEL; GROWTH MODEL]

SARUKHAN, J. AND M. GADGIL. 1974. STUDIES ON PLANT DEMOGRAPHY: RANUNCULUS REPENS L., R. BULBUSUS L. AND R. ACRIS L. III. A MATHEMATICAL MODEL INCORPORATING MULTIPLE MODES OF REPRODUCTION. J. ECOL. 62(3), 921-936. [LOTKA'S MODEL; LESLIE MATRIX; MATRIX; EXPONENTIAL GROWTH]

SCHAEFER, M. B. 1957. A STUDY OF THE DYNAMICS OF THE FISHERY FOR YELLOWFIN TUNA IN THE EASTERN TROPICAL PACIFIC OCEAN. BULL. INTER-AMER. TROP. TUNA COMM. 2(6), 247-268. [SCHAEFER'S MODEL]

SEAL, H. L. 1945. THE MATHEMATICS OF A POPULATION COMPOSED OF K STATIONARY STRATA EACH RECRUITED FROM THE STRATUM BELOW AND SUPPORTED AT THE LOWEST LEVEL BY A UNIFORM ANNUAL NUMBER OF ENTRANTS. BIOMETRIKA 33(3), 226-230. [MODEL]

SHANBHAG, D. N. 1972. ON A VECTOR-VALUED BIRTH AND DEATH PROCESS. BIOMETRICS 28(2), 417-425.

SHEPPARD, P. M., W. W. MACDONALD, R. J. TONN AND B. GRAB. 1969. THE DYNAMICS OF AN ADULT POPULATION OF AEDES AEGYPTI IN RELATION TO DENGUE HAEMORRHAGIC FEVER IN BANGKOK. J. ANIM. ECOL. 38(3), 661-702.

SHINOZAKI, K. AND M. TACHI. 1953. A STUDY ON POPULATION CURVE. ARCH. POPUL. ASSOC. JAPAN 2(1), 34-42.

SKELLAM, J. G. 1955. THE MATHEMATICAL APPROACH TO POPULATION DYNAMICS. PP. 31-45. IN, THE NUMBERS OF MAN AND ANIMALS. (J. B. CRAGG AND N. W. PIRIE, ED.). OLIVER AND BOYD, LONDON, ENGLAND.

SKELLAM, J. G. 1973. THE FORMULATION AND INTERPRETATION OF MATHEMATICAL MODELS OF DIFFUSIONARY PROCESSES IN POPULATION BIOLOGY. PP. 63-85. IN, THE MATHEMATICAL THEORY OF THE DYNAMICS OF BIOLOGICAL POPULATIONS. (M. S. BARTLETT AND R. W. HIORNS, EDS.). ACADEMIC PRESS, NEW YORK, NEW YORK. [MONTE CARLO METHODS; VECTOR; BIVARIATE NORMAL DISTRIBUTION; RANDOM WALK; PROBABILITY; CUMULANTS; NORMAL DISTRIBUTION; COMPUTER; COMPETITION]

SLOBODKIN, L. B. 1955. CONDITIONS FOR POPULATION EQUILIBRIUM. ECOLOGY 36(3), 530-533.

SLUITER, J. W., P. F. VAN HEERDT AND J. J. BEZEM. 1958. POPULATION STATISTICS OF THE BAT MYOTIS MYSTACINUS, BASED ON THE MARKING-RECAPTURE METHOD. ARCHIVES NEERLANDIAISES DE ZOOLOGIE 12, 63-88. [RANK CORRELATION; LIFE TABLE]

SMITH, F. E. 1952. EXPERIMENTAL METHODS IN POPULATION DYNAMICS: A CRITIQUE. ECOLOGY 33(4), 441-450.

SMITH, F. E. 1963. POPULATION DYNAMICS IN DAPHNIA MAGNA AND A NEW MODEL FOR POPULATION GROWTH. ECOLOGY 44(4), 651-663. [LOGISTIC; TIME LAG MODEL]

SMITH, T. D. 1973. A FUNCTION DESCRIBING EQUILIBRIUM GROWTH RATE IN DENSITY DEPENDENT LESLIE MATRIX POPULATION MODELS. (ABSTRACT). BIOMETRICS 29(3), 612.

SMITH, T. G. 1973. POPULATION DYNAMICS OF THE RINGED SEAL IN THE CANADIAN EASTERN ARCTIC. FISH. RES. BOARD CAN. BULL. 181. VIII, 55 PP. [LIFE TABLE; MODEL; LESLIE MATRIX; COMPUTER; INTRINSIC RATE OF INCREASE]

SOLOMON, M. E. 1969. INVESTIGATING THE REGULATORY ASPECT OF INSECT POPULATION DYNAMICS. PP. 87-94. IN, INSECT ECOLOGY AND THE STERILE-MALE TECHNIQUE. INTERNATIONAL ATOMIC ENERGY AGENCY, VIENNA, REPORT STI/PUB/223.

SOMMANI, P. 1972. A STUDY ON THE POPULATION DYNAMICS OF STRIPED BASS (MORONE SAXATILIS WALBAUM) IN THE SAN FRANCISCO BAY ESTUARY. PH.D. DISSERTATION, UNIVERSITY OF WASHINGTON, SEATTLE, WASHINGTON. 133 PP. [LESLIE MATRIX]

SOUTHWOOD, T. R. E. 1967. THE INTERPRETATION OF POPULATION CHANGE. J. ANIM. ECOL. 36(3), 519-529. [KEY FACTOR ANALYSIS]

STIVEN, A. E. 1962. THE EFFECT OF TEMPERATURE AND FEEDING ON THE INTRINSIC RATE OF INCREASE OF THREE SPECIES OF HYDRA. ECOLOGY 43(2), 325-328. [GEOMETRIC SERIES]

STREIFER, W. AND C. A. ISTOCK. 1973. A CRITICAL VARIABLE FORMULATION OF POPULATION DYNAMICS. ECOLOGY 54(2), 392-398. [MODEL; LOGISTIC]

SYKES, Z. M. 1969. ON DISCRETE STABLE POPULATION THEORY. BIOMETRICS 25(2), 285-293.

SYKES, Z. M. 1969. SOME STOCHASTIC VERSIONS OF THE MATRIX MODEL FOR POPULATION DYNAMICS. J. AMER. STAT. ASSOC. 64(325), 111-130.

TAYLOR, H. M., R. S. GOURLEY, C. E. LAWRENCE AND R. S. KAPLAN. 1974. NATURAL SELECTION OF LIFE HISTORY ATTRIBUTES: AN ANALYTICAL APPROACH. THEOR. POPUL. BIOL. 5(1), 104-122. [LIFE HISTORY MODEL; RATE OF INCREASE; ENERGY UTILIZATION]

USHER, M. B. 1972. DEVELOPMENT IN THE LESLIE MATRIX MODEL. PP. 29-60. IN, MATHEMATICAL MODELS IN ECOLOGY. (J. N. R. JEFFERS, ED.). BLACKWELL SCIENTIFIC PUBLICATIONS, OXFORD, ENGLAND.

USHER, M. B. AND M. H. WILLIAMSON. 1970. A DETERMINISTIC MATRIX MODEL FOR HANDLING THE BIRTH, DEATH, AND MIGRATION PROCESSES OF SPATIALLY DISTRIBUTED POPULATIONS. BIOMETRICS 26(1), 1-12.

UTIDA, S. 1957. POPULATION FLUCTUATION, AN EXPERIMENTAL AND THEORETICAL APPROACH. COLD SPRING HARBOR SYMPOSIA ON QUANTITATIVE BIOLOGY 22, 139-150. [LOGISTIC; LOTKA-VOLTERRA; HOST-PARASITE MODEL; PEARL-VERHULST COEFFICIENT]

VAJADA, S. 1947. THE STRATIFIED SEMI-STATIONARY POPULATION. BIOMETRIKA 34(3/4), 243-254. [MODEL]

VAN KAMPEN, N. G. 1973. BIRTH AND DEATH PROCESSES IN LARGE POPULATIONS. BIOMETRIKA 60(2), 419-420. [EPIDEMIC]

VANDERMEER, J. H. 1975. ON THE CONSTRUCTION OF THE POPULATION PROJECTION MATRIX FOR A POPULATION GROUPED IN UNEQUAL STAGES. BIOMETRICS 31(1), 239-242. [LEFKOVITCH MATRIX; MODEL]

VARIOUS AUTHORS. 1966. STATISTICS IN THE STUDY OF BIRD POPULATIONS. THE STATISTICIAN 16(2), 119-182.

VARLEY, G. C. AND G. R. GRADWELL. 1970. RECENT ADVANCES IN INSECT POPULATION DYNAMICS. ANNU. REV. ENTOMOL. 15, 1-24. [LIFE TABLE; KEY FACTOR ANALYSIS; MODEL; MULTIVARIATE ANALYSIS]

VARLEY, G. C. AND G. R. GRADWELL. 1971. THE USE OF MODELS AND LIFE TABLES IN ASSESSING THE ROLE OF NATURAL ENEMIES. PP. 93-112. IN, BIOLOGICAL CONTROL. (C. B. HUFFAKER, ED.). PLENUM PRESS, NEW YORK, NEW YORK.

WATT, K. E. F. 1962. THE CONCEPTUAL FORMULATION AND MATHEMATICAL SOLUTION OF PRACTICAL PROBLEMS IN POPULATION INPUT-OUTPUT DYNAMICS. PP. 191-203. IN, THE EXPLOITATION OF NATURAL ANIMAL POPULATIONS. (E. D. LECREN AND M. W. HOLGATE, EDS.). BLACKWELL SCIENTIFIC PUBLICATIONS, OXFORD, ENGLAND. [COMPUTER]

WATT, K. E. F. 1963. HOW CLOSELY DOES THE MODEL MIMIC REALITY? PP. 109-111. IN, THE DYNAMICS OF EPIDEMIC SPRUCE BUDWORM POPULATIONS. (R. F. MORRIS, ED.). MEMOIRS OF THE ENTOMOLOGICAL SOCIETY OF CANADA NO. 31. [REGRESSION ANALYSIS]

WATT, K. E. F. 1963. WHAT IS THE OPTIMUM STRATEGY FOR STUDYING NATURAL POPULATIONS OF THE SPRUCE BUDWORM? PP. 111-115. IN, THE DYNAMICS OF EPIDEMIC SPRUCE BUDWORM POPULATIONS. (R. F. MORRIS, ED.). MEMOIRS OF THE ENTOMOLOGICAL SOCIETY OF CANADA NO. 31. [MODEL; REGRESSION; CORRELATION COEFFICIENT; VARIANCE]

WATT, K. E. F. 1969. METHODS OF DEVELOPING LARGE-SCALE SYSTEMS MODELS. PP. 35-51. IN, FOREST INSECT POPULATION DYNAMICS. U. S. FOR. SERV. NORTHEASTERN FOREST EXPT. STN. RES. PAPER NE-125.

WIGHT, H. M., R. G. HEATH AND A. D. GEIS 1965. A METHOD FOR ESTIMATING FALL ADULT SEX RATIOS FROM PRODUCTION AND SURVIVAL DATA. J. WILDL. MANAGE. 29(1), 185-195. [SURVIVAL RATE]

WILLIAMS, F. M. 1971. DYNAMICS OF MICROBIAL POPULATIONS. PP. 197-267. IN, SYSTEMS ANALYSIS AND SIMULATION IN ECOLOGY, VOLUME I. (B. C. PATTEN, ED.). ACADEMIC PRESS, NEW YORK, NEW YORK.

WILLIAMSON, M. H. 1959. SOME EXTENSIONS OF THE USE OF MATRICES IN POPULATION THEORY. BULL. MATH. BIOPHYS. 21, 13-17.

POPULATION ESTIMATION (CAPTURE-RECAPTURE METHODS)

AASEN, O. 1958. ESTIMATION OF THE STOCK STRENGTH OF THE NORWEGIAN HERRING. J. CONS. PERM. INT. EXPLOR. MER 24(1), 95-110. [TAG-RECAPTURE]

ADAMS, L. 1951. CONFIDENCE LIMITS FOR THE PETERSEN OR LINCOLN INDEX USED IN ANIMAL POPULATION STUDIES. J. WILDL. MANAGE. 15(1), 13-19. [POISSON DISTRIBUTION; BINOMIAL]

ALLEN, K. R. 1963. A PRELIMINARY STUDY OF THE EFFICIENCY OF SOME TAGGING EXPERIMENTS IN NEW ZEALAND. PP. 328-329. IN, NORTH ATLANTIC FISH MARKING SYMPOSIUM. INTERN. COMM. NORTHWEST ATLANTIC FISH. SPEC. PUBL. NO. 4. [MORTALITY; SURVIVAL; TAG-RECAPTURE]

ANDRZEJEWSKI, R. AND T. WIERZBOWSKA. 1961. AN ATTEMPT AT ASSESSING THE DURATION OF RESIDENCE OF SMALL RODENTS IN A DEFINED FOREST AREA AND THE RATE OF INTERCHANGE BETWEEN INDIVIDUALS. ACTA THERIOL. 5(12), 153-172. [DENSITY FUNCTION; CHI-SQUARE; RATE OF DISAPPEARANCE; TAG-RECAPTURE]

ANDRZEJEWSKI, R., G. BUJALSKA, L. RYSZKOWSKI AND J. USTYNIUK. 1966. ON A RELATION BETWEEN THE NUMBER OF TRAPS IN A POINT OF CATCH AND TRAPPABILITY OF SMALL RODENTS. ACTA THERIOL. 11(13), 343-349.

ANDRZEJEWSKI, R., K. PETRUSEWICZ AND W. WALKOWA. 1959. PRELIMINARY REPORT ON RESULTS OBTAINED WITH A LIVING TRAP IN A CONFINED POPULATION OF MICE. BULL. DE L'ACAD. POL. DES SCI. CL. II, SERIES DES SCIENCES BIOLOGIQUES 7(9), 367-370. [CORRELATION INDEX]

ARNASON, A. N. 1972. PARAMETER ESTIMATES FROM MARK-RECAPTURE EXPERIMENTS ON TWO POPULATIONS SUBJECT TO MIGRATION AND DEATH. RES. POPUL. ECOL.-KYOTO 13(2), 97-113.

BAILEY, N. T. J. 1951. ON ESTIMATING THE SIZE OF MOBILE POPULATIONS FROM RECAPTURE DATA. BIOMETRIKA 38(3/4), 293-306.

BAILEY, N. T. J. 1952. IMPROVEMENTS IN THE INTERPRETATION OF RECAPTURE DATA. J. ANIM. ECOL. 21(1), 120-127.

BARNDORFF-NIELSEN, O. 1972. ESTIMATION PROBLEMS IN CAPTURE-RECAPTURE ANALYSIS. DAN. REV. GAME BIOL. 6(6), 3-22. [MAXIMUM LIKELIHOOD; DARROCH'S MODEL]

BAYLIFF, W. H. 1966. POPULATION DYNAMICS OF THE ANCHOVETA, CETENGRAULIS MYSTICETUS, IN THE GULF OF PANAMA, AS DETERMINED BY TAGGING EXPERIMENTS. BULL. INTER-AMER. TROP. TUNA COMM. 11(4), 175-288. [TAG-RECAPTURE; VARIANCE; ROBSON-CHAPMAN EQUATION; CHAPMAN'S FORMULA; BEVERTON-HOLT EQUATION; COEFFICIENT OF MORTALITY]

BAYLIFF, W. H. AND B. J. ROTHSCHILD. 1974. MIGRATIONS OF YELLOWFIN TUNA TAGGED OFF THE SOUTHERN COAST OF MEXICO IN 1960 AND 1969. BULL. INTER-AMER. TROP. TUNA COMM. 16(1), 3-43. [JONES METHOD; TAG-RECAPTURE; VARIANCE-COVARIANCE METHOD; DISPERSION MODEL]

BAYLIFF, W. H. AND L. M. MOBRAND. 1972. ESTIMATES OF THE RATES OF SHEDDING OF DART TAGS FROM YELLOWFIN TUNA. BULL. INTER-AMER. TROP. TUNA COMM. 15(5), 411-452. [MODEL; TAG-RECAPTURE]

BEVAN, D. E. 1962. ESTIMATION BY TAGGING OF THE SIZE OF MIGRATING SALMON POPULATIONS IN COASTAL WATERS. PP. 377-449; PLUS 20 PP. FIGURES. IN, STUDIES OF ALASKA RED SALMON. (T. S. Y. KOO, ED.). UNIVERSITY OF WASHINGTON PRESS, SEATTLE, WASHINGTON. [TAG-RECAPTURE; MODEL; MORTALITY RATE]

BISHOP, J. A. AND P. M. SHEPPARD. 1973. AN EVALUATION OF TWO CAPTURE-RECAPTURE MODELS USING THE TECHNIQUE OF COMPUTER SIMULATION. PP. 235-252. IN, THE MATHEMATICAL THEORY OF THE DYNAMICS OF BIOLOGICAL POPULATIONS. (M. S. BARTLETT AND R. W. HIORNS, EDS.). ACADEMIC PRESS, NEW YORK, NEW YORK. [COMPUTER; SIMULATION; JOLLY'S MODEL; FISHER AND FORD'S DETERMINISTIC MODEL; BAILEY'S CORRECTION FOR BIAS]

BORROR, D. J. 1948. ANALYSIS OF REPEAT RECORDS OF BANDED WHITE-THROATED SPARROWS. ECOL. MONOGR. 18(3), 411-430. [PROBABILITY; TAG-RECAPTURE ESTIMATION; POISSON DISTRIBUTION]

BUCK, D. H. AND C. F. THOITS, III. 1965. AN EVALUATION OF PETERSEN ESTIMATION PROCEDURES EMPLOYING SEINES IN 1-ACRE PONDS. J. WILDL. MANAGE. 29(3), 598-621. [CUMULATIVE BINOMIAL DISTRIBUTION FUNCTION]

BURNHAM, K. P. AND W. S. OVERTON. 1969. A SIMULATION STUDY OF LIVETRAPPING AND ESTIMATION OF POPULATION SIZE. OREGON STATE UNIVERSITY, CORVALLIS, OREGON, DEPT. STAT. TECH. REPT. NO. 14. 152 PP. [MODEL; SIMULATION; COMPUTER; MATRIX; PROBABILITY; CAPTURE-RECAPTURE]

CAROTHERS, A. D. 1971. AN EXAMINATION AND EXTENSION OF LESLIE'S TEST OF EQUAL CATCHABILITY. BIOMETRICS 27(3), 615-630.

CAROTHERS, A. D. 1973. CAPTURE-RECAPTURE METHODS APPLIED TO A POPULATION WITH KNOWN PARAMETERS. J. ANIM. ECOL. 42(1), 125-146.

CAROTHERS, A. D. 1973. THE EFFECTS OF UNEQUAL CATCHABILITY ON JOLLY-SEBER ESTIMATES. BIOMETRICS 29(1), 79-100.

CHAPMAN, D. G. 1948. A MATHEMATICAL STUDY OF CONFIDENCE LIMITS OF SALMON POPULATIONS CALCULATED FROM SAMPLE TAG RATIOS. INTERN. PACIFIC SALMON FISHERIES COMM. BULL. 2. PP. 67-85. [TAG-RECAPTURE ESTIMATION; PROBABILITY; MAXIMUM LIKELIHOOD; POISSON DISTRIBUTION; BINOMIAL DISTRIBUTION; NORMAL]

CHAPMAN, D. G. 1952. INVERSE, MULTIPLE AND SEQUENTIAL SAMPLE CENSUSES. BIOMETRICS 8(4), 286-306.

CHAPMAN, D. G. AND W. S. OVERTON. 1966. ESTIMATING AND TESTING DIFFERENCES BETWEEN POPULATION LEVELS BY THE SCHNABEL ESTIMATION METHOD. J. WILDL. MANAGE. 30(1), 173-180.

CHAPMAN, D. G., B. D. FINK AND E. B. BENNETT. 1965. A METHOD FOR ESTIMATING THE RATE OF SHEDDING OF TAGS FROM YELLOWFIN TUNA. BULL. INTER-AMER. TROP. TUNA COMM. 10(5), 335-342.

CHATTERJEE, S. 1973. ESTIMATING WILDLIFE POPULATIONS BY THE CAPTURE-RECAPTURE METHOD. PP. 25-33. IN, STATISTICS BY EXAMPLE: FINDING MODELS. (F. MOSTELLER, W. H. KRUSKAL, R. F. LINK, R. S. PIETERS AND G. R. RISING, EDS.). ADDISON-WESLEY PUBLISHING CO., READING, MASSACHUSETTS. [PROBABILITY; MAXIMUM LIKELIHOOD ESTIMATION]

COCHRAN, W. G. 1949. THE PRESENT STATUS OF BIOMETRY. PAPER GIVEN BEFORE THE JOINT MEETING OF THE I. M. S. AND THE BIOMETRIC SOCIETY AT BERN, SWITZERLAND. 19 PP. [TAG-RECAPTURE ESTIMATION; HISTORY]

CONWAY, G. R., M. TRPIS AND G. A. H. MCCLELLAND. 1974. POPULATION PARAMETERS OF THE MOSQUITO AEDES AEGYPTI (L.) ESTIMATED BY MARK-RELEASE-RECAPTURE IN A SUBURBAN HABITAT IN TANZANIA. J. ANIM. ECOL. 43(2), 289-304. [MODEL; SURVIVAL RATE; FISHER AND FORD METHOD; MAXIMUM LIKELIHOOD ESTIMATION; LINCOLN INDEX; BAILEY'S METHOD]

COOK, L. M., L. P. BROWER AND H. J. CROZE. 1967. THE ACCURACY OF A POPULATION ESTIMATION FROM MULTIPLE RECAPTURE DATA. J. ANIM. ECOL. 36(1), 57-60.

CORMACK, R. M. 1966. A TEST FOR EQUAL CATCHABILITY. BIOMETRICS 22(2), 330-342.

CORMACK, R. M. 1968. THE STATISTICS OF CAPTURE-RECAPTURE METHODS. OCEANOGR. MAR. BIOL. ANNU. REV. 6, 455-506.

CORMACK, R. M. 1972. THE LOGIC OF CAPTURE-RECAPTURE ESTIMATES. BIOMETRICS 28(2), 337-343.

CORMACK, R. M. 1973. COMMON SENSE ESTIMATES FROM CAPTURE-RECAPTURE STUDIES. PP. 225-234. IN, THE MATHEMATICAL THEORY OF THE DYNAMICS OF BIOLOGICAL POPULATIONS. (M. S. BARTLETT AND R. W. HIORNS, EDS.). ACADEMIC PRESS, NEW YORK, NEW YORK. [MODEL; MAXIMUM LIKELIHOOD]

CRAIG, C. C. 1953. ON THE UTILIZATION OF MARKED SPECIMENS IN ESTIMATING POPULATIONS OF FLYING INSECTS. BIOMETRIKA 40(1/2), 170-176.

CRUMPACKER, D. W. AND J. S. WILLIAMS. 1973. DENSITY, DISPERSION, AND POPULATION STRUCTURE IN DROSOPHILA PSEUDOOBSCURA. ECOL. MONOGR. 43(4), 499-538. [TAG-RECAPTURE; MAXIMUM LIKELIHOOD; DENSITY MODEL; DISPERSION MODEL]

DARROCH, J. N. 1958. THE MULTIPLE-RECAPTURE CENSUS. I. ESTIMATION OF A CLOSED POPULATION. BIOMETRIKA 45(3/4), 343-359.

DARROCH, J. N. 1959. THE MULTIPLE-RECAPTURE CENSUS. II. ESTIMATION WHEN THERE IS IMMIGRATION OR DEATH. BIOMETRIKA 46(3/4), 336-351.

DARROCH, J. N. 1961. THE TWO-SAMPLE CAPTURE-RECAPTURE CENSUS WHEN TAGGING AND SAMPLING ARE STRATIFIED. BIOMETRIKA 48(3/4), 241-260.

DAVIS, W. S. 1964. GRAPHIC REPRESENTATION OF CONFIDENCE INTERVALS FOR PETERSEN POPULATION ESTIMATES. TRANS. AMER. FISH. SOC. 93(3), 227-232. [TAG-RECAPTURE ESTIMATION]

DELURY, D. B. 1958. THE ESTIMATION OF POPULATION SIZE BY A MARKING AND RECAPTURE PROCEDURE. J. FISH. RES. BOARD CAN. 15(1), 19-25.

DICE, L. R. AND P. J. CLARK. 1953. THE STATISTICAL CONCEPT OF HOME RANGE AS APPLIED TO THE RECAPTURE RADIUS OF THE DEERMOUSE (PEROMYSCUS). CONTRIB. LAB. VERTEBR. BIOL., UNIVERSITY OF MICHIGAN, ANN ARBOR, MICHIGAN. NO. 62. 15 PP.

EBERHARDT, L. L. 1969. POPULATION ESTIMATES FROM RECAPTURE FREQUENCIES. J. WILDL. MANAGE. 33(1), 28-39.

EDWARDS, W. R. AND L. EBERHARDT. 1967. ESTIMATING COTTONTAIL ABUNDANCE FROM LIVE TRAPPING DATA. J. WILDL. MANAGE. 31(1), 87-96. [SCHNABEL METHOD OF ESTIMATION ; POISSON DISTRIBUTION;LOG TRANSFORMATION;MULTIPLE CENSUS; MAXIMUM LIKELIHOOD; LINCOLN INDEX; GEOMETRIC MODEL; REGRESSION; GEOMETRIC DISTRIBUTION; CONFIDENCE INTERVAL]

FELLER, W. 1952. A SAMPLING PROBLEM. PP. 37-38. IN, INTRODUCTION TO PROBABILITY THEORY AND ITS APPLICATIONS. JOHN WILEY AND SONS, NEW YORK, NEW YORK. [FISH; HYPERGEOMETRIC DISTRIBUTION; PROBABILITY; MAXIMUM LIKELIHOOD]

FIENBERG, S. E. 1972. THE MULTIPLE RECAPTURE CENSUS FOR CLOSED POPULATIONS AND INCOMPLETE 2K CONTINGENCY TABLES. BIOMETRIKA 59(3), 591-603. [TAG-RECAPTURE; MODEL; MAXIMUM LIKELIHOOD]

FREEMAN, P. R. 1972. SEQUENTIAL ESTIMATION OF POPULATION SIZE. (ABSTRACT). BIOMETRICS 28(4), 1158-1159.

FREEMAN, P. R. 1972. SEQUENTIAL ESTIMATION OF THE SIZE OF A POPULATION. BIOMETRIKA 59(1), 9-17. [BAYESIAN INFERENCE; TAG-RECAPTURE]

FREEMAN, P. R. 1973. A NUMERICAL COMPARISON BETWEEN SEQUENTIAL TAGGING AND SEQUENTIAL RECAPTURE. BIOMETRIKA 60(3), 499-508. [SEQUENTIAL ESTIMATION; TAG-RECAPTURE; DYNAMIC PROGRAMMING]

FREEMAN, P. R. 1973. SEQUENTIAL RECAPTURE. BIOMETRIKA 60(1), 141-153. [BAYESIAN INFERENCE; DECISION THEORY; TAG-RECAPTURE; DYNAMIC PROGRAMMING]

GASKELL, T. J. AND B. J. GEORGE. 1972. A BAYESIAN MODIFICATION OF THE LINCOLN INDEX. J. APPL. ECOL. 9(2), 377-384. [COMPUTER; SIMULATION]

GILBERT, N. E. G. 1956. LIKELIHOOD FUNCTION FOR CAPTURE-RECAPTURE SAMPLES. BIOMETRIKA 43(3/4), 488-489.

GILBERT, R. O. 1971. APPROXIMATIONS OF THE BIAS IN JOLLY'S STOCHASTIC CAPTURE-RECAPTURE MODEL. (ABSTRACT). BIOMETRICS 27(2), 480.

GILBERT, R. O. 1972. PARTITIONING OF THE BIAS OF THE JOLLY-SEBER CAPTURE-RECAPTURE ESTIMATE OF POPULATION SIZE WHEN UNEQUAL CATCHABILITY IS PRESENT. (ABSTRACT). BIOMETRICS 28(4), 1161.

GILBERT, R. O. 1973. APPROXIMATIONS OF THE BIAS IN THE JOLLY-SEBER CAPTURE-RECAPTURE MODEL. BIOMETRICS 29(3), 501-526. [SIMULATION; TAYLOR SERIES; MONTE CARLO; COMPUTER]

GOODMAN, L. A. 1953. SEQUENTIAL SAMPLING TAGGING FOR POPULATION SIZE PROBLEMS. ANN. MATH. STAT. 24(1), 56-69.

GRUM, L. 1971. REMARKS ON THE DIFFERENTIATION IN CARABIDAE MOBILITY. EKOLOGIYA POLSKA-SERIA 19(3), 47-56. [MODEL; ESTIMATION; TAG-RECAPTURE]

GULLAND, J. A. 1955. ON THE ESTIMATION OF POPULATION PARAMETERS FROM MARKED MEMBERS. BIOMETRIKA 42(1/2), 269-270. [POISSON DISTRIBUTION; VARIANCE]

HAMMERSLEY, J. M. 1953. CAPTURE-RECAPTURE ANALYSIS. BIOMETRIKA 40(3/4), 265-278.

HANSSON, L. 1974. INFLUENCE AREA OF TRAP STATIONS AS A FUNCTION OF NUMBER OF SMALL MAMMALS EXPOSED PER TRAP. ACTA THERIOL. 19(2), 19-25.

HOLGATE, P. 1966. CONTRIBUTIONS TO THE MATHEMATICS OF ANIMAL TRAPPING. BIOMETRICS 22(4), 925-936. [PROBABILITY; BETA DISTRIBUTION; TAG-RECAPTURE ESTIMATION; GEOMETRIC DISTRIBUTION; GAMMA DISTRIBUTION; MAXIMUM LIKELIHOOD; TRUNCATED NEGATIVE BINOMIAL DISTRIBUTION]

HOLST, L. 1971. A NOTE ON FINDING THE SIZE OF A FINITE POPULATION. BIOMETRIKA 58(1), 228-229. [TAG-RECAPTURE]

HOLT, S. J. 1963. TAGGING EXPERIMENTS AND THE THEORY OF ADVERTISING. PP. 29-30, IN, NORTH ATLANTIC FISH MARKING SYMPOSIUM. INTERN. COMM. NORTHWEST ATLANTIC FISH. SPEC. PUBL NO. 4. [OPERATIONAL RESEARCH; TAG-RECAPTURE]

ITO, Y. 1973. A METHOD TO ESTIMATE A MINIMUM POPULATION DENSITY WITH A SINGLE RECAPTURE CENSUS. RES. POPUL. ECOL.-KYOTO 14(2), 159-168.

ITO, Y., A. TAKAI, K. MIYASHITA AND K. NAKAMURA. 1963. ESTIMATION OF DENSITY, SURVIVAL RATE AND DILUTION RATE OF MECOSTETHUS MAGISTER (ORTHOPTERA: ACRIDIDAE) POPULATIONS BY THE MARK AND RECAPTURE METHOD. RES. POPUL. ECOL.-KYOTO 5(1), 51-64. [LESLIE MATRIX; BAILEY'S TRIPLE-CATCH METHOD; JACKSON'S COMPOUND POSITIVE AND NEGATIVE METHOD]

ITO, Y., M. MURAI AND T. TERUYA. 1974. AN ESTIMATION OF POPULATION DENSITY OF DACUS CUCURBITAE WITH MARK-RECAPTURE METHODS. RES. POPUL. ECOL.-KYOTO 15(2), 213-222. [JACKSON'S METHOD; JOLLY'S METHOD; SURVIVAL RATE; ITO'S METHOD; CORRECTION OF ITO, Y. 1973. RES. POPUL. ECOL.-KYOTO 14(2), 159-168.]

IWAO, S. 1963. ON A METHOD FOR ESTIMATING THE RATE OF POPULATION INTERCHANGE BETWEEN TWO AREAS. RES. POPUL. ECOL.-KYOTO 5(1), 44-50. [TAG-RECAPTURE]

IWAO, S., K. KIRITANI AND N. HOKYO. 1966. APPLICATION OF A MARKING-AND-RECAPTURE METHOD FOR THE ANALYSIS OF LARVAL-ADULT POPULATIONS OF AN INSECT, NEZARA VIRIDULA. RES. POPUL. ECOL.-KYOTO 8(2), 147-160. [MODIFICATION OF METHOD OF LESLIE, CHITTY AND CHITTY]

JACKSON, C. H. N. 1933. ON THE TRUE DENSITY OF TSETSE FLIES. J. ANIM. ECOL. 2(2), 204-209. [TAG-RECAPTURE ESTIMATION]

JACKSON, C. H. N. 1939. THE ANALYSIS OF AN ANIMAL POPULATION. J. ANIM. ECOL. 8(2), 238-246. [TAG-RECAPTURE ESTIMATION]

JACKSON, C. H. N. 1940. THE ANALYSIS OF A TSETSE-FLY POPULATION. ANN. EUGEN. 10(4), 332-369. [TAG-RECAPTURE ESTIMATION]

JACKSON, C. H. N. 1944. THE ANALYSIS OF TSETSE-FLY POPULATION. II. ANN. EUGEN. 12(3), 176-205. [BIRTH RATE; DEATH; TAG-RECAPTURE ESTIMATION]

JACKSON, C. H. N. 1948. THE ANALYSIS OF A TSETSE-FLY POPULATION. III. ANN. EUGEN. 14(2), 91-108. [CAPTURE-RECAPTURE; BIRTH RATE; DEATH RATE]

JANION, M., L. RYSZKOWSKI AND T. WIERZBOWSKA. 1968. ESTIMATE OF NUMBER OF RODENTS WITH VARIABLE PROBABILITY OF CAPTURE. ACTA THERIOL. 13(16), 285-294. [REMOVAL METHOD]

JANION, S. M. AND T. WIERZBOWSKA. 1969/1970. ESTIMATION OF THE NUMBER OF RODENTS ACCORDING TO THE PROBABILITY OF CAPTURES AND THE TIME OF RESIDENCY. PP. 71-74, IN, ENERGY FLOW THROUGH SMALL MAMMAL POPULATIONS. (K. PETRUSEWICZ AND L. RYSZKOWSKI, EDS.). PWN-POLISH SCIENTIFIC PUBLISHERS, WARSAW, POLAND. [TAG-RECAPTURE ESTIMATION; NEGATIVE BINOMIAL DISTRIBUTION]

JANION, S. M. AND T. WIERZBOWSKA. 1970. TRAPABILITY OF RODENTS DEPENDING ON POPULATION DENSITY. ACTA THERIOL. 15(13), 199-207. [PROBABILITY]

JOLLY, G. M. 1963. ESTIMATES OF POPULATION PARAMETERS FROM MULTIPLE RECAPTURE DATA WITH BOTH DEATH AND DILUTION-DETERMINISTIC MODEL. BIOMETRIKA 50(1/2), 113-128.

JOLLY, G. M. 1965. EXPLICIT ESTIMATES FROM CAPTURE-RECAPTURE DATA WITH BOTH DEATH AND IMMIGRATION-STOCHASTIC MODEL. BIOMETRIKA 52(1/2), 225-247.

JOLLY, G. M. 1971. THE ESTIMATION OF MEAN PARAMETER VALUES FROM CAPTURE-RECAPTURE CHAINS OF SAMPLES. (ABSTRACT). BIOMETRICS 27(1), 252-253.

JONES, R. 1959. A METHOD OF ANALYSIS OF SOME TAGGED HADDOCK RETURNS. J. CONS. PERM. INT. EXPLOR. MER 25(1), 58-72. [TAG-RECAPTURE; DISPERSION MODEL]

JONES, R. 1964. A REVIEW OF METHODS OF ESTIMATING POPULATION SIZE FROM MARKING EXPERIMENTS. RAPP. PROCES-VERB. REUNIONS J. CONS. PERM. INT. EXPLOR. MER 155, 202-209. [TAG-RECAPTURE; MORTALITY RATE; INVERSE SAMPLING; DIRECT SAMPLING; PETERSEN INDEX; SCHNABEL METHOD; MULTIPLE RECAPTURES; TYPE A AND B ERRORS]

KENYON, K. W., V. B. SCHEFFER AND D. G. CHAPMAN. 1954. A POPULATION STUDY OF THE ALASKA FUR-SEAL HERD. U. S. FISH WILDL. SERV. SPEC. SCI. REPT. WILDL. 12. 77 PP. [TAG-RECAPTURE]

KETCHEN, K. S. 1953. THE USE OF CATCH-EFFORT AND TAGGING DATA IN ESTIMATING A FLATFISH POPULATION. J. FISH. RES. BOARD CAN. 10(8), 459-485. [DELURY'S METHOD; TAG-RECAPTURE ESTIMATION]

LAMBOU, V. W. AND R. J. MUNCY. 1961. ESTIMATING THE NUMBER OF MARKED ANIMALS WHICH HAVE RETAINED THEIR IDENTITY FROM MULTIPLE MARKED ANIMALS AND ITS APPLICATION TO THE PETERSON METHOD. PROC. ANN. CONF. S. E. ASSOC. GAME AND FISH COMMISSIONERS 15, 161-173. [TAG-RECAPTURE; MODEL; PROBABILITY; SAMPLING ERROR; CONFIDENCE INTERVAL; VARIANCE]

LE CREN, E. D. 1965. A NOTE ON THE HISTORY OF MARK-RECAPTURE POPULATION ESTIMATES. J. ANIM. ECOL. 34(2), 453-454.

LENARZ, W. H., F. J. MATTER, III, J. S. BECKETT, A. C. JONES AND J. M. MASON, JR. 1973. ESTIMATION OF RATES OF TAG SHEDDING BY NORTHWEST ATLANTIC BLUEFIN TUNA. U. S. NATL. MAR. FISH. SERV. FISH. BULL. 71(4), 1103-1105. [BAYLIFF-MOBRAND EQUATION]

LESLIE, P. H. 1952. THE ESTIMATION OF POPULATION PARAMETERS FROM DATA OBTAINED BY MEANS OF THE CAPTURE-RECAPTURE METHOD. II. THE ESTIMATION OF TOTAL NUMBERS. BIOMETRIKA 39(3/4), 363-388.

LESLIE, P. H. 1958. STATISTICAL APPENDIX. (ORIANS, G. H., A CAPTURE-RECAPTURE ANALYSIS OF A SHEARWATER POPULATION.). J. ANIM. ECOL. 27(1), 84-86. [PROBABILITY]

LESLIE, P. H. AND D. CHITTY. 1951. THE ESTIMATION OF POPULATION PARAMETERS FROM DATA OBTAINED BY MEANS OF THE CAPTURE-RECAPTURE METHOD I. THE MAXIMUM LIKELIHOOD EQUATIONS FOR ESTIMATING THE DEATH-RATE. BIOMETRIKA 38(3/4), 269-292.

LESLIE, P. H., D. CHITTY AND H. CHITTY. 1953. THE ESTIMATION OF POPULATION PARAMETERS FROM DATA OBTAINED BY MEANS OF THE CAPTURE-RECAPTURE METHOD. III. AN EXAMPLE OF THE PRACTICAL APPLICATIONS OF THE METHOD. BIOMETRIKA 40(1/2), 137-169.

LINCOLN, F. C. 1930. CALCULATING WATERFOWL ABUNDANCE ON THE BASIS OF BANDING RETURNS. U. S. DEPT. AGRICULTURE, CIRCULAR NO. 118, 4 PP. [TAG-RECAPTURE]

MANLY, B. F. J. 1969. SOME PROPERTIES OF A METHOD OF ESTIMATING THE SIZE OF MOBILE ANIMAL POPULATIONS. BIOMETRIKA 56(2), 407-410. [TAG-RECAPTURE; LIKELIHOOD FUNCTION; JOLLY'S METHOD]

MANLY, B. F. J. 1970. A SIMULATION STUDY OF ANIMAL POPULATION ESTIMATION USING THE CAPTURE-RECAPTURE METHOD. J. APPL. ECOL. 7(1), 13-39.

MANLY, B. F. J. 1971. ESTIMATES OF A MARKING EFFECT WITH CAPTURE-RECAPTURE SAMPLING. J. APPL. ECOL. 8(1), 181-189.

MANLY, B. F. J. 1971. A SIMULATION STUDY OF JOLLY'S METHOD FOR ANALYSING CAPTURE-RECAPTURE DATA. BIOMETRICS 27(2), 415-424.

MANLY, B. F. J. 1973. A NOTE ON THE ESTIMATION OF SELECTIVE VALUES FROM RECAPTURES OF MARKED ANIMALS WHEN SELECTION PRESSURES REMAIN CONSTANT OVER TIME. RES. POPUL. ECOL.-KYOTO 14(2), 151-158.

MANLY, B. F. J. AND G. A. F. SEBER. 1973. ANIMAL LIFE TABLES FROM CAPTURE-RECAPTURE DATA. BIOMETRICS 29(3), 487-500. [SIMULATION METHODS; PROBABILITY; COMPUTER]

MANLY, B. F. J. AND M. J. PARR. 1968. A NEW METHOD OF ESTIMATING POPULATION SIZE, SURVIVORSHIP, AND BIRTH RATE FROM CAPTURE-RECAPTURE DATA. TRANS. BRITISH ENTOMOLOGICAL AND NATURAL HISTORY SOCIETY 18(5), 81-89.

MORAN, P. A. P. 1951. A MATHEMATICAL THEORY OF ANIMAL TRAPPING. BIOMETRIKA 38(3/4), 307-311.

MORAN, P. A. P. 1952. THE ESTIMATION OF DEATH-RATES FROM CAPTURE-MARK-RECAPTURE SAMPLING. BIOMETRIKA 39(1/2), 181-188.

MYHRE, R. J. 1966. LOSS OF TAGS FROM PACIFIC HALIBUT AS DETERMINED BY DOUBLE-TAG EXPERIMENTS. INTERN. PACIFIC HALIBUT COMM. REPT. NO. 41. 31 PP. [MODEL]

NAIR, K. R. 1936. NOTE ON KARL PEARSON'S PAPER: "ON A METHOD OF ASCERTAINING LIMITS TO THE ACTUAL NUMBER OF MARKED MEMBERS IN A POPULATION OF GIVEN SIZE FROM A SAMPLE". BIOMETRIKA 28(3/4), 442-443. [BINOMIAL EXPANSION PROOF; TAG-RECAPTURE]

NAKAMURA, K., Y. ITO, K. MIYASHITA AND A. TAKAI. 1967. THE ESTIMATION OF POPULATION DENSITY OF THE GREEN RICE LEAFHOPPER, NEPHOTETTIX CINCTICEPS UHLER, IN SPRING FIELD BY THE CAPTURE-RECAPTURE METHOD. RES. POPUL. ECOL.-KYOTO 9(2), 113-129. [BAILEY'S TRIPLE-CATCH METHOD; ESTIMATION OF DEATH RATE]

NIXON, C. M., W. R. EDWARDS AND L. EBERHARDT. 1967. ESTIMATING SQUIRREL ABUNDANCE FROM LIVETRAPPING DATA. J. WILDL. MANAGE. 31(1), 96-101. [SCHNABEL METHOD OF ESTIMATION; MAXIMUM LIKELIHOOD; GEOMETRIC DISTRIBUTION; LOG TRANSFORMATION; POISSON DISTRIBUTION; REGRESSION]

NOSE, Y. 1961. AN ANALYSIS OF THE PETERSEN-TYPE FISH POPULATION ESTIMATE. BULL. JAPANESE SOC. SCI. FISH. 27(8), 763-773. [TAG-RECAPTURE]

NOSE, Y. 1961. ON THE PETERSEN-TYPE ESTIMATE OF FISH POPULATION UNDER THE EXISTENCE OF RECRUITMENT. BULL. JAPANESE SOC. SCI. FISH. 27(10), 881-892. [TAG-RECAPTURE]

OVERTON, W. S. 1965. A MODIFICATION OF THE SCHNABEL ESTIMATOR TO ACCOUNT FOR REMOVAL OF ANIMALS FROM THE POPULATION. J. WILDL. MANAGE. 29(2), 392-395.

PALOHEIMO, J. E. 1963. ESTIMATION OF CATCHABILITIES AND POPULATION SIZES OF LOBSTERS. J. FISH. RES. BOARD CAN. 20(1), 59-88. [TAG-RECAPTURE; DELURY'S METHOD; POISSON DISTRIBUTION]

PARKER, R. A. 1955. A METHOD FOR REMOVING THE EFFECT OF RECRUITMENT ON PETERSEN-TYPE POPULATION ESTIMATES. J. FISH. RES. BOARD CAN. 12(3), 447-450.

PARR, M. J., T. J. GASKELL AND B. J. GEORGE. 1968. CAPTURE-RECAPTURE METHODS OF ESTIMATING ANIMAL NUMBERS. J. BIOL. EDUC. 2(2), 95-117.

PATHAK, P. K. 1964. ON ESTIMATING THE SIZE OF A POPULATION AND ITS INVERSE BY CAPTURE MARK METHOD. SANKHYA 26A(1), 75-80. [VARIANCE; PROBABILITY]

PAULIK, G. J. 1961. DETECTION OF INCOMPLETE REPORTING OF TAGS. J. FISH. RES. BOARD CAN. 18(5), 817-832. [BINOMIAL DISTRIBUTION; TAG-RECAPTURE ESTIMATION; POWER OF TEST; MODEL; SAMPLE SIZE; POISSON DISTRIBUTION]

PAULIK, G. J. 1963. DETECTION OF INCOMPLETE REPORTING OF TAGS. PP. 238-247. IN: NORTH ATLANTIC FISH MARKING SYMPOSIUM. INTERN. COMM. NORTHWEST ATLANTIC FISH. SPEC. PUBL. NO. 4. [SAMPLE SIZE; TAG-RECAPTURE]

PEARSON, K. 1928. ON A METHOD OF ASCERTAINING LIMITS TO THE ACTUAL NUMBER OF MARKED MEMBERS IN A POPULATION OF GIVEN SIZE FROM A SAMPLE. BIOMETRIKA 20A(1/2), 148-174. [TAG-RECAPTURE; INVERSE PROBABILITY; HYPERGEOMETRIC SERIES]

PHILLIPS, B. F. AND N. A. CAMPBELL. 1970. COMPARISON OF METHODS OF ESTIMATING POPULATION SIZE USING DATA ON THE WHELK DICATHAIS AEGROTA (REEVE). J. ANIM. ECOL. 39(3), 753-759. [TAG-RECAPTURE; SCHNABEL METHOD; SCHUMACHER AND ESCHMEYER METHOD; DELURY'S METHOD; EDWARDS AND EBERHARDT METHOD]

POLLOCK, K. H. 1973. DISTRIBUTION OF THE RECAPTURE VECTOR IN A TAG-RECAPTURE MODEL ALLOWING MORTALITY. (ABSTRACT). TWENTY-FIFTH ANN. REPT. BIOMETRICS UNIT, CORNELL UNIVERSITY, ITHACA, NEW YORK. P. 93. (BU-475-M). [PROBABILITY DISTRIBUTION]

POLLOCK, K. H. 1974. THE ASSUMPTION OF EQUAL CATCHABILITY OF ANIMALS IN TAG-RECAPTURE EXPERIMENTS. PH.D. DISSERTATION, CORNELL UNIVERSITY, ITHACA, NEW YORK. 89 PP. (DISS. ABSTRS. 35(6), 2583B).

POLLOCK, K. H. 1974. A TAG-RECAPTURE MODEL ALLOWING FOR UNEQUAL SURVIVAL AND CATCHABILITY. (ABSTRACT). BIOMETRICS 30(2), 384.

POPE, J. A. 1963. THE DESIGN AND ANALYSIS OF CAPTURE-RECAPTURE EXPERIMENTS. PP. 342-347. IN, NORTH ATLANTIC FISH MARKING SYMPOSIUM. INTERN. COMM. NORTHWEST ATLANTIC FISH. SPEC. PUBL. NO. 4. [MORTALITY ESTIMATES; RECRUITMENT ESTIMATES]

RAFAIL, S. Z. 1971. ESTIMATION OF ABUNDANCE OF FISH POPULATIONS BY CAPTURE-RECAPTURE EXPERIMENTS. MARINE BIOL. 10(1), 1-7. [PROBABILITY]

RAFAIL, S. Z. 1971. ESTIMATION OF SOME PARAMETERS OF LARGE FISH POPULATIONS BY CAPTURE-RECAPTURE EXPERIMENTS. MARINE BIOL. 10(1), 8-12.

RAFAIL, S. Z. 1972. THE APPLICATION OF A MULTIPLE-RECAPTURE TECHNIQUE FOR THE STUDY OF RELATIVELY SMALL FISH POPULATIONS. MARINE BIOL. 16(3), 253-260.

RAFAIL, S. Z. 1972. THE APPLICATION OF MULTIPLE-RECAPTURE TECHNIQUE FOR THE STUDY OF SOME ASPECTS OF DYNAMICS OF LARGE FISH POPULATIONS. MARINE BIOL. 17(3), 251-255. [TAG-RECAPTURE; SAMPLING; MORTALITY RATE]

RAFAIL, S. Z. 1972. A FURTHER CONTRIBUTION TO THE STUDY OF FISH POPULATIONS BY CAPTURE-RECAPTURE EXPERIMENTS. MARINE BIOL. 14(4), 338-340.

READSHAW, J. L. 1968. ESTIMATES OF THE SIZE OF WINTER FLOCKS OF THE PIED CURRAWONG, STREPERA GRACULINA (SHAW), FROM MARK-RECAPTURE DATE - A NEW APPROACH. AUST. J. ZOOL. 16(1), 27-35.

ROBERTS, H. V. 1967. INFORMATIVE STOPPING RULES AND INFERENCES ABOUT POPULATION SIZE. J. AMER. STAT. ASSOC. 62(319), 763-775. [MAXIMUM LIKELIHOOD; PROBABILITY; FISH TAGGING]

ROBSON, D. S. 1952. INTERVAL ESTIMATION OF FISH POPULATIONS BY MEANS OF MARKED MEMBERS. (ABSTRACT). BIOMETRICS 8(4), 391. [HYPERGEOMETRIC DISTRIBUTION; CONFIDENCE INTERVAL; BINOMIAL DISTRIBUTION]

ROBSON, D. S. 1963. MAXIMUM LIKELIHOOD ESTIMATION OF A SEQUENCE OF ANNUAL SURVIVAL RATES FROM A CAPTURE-RECAPTURE SERIES. PP. 330-335. IN, NORTH ATLANTIC FISH MARKING SYMPOSIUM. INTERN. COMM. NORTHWEST ATLANTIC FISH. SPEC. PUBL. NO. 4. [POISSON FUNCTION; MATRIX; MAXIMUM LIKELIHOOD]

ROBSON, D. S. 1969. MARK-RECAPTURE METHODS IN POPULATION ESTIMATION. PP. 120-146. IN, NEW DEVELOPMENTS IN SURVEY SAMPLING. (N. L. JOHNSON AND H. SMITH, JR., EDS.). WILEY-INTERSCIENCE, NEW YORK, NEW YORK.

ROBSON, D. S. AND H. A. REGIER. 1964. SAMPLE SIZE IN PETERSEN MARK-RECAPTURE EXPERIMENTS. TRANS. AMER. FISH. SOC. 93(3), 215-226.

ROBSON, D. S. AND H. A. REGIER. 1966. ESTIMATES OF TAG LOSS FROM RECOVERIES OF FISH TAGGED AND PERMANENTLY MARKED. TRANS. AMER. FISH. SOC. 95(1), 56-59.

ROBSON, D. S. AND W. A. FLICK. 1965. A NONPARAMETRIC STATISTICAL METHOD FOR CULLING RECRUITS FROM A MARK-RECAPTURE EXPERIMENT. BIOMETRICS 21(4), 936-947.

ROFF, D. A. 1973. AN EXAMINATION OF SOME STATISTICAL TESTS USED IN THE ANALYSIS OF MARK-RECAPTURE DATA. OECOLOGIA 12(1), 35-54. [GEOMETRIC DISTRIBUTION; CHI-SQUARE; POISSON DISTRIBUTION; LESLIE MATRIX; REGRESSION TESTS; COMPUTER]

ROFF, D. A. 1973. ON THE ACCURACY OF SOME MARK-RECAPTURE ESTIMATORS. OECOLOGIA 12(1), 15-34. [PETERSEN INDEX; BAILEY'S MODEL; JOLLY AND SEBER METHOD; SIMULATION; COMPUTER]

SAMUEL, E. 1968. SEQUENTIAL MAXIMUM LIKELIHOOD ESTIMATION OF THE SIZE OF A POPULATION. ANN. MATH. STAT. 39(3), 1057-1068. [TAG-RECAPTURE]

SAMUEL, E. 1969. COMPARISON OF SEQUENTIAL RULES FOR ESTIMATION OF THE SIZE OF A POPULATION. BIOMETRICS 25(3), 517-527.

SCHAEFER, M. B. 1951. ESTIMATION OF SIZE OF ANIMAL POPULATIONS BY MARKING EXPERIMENTS. U. S. FISH WILDL. SERV. FISH. BULL. 52, 191-203.

SCHAEFER, M. B. 1951. A STUDY OF THE SPAWNING POPULATIONS OF SOCKEYE SALMON IN THE HARRISON RIVER SYSTEM, WITH SPECIAL REFERENCE TO THE PROBLEM OF ENUMERATION BY MEANS OF MARKED MEMBERS. INTERN. PACIFIC SALMON FISHERIES COMM. BULL. 4, V, 207 PP. [TAG-RECAPTURE; CONFIDENCE INTERVAL; ESTIMATION]

SCHAEFER, M. B., B. M. CHATWIN AND G. C. BROADHEAD. 1961. TAGGING AND RECOVERY OF TROPICAL TUNAS, 1955-1959. BULL. INTER-AMER. TROP. TUNA COMM. 5(5), 343-416. [TAG-RECAPTURE; ANALYSIS OF VARIANCE]

SCHNABEL, Z. E. 1938. THE ESTIMATION OF THE TOTAL FISH POPULATION OF A LAKE. AMER. MATH. MONTHLY 45(6), 348-352. [TAG-RECAPTURE ESTIMATION; MAXIMUM LIKELIHOOD]

SCHUMACHER, F. X. AND R. W. ESCHMEYER 1943. THE ESTIMATE OF FISH POPULATION IN LAKES OR PONDS. J. TENN. ACAD. SCI. 18(3), 228-249. [TAG-RECAPTURE ESTIMATION]

SEBER, G. A. F. 1962. THE MULTI-SAMPLE SINGLE RECAPTURE CENSUS. BIOMETRIKA 49(3/4), 339-350.

SEBER, G. A. F. 1965. A NOTE ON THE MULTIPLE-RECAPTURE CENSUS. BIOMETRIKA 52(1/2), 249-259.

SEBER, G. A. F. 1970. THE EFFECTS OF TRAP RESPONSE ON TAG RECAPTURE ESTIMATES. BIOMETRICS 26(1), 13-22.

SMIRNOV, V. S. 1967. THE ESTIMATION OF ANIMAL NUMBERS BASED ON THE ANALYSIS OF POPULATION STRUCTURE. PP. 199-224. IN, SECONDARY PRODUCTIVITY OF TERRESTRIAL ECOSYSTEMS, VOLUME I. (K. PETRUSEWICZ, ED.) INSTITUTE OF ECOLOGY, POLISH ACAD. SCI., WARSAW, POLAND. [GEOMETRIC SERIES; PROBABILITY; POISSON DISTRIBUTION; KELKER'S EQUATION; TAG-RECAPTURE ESTIMATION]

SMITH, H. D., C. D. JORGENSEN AND H. D. TOLLEY. 1972. ESTIMATION OF SMALL MAMMAL USING RECAPTURE METHODS: PARTITIONING OF ESTIMATOR VARIABLES. ACTA THERIOL. 17(5), 57-66. [MODEL; CAPTURE-RECAPTURE; PROBABILITY; CONFIDENCE INTERVAL; BINOMIAL MODEL]

SONLEITNER, F. J. 1965. APPLICATION AND COMPARISON OF MARK-CAPTURE MODELS. PROC. 12TH INTERN. CONGR. ENTOMOL., LONDON, 1964. P. 399.

SONLEITNER, F. J. 1973. MARK-RECAPTURE ESTIMATES OF OVER-WINTERING SURVIVAL OF THE QUEENSLAND FRUIT FLY, DACUS TRIONI, IN FIELD CAGES. RES. POPUL. ECOL.-KYOTO 14(2), 188-208.

SONLEITNER, F. J. AND M. A. BATEMAN. 1963. MARK-RECAPTURE ANALYSIS OF A POPULATION OF QUEENSLAND FRUIT-FLY, DACUS TRYONI (FROGG.) IN AN ORCHARD. J. ANIM. ECOL. 32(2), 259-269. [VARIANCE; DETERMINISTIC MODEL OF LESLIE AND CHITTY; SAMPLE SIZE; THREE-POINT SAMPLING MODEL]

STEINHORST, R. K. AND D. M. SWIFT. 1973. ESTIMATING SMALL MAMMAL DENSITIES FROM ASSESSMENT LINE DATA. (ABSTRACT). BIOMETRICS 29(3), 612-613. [TAG-RECAPTURE; LAG REGRESSION; REGRESSION; MOVING AVERAGES]

STRANDGAARD, H. 1967. RELIABILITY OF THE PETERSEN METHOD TESTED ON A ROE-DEER POPULATION. J. WILDL. MANAGE. 31(4), 643-651.

TAKAHASI, K. 1960. MODEL FOR THE ESTIMATION OF THE SIZE OF A POPULATION BY USING CAPTURE-RECAPTURE METHOD. ANN. INST. STAT. MATH. (TOKYO) 12(1), 237-248.

TANAKA, R. 1951. ESTIMATION OF VOLE AND MOUSE POPULATIONS ON MT. ISHIZUCHI AND ON THE UPLANDS OF SOUTHERN SHIKOKU. J. MAMMAL. 32(4), 450-458. [TAG-RECAPTURE; HAYNE'S METHOD; FORMULA]

TANAKA, R. 1952. THEORETICAL JUSTIFICATION OF THE MARK-AND-RELEASE INDEX FOR SMALL MAMMALS. BULL. KOCHI WOMEN'S COLLEGE 1, 38-47.

TANAKA, R. 1956. ON DIFFERENTIAL RESPONSE TO LIVE TRAPS OF MARKED AND UNMARKED SMALL MAMMALS. ANNOTATIONES ZOOLOGICAE JAPONENSES 29(1), 44-51. [MAXIMUM LIKELIHOOD; STANDARD ERROR]

TANAKA, R. 1959. A CRITICISM ON SOME EXAMPLES OFFERING A KILL-EFFICIENCY BY THE MARK-AND-RELEASE CENSUS OF PRE-AND POST-POISONING VOLE POPULATIONS. BULL. KOCHI WOMEN'S UNIV. 7(3), 7-13. [MAXIMUM LIKELIHOOD; LATIN SQUARE]

TANAKA, R. 1961. A FIELD STUDY OF EFFECT OF TRAP SPACING UPON ESTIMATES OF RANGES AND POPULATIONS IN SMALL MAMMALS BY MEANS OF A LATIN SQUARE ARRANGEMENT OF QUADRATS. BULL. KOCHI WOMEN'S UNIV. 9(5), 8-16. [PROBABILITY]

TANAKA, R. 1970. A FIELD STUDY OF THE EFFECT OF PREBAITING ON CENSUSING BY THE CAPTURE-RECAPTURE METHOD IN A VOLE POPULATION. RES. POPUL. ECOL.-KYOTO 12(1), 111-125. [EBERHARDT'S EQUATION]

TANAKA, R. 1972. INVESTIGATION INTO THE EDGE EFFECT BY USE OF CAPTURE-RECAPTURE DATA IN A VOLE POPULATION. RES. POPUL. ECOL.-KYOTO 13(2), 127-151. [DICE'S ASSESSMENT LINE; BIVARIATE NORMAL DISTRIBUTION]

TANAKA, R. 1974. AN APPROACH TO THE EDGE EFFECT IN PROOF OF THE VALIDITY OF DICE'S ASSESSMENT LINES IN SMALL MAMMAL CENSUSING. RES. POPUL. ECOL.-KYOTO 15(2), 121-137. [TAG-RECAPTURE; JOLLY'S EQUATIONS; RANDOM WALK; HOME RANGE; WIERZBOWSKA'S METHOD]

TANAKA, R. AND M. KANAMORI. 1967. NEW REGRESSION FORMULA TO ESTIMATE THE WHOLE POPULATION FOR RECAPTURE-ADDICTED SMALL MAMMALS. RES. POPUL. ECOL.-KYOTO 9(2), 83-94.

TANAKA, S. 1956. ESTIMATION OF WHALE POPULATION ON THE BASIS OF TAGGING EXPERIMENTS. BULL. JAPANESE SOC. SCI. FISH. 22(6), 330-337. [POISSON DISTRIBUTION; DELURY'S METHOD; MAXIMUM LIKELIHOOD; BINOMIAL DISTRIBUTION; TAG-RECAPTURE]

TANAKA, S. 1971. ESTIMATION OF THE COEFFICIENT OF TAG SHEDDING FROM DRIFTING SEAWEED. BULL. JAPANESE SOC. SCI. FISH. 37(11), 1067-1072. [MAXIMUM LIKELIHOOD; ROBSON AND REIGER METHOD]

TANTON, M. T. 1965. PROBLEMS OF LIVE-TRAPPING AND POPULATION ESTIMATION FOR THE WOOD MOUSE, APODEMUS SYLVATICUS (L.). J. ANIM. ECOL. 34(1), 1-22. [TAG-RECAPTURE; MIGRATION DIRECTION MODEL; POISSON DISTRIBUTION; DARROCH'S METHOD; TRUNCATED NEGATIVE BINOMIAL DISTRIBUTION; FREQUENCY OF CAPTURE; GAMMA DISTRIBUTION]

TAYLOR, S. M. 1966. RECENT QUANTITATIVE WORK ON BRITISH BIRD POPULATIONS. THE STATISTICIAN 16(2), 119-170. [CAPTURE-RECAPTURE; MAXIMUM LIKELIHOOD; PROBABILITY]

TEPPER, E. E. 1967. STATISTICAL METHODS IN USING MARK-RECAPTURE DATA FOR POPULATION ESTIMATION. U. S. DEPARTMENT OF INTERIOR, OFFICE OF LIBRARY SERVICES, BIBLIOGRAPHIC SERIES BIBLIO. NO. 4. II, 65 PP.

WELCH, H. E. 1960. TWO APPLICATIONS OF A METHOD OF DETERMINING THE ERROR OF POPULATION ESTIMATES OF MOSQUITO LARVAE BY THE MARK AND RECAPTURE TECHNIQUE. ECOLOGY 41(1), 228-229.

WHITE, E. G. 1971. A VERSATILE FORTRAN COMPUTER PROGRAM FOR THE CAPTURE-RECAPTURE STOCHASTIC MODEL OF G. M. JOLLY. J. FISH. RES. BOARD CAN. 28(3), 443-445.

WHITE, E. G. 1975. IDENTIFYING POPULATION UNITS THAT COMPLY WITH CAPTURE-RECAPTURE ASSUMPTIONS IN AN OPEN COMMUNITY OF ALPINE GRASSHOPPERS. RES. POPUL. ECOL.-KYOTO 16(2), 153-187. [COMPUTER; JOLLY'S STOCHASTIC MODEL]

WIERZBOWSKA, T. 1972. STATISTICAL ESTIMATION OF HOME RANGE SIZE OF SMALL RODENTS. EKOLOGIYA POLSKA-SERIA A 20(49), 781-831. [MODEL; PROBABILITY; TAG-RECAPTURE]

WILBUR, H. M. AND J. M. LANDWEHR. 1974. THE ESTIMATION OF POPULATION SIZE WITH EQUAL AND UNEQUAL RISKS OF CAPTURE. ECOLOGY 55(6), 1339-1348. [TAG-RECAPTURE; NEGATIVE BINOMIAL DISTRIBUTION; POISSON DISTRIBUTION]

WITTES, J. T. 1972. ON THE BIAS AND ESTIMATED VARIANCE OF CHAPMAN'S TWO-SAMPLE CAPTURE-RECAPTURE POPULATION ESTIMATE. BIOMETRICS 28(2), 592-597.

WITTES, J. T. 1974. APPLICATIONS OF A MULTINOMIAL CAPTURE-RECAPTURE MODEL TO EPIDEMIOLOGICAL DATA. J. AMER. STAT. ASSOC. 69(345), 93-97.

WITTES, J. T. 1974. CAPTURE-RECAPTURE MODELS IN MEDICINE AND ALLIED FIELDS. (ABSTRACT). BIOMETRICS 30(2), 386-387.

WITTES, J. T. AND V. W. SIDEL. 1968. A GENERALIZATION OF THE SIMPLE CAPTURE-RECAPTURE MODEL WITH APPLICATIONS TO EPIDEMIOLOGICAL RESEARCH. J. CHRONIC DIS. 21(5), 287-301. [MAXIMUM LIKELIHOOD ESTIMATION; FISH]

YOUNG, H. 1961. A TEST FOR RANDOMNESS IN TRAPPING. BIRD-BANDING 32(3), 160-162. [RUN THEORY]

YOUNGS, W. D. 1972. ESTIMATION OF NATURAL AND FISHING MORTALITY RATES FROM TAG RECAPTURES. TRANS AMER. FISH. SOC. 101(3), 542-545. [REGRESSION]

POPULATION ESTIMATION (OTHER METHODS)

ALLEN, K. R. 1966. SOME METHODS FOR ESTIMATING EXPLOITED POPULATIONS. J. FISH. RES. BOARD CAN. 23(10), 1553-1574. [MORTALITY RATE; Q-METHOD; TAG-RECAPTURE ESTIMATION]

ALLEN, W. E. 1930. METHODS IN QUANTITATIVE RESEARCH ON MARINE MICROPLANKTON. BULL. SCRIPPS INST. OCEANOGR. UNIV. CALIF. 2(8), 319-329.

ANDERSON, K. P. 1964. STATISTICAL ASPECTS OF ABUNDANCE ESTIMATES. RAPP. PROCES-VERB. REUNIONS J. CONS. PERM. INT. EXPLOR. MER 155, 15-18. [NORMAL DISTRIBUTION; LOG-NORMAL DISTRIBUTION; NEGATIVE BINOMIAL DISTRIBUTION; POISSON DISTRIBUTION; STOCHASTIC VARIABLE; REGRESSION; VARIANCE]

ANDRZEJEWSKI, R. AND W. JEZIERSKI. 1966. STUDIES ON THE EUROPEAN HARE. XI. ESTIMATION OF POPULATION DENSITY AND ATTEMPT TO PLAN THE YEARLY TAKE OF HARES. ACTA THERIOL. 11(21), 433-448.

ANONYMOUS. 1958. A COMPARISON OF THE PRECISION OF VARIOUS ESTIMATORS FOR ESTIMATING WATERFOWL KILL FROM SURVEY DATA. INST. OF STAT., NORTH CAROLINA STATE COLLEGE, RALEIGH, NORTH CAROLINA. U. S. FISH AND WILDLIFE SERVICE CONTRACT NO. 14-16-008-508. MIMEO V, 51 PP.

ANONYMOUS. 1967. REPORT ON THE FOURTH SESSION OF THE ADVISORY COMMITTEE ON MARINE RESOURCES RESEARCH 16-21 JANUARY 1967. REPORT OF AGMRR WORKING PARTY ON DIRECT AND SPEEDIER ESTIMATION OF FISH ABUNDANCE. FAO FISHERIES REPORT NO. 41 (SUPPL. 1). I, 39 PP. [EQUATIONS OF ACOUSTIC FISH DETECTION]

ARCHIBALD, E. E. A. 1950. PLANT POPULATIONS. II. THE ESTIMATION OF THE NUMBER OF INDIVIDUALS PER UNIT AREA OF SPECIES IN HETEROGENEOUS PLANT POPULATIONS. ANN. BOT. (LONDON) NEW SERIES 14(53), 7-21. [DISTRIBUTION; POISSON DISTRIBUTION; THOMAS'S DISTRIBUTION; COEFFICIENT OF DISPERSION]

ARNASON, A. N. 1973. THE ESTIMATION OF POPULATION SIZE, MIGRATION RATES AND SURVIVAL IN A STRATIFIED POPULATION. RES. POPUL. ECOL.-KYOTO 15(1), 1-8. [TAG-RECAPTURE; MATRIX; CONDITIONAL LIKELIHOOD; MOMENT EQUATIONS]

ATWOOD, E. L. 1958. A PROCEDURE FOR REMOVING THE EFFECT OF RESPONSE BIAS ERRORS FROM WATERFOWL HUNTER QUESTIONNAIRE RESPONSES.(ABSTRACT). BIOMETRICS 14(1), 132-133.

BARBEHENN, K. R. 1974. ESTIMATING DENSITY AND HOME RANGE SIZE WITH REMOVAL GRIDS: THE RODENTS AND SHREWS OF GUAM. ACTA THERIOL. 19(14), 191-234. [MODEL]

BEALL, G. 1939. METHODS OF ESTIMATING THE POPULATION OF INSECTS IN A FIELD. BIOMETRIKA 30(3/4), 422-439.

BENNETT, C., JR. 1967. A NEW METHOD FOR ESTIMATING NUMBERS OF DUCK BROODS. J. WILDL. MANAGE. 31(3), 555-562. [MODEL; LOGARITHMIC SERIES]

BERGERUD, A. T. 1968. NUMBERS AND DENSITIES. PP. 21-42. IN: A PRACTICAL GUIDE TO THE STUDY OF THE PRODUCTIVITY OF LARGE HERBIVORES. (F. B. GOLLEY AND H. K. BUECHNER, EDS.). BLACKWELL SCIENTIFIC PUBLICATIONS, OXFORD, ENGLAND. [RANDOM SAMPLING; SYSTEMATIC SAMPLING; VARIANCE; STRATIFICATION; INDEX OF DISPERSION; TAG-RECAPTURE; SEX AND AGE RATIOS]

BLISS, C. I. 1941. STATISTICAL PROBLEMS IN ESTIMATING POPULATIONS OF JAPANESE BEETLE LARVAE. J. ECON. ENTOMOL. 34(2), 221-232. [POISSON DISTRIBUTION; BLOCKS; TRANSFORMATION]

BOGUSLAVSKY, G. W. 1956. STATISTICAL ESTIMATION OF THE SIZE OF A SMALL POPULATION. SCIENCE 124(3216), 317-318.

BRAATEN, D. L. 1969. ROBUSTNES OF THE DELURY POPULATION ESTIMATOR J. FISH. RES. BOARD CAN. 26(2), 339-355. [MONTE CARLO]

CALHOUN, J. B. AND J. U. CASBY, 1958. CALCULATION OF HOME RANGE AND DENSITY OF SMALL MAMMALS. U. S. PUBLIC HEALTH SERVICE, PUBLIC HEALTH MONOGR. NO. 55. IV, 24 PP. [NORMAL DISTRIBUTION]

CAUGHLEY, G. 1974. BIAS IN AERIAL SURVEY. J. WILDL. MANAGE. 38(4), 921-933. [MODEL; MULTIPLE REGRESSION MODEL]

CAUGHLEY, G. AND J. GODDARD. 1972. IMPROVING THE ESTIMATES FROM INACCURATE CENSUSES. J. WILDL. MANAGE. 36(1), 135-140. [COMPUTER]

CHAPMAN, D. G. 1954. THE ESTIMATION OF BIOLOGICAL POPULATIONS. ANN. MATH. STAT. 25(1), 1-15.

CHAPMAN, D. G. 1955. POPULATION ESTIMATION BASED ON CHANGE OF COMPOSITION CAUSED BY A SELECTIVE REMOVAL. BIOMETRIKA 42(3/4), 279-290.

CHAPMAN, D. G. 1964. A CRITICAL STUDY OF PRIBILOF FUR SEAL POPULATION ESTIMATES. U. S. FISH WILDL. SERV. FISH. BULL. 63(3), 657-669. [SAMPLING]

CHAPMAN, D. G. 1971. ESTIMATION OF POPULATION PARAMETERS OF ANTARCTIC BALEEN WHALES. QUANTITATIVE SCIENCE PAPER NO. 22. CENTER FOR QUANTITATIVE SCIENCE IN FORESTRY, FISHERIES AND WILDLIFE, UNIVERSITY OF WASHINGTON, SEATTLE, WASHINGTON. MIMEO. 23 PP.

HAPMAN, D. G. AND C. O. JUNGE, JR. 956. THE ESTIMATION OF THE SIZE F A STRATIFIED ANIMAL POPULATION. NN. MATH. STAT. 27(2), 375-389.

HAPMAN, D. G., W. S. OVERTON AND . L. FINKNER. 1959. METHODS OF STIMATING DOVE KILL. INST. OF TAT., NORTH CAROLINA STATE OLLEGE, RALEIGH, NORTH CAROLINA. IMEO SERIES NO. 243. 62 PP.

CHATTERJEE, S. 1973. ESTIMATING THE SIZE OF WILDLIFE POPULATIONS. PP. 99-104. IN, STATISTICS BY EXAMPLE: EXPLORING DATA. (F. MOSTELLER, W. H.KRUSKAL, R. F. LINK, R. S. PIETERS AND G. R. RISING, EDS.). ADDISON-WESLEY PUBLISHING CO., READING, MASSACHUSETTS.

COOPER, G. P. AND K. F. LAGLER 1956. APPRAISAL OF METHODS OF FISH POPULATION STUDY - PART III. THE MEASUREMENT OF FISH POPULATION SIZE. TRANS. N. AMER. WILDL. CONF. 21, 281-297. [DELURY'S METHOD ; SCHNABEL METHOD OF ESTIMATION ; PETERSEN INDEX ; TAG-RECAPTURE ESTIMATION]

COX, E. L. 1949. MATHEMATICAL BASES OF EXPERIMENTAL SAMPLING FOR ESTIMATION OF SIZE OF CERTAIN BIOLOGICAL POPULATIONS. STATISTICAL LABORATORY REPORT, VIRGINIA AGRICULTURAL EXPERIMENT STATION, BLACKSBURG, VIRGINIA. 42 PP. [TAG-RECAPTURE ESTIMATION]

CRAIG, C. C. 1953. ON A METHOD OF ESTIMATING BIOLOGICAL POPULATIONS IN THE FIELD. BIOMETRIKA 40(1/2), 216-218.

CUSHING, D. H. 1964. THE COUNTING OF FISH WITH AN ECHO-SOUNDER. RAPP. PROCES-VERB. REUNIONS J. CONS. PERM. INT. EXPLOR. MER 155, 190-195. [TARGET STRENGTH FUNCTION; SIZE OF TARGET]

CUSHING, D. H. 1968. DIRECT ESTIMATION OF A FISH POPULATION ACOUSTICALLY. J. FISH. RES. BOARD CAN. 25(11), 2349-2364. [EQUATIONS]

CUSHING, D. H. 1969. THE ESTIMATION OF FISH ABUNDANCE. PP. 57-61. IN, MANUAL OF METHODS FOR FISH STOCK ASSESSMENT PART V. THE USE OF ACOUSTIC INSTRUMENTS IN FISH DETECTION AND FISH ABUNDANCE ESTIMATION. (B. B. PARRISH, ED.). FAO FISHERIES TECHNICAL PAPER NO. 83. [EQUATIONS]

DAVIS, D. E. 1965. ESTIMATING THE NUMBERS OF GAME POPULATIONS. PP. 89-118. IN, WILDLIFE INVESTIGATIONAL TECHNIQUES (2ND REVISED EDITION). (H. S. MOSBY, ED.). THE WILDLIFE SOCIETY, WASHINGTON, D. C.

DELURY, D. B. 1947. ON THE ESTIMATION OF BIOLOGICAL POPULATIONS. BIOMETRICS 3(4), 145-167.

DELURY, D. B. 1951. ON THE PLANNING OF EXPERIMENTS FOR THE ESTIMATION OF FISH POPULATIONS. J. FISH. RES. BOARD CAN. 8(4), 281-307.

DELURY, D. B. 1954. ON THE ASSUMPTIONS UNDERLYING ESTIMATES OF MOBILE POPULATIONS. PP. 287-293. IN, STATISTICS AND MATHEMATICS IN BIOLOGY. (O. KEMPTHORNE, T. A. BANCROFT, J. W. GOWEN AND J. L. LUSH, EDS.) IOWA STATE COLLEGE PRESS, AMES, IOWA.

DICE, L. R. 1938. SOME CENSUS METHODS FOR MAMMALS. J. WILDL. MANAGE. 2(3), 119-130. [HOME RANGE]

DICKIE, L. M. 1955. FLUCTUATIONS IN ABUNDANCE OF THE GIANT SCALLOP, PLACOPECTEN MAGELLANICUS (GMELIN), IN THE DIGBY AREA OF THE BAY OF FUNDY. J. FISH. RES. BOARD CAN. 12(6), 797-857. [DELURY'S METHOD; ESTIMATION]

DOI, T. 1960. ON THE TEN-DAY FLUCTUATION OF INDEX TO SIZE OF FISH POPULATION IN THE WATERS ADJACENT TO THE BUNGO STRAIT. BULL. TOKAI REG. FISH. RES. LAB. 27, 15-21. [EQUATIONS]

DOI, T. 1964. ESTIMATION OF THE SIZE OF FIN WHALE POPULATION IN THE ANTARCTIC. BULL. TOKAI REG. FISH. RES. LAB. 39, 1-11. [CATCHABILITY COEFFICIENT]

DOI, T. 1965. ON THE PREDICTION OF FUTURE CATCHES OF YELLOWTAILS CAUGHT BY SET NET AT KUKI IN MIE PERFECTURE FROM TECHNIQUES OF TIME SERIES. BULL. TOKAI REG. FISH. RES. LAB. 44, 1-15.

DOI, T. 1970. RE-EVALUATION OF POPULATION STUDIES BY SIGHTING OBSERVATION OF WHALE. BULL. TOKAI REG. FISH. RES. LAB. 63, 1-10. [PROBABILITY; EQUATIONS]

DOI, T. 1971. DIAGNOSIS METHODS OF SPERM WHALE POPULATION. BULL. TOKAI REG. FISH. RES. LAB. 64, 89-143. [MODEL; ESTIMATION]

DOI, T. 1971. FURTHER DEVELOPMENT OF SIGHTING THEORY ON WHALE. BULL. TOKAI REG. FISH. RES. LAB. 68, 1-29. [PROBABILITY; EQUATIONS]

DOI, T. 1974. FURTHER DEVELOPMENT OF WHALE SIGHTING THEORY. PP. 359-368. IN, THE WHALE PROBLEM: A STATUS REPORT. (W. E. SCHEVILL, ED.). HARVARD UNIVERSITY PRESS, CAMBRIDGE, MASSACHUSETTS. [ESTIMATION]

ENGLISH, T. S. 1964. A THEORETICAL MODEL FOR ESTIMATING THE ABUNDANCE OF PLANKTONIC FISH EGGS. RAPP. PROCES-VERB. REUNIONS J. CONS. PERM. INT. EXPLOR. MER 155, 174-182. [ANALYSIS OF VARIANCE; PARTIALLY HIERARCHICAL MODEL]

ERICKSON, R. O. AND J. R. STEHN. 1945. A TECHNIQUE FOR ANALYSIS OF POPULATION DENSITY DATA. AMER. MIDL. NAT. 33(3), 781-787. [POISSON DISTRIBUTION; ARTIFICIAL POPULATION; CHI-SQUARE; DOUBLE POISSON DISTRIBUTION]

FORBES, S. T. AND O. NAKKEN. (EDS.). 1972. MANUAL OF METHODS FOR FISHERIES RESOURCE SURVEY AND APPRAISAL PART 2. THE USE OF ACOUSTIC INSTRUMENTS FOR FISH DETECTION AND ABUNDANCE ESTIMATION. FAO MANUALS IN FISHERIES SCIENCE NO. 5. IX, 138 PP. [MODEL; EQUATIONS]

FORD, J. S. 1974. ECHO INTEGRATOR ABSOLUTE CONSTANTS PUT INTO REAL NUMBERS. FISH. RES. BOARD CAN. TECH. REPT. NO. 467. 33 PP. [POPULATION ESTIMATION]

GATES, C. E. AND W. B. SMITH 1972. ESTIMATION OF DENSITY OF MOURNING DOVES FROM AURAL INFORMATION. BIOMETRICS 28(2), 345-359. [SAMPLING; SIMULATION; TIME DEPENDENT DATA; COMPUTER]

GATES, C. E. AND W. B. SMITH. 1970. ESTIMATION OF DENSITY OF MOURNING DOVES FROM AURAL INFORMATION. (ABSTRACT). BIOMETRICS 26(3), 601. [SAMPLING; COMPUTER; SIMULATION]

GERRARD, D. J. AND H. C. CHIANG. 1970. DENSITY ESTIMATION OF CORN ROOTWORM EGG POPULATIONS BASED UPON FREQUENCY OF OCCURRENCE. ECOLOGY 51(2), 237-245. [SAMPLING; MODEL; CONTAGIOUS DISTRIBUTION; PROBABILITY; VARIANCE; POISSON DISTRIBUTION]

GERRARD, D. J. AND R. D. COOK. 1972. INVERSE BINOMIAL SAMPLING AS A BASIS FOR ESTIMATING NEGATIVE BINOMIAL POPULATION DENSITIES. BIOMETRICS 28(4), 971-980. [INSECT]

GLASGOW, J. P. 1953. THE EXTERMINATION OF ANIMAL POPULATIONS BY ARTIFICIAL PREDATION AND THE ESTIMATION OF POPULATIONS. J. ANIM. ECOL. 22(1), 32-46. [GEOMETRIC DISTRIBUTION; CAPTURE-RECAPTURE; BIRTH RATE; DEATH RATE; FISHER'S LATTICE]

GORDON, R. D. 1938. NOTE ON ESTIMATING BACTERIAL POPULATIONS BY THE DILUTION METHOD. PROC. NATL. ACAD. SCI. U.S.A. 24(5), 212-215. [EQUATIONS]

GRAY, H. E. AND A. E. TRELOAR. 1935. ON THE ENUMERATION OF INSECT POPULATIONS BY THE METHOD OF NET COLLECTION. ECOLOGY 14(4), 356-367.

GRIFFITHS, R. C. 1960. A STUDY OF MEASURES OF POPULATION DENSITY AND OF CONCENTRATION OF FISHING EFFORT IN THE FISHERY FOR YELLOWFIN TUNA, NEOTHUNNUS MACROPTERUS, IN THE EASTERN TROPICAL PACIFIC OCEAN, FROM 1951 TO 1956. BULL. INTER-AMER. TROP. TUNA COMM. 4(3), -98. [INDEX OF DENSITY; INDEX OF CONCENTRATION OF EFFORT; GULLAND'S EQUATION; REGRESSION]

GRODZINSKI, W., Z. PUCEK AND L. RYSZKOWSKI. 1966. ESTIMATION OF RODENT NUMBERS BY MEANS OF PREBAITING AND INTENSIVE REMOVAL. ACTA THERIOL. 11(10), 297-314.

GULLAND, J. A. 1964. CATCH PER UNIT EFFORT AS A MEASURE OF ABUNDANCE. RAPP. PROCES-VERB. REUNIONS J. CONS. PERM. INT. EXPLOR. MER 155, 8-14. [INDEX OF ABUNDANCE]

GULLAND, J. A. 1966. ANALYSIS OF TARGET DISCRIMINATION WITH SECTOR SCANNING EQUIPMENT. J. CONS. PERM. INT. EXPLOR. MER 30(3), 343-345. [EQUATIONS]

HAMLEY, J. M. 1972. USE OF THE DELURY METHOD TO ESTIMATE GILLNET SELECTIVITY. J. FISH. RES. BOARD CAN. 29(11), 1636-1638.

HANSON, W. R. 1963. CALCULATION OF PRODUCTIVITY, SURVIVAL, AND ABUNDANCE OF SELECTED VERTEBRATES FROM SEX AND AGE RATIOS. WILDL. MONOGR. NO. 9. 60 PP.

HANSON, W. R. 1968. ESTIMATING THE NUMBER OF ANIMALS: A RAPID METHOD FOR UNIDENTIFIED INDIVIDUALS. SCIENCE 162(3854), 675-676.

HANSSON, L. 1969. HOME RANGE, POPULATION STRUCTURE AND DENSITY ESTIMATES AT REMOVAL CATCHES WITH EDGE EFFECT. ACTA THERIOL. 14(11), 153-160.

HARTLEY, H. O., P. G. HOMEYER AND E. L. KOZICKY. 1955. THE USE OF LOG TRANSFORMATIONS IN ANALYZING FALL ROADSIDE PHEASANT COUNTS. J. WILDL. MANAGE. 19(4), 495-496. [ANALYSIS OF VARIANCE]

HAYNE, D. W. 1949. TWO METHODS FOR ESTIMATING POPULATION FROM TRAPPING RECORDS. J. MAMMAL. 30(4), 399-411.

HEIP, C. 1974. A RAPID METHOD TO EVALUATE NEMATODE DENSITY. NEMATOLOGICA 20(2), 266-268. [GEOMETRIC SERIES; ESTIMATION]

HOLGATE, P. 1973. AN ESTIMATOR FOR THE SIZE OF AN ANIMAL POPULATION. BIOMETRIKA 60(1), 135-140. [STOCHASTIC MODEL; LESLIE MATRIX; POISSON DISTRIBUTION]

HOUSER, A. AND J. E. DUNN. 1967. ESTIMATING THE SIZE OF THE THREADFIN SHAD POPULATION IN BULL SHOALS RESERVOIR FROM MIDWATER TRAWL CATCHES. TRANS. AMER. FISH. SOC. 96(2), 176-184. [MAXIMUM LIKELIHOOD; STRATIFIED SAMPLING; GAMMA DISTRIBUTION; PROBABILITY DISTRIBUTION; POISSON DISTRIBUTION; NEGATIVE BINOMIAL DISTRIBUTION; VARIANCE]

HUNTER, W. R. AND D. C. GRANT. 1966. ESTIMATES OF POPULATION DENSITY AND DISPERSAL IN THE NATICID GASTROPOD, POLINICES DUPLICATUS, WITH A DISCUSSION OF COMPUTATIONAL METHODS. BIOL. BULL. 131(2), 292-307. [TAG-RECAPTURE METHODS]

IMAIZUMI, Y. 1973. ON A NEW METHOD OF ANALYSIS OF LOCAL POPULATIONS EMPLOYING THE BINOMIAL DISTRIBUTION, WITH ITS APPLICATION TO THE RED BACKED VOLES OF HOKKAIDO. J. MAMMAL SOC. JAPAN 5(6), 213-223.

INOUE, T. 1973. A QUANTITATIVE ANALYSIS OF DISPERSAL IN A HORSE-FLY, TABANUS IYOENSIS SHIRAKI, AND ITS APPLICATION TO ESTIMATE THE POPULATION SIZE. RES. POPUL. ECOL.-KYOTO 14(2), 209-233.

ISHII, T. 1967. STUDIES ON ESTIMATING PARAMETERS OF A FISH POPULATION SUPPLIED BY SEQUENTIAL RECRUITMENT-I. THE EFFECT ON ESTIMATES FOR PACIFIC YELLOWFIN TUNA. BULL. JAPANESE SOC. SCI. FISH. 33(6), 513-523. [BEVERTON-HOLT MODEL]

ISHII, T. 1967. STUDIES ON ESTIMATING PARAMETERS OF A FISH POPULATION SUPPLIED BY SEQUENTIAL RECRUITMENT-II. A METHOD OF SIMULTANEOUS ESTIMATION BY MINIMIZING THE TRACING INDEX. BULL. JAPANESE SOC. SCI. FISH. 33(8), 738-745. [EQUATIONS]

ISHII, T. 1968. STUDIES ON ESTIMATING PARAMETERS OF A FISH POPULATION SUPPLIED BY SEQUENTIAL RECRUITMENT-III. SIMULTANEOUS ESTIMATION OF PARAMETERS OF PACIFIC BIGEYE TUNA BY THE TRACING METHOD. BULL. JAPANESE SOC. SCI. FISH. 34(6), 488-494. [TRACING INDEX; RECRUITMENT INDEX]

ISHII, T. 1969. STUDIES ON ESTIMATING PARAMETERS OF A FISH POPULATION SUPPLIED BY SEQUENTIAL RECRUITMENT-V. SIMULTANEOUS ESTIMATION OF PARAMETERS WITH THE TRANSFER EFFECT OF PACIFIC YELLOWFIN TUNA. BULL. JAPANESE SOC. SCI. FISH. 35(6), 537-545. [TRACING INDEX]

ISHII, T. 1969. STUDIES ON ESTIMATING PARAMETERS OF FISH POPULATION SUPPLIED BY SEQUENTIAL RECRUITMENT-IV. VARIABILITY OF ESTIMATED VALUES OF PARAMETERS BY THE TRACING METHOD. BULL. JAPANESE SOC. SCI. FISH. 35(3), 265-272. [TRACING INDEX]

ISHII, T. 1974. STUDIES ON ESTIMATING PARAMETERS OF FISH POPULATION SUPPLIED BY SEQUENTIAL RECRUITMENT-VI. SIMULTANEOUS ESTIMATION OF PARAMETERS WITH THE MEAN RECRUITING AGE OF PACIFIC BIGEYE TUNA. BULL. JAPANESE SOC. SCI. FISH. 40(1), 51-56. [MODEL]

JANION, M. S. AND T. WIERZBOWSKA. 1972. ESTIMATING THE DENSITY OF RODENTS BY MEANS OF STAINED FOOD. ACTA THERIOL. 17(35), 467-474.

JORGENSEN, C. D., H. D. SMITH AND D. T. SCOTT. 1972. EVALUATION OF TECHNIQUES FOR ESTIMATING POPULATION SIZES OF DESERT RODENTS RESEARCH MEMORANDUM RM 72-10, DESERT BIOME, IBP, BRIGHAM YOUNG UNIVERSITY, PROVO, UTAH, 196 PP. [TAG-RECAPTURE ESTIMATION; STOCHASTIC]

JUNGE, C. O., JR. 1963. A QUANTITATIVE EVALUATION OF THE BIA IN POPULATION ESTIMATES BASED ON SELECTIVE SAMPLES. PP. 26-28. IN NORTH ATLANTIC FISH MARKING SYMPOSIUM, INTERN. COMM. NORTHWEST ATLANTIC FISH., SPECIAL PUBL. 4. [DETERMINISTIC MODEL; CORRELATION COEFFICIENT; TAG-RECAPTURE]

KAUFMAN, D. W., G. C. SMITH, R. M. JONES, J. B. GENTRY AND M. H. SMITH. 1971. USE OF ASSESSMENT LINES TO ESTIMATE DENSITY OF SMALL MAMMALS. ACTA THERIOL. 16(9), 127-147. [REMOVAL TRAPPING; AREA OF EFFECT; DENSITY]

KAWAKAMI, T. 1970. ON THE METHOD OF INFERENCE OF STANDING CROP AND GREGARIOUS STATE OF NEKTON IN A FISHING GROUND BY THE SIMULTANEOUS USE OF TWO ECHO-SOUNDERS OF DIFFERENT DIRECTIVITIES. BULL. JAPANESE SOC. SCI. FISH. 36(12), 1203-1207. [EQUATIONS]

KELKER, G. H. 1942. SEX-RATIO EQUATIONS AND FORMULAS FOR DETERMINING WILDLIFE POPULATIONS. PROC. UTAH ACAD. SCI., ARTS LETT. 19, 189-198.

KIMURA, M. AND J. F. CROW. 1962. THE MEASUREMENT OF EFFECTIVE POPULATION NUMBER. MRC TECHN. SUMM. REPORT-335, U.S. ARMY, MADISON, WISCONSIN. 23 PP.

KING, E. W. AND J. R. HOLMAN. 1966. THE DETERMINATION OF POPULATION SIZE BY MEASUREMENT OF RESPONSE TO THE PHYSICAL ENVIRONMENT. ANN. ENTOMOL. SOC. AMER. 59(6), 1200-1205. [MODEL; REGRESSION; LOG TRANSFORMATION]

KOZICKY, E. L., H. O. HARTLEY AND G. O. HENDRICKSON. 1954. A PROPOSED COMPARISON OF FALL ROADSIDE PHEASANT COUNTS AND FLUSHING RATES. PROC. IOWA ACAD. SCI. 61, 528-534. [DESIGN; LATIN SQUARE]

UDRIN, A. I. 1971. SOME QUANTITATIVE INDICES USED IN TREATMENT OF ANIMAL POPULATIONS. OOLOGISCHESKII ZHURNAL 50(12), 861-1864.

UROKI, T., K. KAWAGUCHI, W. AKAMOTO AND H. WATANABE. 1971. NEW TELEMETRIC APPARATUS TO ETECT FISH LOCATION AND ITS URROUNDING WATER TEMPERATURE. JLL. JAPANESE SOC. SCI. FISH. 7(10), 964-972. [EQUATIONS]

KUTKUHN, J. H. 1958. NOTES ON THE PRECISION OF NUMERICAL AND VOLUMETRIC PLANKTON ESTIMATES FROM SMALL-SAMPLE CONCENTRATES. LIMNOL. OCEANOGR. 3(1), 69-83. [POISSON DISTRIBUTION; NEGATIVE BINOMIAL DISTRIBUTION; SAMPLE SIZE; VARIANCE]

LEBEDEVA, L. P. 1972. SOUND SCATTERING BY FISH. J. ICHTHYOLOGY 12(1), 144-149. (ENGLISH TRANSL.). [EQUATIONS]

LESLIE, P. H. AND D. H. S. DAVIS. 1939. AN ATTEMPT TO DETERMINE THE ABSOLUTE NUMBER OF RATS ON A GIVEN AREA. J. ANIM. ECOL. 8(1), 94-113. [PROBABILITY]

LEWIS, J. C. AND J. W. FARRAR. 1968. AN ATTEMPT TO USE THE LESLIE CENSUS METHOD ON DEER. J. WILDL. MANAGE. 32(4), 760-764.

LORD, G. E. 1973. POPULATION AND PARAMETER ESTIMATION IN THE ACOUSTIC ENUMERATION OF A MIGRATING FISH POPULATION. BIOMETRICS 29(4), 713-725. [POISSON DISTRIBUTION; SAMPLE SIZE; MAXIMUM LIKELIHOOD; SAMPLING]

MACARTHUR, R. H. AND A. T. MACARTHUR. 1974. ON THE USE OF MIST NETS FOR POPULATION STUDIES OF BIRDS. PROC. NATL. ACAD. SCI. U.S.A. 71(8), 3230-3233. [TAG-RECAPTURE; ESTIMATION PROCEDURE]

MACLULICH, D. A. 1951. A NEW TECHNIQUE OF ANIMAL CENSUS, WITH EXAMPLES. J. MAMMAL. 32(3), 318-328. [POISSON DISTRIBUTION]

MACNEILL, I. B. 1971. ON THE ESTIMATION OF FISH POPULATION DISTRIBUTIONS USING ACOUSTIC METHODS. PP. 553-581. IN, STATISTICAL ECOLOGY, VOLUME 1, SPATIAL PATTERNS AND STATISTICAL DISTRIBUTIONS. (G. P. PATIL, E. C. PIELOU AND W. E. WATERS, EDS.). PENNSYLVANIA STATE UNIVERSITY PRESS, UNIVERSITY PARK, PENNSYLVANIA.

MACNEILL, I. B. 1971. QUICK STATISTICAL METHODS FOR ANALYZING THE SEQUENCES OF FISH COUNTS PROVIDED BY DIGITAL ECHO COUNTERS. J. FISH. RES. BOARD CAN. 28(7), 1035-1042. [EQUATIONS; VARIANCE; CONFIDENCE INTERVAL; VOLUME OF SONIFIED LAYER]

MACNEILL, I. B. 1972. CENSORED OVERLAPPING POISSON COUNTS. BIOMETRIKA 59(2), 427-434. [ECHO COUNTING]

MARTEN, G. G. 1970. A REGRESSION METHOD FOR MARK-RECAPTURE ESTIMATION OF POPULATION SIZE WITH UNEQUAL CATCHABILITY. ECOLOGY 51(2), 291-295.

MARTEN, G. G. 1972. CENSUSING MOUSE POPULATIONS BY MEANS OF TRACKING. ECOLOGY 53(5), 859-867. [TAG-RECAPTURE; VARIANCE]

MARTEN, G. G. 1972. THE REMOTE SENSING APPROACH TO CENSUSING. RES. POPUL. ECOL.-KYOTO 14(1), 36-57.

MCLAREN, I. A. 1966. ANALYSIS OF AN AERIAL CENSUS OF RINGED SEAL. J. FISH. RES. BOARD CAN. 23(5), 769-773. [ESTIMATION]

MENHINICK, E. F. 1963. ESTIMATION OF INSECT POPULATION DENSITY IN HERBACEOUS VEGETATION WITH EMPHASIS ON REMOVAL SWEEPING. ECOLOGY 44(3), 617-621. [TAG-RECAPTURE; HAYNE'S METHOD]

MIDTTUN, L. M. 1969. PHYSICAL PROPERTIES OF SOUND IN WATER; ITS PROPAGATION AND REFLECTION FROM TARGETS. PP. 1-16. IN, MANUAL OF METHODS FOR FISH STOCK ASSESSMENT PART V. THE USE OF ACOUSTIC INSTRUMENTS IN FISH DETECTION AND FISH ABUNDANCE ESTIMATION. (B. B. PARRISH, ED.). FAO FISHERIES TECHNICAL PAPER NO. 83. [EQUATIONS]

MITCHELL, B. 1963. ECOLOGY OF TWO CARABID BEETLES, BEMBIDION LAMPROS (HERBST) AND TRECHUS QUADRISTRIATUS (SCHRANK). II. STUDIES ON POPULATIONS OF ADULTS IN THE FIELD, WITH SPECIAL REFERENCE TO THE TECHNIQUE OF PITFALL TRAPPING. J. ANIM. ECOL. 32(3), 377-392. [MATHEMATICS AND ANALYSIS OF PITFALL TRAPPING]

MOOSE, P. H. AND J. E. EHRENBERG. 1971. AN EXPRESSION FOR THE VARIANCE OF ABUNDANCE ESTIMATES USING A FISH ECHO INTEGRATOR. J. FISH. RES. BOARD CAN. 28(9), 1293-1301. [MODEL; POISSON DISTRIBUTION; VARIANCE]

MOTTLEY, C. M. 1949. THE STATISTICAL ANALYSIS OF CREEL-CENSUS DATA. TRANS. AMER. FISH. SOC. 76, 290-300. [ANALYSIS OF COVARIANCE; DELURY'S METHOD]

MURPHY, G. I. 1960. ESTIMATING ABUNDANCE FROM LONGLINE CATCHES. J. FISH. RES. BOARD CAN. 17(1), 33-40. [EQUATIONS]

MURPHY, G. I. 1964. A SOLUTION OF THE CATCH EQUATION. J. FISH. RES. BOARD CAN. 22(1), 191-202.

NELSON, L., JR. AND F. W. CLARK. 1973. CORRECTION FOR SPRUNG TRAPS IN CATCH/EFFORT CALCULATIONS OF TRAPPING RESULTS. J. MAMMAL. 54(1), 295-298.

NELSON, W. 1967. ESTIMATION FOR A POPULATION DEPLETED BY SAMPLING. AMER. STATISTICIAN 21(4), 19-21.

NELSON, W. C. 1959. METHODS OF ESTIMATING ABSOLUTE POPULATIONS OF MOBILE ANIMALS: A REVIEW OF THE LITERATURE. COLORADO DEPARTMENT OF GAME AND FISH, DENVER, COLORADO. MIMEO. II, 31 PP.

NEYMAN, J. 1949. ON THE PROBLEM OF ESTIMATING THE NUMBER OF SCHOOLS OF FISH. UNIV. CALIF. PUBL. STAT 1(3), 21-36.

NORTON, H. W., T. G. SCOTT, W. R. HANSON AND W. D. KLIMSTRA. 1961. WHISTLING-COCK INDICES AND BOBWHIT POPULATIONS IN AUTUMN. J. WILDL. MANAGE. 25(4), 398-403. [ANALYSI OF COVARIANCE]

NOSE, Y. 1959. ON THE CONFIDENCE LIMITS CORRESPONDING TO THE ESTIMATE OBTAINED BY DELURY'S LOGARITHMIC CATCH-EFFORT METHOD. BULL. JAPANESE SOC. SCI. FISH. 24(12), 953-956.

OVERTON, W. S. AND D. E. DAVIS 1969. ESTIMATING THE NUMBERS OF ANIMALS IN WILDLIFE POPULATIONS. PP. 403-455. IN, WILDLIFE MANAGEMENT TECHNIQUES (3RD REVISED EDITION.). (R. H. GILES, JR., ED.). THE WILDLIFE SOCIETY, WASHINGTON, D. C.

PARKER, R. A. 1963. ON THE ESTIMATION OF POPULATION SIZE, MORTALITY, AND RECRUITMENT. BIOMETRICS 19(2), 318-323.

PARKER, R. R. 1957. TWO PROPOSED METHODS OF ESTIMATING ANIMAL POPULATIONS. (ABSTRACT) PROC. ALASKA SCI. CONF. 5, 39.

PAULIK, G. J. AND D. S. ROBSON. 1969. STATISTICAL CALCULATIONS FOR CHANGE-IN-RATIO ESTIMATORS OF POPULATION PARAMETERS. J. WILDL. MANAGE. 33(1), 1-27.

PESENKO, YU. A. 1972. METHOD OF COUNTING INSECT POPULATIONS. EKOLOGIYA 3(1), 89-95. (ENGLISH TRANSL. PP. 68-73). [ESTIMATION; OPTIMAL SAMPLING UNIT; SAMPLE SIZE]

PETRIDES, G. A. 1954. ESTIMATING THE PERCENTAGE KILL IN RINGNECKED PHEASANTS AND OTHER GAME SPECIES. J. WILDL. MANAGE. 18(3), 294-297.

PHILLIPS, J. F. V. 1931. QUANTITATIVE METHODS IN THE STUDY OF NUMBERS OF TERRESTRIAL ANIMALS IN BIOTIC COMMUNITIES: A REVIEW, WITH SUGGESTIONS. ECOLOGY 12(4), 633-649.

POPE, J. G. 1972. AN INVESTIGATION OF THE ACCURACY OF VIRTUAL POPULATION ANALYSIS USING COHORT ANALYSIS. INTERN. COMM. NORTHWEST ATLANTIC FISH. RES. BULL. 9, 65-74. [GULLAND'S VIRTUAL POPULATION ANALYSIS; VARIANCE; MODEL]

POWER, G. AND L. H. SAUNDERS. 1973. AN ERROR IN LOGIC IN ESTIMATING STREAM POPULATIONS OF PREMIGRANT TROUT FROM STREAM SURVIVAL RATES AND LAKE POPULATIONS. J. FISH. RES. BOARD CAN. 30(7), 1033-1034.

PRUESS, K. P. AND N. C. PRUESS. 1971. TELESCOPIC OBSERVATION OF THE MOON AS A MEANS FOR OBSERVING MIGRATION OF THE ARMY CUTWORM, CHORIZAGROTIS AUXILIARIS (LEPIDOPTERA: NOCTUIDAE). ECOLOGY 52(6), 999-1007. [MODEL]

REGIER, H. A. AND D. S. ROBSON 1967. ESTIMATING POPULATION NUMBER AND MORTALITY RATES. PP. 31-66. IN, THE BIOLOGICAL BASIS OF FRESHWATER FISH PRODUCTION. (S. D. GERKING, ED.). BLACKWELL SCIENTIFIC PUBLICATIONS, OXFORD, ENGLAND.

REID, E. H., J. L. KOVNER AND S. C. MARTIN. 1963. A PROPOSED METHOD FOR DETERMINING CATTLE NUMBERS IN RANGE EXPERIMENTS. J. RANGE MANAGE. 16(4), 184-187. [CONFIDENCE INTERVAL; VARIANCE; T-TEST; MULTIPLE CORRELATION]

RICKER, W. E. 1971. DERZHAVIN'S BIOSTATISTICAL METHOD OF POPULATION ANALYSIS. J. FISH. RES. BOARD CAN. 28(10), 1666-1672. [VIRTUAL POPULATION ESTIMATES]

ROBEL, R. J., D. J. DICK AND G. F. KRAUSE. 1969. REGRESSION COEFFICIENTS USED TO ADJUST BOBWHITE QUAIL WHISTLE COUNT DATA. J. WILDL. MANAGE. 33(3), 662-668.

ROBSON, D. S. AND H. A. REGIER. 1968. ESTIMATION OF POPULATION NUMBER AND MORTALITY RATES. PP. 124-158. IN, METHODS FOR ASSESSMENT OF FISH PRODUCTION IN FRESH WATERS.(W. E. RICKER, ED.). BLACKWELL SCIENTIFIC PUBLICATIONS, OXFORD, ENGLAND. [TAG-RECAPTURE; REMOVAL METHOD]

ROESSLER, M. 1965. AN ANALYSIS OF THE VARIABILITY OF FISH POPULATIONS TAKEN BY OTTER TRAWL IN BISCAYNE BAY, FLORIDA. TRANS. AMER. FISH. SOC. 94(4), 311-318. [STANDARD ERROR; POISSON DISTRIBUTION; NEGATIVE BINOMIAL DISTRIBUTION; DIVERSITY INDEX; LOGARITHMIC SERIES]

RUPP, R. S. 1966. GENERALIZED EQUATION FOR THE RATIO METHOD OF ESTIMATING POPULATION ABUNDANCE. J. WILDL. MANAGE. 30(3), 523-526.

SAVILLE, A. 1964. ESTIMATION OF THE ABUNDANCE OF FISH STOCK FROM EGG AND LARVAL SURVEYS. RAPP. PROCES-VERB. REUNIONS J. CONS. PERM. INT. EXPLOR. MER 155, 164-170. [STRATIFICATION; TRIGAMMA FUNCTION; NEGATIVE BINOMIAL DISTRIBUTION]

SEBER, G. A. F. AND E. D. LE CREN. 1967. ESTIMATING POPULATION PARAMETERS FROM CATCHES LARGE RELATIVE TO THE POPULATION. J. ANIM. ECOL. 36(3), 631-643.

SEBER, G. A. F. AND J. F. WHALE. 1970. THE REMOVAL METHOD FOR TWO AND THREE SAMPLES. BIOMETRICS 26(3), 393-400.

SEIERSTAD, S., A. SEIERSTAD AND I. MYSTERUD. 1965. STATISTICAL TREATMENT OF THE 'INCONSPICUOUSNESS PROBLEM' IN ANIMAL POPULATION SURVEYS. NATURE 206(4979), 22-23.

SEIERSTAD, S., A. SEIERSTAD AND I. MYSTERUD. 1967. ESTIMATION OF SURVEY EFFICIENCY FOR ANIMAL POPULATIONS WITH UNIDENTIFIABLE INDIVIDUALS. NATURE 213(5075), 524-525.

SHIBATA, K. 1970. ANALYSIS OF ECHO-SOUNDER RECORDS ACOUSTIC INFORMATION OF FISH SIZE. BULL. JAPANESE SOC. SCI. FISH. 36(5), 462-468. [EQUATIONS]

SHIBATA, K. 1971. STUDIES ON ECHO COUNTING FOR ESTIMATION OF FISH STOCKS - I. OVERLAP COUNTING AND READING OF S-TYPE ECHO COUNTER. BULL. JAPANESE SOC. SCI. FISH. 37(8), 711-719. [EQUATIONS]

SITTLER, O. D. 1965. THEORETICAL BASIS FOR ESTIMATING DEER POPULATION FROM AUTOMATICALLY COLLECTED DATA. J. WILDL. MANAGE. 29(2), 381-387. [BEHAVIOR MODEL]

SKELLAM, J. G. 1962. ESTIMATION OF ANIMAL POPULATIONS BY EXTRACTION PROCESSES CONSIDERED FROM THE MATHEMATICAL STANDPOINT. PP. 26-36. IN. PROGRESS IN SOIL ZOOLOGY. (P. W. MURPHY, ED.). BUTTERWORTHS, LONDON, ENGLAND.

SMIRNOV, V. S. 1969. FORMOSOV'S FORMULA FOR THE REGISTRATION OF ANIMALS DENSITY BY TRACKS, ITS MATHEMATICAL INTERPRETATION AND APPLICATION. TRUDY INST. EKOL. RASTENII ZHIVOTNYKN (SVERDLOVSK) 71, 156-164.

SMITH, M. H., R. BLESSING, J. G. CHELTON, J. B. GENTRY, F. B. GOLLEY AND J. T. MCGINNIS. 1971. DETERMINING DENSITY FOR SMALL MAMMAL POPULATIONS USING A GRID AND ASSESSMENT LINES. ACTA THERIOL. 16(8), 105-125.

STORMER, F. A., T. W. HOEKSTRA, C. M. WHITE AND C. M. KIRKPATRICK. 1974. ASSESSMENT OF POPULATION LEVELS OF WHITE-TAILED DEER ON NAD CRANE. RESEARCH BULL. NO. 910. AGRICULTURAL EXPERIMENT STATION, PURDUE UNIVERSITY, WEST LAFAYETTE, INDIANA. 11 PP. [PELLET GROUP INDEX; POPULATION ESTIMATION; EBERHARDT'S EQUATION]

TANAKA, H., M. IKUZAWA AND H. SUGIYAMA. 1953. SOME METHODS OF ESTIMATING THE POPULATION SIZE OF RATS OR VOLES LIVING IN A SPECIFIED AREA. J. OSAKA CITY MED. CENTER 3(1), 1-13.

TANAKA, H., M. IKUZAWA AND H. SUGIYAMA. 1954. SUMMARY OF SOM STATISTICAL STUDIES OF THE ESTIMATION OF POPULATION SIZE OF RATS. JAPANESE J. SANIT. ZOOL. 4 200-208.

TANAKA, R. 1954. A REVISED METHOD FOR ESTIMATING REDUCTION OF NATURAL POPULATIONS EFFECTED BY POISONING. JAPANESE J. SANIT. ZOOL. 4(3), 186-193. [MAXIMUM LIKELIHOOD; HAYNE'S METHOD; TAG-RECAPTURE ESTIMATION]

TANAKA, R. 1960. EVIDENCE AGAINST RELIABILITY OF THE TRAP-NIGHT INDEX AS A RELATIVE MEASURE OF POPULATION IN SMALL MAMMALS. JAPANESE J. ECOL. 10(3), 102-106. [REMOVAL METHOD OF ESTIMATION; REGRESSION]

TANAKA, R. 1963. EXAMINATION OF THE ROUTINE CENSUS EQUATION BY CONSIDERING MULTIPLE COLLISIONS WITH A SINGLE-CATCH TRAP IN SMALL MAMMALS. JAPANESE J. ECOL. 13(1), 16-21. [REMOVAL METHOD OF ESTIMATION]

TANAKA, R. AND M. KANAMORI. 1969. INQUIRY INTO EFFECTS OF PREBAITING ON REMOVAL CENSUS IN A VOLE POPULATION. RES. POPUL. ECOL.-KYOTO 11(1), 1-13.

TAUTI, M. 1953. HOW TO READ IMAGES OF FISH SCHOOLS. BULL. JAPANESE SOC. SCI. FISH. 19(4), 372-375.

TAYLOR, L. R. 1963. ANALYSIS OF THE EFFECT OF TEMPERATURE ON INSECTS IN FLIGHT. J. ANIM. ECOL. 32(1), 99-117. [THEORY OF TRAP CAPTURES]

TAYLOR, R. H. AND R. M. WILLIAMS. 1956. THE USE OF PELLET COUNTS FOR ESTIMATING THE DENSITY OF POPULATIONS OF THE WILD RABBIT, ORYCTOLAGUS CUNICULUS (L.). N. Z. J. SCI. TECHNOL. SECT. B 38(3), 236-256.

THOMPSON, K. H. 1962. ESTIMATION OF THE PROPORTION OF VECTORS IN A NATURAL POPULATION OF NSECTS. BIOMETRICS 18(4), 68-578.

HORNE, R. E., J. E. REEVES, AND A. . MILLIKAN. 1971. ESTIMATION F THE PACIFIC HAKE (MERLUCCIUS RODUCTUS) POPULATION IN PORT USAN, WASHINGTON, USING AN ECHO NTEGRATOR. J. FISH. RES. BOARD AN. 28(9), 1275-1284. STRATIFIED SAMPLING; VARIANCE; QUATIONS]

TRUSKANOV, M. D. AND M. N. SCHERBINO. 1966. METHODS OF DIRECT CALCULATION OF FISH CONCENTRATIONS BY MEANS OF HYDROACOUSTIC APPARATUS. INTERN. COMM. NORTHWEST ATLANTIC FISH. RES. BULL. 3, 70-80. [EQUATIONS]

VAUGHAN, A. E. 1955. ESTIMATION OF A BIOLOGICAL POPULATION WHICH IS SUBJECT TO A BIASED MORTALITY. DISSERTATION. STANFORD UNIVERSITY, STANFORD, CALIFORNIA.

WADLEY, F. M. 1954. LIMITATIONS ON THE "ZERO METHOD" OF POPULATION COUNTS. SCIENCE 119(3098), 689-690.

WALLIN, L. 1973. RELATIVE ESTIMATES OF SMALL-MAMMAL POPULATIONS IN RELATION TO THE SPATIAL PATTERN OF TRAPPABILITY. OIKOS 24(2), 282-286. [POISSON DISTRIBUTION; REGRESSION]

WALOFF, N. AND K. BAKKER 1963. THE FLIGHT ACTIVITY OF MIRIDAE (HETEROPTERA) LIVING ON BROOM, SAROTHAMNUS SCOPARIUS (L.) WIMM. J. ANIM. ECOL. 32(3), 461-480. [POPULATION ESTIMATION; DEMPSTER'S METHOD]

WEBB, W. L. 1942. NOTES ON A METHOD FOR CENSUSING SNOWSHOE HARE POPULATIONS. J. WILDL. MANAGE. 6(1), 67-69. [STRIP CENSUS]

WIDRIG, T. M. 1954. METHOD OF ESTIMATING FISH POPULATIONS, WITH APPLICATION TO PACIFIC SARDINE. U. S. FISH WILDL. SERV. FISH. BULL. 56, 141-166.

WILSON, L. F. AND D. J. GERRARD. 1971. A NEW PROCEDURE FOR RAPIDLY ESTIMATING EUROPEAN PINE SAWFLY (HYMENOPTERA: DIPRIONIDAE) POPULATION LEVELS IN YOUNG PINE PLANTATIONS. CAN. ENTOMOL. 103(9), 1315-1322. [NEGATIVE BINOMIAL SERIES; MAXIMUM LIKELIHOOD; CONFIDENCE INTERVAL]

WILSON, R. W., JR. 1969. CONSIDERATIONS OF THE RELATION OF VARIOUS ERRORS TO ESTIMATES OF POPULATION CHARACTERISTICS. PP. 101-111, IN, FOREST INSECT POPULATION DYNAMICS. U. S. FOR. SERV. NORTHEASTERN FOREST EXPT. STN. RES. PAPER NE-125.

WOLF, W. W., A. N. KISHABA AND H. H. TOBA. 1971. PROPOSED METHOD FOR DETERMINING DENSITY OF TRAPS REQUIRED TO REDUCE AN INSECT POPULATION. J. ECON. ENTOMOL. 64(4), 872-877. [PROBABILITY; TRAP-DENSITY FUNCTION; TAG-RECAPTURE]

ZIPPIN, C. 1956. AN EVALUATION OF THE REMOVAL METHOD OF ESTIMATING ANIMAL POPULATIONS. BIOMETRICS 12(2), 163-189.

ZIPPIN, C. 1958. THE REMOVAL METHOD OF POPULATION ESTIMATION. J. WILDL. MANAGE. 22(1), 82-90.

REPRODUCTION

ALLEN, K. R. 1968. SIMPLIFICATION OF A METHOD COMPUTING RECRUITMENT RATES. J. FISH. RES. BOARD CAN. 25(12), 2701-2702.

BEZEM, J. J. 1960. MATHEMATICAL APPENDIX. (SLUITER, J. W., REPRODUCTIVE RATE OF THE BAT RHINOL KONINKL. NEDERL. AKADEMIE VAN WETENSCHAPPEN-AMSTERDAM. PROC. SERIES C 63(3), 392-393.

BROWN, J. L. 1969. THE BUFFER EFFECT AND PRODUCTIVITY IN TIT POPULATIONS. AMER. NAT. 103(932), 347-354. [EXCHANGE RATIO; EQUATIONS]

BUJALSKA, G. 1970. REPRODUCTION STABILIZING ELEMENTS IN AN ISLAND POPULATION OF CLETHRIONOMYS GLAREOLUS (SCHREBER, 1780). ACTA THERIOL. 15(25), 381-412. [INDEX OF VARIATION; HOME RANGE; NUMBER OF CAPTURES OF FEMALES]

CASWELL, H. 1972. ON INSTANTANEOUS AND FINITE BIRTH RATES. LIMNOL. OCEANOGR. 17(5), 787-791.

CAUGHLEY, G. 1970. A COMMENT ON VANDERMEER'S "PSEUDO-REPRODUCTIVE VALUE." AMER. NAT. 104(936), 214-215.

CAUGHLEY, G. 1974. PRODUCTIVITY, OFFTAKE, AND RATE OF INCREASE. J. WILDL. MANAGE. 38(3), 566-567. [HUNTING AND NATURAL MORTALITY]

CAUGHLEY, G. AND L. C. BIRCH. 1971. RATE OF INCREASE. J. WILDL. MANAGE. 35(4), 658-663.

CODY, M. L. 1966. A GENERAL THEORY OF CLUTCH SIZE. EVOLUTION 20(2), 174-184.

CODY, M. L. 1971. ECOLOGICAL ASPECTS OF REPRODUCTION. PP. 461-512. IN, AVIAN BIOLOGY, VOLUME 1, (D. S. FARNER AND J. R. KING, EDS.). ACADEMIC PRESS, NEW YORK, NEW YORK.

COHEN, D. 1966. OPTIMIZING REPRODUCTION IN A RANDOMLY VARYING ENVIRONMENT. J. THEOR. BIOL. 12(1), 119-129.

COHEN, D. 1967. OPTIMIZING REPRODUCTION IN A RANDOMLY VARYING ENVIRONMENT WHEN A CORRELATION MAY EXIST BETWEEN THE CONDITIONS AT THE TIME A CHOICE HAS TO BE MADE AND THE SUBSEQUENT OUTCOME. J. THEOR. BIOL. 16(1), 1-14. [MODEL; INFORMATION; LOG TRANSFORMATION]

COHEN, D. 1968. A GENERAL MODEL OF OPTIMAL REPRODUCTION IN A RANDOMLY VARYING ENVIRONMENT. J. ECOL. 56(1), 219-228.

COLE, L. C. 1960. A NOTE ON POPULATION PARAMETERS IN CASES OF COMPLEX REPRODUCTION. ECOLOGY 41(2), 372-375.

CONWAY, G. R. 1969. A BASIC MODEL OF INSECT REPRODUCTION AND ITS IMPLICATIONS FOR PEST CONTROL. PH.D. DISSERTATION, UNIVERITY OF CALIFORNIA, DAVIS, CALIFORNIA. 256 PP. (DISS. ABSTR. 31(6), 3457-B).

DAVIS, D. E. 1973. COMMENTS ON R. BULL. ECOL. SOC. AMER. 54(3), 14-15; 26.

EDMONDSON, W. T. 1968. A GRAPHICAL MODEL FOR EVALUATING THE USE OF THE EGG RATIO FOR MEASURING BIRTH AND DEATH RATES. OECOLOGIA 1(1), 1-37.

FAGEN, R. M. 1972. AN OPTIMAL LIFE-HISTORY STRATEGY IN WHICH REPRODUCTIVE EFFORT DECREASES WITH AGE. AMER. NAT. 106(948), 258-261. [MODEL; MALTHUSIAN PARAMETER; GADGIL AND BOSSERT MODEL]

GIESEL, J. T. 1974. FITNESS AN POLYMORPHISM FOR NET FECUNDITY DISTRIBUTION IN ITEROPAROUS POPULATIONS. AMER. NAT. 108(961) 321-331. [MODEL; GENETICS; DEMOGRAPHY; SIMULATION; RANDOM NUMBER GENERATOR; COMPUTER]

GOODMAN, D. 1974. NATURAL SELECTION AND A COST CEILING ON REPRODUCTIVE EFFORT. AMER. NAT. 108(961), 247-268. [COLE'S RESULT; BIRDS; LIFE TABLE; MODEL; FISHER'S REPRODUCTIVE VALUE; FITNESS; SIMULATION]

GREEN, R. AND P. R. PAINTER. 1975. SELECTION FOR FERTILITY AND DEVELOPMENT TIME. AMER. NAT. 109(965), 1-10. [INTRINSIC RATE OF INCREASE; VOLTERRA'S EQUATIONS]

HANNESSON, R. 1974. RELATION BETWEEN REPRODUCTIVE POTENTIAL AND SUSTAINED YIELD OF FISHERIES. J. FISH. RES. BOARD CAN. 31(3), 359-362. [GROWTH EQUATIONS]

HOKYO, N. AND K. KIRITANI. 1967. A METHOD FOR ESTIMATING NATURAL SURVIVAL RATE AND MEAN FECUNDITY OF AN ADULT INSECT POPULATION BY DISSECTING THE FEMALE REPRODUCTIVE ORGANS. RES. POPUL. ECOL.-KYOTO 9(2), 130-142. [MACDONALD'S METHOD]

HOWE, R. W. 1953. THE RAPID DETERMINATION OF THE INTRINSIC RATE OF INCREASE OF AN INSECT POPULATION. ANN. APPL. BIOL. 40(1), 134-151.

KANDLER, R. AND S. DUTT. 1958. FECUNDITY OF BALTIC HERRING. RAPP. PROCES-VERB. REUNIONS J. CONS. PERM. INT. EXPLOR. MER 143(2), 99-108. [REGRESSION]

LESLIE, P. H. 1948. ON THE DISTRIBUTION IN TIME OF THE BIRTHS IN SUCCESSIVE GENERATIONS. J. ROYAL STAT. SOC. SER. A 111(1), 44-53. [MATRIX; VECTOR]

MATHISEN, O. A. AND T. GUNNEROD. 1969. VARIANCE COMPONENTS IN THE ESTIMATION OF POTENTIAL EGG DEPOSITION OF SOCKEYE SALMON ESCAPEMENT. J. FISH. RES. BOARD CAN. 26(3), 655-670.

MAYFIELD, H. 1961. NESTING SUCCESS CALCULATED FROM EXPOSURE. WILSON BULL. 73(3), 255-261. [PROBABILITY]

MOUNTFORD, M. D. 1968. THE SIGNIFICANCE OF LITTER-SIZE. J. ANIM. ECOL. 37(2), 363-367. [NUMERICAL MODELS]

MOUNTFORD, M. D. 1970. TEST OF CHANGE IN EGG-SHELL INDEX. J. APPL. ECOL. 7(1), 113-115. [PROBABILITY DISTRIBUTION OF MAXIMUM; MAXIMUM DIFFERENCE]

MOUNTFORD, M. D. 1973. THE SIGNIFICANCE OF CLUTCH-SIZE. PP. 315-323. IN, THE MATHEMATICAL THEORY OF THE DYNAMICS OF BIOLOGICAL POPULATIONS. (M. S. BARTLETT AND R. W. HIORNS, EDS.). ACADEMIC PRESS, NEW YORK, NEW YORK. [MODEL; PROBABILITY GENERATING FUNCTION; SIGMOID FUNCTION; SURVIVORSHIP; MONTE CARLO STUDIES]

O'DONALD, P. 1972. NATURAL SELECTION OF REPRODUCTIVE RATES AND BREEDING TIMES AND ITS EFFECT ON SEXUAL SELECTION. AMER. NAT. 106(949), 368-379. [MODEL; BIRDS]

PETRUSEWICZ, K. 1969/1970. ESTIMATION OF NUMBER OF NEW-BORN ANIMALS. PP. 181-185. IN, ENERGY FLOW THROUGH SMALL MAMMAL POPULATIONS. (K. PETRUSEWICZ AND L. RYSZKOWSKI, EDS.). PWN-POLISH SCIENTIFIC PUBLISHERS, WARSAW, POLAND.

RICKLEFS, R. E. 1973. FECUNDITY, MORTALITY AND AVIAN DEMOGRAPHY. PP. 366-435. IN, BREEDING BIOLOGY OF BIRDS. (D. S. FARNER, ED.). NATL. ACAD. SCI., WASHINGTON, D. C. [GROWTH FORMULA; LIFE TABLE]

ROYAMA, T. 1969. A MODEL FOR THE GLOBAL VARIATION OF CLUTCH SIZE IN BIRDS. OIKOS 20(2), 562-567.

SCHAFFER, W. M. 1974. OPTIMAL REPRODUCTIVE EFFORT IN FLUCTUATING ENVIRONMENTS. AMER. NAT. 108(964), 783-790. [MODEL; FITNESS; LIFE HISTORY FUNCTIONS]

SHIYOMI, M. 1967. A STATISTICAL MODEL OF THE REPRODUCTION OF APHIDS. RES. POPUL. ECOL.-KYOTO 9(2), 167-176. [POISSON DISTRIBUTION; BINOMIAL DISTRIBUTION; NEGATIVE BINOMIAL DISTRIBUTION; MOMENTS]

SLUITER, J. W. 1960. REPRODUCTIVE RATE OF THE BAT RHINOLOPHUS HIPPOSIDEROS. (MATHEMATICAL APPENDIX BY J. J. BEZEM). KONINKL. NEDERL. AKADEMIE VAN WETENSCHAPPEN-AMSTERDAM. PROC. SERIES C 63(3), 383-393. [SURVIVAL RATE]

SMITH, C. C. AND S. D. FRETWELL. 1974. THE OPTIMAL BALANCE BETWEEN SIZE AND NUMBER OF OFFSPRING. AMER. NAT. 108(962), 499-506. [MODEL; FITNESS FUNCTION]

STROSS, R. G., J. C. NEESS AND A. D. HASLER. 1961. TURNOVER TIME AND PRODUCTION OF PLANKTONIC CRUSTACEA IN LIMED AND REFERENCE PORTION OF A BOG LAKE. ECOLOGY 42(2), 237-245. [RECRUITMENT MODEL]

THOMPSON, W. R. 1931. ON THE REPRODUCTION OF ORGANISMS WITH OVERLAPPING GENERATIONS. BULL. ENTOMOL. RES. 22(1), 147-172. [MODEL]

TURNER, H. N. 1968. VITAL STATISTICS AND MEASUREMENTS OF REPRODUCTION RATE IN SHEEP. PP. 172-182. IN, A PRACTICAL GUIDE TO THE STUDY OF THE PRODUCTIVITY OF LARGE HERBIVORES. (F. B. GOLLEY AND H. K. BUECHNER, EDS.). BLACKWELL SCIENTIFIC PUBLICATIONS, OXFORD, ENGLAND. [LIFE TABLE; MODEL; INTRINSIC RATE OF INCREASE]

UNDERWOOD, A. J. 1974. ON MODELS FOR REPRODUCTIVE STRATEGY IN MARINE BENTHIC INVERTEBRATES. AMER. NAT. 108(964), 874-878. [VANCE MODELS]

VANCE, R. R. 1973. MORE ON REPRODUCTIVE STRATEGIES IN MARINE BENTHIC INVERTEBRATES. AMER. NAT. 107(955), 353-361. [MODEL]

VANCE, R. R. 1973. ON REPRODUCTIVE STRATEGIES IN MARINE BENTHIC INVERTEBRATES. AMER. NAT. 107(955), 339-352. [MODEL]

VANCE, R. R. 1974. REPLY TO UNDERWOOD. AMER. NAT. 108(964), 879-880. [VANCE MODELS]

VANDERMEER, J. H. 1968. REPRODUCTIVE VALUE IN A POPULATION OF ARBITRARY AGE DISTRIBUTION. AMER. NAT. 102(928), 586-589. [FISHER'S REPRODUCTIVE VALUE; EQUATIONS; SHANNON-WIENER FUNCTION]

WATT, K. E. F. 1960. THE EFFECT OF POPULATION DENSITY ON FECUNDITY IN INSECTS. CAN. ENTOMOL. 92(9), 674-695. [MODEL]

YOSHIHARA, T. 1956. ON SOME THEORETICAL CONSIDERATIONS OF THE REPRODUCTION CURVE. RES. POPUL. ECOL.-KYOTO 3, 1-7. [LOGISTIC EQUATION]

SAMPLING

ABRAHAMSEN, G. 1969. SAMPLING DESIGN IN STUDIES OF POPULATION DENSITIES IN ENCHYTRAEIDAE (OLIGOCHAETA). OIKOS 20(1), 54-66.

ABRAMSON, N. AND J. TOLLADAY. 1959. THE USE OF PROBABILITY SAMPLING FOR ESTIMATING ANNUAL NUMBER OF ANGLER DAYS. CALIF. FISH GAME 45(4), 303-311. [SIMPLE RANDOM SAMPLING; STRATIFIED SAMPLING; OPTIMUM ALLOCATION]

ABRAMSON, N. J. 1968. A PROBABILITY SEA SURVEY PLAN FOR ESTIMATING RELATIVE ABUNDANCE OF OCEAN SHRIMP. CALIF. FISH GAME 54(4), 257-268. [STRATIFIED TWO-STAGE SAMPLING; PROPORTIONAL ALLOCATION; VARIANCE]

ADAMS, L., W. G. O'REGAN AND D. J. DUNAWAY. 1962. ANALYSIS OF FORAGE CONSUMPTION BY FECAL EXAMINATION. J. WILDL. MANAGE. 26(1), 108-111. [REGRESSION; COEFFICIENT OF VARIATION]

ALDRICH, R. C. AND A. T. DROOZ. 1967. ESTIMATED FRASER FIR MORTALITY AND BALSAM WOOLLY APHID INFESTATION TREND USING AERIAL COLOR PHOTOGRAPHY. FOREST SCI. 13(3), 300-313. [STRATIFIED-SYSTEMATIC SAMPLE; INDEX OF MORTALITY]

ANDERSON, K. P. 1965. MANUAL OF SAMPLING AND STATISTICAL METHODS FOR FISHERIES BIOLOGY. PART II - STATISTICAL METHODS, CHP. 5. FAO FISHERIES TECHNICAL PAPER NO. 26(SUPPL.), VI, 25 PP.

ANDREYEV, V. L. AND V. YE MOLOTKOV. 1972. THE OPTIMUM NUMBER OF SAMPLES IN STUDY OF MULTICOMPONENT BIOLOGICAL SYSTEMS. HYDROBIOL. J. 8(5), 96-100. (ENGLISH TRANSL.). [MATRIX; COVARIANCE]

ANONYMOUS. 1956. NATIONAL SURVEY OF FISHING AND HUNTING: TECHNICAL SUPPLEMENT. U. S. FISH WILDL. SERV. CIRCULAR 44-SUPP. MISC. PAGES. [SAMPLING; STRATIFIED SAMPLING; PROPORTIONAL ALLOCATION]

ARNOLD, J. C. 1970. A MARKOVIAN SAMPLING POLICY APPLIED TO WATER QUALITY MONITORING OF STREAMS. BIOMETRICS 26(4), 739-747. [MODEL; MATRIX; PROBABILITY; COMPUTER]

ARVANITIS, L. G. 1966. DECISION RULES FOR DESIGN OF FOREST SAMPLING SYSTEMS: A CONTRIBUTION TO METHODOLOGY BASED ON COMPUTER SIMULATION. DISS. ABSTR. 27B(658-659) O.R.S. UNIVERSITY OF CALIFORNIA, BERKELEY, CALIFORNIA.

ARVANITIS, L. G. AND W. G. O'REGAN. 1972. CLUSTER OR SATELLITE SAMPLING IN FORESTRY: A MONTE CARLO COMPUTER SIMULATION STUDY. PP. 191-205. IN, 3RD CONFERENCE ADVISORY GROUP OF FOREST STATISTICIANS. INSTITUT NATIONAL DE LA RECHERCHE AGRONOMIQUE, PARIS, FRANCE. I.N.R.A. PUBL. 72-3. [OPTIMUM ALLOCATION; SIMULATION; VARIANCE COMPONENTS]

AVERY, T. E. 1975. PROBABILITY, SAMPLING, AND ESTIMATION. PP. 11-43. IN, NATURAL RESOURCES MEASUREMENTS. 2ND EDITION. MCGRAW-HILL BOOK CO., NEW YORK, NEW YORK.

BANERJEE, B. 1970. A MATHEMATICAL MODEL ON SAMPLING DIPLOPODS USING PITFALL TRAPS. OECOLOGIA 4(9), 102-105.

BARKLEY, F. A. 1934. THE STATISTICAL THEORY OF POLLEN ANALYSIS. ECOLOGY 15(3), 283-289. [CORRELATION; BROWN(SPEARMAN)PROPHECY FORMULA; SAMPLE SIZE]

BARKLEY, R. A. 1964. THE THEORETICAL EFFECTIVENESS OF TOWED-NET SAMPLERS AS RELATED TO SAMPLER SIZE AND TO SWIMMING SPEED OF ORGANISMS. J. CONS. PERM. INT. EXPLOR. MER 29(2), 146-157. [EQUATIONS]

BARKLEY, R. A. 1972. SELECTIVITY OF TOWED-NET SAMPLERS. U. S. NATL. MAR. FISH. SERV. FISH. BULL. 70(3), 799-820. [SAMPLING; PROBABILITY; MODEL]

BEALL, G. 1935. STUDY OF ARTHROPOD POPULATIONS BY THE METHOD OF SWEEPING. ECOLOGY 16(2), 216-225. [LEXIS' COEFFICIENT OF DISTURBANCE; CHARLIER COEFFICIENT OF DISTURBANCE; SAMPLE SIZE]

BELL, R. H. V., J. J. R. GRIMSDELL, L. P. VAN LARIEREN AND J. A. SAYER. 1973. CENSUS OF THE KAFUE LECHWE BY AERIAL STRATIFIED SAMPLING. EAST AFR. WILDL. J. 11(1), 55-74. [JOLLY'S METHOD; SAMPLING ERROR]

BELYAREV, B. N. 1964. SELECTION OF THE SAMPLING INTERVAL AND ESTIMATE OF THE LOSS OF INFORMATION BY USING SAMPLING INSTEAD OF CONTINUOUS OCEANOGRAPHIC MEASUREMENTS. OKEANOLOGIYA 4(3), 497-504.

BERTHET, P. 1971. MITES. PP. 186-208. IN, METHODS OF STUDY IN QUANTITATIVE SOIL ECOLOGY: POPULATION, PRODUCTION AND ENERGY FLOW. (J. PHILLIPSON, ED.). BLACKWELL SCIENTIFIC PUBLICATIONS, OXFORD, ENGLAND. [SAMPLE SIZE; DENSITY; RELATIVE ERROR]

BEST, E. A. AND H. D. BOLES. 1956. AN EVALUATION OF CREEL CENSUS METHODS. CALIF. FISH GAME 42(2), 109-115. [STRATIFIED SAMPLING; RANDOM SAMPLING; TABLE OF RANDOM NUMBERS]

BICKFORD, C. A. 1952. THE SAMPLING DESIGN USED IN THE FOREST SURVEY OF THE NORTHEAST. J. FORESTRY 50(4), 290-293. [STRATIFIED SAMPLING; DOUBLE SAMPLING; SAMPLING ERROR]

BLANKENSHIP, L. H., A. B. HUMPHREY AND D. MACDONALD. 1971. A NEW STRATIFICATION FOR MOURNING DOVE CALL-COUNT ROUTES. J. WILDL. MANAGE. 35(2), 319-326. [STRATIFIED RANDOM SAMPLING; MODEL]

BLYTHE, R. H., JR. 1945. THE ECONOMICS OF SAMPLE SIZE APPLIED TO THE SCALING OF SAWLOGS. BIOMETRICS 1(5), 67-70. [SAMPLING ERROR; COST FUNCTION; SAMPLE SIZE]

BONHAM, C. D. AND R. E. FYE. 1971. AN EMPIRICAL MODEL FOR PREDICTING BOLL WEEVIL DISTRIBUTION ON COTTON PLANTS. J. ECON. ENTOMOL. 64(2), 539-540.

BOURDEAU, P. F. 1953. A TEST OF RANDOM VERSUS SYSTEMATIC ECOLOGICAL SAMPLING. ECOLOGY 34(3), 499-512. [STRATIFIED SAMPLING; ANALYSIS OF VARIANCE; CORRELATION]

BOYCE, F. M. 1974. MIXING WITHIN EXPERIMENTAL ENCLOSURES: A CAUTIONARY NOTE ON THE LIMNOCORRAL J. FISH. RES. BOARD CAN. 31(8), 1400-1405. [MODEL]

BRINKHURST, R. O., K. E. CHUA AND E. BATOOSINGH. 1969. MODIFICATIONS IN SAMPLING PROCEDURES AS APPLIED TO STUDIES ON THE BACTERIA AND TUBIFICID OLIGOCHAETES INHABITING AQUATIC SEDIMENTS. J. FISH. RES. BOARD CAN. 26(10), 2581-2593.

BROWN, D. 1954. CHP. 2. THE THEORY OF SAMPLING. PP. 8-18. IN, METHODS BUREAU OF SURVEYING AND MEASURING VEGETATION. COMMONWEALTH BUREAU OF PASTURES AND FIELD CROPS, HURLEY, BERKS, ENGLAND. BULL. 42.

BUCHANAN-WOLLASTON, H. J. 1931. SOME REMARKS ON THE GRADUATION OF MEASUREMENT DATA. J. CONS. PERM. INT. EXPLOR. MER 6(1), 47-63. [FISHERIES]

CAILLOUET, C. W., JR. AND J. B. HIGMAN. 1973. SAMPLE SIZE IN SPORT FISHERY SURVEYS. TRANS. AMER. FISH. SOC. 102(2), 466-468. [REGRESSION; POWER FUNCTIONS]

CARLANDER, K. D., C. J. DICOSTANZO AND R. J. JESSEN. 1958. SAMPLING PROBLEMS IN CREEL CENSUS. PROG. FISH-CULT. 20(2), 73-81. [PROBABILITY SAMPLING; STRATIFIED SAMPLING]

CASSIE, R. M. 1968. SAMPLE DESIGN. PP. 105-121. IN, ZOOPLANKTON SAMPLING. (TRANTER, D. J. AND J. H. FRASER, EDS.). UNESCO, PARIS, FRANCE. [PRECISION; ACCURACY; COVARIANCE; NORMAL DISTRIBUTION; TRANSFORMATION; POISSON DISTRIBUTION; VARIANCE; DOUBLE POISSON DISTRIBUTION OF NEYMAN AND THOMAS; NEGATIVE BINOMIAL DISTRIBUTION; STRATIFIED SAMPLING; RANDOM SAMPLING; SAMPLE SIZE; REGRESSION SAMPLING; MULTIPLE REGRESSION; NORMAL DISTRIBUTION]

CASSIE, R. M. 1971. SAMPLING AND STATISTICS. PP. 174-209. IN, A MANUAL ON METHODS FOR THE ASSESSMENT OF SECONDARY PRODUCTIVITY IN FRESH WATERS. (W. T. EDMONDSON AND G. G. WINBERG, EDS.). BLACKWELL SCIENTIFIC PUBLICATIONS, OXFORD, ENGLAND. [REGRESSION SAMPLING; RANDOM SAMPLING; FIDUCIAL LIMITS; SAMPLE SIZE; SYSTEMATIC SAMPLING; BINOMIAL DISTRIBUTION; POISSON DISTRIBUTION; TESTS; NON-RANDOM DISTRIBUTION; MODEL; STRATIFIED SAMPLING; REGRESSION]

CHACKO, V. J. 1969. PROBLEMS IN STATISTICAL ASSESSMENT OF FOREST RESOURCES AND PRODUCTS. BULL. INTERN. STAT. INST. 42(1), 327-334. [INVENTORIES]

CHAPMAN, D. G. AND A. M. JOHNSON. 1968. ESTIMATION OF FUR SEAL PUP POPULATIONS BY RANDOMIZED SAMPLING. TRANS. AMER. FISH. SOC. 97(3), 264-270. [TAG-RECAPTURE ESTIMATION; VARIANCE; CLUSTER ANALYSIS; SUBSAMPLING]

CHEPURNOVA, E. A. AND M. N. LEBEDEVA. 1972. STATISTICAL ANALYSIS OF DATA OBTAINED BY COUNTING BACTERIAL COLONIES ON DISHES. HYDROBIOL. J. 8(1), 86-90. (ENGLISH TRANSL.). [SAMPLE SIZE; ESTIMATION]

COCHRAN, W. G., F. MOSTELLER AND J. W. TUKEY. 1954. PRINCIPLES OF SAMPLING. J. AMER. STAT. ASSOC. 49(265), 13-35. [PROBABILITY SAMPLING; SUBSAMPLING; STRATIFIED SAMPLING; PROPORTIONAL ALLOCATION; SYSTEMATIC ERRORS]

COLE, W. E. 1960. SEQUENTIAL SAMPLING IN SPRUCE BUDWORM CONTROL PROJECTS. FOREST SCI. 6(1), 51-59.

COTTAM, G., J. T. CURTIS AND A. J. CATANA, JR. 1957. SOME SAMPLING CHARACTERISTICS OF A SERIES OF AGGREGATED POPULATIONS. ECOLOGY 38(4), 610-622. [POISSON DISTRIBUTION; INDEX OF AGGREGATION]

COTTAM, G., J. T. CURTIS AND B. W. HALE. 1953. SOME SAMPLING CHARACTERISTICS OF RANDOMLY DISPERSED INDIVIDUALS. ECOLOGY 34(4), 741-757. [POISSON DISTRIBUTION; CHI-SQUARE TEST; ANGLE METHOD; ARTIFICIAL POPULATION]

CUNIA, T. 1965. CONTINUOUS FOREST INVENTORY, PARTIAL REPLACEMENT OF SAMPLES AND MULTIPLE REGRESSION. FOREST SCI. 11(4), 480-502. [COMPUTER; SAMPLING; STANDARD ERROR; SIMPLE RANDOM SAMPLING; VECTOR; MATRIX; COVARIANCE]

CUNIA, T. 1965. SOME THEORY ON RELIABILITY OF VOLUME ESTIMATES IN A FOREST INVENTORY SAMPLE. FOREST SCI. 11(1), 115-128. [CLUSTER SAMPLING; SUBPLOTS; SYSTEMATIC SUBPLOTS; MODEL; OPTIMUM SAMPLE SIZE; SIMPLE RANDOM SAMPLING; MULTIPLE REGRESSION; VECTOR; MATRIX; COVARIANCE; VARIANCE]

CUNIA, T. AND R. B. CHEVROU. 1969. SAMPLING WITH PARTIAL REPLACEMENT ON THREE OR MORE OCCASIONS. FOREST SCI. 15(2), 204-224. [MULTIPLE REGRESSION; FOREST INVENTORY; SIMPLE RANDOM SAMPLING; MATRIX; VECTOR; COVARIANCE]

DAVIS, D. E. AND C. ZIPPIN. 1954 PLANNING WILDLIFE EXPERIMENTS INVOLVING PERCENTAGES. J. WILDL. MANAGE. 18(2), 170-178. [TYPE I AND II ERRORS; SAMPLE SIZE; CHART]

DELOYA, M. C. AND A. B. VILLA SALAS. 1972. EVALUACION ESTADISTICA DE CUATRO TAMANOS DE SITIOS CIRCULARES EN INVENTAIOSFORESTALES. PP. 59-82. IN, 3RD CONFERENCE ADVISORY GROUP OF FOREST STATISTICIANS, INSTITUT NATIONAL DE LA RECHERCHE AGRONOMIQUE, PARIS, FRANCE. I.N.R.A. PUBL. 72-3. [FOREST INVENTORY; SAMPLE SIZE; SYSTEMATIC SAMPLING; VARIANCE COMPONENTS; OPTIMUM ALLOCATION; FOREST VOLUMES]

DEMARS, C. J., JR. 1970. FREQUENCY DISTRIBUTIONS, DATA TRANSFORMATIONS, AND ANALYSIS OF VARIATIONS USED IN DETERMINATION OF OPTIMUM SAMPLE SIZE AND EFFORT FOR BROODS OF THE WESTERN PINE BEETLE. PP. 42-65. IN, STUDIES ON THE POPULATION DYNAMICS OF THE WESTERN PINE BEETLE DENDROCTONUS BREVICOMIS LECONTE (COLEOPTERA: SCOLYTIDAE). (R. W. STARK AND D. L. DAHLSTEN, EDS.). UNIVERSITY OF CALIFORNIA, DIVISION OF AGRICULTURAL SCIENCES, BERKELEY, CALIFORNIA. [COMPUTER; ANALYSIS OF VARIANCE; LOG-NORMAL DISTRIBUTION; FREQUENCY DISTRIBUTION; POISSON DISTRIBUTION; NEYMAN'S TYPE A DISTRIBUTION; BINOMIAL DISTRIBUTION; NORMAL; NEGATIVE BINOMIAL DISTRIBUTION; SAMPLING; OPTIMUM ALLOCATION]

DEMPSTER, J. P. 1961. THE ANALYSIS OF DATA OBTAINED BY REGULAR SAMPLING OF AN INSECT POPULATION. J. ANIM. ECOL. 30(2), 429-432. [EQUATIONS; ESTIMATING MORTALITY]

DUNNET, G. M., A. E. HARVIE AND T. J. SMITH. 1973. ESTIMATING THE PROPORTIONS OF VARIOUS LEAVES IN THE DIET OF THE OPOSSUM, TRICHOSURUS VULPECULA KERR, BY FAECAL ANALYSIS. J. APPL. ECOL. 10(3), 737-745. [INDEX OF RESISTANCE TO DIGESTION]

EBERHARDT, L. L. 1963. PROBLEMS IN ECOLOGICAL SAMPLING. NORTHWEST SCI. 37(4), 144-154.

EK, A. R. 1971. A COMPARISON OF SOME ESTIMATORS IN FOREST SAMPLING. FOREST SCI. 17(1), 2-13. [COMPUTER]

FINNEY, D. J. 1946. FIELD SAMPLING FOR THE ESTIMATION OF WIREWORM POPULATIONS. BIOMETRICS 2(1), 1-7. [SAMPLING; PRECISION]

FINNEY, D. J. 1948. RANDOM AND SYSTEMATIC SAMPLING IN TIMBER SURVEYS. J. FORESTRY 22(1), 1-36.

FINNEY, D. J. 1950. AN EXAMPLE OF PERIODIC VARIATION IN FOREST SAMPLING. J. FORESTRY 23(2), 96-111.

FINNEY, D. J. 1953. THE ESTIMATION OF ERROR IN THE SYSTEMATIC SAMPLING OF FORESTS. J. INDIAN SOC. AGRIC. STAT. 5(1), 6-16.

FINNEY, D. J. AND H. PALCA. 1949. THE ELIMINATION OF BIAS DUE TO EDGE-EFFECTS IN FOREST SAMPLING. J. FORESTRY 23(1), 31-47.

FOOTE, L. E., H. S. PETERS AND A. L. FINKNER. 1958. DESIGN TESTS FOR MOURNING DOVE CALL-COUNT SAMPLING IN SEVEN SOUTHEASTERN STATES. J. WILDL. MANAGE. 22(4), 402-408. [STRATIFIED RANDOM SAMPLING]

FRAYER, W. E. 1966. WEIGHTED REGRESSION IN SUCCESSIVE FOREST INVENTORIES. FOREST SCI. 12(4), 464-472. [MODEL; ASSUMPTIONS]

FRAYER, W. E. AND G. M. FURNIVAL. 1967. AREA CHANGE ESTIMATES FROM SAMPLING WITH PARTIAL REPLACEMENT. FOREST SCI. 13(1), 72-77. [STRATIFIED SAMPLING; SUBSAMPLING; VARIANCE]

FRAYER, W. E., R. C. VANAKEN AND R. D. SULLIVAN. 1971. APPLICATION OF SAMPLING WITH PARTIAL REPLACEMENT TO TIMBER INVENTORIES, CENTRAL ROCKY MOUNTAINS. FOREST SCI. 17(2), 160-162. [MULTIPLE REGRESSION MODEL; MODEL]

FREESE, F. 1962. ELEMENTARY FOREST SAMPLING. AGRICULTURE HANDBOOK NO. 232. U. S. FOREST SERVICE. WASHINGTON, D. C. IV, 91 PP.

FUJITA, H. 1956. THE COLLECTION EFFICIENCY OF A PLANKTON NET. RES. POPUL. ECOL.-KYOTO 3, 8-15.

GAISER, R. N. 1951. RANDOM SAMPLING WITHIN CIRCULAR PLOTS BY MEANS OF POLAR COORDINANTS. J. FORESTRY 49(12), 916-917.

GATES, C. E., T. L. CLARK AND K. E. GABLE. 1975. OPTIMIZING MOURNING DOVE BREEDING POPULATION SURVEYS IN TEXAS. J. WILDL. MANAGE. 39(2), 237-242. [VARIANCE MINIMIZATION; MODEL; COST FUNCTION]

GAUFIN, A. R., E. K. HARRIS AND H. J. WALTER. 1956. A STATISTICAL EVALUATION OF STREAM BOTTOM SAMPLING DATA OBTAINED FROM THREE STANDARD SAMPLERS. ECOLOGY 37(4), 643-648. [PROBABILITY]

GERARD, G. AND P. BERTHET. 1971. SAMPLING STRATEGY IN CENSUSING PATCHY POPULATIONS. PP. 59-65. IN, STATISTICAL ECOLOGY, VOLUME 1. SPATIAL PATTERNS AND STATISTICAL DISTRIBUTIONS. (G. P. PATIL, E. C. PIELOU AND W. E. WATERS, ED.). PENNSYLVANIA STATE UNIVERSITY PRESS, UNIVERSITY PARK, PENNSYLVANIA.

GEVORKIANTZ, S. R. 1954. PROBLEMS IN FOREST INVENTORY FROM THE FORESTER'S POINT OF VIEW. PP. 251-262. IN, STATISTICS AND MATHEMATICS IN BIOLOGY. (O. KEMPTHORNE, T. A. BANCROFT, J. W. GOWEN AND J. L. LUSH, EDS.). IOWA STATE COLLEGE PRESS, AMES, IOWA.

GHENT, A. W. 1969. STUDIES OF REGENERATION IN FOREST STANDS DEVASTATED BY THE SPRUCE BUDWORM. IV. PROBLEMS OF STOCKED-QUADRAT SAMPLING. FOREST SCI. 15(4), 417-429. [BINOMIAL DISTRIBUTION; POISSON DISTRIBUTION; MODEL; CONTAGIOUS DISTRIBUTION]

GLEASON, H. A. 1922. ON THE RELATION BETWEEN SPECIES AND AREA. ECOLOGY 3(2), 158-162. [ARRHENIUS'S EQUATION]

GLEASON, H. A. 1925. SPECIES AND AREA. ECOLOGY 6(1), 66-74. [ARRHENIUS'S EQUATION; EQUATIONS]

GOFF, F. G. 1968. USE OF SIZE STRATIFICATION AND DIFFERENTIAL WEIGHTING TO MEASURE FOREST TRENDS. AMER. MIDL. NAT. 79(2), 402-418. [EQUATIONS]

GOOD, I. J. AND G. H. TOULMIN. 1956. THE NUMBER OF NEW SPECIES, AND THE INCREASE IN POPULATION COVERAGE, WHEN A SAMPLE IS INCREASED. BIOMETRIKA 43(1/2), 45-63.

GREEN, R. H. 1970. ON FIXED PRECISION LEVEL SEQUENTIAL SAMPLING. RES. POPUL. ECOL.-KYOTO 12(2), 249-251. [MOLLUSCA]

GROSENBAUGH, L. R. 1965. THREE-PEE SAMPLING THEORY AND PROGRAM "THRP" FOR COMPUTER GENERATION OF SELECTION CRITERIA. U. S. FOR. SERV. PACIFIC SOUTHWEST FOREST EXPT. STN. RES. PAPER PSW-21. 53 PP.

GROSENBAUGH, L. R. 1967. THE GAINS FROM SAMPLE-TREE SELECTION WITH UNEQUAL PROBABILITIES. J. FORESTRY 65(3), 203-206. [MODEL; PROBABILITY; ARTIFICIAL POPULATION; COMPONENTS OF VARIANCE]

GULLAND, J. 1957. SAMPLING PROBLEMS AND METHODS IN FISHERIES RESEARCH. FAO FISH. BULL. 10(4), 157-181.

GULLAND, J. A. 1958. SAMPLING OF SEMI-OCEANIC STOCKS OF FISH. PP. 71-76. IN, SOME PROBLEMS FOR BIOLOGICAL FISHERY SURVEY AND TECHNIQUES FOR THEIR SOLUTION. INTERN. COMM. FOR THE NORTHWEST ATLANTIC FISHERIES, SPECIAL PUBL. NO. 1.

GULLAND, J. A. 1966. MANUAL OF SAMPLING AND STATISTICAL METHODS FOR FISHERIES BIOLOGY, PART I. SAMPLING METHODS. FAO MANUALS IN FISHERIES SCIENCE NO. 3. MISC. PAGES.

HANSEN, V. KR. AND K. P. ANDERSEN. 1962. SAMPLING SMALLER ZOOPLANKTON. RAPP. PROCES-VERB. REUNIONS J. CONS. PERM. INT. EXPLOR. MER 153, 39-47. [VARIANCE; LINEAR REGRESSION MODEL CATCH EQUATION]

HANSON, W. R. AND F. GRAYBILL. 1956. SAMPLE SIZE IN FOOD-HABITS ANALYSES. J. WILDL. MANAGE. 20(1), 64-68.

HARRIS, E. K. 1957. FURTHER RESULTS IN THE STATISTICAL ANALYSIS OF STREAM SAMPLING. ECOLOGY 38(3), 463-468.

HARRIS, R. W. 1951. USE OF AERIAL PHOTOGRAPHS AND SUB-SAMPLING IN RANGE INVENTORIES. J. RANGE MANAGE. 4(4), 270-278. [SAMPLING; OPTIMUM ALLOCATION]

HASEL, A. A. 1938. SAMPLING ERROR IN TIMBER SURVEYS. J. AGRIC. RES. 57(10), 713-736.

HASEL, A. A. 1942. ESTIMATION OF VOLUME IN TIMBER STANDS BY STRIP SAMPLING. ANN. MATH. STAT. 13(2), 179-206.

HASEL, A. A. 1954. PROBLEMS IN FOREST INVENTORY: FROM THE STATISTICAL POINT OF VIEW. PP. 263-272. IN. STATISTICS AND MATHEMATICS IN BIOLOGY. (O. KEMPTHORNE, T. A. BANCROFT, J. W. GOWEN AND J. L. LUSH, EDS.). IOWA STATE COLLEGE PRESS, AMES, IOWA. [SAMPLING]

HASEL, A. A. AND A. POLI. 1949. A NEW APPROACH TO FOREST OWNERSHIP SURVEYS. LAND ECONOMICS 25(1), 1-10. [SAMPLING]

HAYNE, D. W. 1951. A STUDY OF BIAS IN THE SELECTION OF A "REPRESENTATIVE" SAMPLE OF SMALL FISH. PAPERS MICH. ACAD. SCI., ARTS AND LETTERS 37, 133-141. [SAMPLING]

HAYNE, D. W. AND L. EBERHARDT. 1954. NATURE OF THE BIAS OF ESTIMATES COMPUTED FROM VOLUNTARY REPORTS. MICH. DEPT. CONSERV. GAME DIV. REPT. NO. 2325. 8 PP. [MODEL]

HEADY, H. F. 1955. TECHNIQUES USEFUL IN RANGE RESEARCH. J. RANGE MANAGE. 8(3), 114-116. [SAMPLING]

HEALY, M. J. R. 1962. SOME BASIC STATISTICAL TECHNIQUES IN SOIL ZOOLOGY. PP. 3-9. IN. PROGRESS IN SOIL ZOOLOGY. (P. W. MURPHY, ED.). BUTTERWORTHS, LONDON, ENGLAND. [SAMPLING; NEGATIVE BINOMIAL DISTRIBUTION; POISSON DISTRIBUTION; NORMAL; SAMPLE SIZE]

HENNEMUTH, R. C. 1957. AN ANALYSIS OF METHODS OF SAMPLING TO DETERMINE THE SIZE COMPOSITION OF COMMERCIAL LANDINGS OF YELLOWFIN TUNA (NEOTHUNNUS MACROPTERUS) AND SKIPJACK (KATSUWONUS PELAMIS). BULL. INTER-AMER. TROP. TUNA COMM. 2(5), 174-225. [STRATIFIED SAMPLING; SAMPLE SIZE; VARIANCE; ANALYSIS OF VARIANCE]

HICKS, R. H. AND L. D. CALVIN. 1964. AN EVALUATION OF THE PUNCH CARD METHOD OF ESTIMATING SALMON-STEELHEAD SPORT CATCH. OREGON AGRIC. EXPT. STN. TECH. BULL. 81. 75 PP. [STRATIFICATION; COMPUTER; ESTIMATION; VARIANCE; SUBSAMPLING]

HILLBRICHT-ILKOWSKA, A. AND T. WEGLENSKA. 1970. THE EFFECT OF SAMPLING FREQUENCY AND THE METHOD OF ASSESSMENT ON THE PRODUCTION VALUES OBTAINED FOR SEVERAL ZOOPLANKTON SPECIES. EKOLOGIYA POLSKA-SERIA A 18(27), 539-557. [VINBERG PRODUCTION MODEL; EDMONDSON'S EQUATION]

HIRANO, T. 1965. AN ESTIMATION ON DISTRIBUTION OF FISH EGGS AND LARVAE DRIFTED BY THE KUROSHIO CURRENT. BULL. TOKAI REG. FISH. RES. LAB. 44, 25-30. [EQUATIONS]

HOEL, P. G. 1943. THE ACCURACY OF SAMPLING METHODS IN ECOLOGY. ANN. MATH. STAT. 14(3), 289-300.

HOLDEN, M. J. AND D. F. S. RAITT. (EDS). 1974. MANUAL OF FISHERIES SCIENCE PART 2 - METHODS OF RESOURCE INVESTIGATION AND THEIR APPLICATION. FAO FISHERIES TECHNICAL PAPER NO. 115(REV. 1). IX, 214 PP. [SAMPLING; NORMAL DISTRIBUTION; CONFIDENCE LIMITS; STRATIFIED SAMPLING; VARIANCE]

HOLT, W. R. 1968. A FURTHER NOTE ON SAMPLE SIZE FOR CHI-SQUARE TEST: THE R X C TABLE. J. ECON. ENTOMOL. 61(3), 853-854.

HOLT, W. R., B. H. KENNEDY AND J. W. PEACOCK. 1967. FORMULAE FOR ESTIMATING SAMPLE SIZE FOR CHI-SQUARE TEST. J. ECON. ENTOMOL. 60(1), 286-288.

INSTITUTE OF STATISTICS. 1950. A COMPARISON OF THE PRECISION OF VARIOUS ESTIMATORS FOR ESTIMATING WATERFOWL KILL FROM SURVEY DATA. UNIVERSITY OF NORTH CAROLINA, RALEIGH, NORTH CAROLINA, MIMEO V, 51 PP.

IVES, W. G. H. 1954. SEQUENTIAL SAMPLING OF INSECT POPULATIONS. FORESTRY CHRON. 30(3), 287-291.

IVES, W. G. H. AND G. L. WARREN. 1965. SEQUENTIAL SAMPLING FOR WHITE GRUBS. CAN. ENTOMOL. 97(6), 596-604.

IVES, W. G. H. AND R. M. PRENTICE. 1959. ESTIMATION OF PARASITISM OF LARCH SAWFLY COCOONS BY BESSA HARVEYI TUSD. IN SURVEY COLLECTIONS. CAN. ENTOMOL. 91(8), 496-500. [CONFIDENCE INTERVAL; VARIANCE; CLUSTER SAMPLING; SAMPLE SIZE]

IWAO, S. 1975. A NEW METHOD OF SEQUENTIAL SAMPLING TO CLASSIFY POPULATIONS RELATIVE TO A CRITICAL DENSITY. RES. POPUL. ECOL.-KYOTO 16(2), 281-288. [SIMPLE RANDOM SAMPLING; CONFIDENCE INTERVAL; BINOMIAL DISTRIBUTION; POISSON DISTRIBUTION; NEGATIVE BINOMIAL DISTRIBUTION; TWO-STAGE SAMPLING]

IWAO, S. AND E. KUNO. 1968. USE OF THE REGRESSION OF MEAN CROWDING ON MEAN DENSITY FOR ESTIMATING SAMPLE SIZE AND THE TRANSFORMATION OF DATA FOR THE ANALYSIS OF VARIANCE. RES. POPUL. ECOL.-KYOTO 10(2), 210-214.

JEBE, E. H. 1959. 142 QUERY: ON A QUAIL ROADSIDE COUNT TECHNIQUE. BIOMETRICS 15(4), 628-631. [ANALYSIS OF VARIANCE; COMPONENTS OF VARIANCE]

JEFFERS, J. N. R. 1969. THE CONTRIBUTION OF STATISTICAL METHODS TO FOREST RESEARCH AND MANAGEMENT. BULL. INTERN. STAT. INST. 42(1), 334-342. [DESIGN; SURVEYS; COMPUTER]

JOHNSON, F. A. 1943. A STATISTICAL STUDY OF SAMPLING METHODS FOR TREE NURSERY INVENTORIES. J. FORESTRY 41(9), 674-679. [RANDOM SAMPLING; SYSTEMATIC SAMPLING; STRATIFIED SAMPLING]

JOHNSON, M. W. AND L. WROBLEWSKI. 1962. ERRORS ASSOCIATED WITH A SYSTEMATIC SAMPLING CREEL CENSUS. TRANS. AMER. FISH. SOC. 91(2), 201-207. [EQUATIONS]

JOLLY, G. M. 1954. THEORY OF SAMPLING. PP. 8-22. IN, METHODS OF SURVEYING AND MEASURING VEGETATION. (D. BROWN). COMMONWEALTH BUREAU OF PASTURES AND FIELD CROPS, HURLEY, BERKS, ENGLAND. BULL. 42.

JOLLY, G. M. 1969. SAMPLING METHODS FOR AERIAL CENSUSES OF WILDLIFE POPULATIONS. PP. 46-49. IN, PROCEEDINGS OF THE WORKSHOP ON THE USE OF LIGHT AIRCRAFT IN WILDLIFE MANAGEMENT IN EAST AFRICA. (W. G. SWANK, R. M. WATSON, G. H. FREEMAN AND T. JONES, EDS.). EAST AFRICAN AGRICULTURAL AND FORESTRY JOURNAL, SPECIAL ISSUE. [STRATIFIED SAMPLING; RATIO ESTIMATOR; VARIANCE]

JOLLY, G. M. 1969. THE TREATMENT OF ERRORS IN AERIAL COUNTS OF WILDLIFE POPULATIONS. PP. 50-55. IN, PROCEEDINGS OF THE WORKSHOP ON THE USE OF LIGHT AIRCRAFT IN WILDLIFE MANAGEMENT IN EAST AFRICA. (W. G. SWANK, R. M. WATSON, G. H. FREEMAN AND T. JONES, EDS.). EAST AFRICAN AGRICULTURAL AND FORESTRY JOURNAL, SPECIAL ISSUE. [BIAS; SAMPLING; VARIANCE]

JOWETT, G. H. AND G. SCURFIELD. 1949. STATISTICAL TEST FOR OPTIMAL CONDITIONS: NOTE ON A PAPER OF EMMETT AND ASHBY. J. ECOL. 37(1), 65-67. [PLANTS; CHI-SQUARE; T-TEST]

JUMBER, J. F., H. O. HARTLEY, E. L. KOZICKY AND A. M. JOHNSON. 1957. A TECHNIQUE FOR SAMPLING MOURNING DOVE PRODUCTION. J. WILDL. MANAGE. 21(2), 226-229. [STRATIFIED SAMPLING]

KAELIN, A. AND C. AUER. 1954. STATISTISCHE METHODEN ZUR UNTERSUCHUNG VON INSEKTEN POPULATIONEN. ZEITSCHR. F. ANGEW. ENTOMOLOGIE. 36(3), 241-282 AND 36(4), 423-461.

KINASHI, K. 1954. FOREST INVENTORY BY SAMPLING METHODS. BULL. KYUSHU UNIV. FORESTS 23. 153 PP. [STRATIFIED SAMPLING; SIMPLE RANDOM SAMPLING; SYSTEMATIC SAMPLING; AUTOCORRELATION; ANALYSIS OF VARIANCE; SUBSAMPLING; DOUBLE SAMPLING]

KNIGHT, F. B. 1967. EVALUATION OF FOREST INSECT INFESTATIONS. ANNU. REV. ENTOMOL. 12, 207-228. [SEQUENTIAL SAMPLING; LIFE TABLE]

KOBAYASHI, S. 1974. THE SPECIES-AREA RELATION I. A MODEL FOR DISCRETE SAMPLING. RES. POPUL. ECOL.-KYOTO 15(2), 223-237. [PROBABILITY; NEGATIVE BINOMIAL DISTRIBUTION; QUADRAT]

KOBAYASHI, S. 1975. THE SPECIES-AREA RELATION II. A SECOND MODEL FOR CONTINUOUS SAMPLING. RES. POPUL. ECOL.-KYOTO 16(2), 265-280.

KULOW, D. L. 1966. COMPARISON OF FOREST SAMPLING DESIGNS. J. FORESTRY 64(7), 469-474.

KUNO, E. 1969. A NEW METHOD OF SEQUENTIAL SAMPLING TO OBTAIN THE POPULATION ESTIMATES WITH A FIXED LEVEL OF PRECISION. RES. POPUL. ECOL.-KYOTO 11(2), 127-136.

KUNO, E. 1971. SAMPLING ERROR AS A MISLEADING ARTIFACT IN "KEY FACTOR ANALYSIS." RES. POPUL. ECOL.-KYOTO 13(1), 28-45. [VARIANCE; COVARIANCE; EXPECTED VALUE]

KUNO, E. 1972. SOME NOTES ON POPULATION ESTIMATION BY SEQUENTIAL SAMPLING. RES. POPUL. ECOL.-KYOTO 14(1) 58-73.

LAMBERT, J. M. 1972. THEORETICAL MODELS FOR LARGE-SCALE VEGETATION SURVEY. PP. 87-109. IN, MATHEMATICAL MODELS IN ECOLOGY. (J. N. R. JEFFERS, ED.), BLACKWELL SCIENTIFIC PUBLICATIONS, OXFORD, ENGLAND.

LAMBOU, V. W. 1961. DETERMINATION OF FISHING PRESSURE FROM FISHERMEN OR PARTY COUNTS WITH A DISCUSSION OF SAMPLING PROBLEMS. PROC. ANN. CONF. S. E. ASSOC. GAME AND FISH COMMISSIONERS 15, 380-401. [BIAS; RANDOM SAMPLING; NEGATIVE BINOMIAL DISTRIBUTION; POISSON DISTRIBUTION; SIMPSON'S RULE; TRAPEZOIDAL RULE; SYSTEMATIC SAMPLING; NORMAL THEORY; STRATIFICATION; SUBSAMPLING]

LAMBOU, V. W. 1963. APPLICATION OF DISTRIBUTION PATTERN OF FISHES IN LAKE BISTINEAU TO DESIGN OF SAMPLING PROGRAMS. PROG. FISH-CULT. 25(2), 79-87. [POISSON DISTRIBUTION; NEGATIVE BINOMIAL DISTRIBUTION; T-TEST; LOG DISTRIBUTION OF FISHER; VARIANCE]

LANDER, R. H. 1956. SEQUENTIAL ANALYSIS IN FISHERY RESEARCH. COPEIA 1956(3), 151-154.

LANG, A. R. G. AND F. M. MELHUISH. 1970. LENGTHS AND DIAMETERS OF PLANT ROOTS IN NON-RANDOM POPULATIONS BY ANALYSIS OF PLANE SURFACES. BIOMETRICS 26(3), 421-431.

LASSITER, R., R. HARKINS AND L. TEBO. 1973. BIOMETRICS. PP. 1-27. IN, BIOLOGICAL AND LABORATORY METHODS FOR MEASURING THE QUALITY OF SURFACE WATERS AND EFFLUENTS. (C. I. WEBER, ED.), NATIONAL ENVIRONMENTAL RESEARCH CENTER, OFFICE OF RESEARCH AND DEVELOPMENT, U.S. EPA, CINCINNATI, OHIO, ENVIRONMENTAL MONITORING SERIES EPA-670/4-73-001. [GRAPH; DESIGN; SAMPLING; SUBSAMPLING; SAMPLE SIZE; TEST OF HYPOTHESIS; REGRESSION; STRATIFIED RANDOM SAMPLING]

LEAK, W. 1966. ANALYSIS OF MULTIPLE SYSTEMATIC REMEASUREMENTS. FOREST SCI. 12(1), 69-73. [REGRESSION; MODEL; PLOT; VARIANCE]

LEDIG, F. T. 1974. AN ANALYSIS OF METHODS FOR SELECTION OF TREES FROM WILD STANDS. FOREST SCI. 20(1), 2-16. [ANALYSIS OF VARIANCE; GENETICS]

LEGLER, E., JR., H. STERN, JR. AND W. S. OVERTON. 1961. A PRELIMINARY EVALUATION OF TELEPHONE AND FIELD SAMPLING FRAMES. TRANS. N. AMER. WILDL. NAT. RESOUR. CONF. 26, 405-417. [DOVES]

LINDSEY, A. A. 1956. SAMPLING METHODS AND COMMUNITY ATTRIBUTES IN FORESTRY ECOLOGY. FOREST SCI. 2(4), 287-296.

LINDSEY, A. A., J. D. BARTON, JR. AND S. R. MILES. 1958. FIELD EFFICIENCIES OF FOREST SAMPLING METHODS. ECOLOGY 39(3), 428-444.

LLOYD, F. T. AND D. F. OLSON. 1974. THE PRECISION AND REPEATABILITY OF A LEAF BIOMASS SAMPLING TECHNIQUE FOR MIXED HARDWOOD STANDS. J. APPL. ECOL. 11(3), 1035-1042. [STATISTICAL MODEL; VARIANCE; ANALYSIS OF VARIANCE; VARIANCE COMPONENTS]

LORD, F. T. 1968. AN APPRAISAL OF METHODS OF SAMPLING APPLE TREES AND RESULTS OF SOME TESTS USING A SAMPLING UNIT COMMON TO INSECT PREDATORS AND THEIR PREY. CAN. ENTOMOL. 100(1), 23-33.

MANLY, B. F. J. 1975. THE ESTIMATION OF THE FITNESS FUNCTION FROM TWO SAMPLES TAKEN FROM A POPULATION. RES. POPUL. ECOL.-KYOTO 16(2), 219-230. [MODEL; PROBABILITY; MAXIMUM LIKELIHOOD ESTIMATION; LEAST SQUARES; MOMENTS; LINEAR REGRESSION MODEL]

MASUYAMA, M. 1953. A RAPID METHOD OF ESTIMATING BASAL AREA IN TIMBER SURVEY--AN APPLICATION OF INTEGRAL GEOMETRY TO AREAL SAMPLING PROBLEMS. SANKHYA 12(3), 291-302. [MOMENT GENERATING FUNCTIONS]

MASUYAMA, M. 1953. RAPID METHODS OF ESTIMATING THE SUM OF SPECIFIED AREAS IN A FIELD OF GIVEN SIZE. REP. STATIST. APPL. RES. UNION JAP. SCI. ENG. 2(4), 113-119.

MASUYAMA, M. 1954. A NOTE ON "RAPID METHODS OF ESTIMATING THE SUM OF SPECIFIED AREAS IN A FIELD OF GIVEN SIZE". REP. STATIST. APPL. RES. UNION JAP. SCI. ENG. 3(2), 32.

MASUYAMA, M. 1957. AN ORNITHOLOGICAL APPLICATION OF LINE-GRID SAMPLING. REP. STATIST. APPL. RES. UNION JAP. SCI. ENG. 5(1), 1-3.

MATERN, B. 1949. FOREST INVENTORIES - SAMPLING TECHNIQUES (ADAPTATION OF MODERN STATISTICAL METHODS TO THE ESTIMATION OF FOREST AREAS, TIMBER VOLUMES, GROWTH, AND DRAIN). UNITED NATIONS ECONOMIC AND SOCIAL COUNCIL, REPORT E/CONF.7/SEC/W.336, 8 PP.

MATERN, B. 1969. SAMPLE SURVEY PROBLEMS. BULL. INTERN. STAT. INST. 42(1), 143-152. [PLANTS; ANIMALS; ESTIMATION]

MCALICE, B. J. 1971. PHYTOPLANKTON SAMPLING WITH THE SEDGWICK-RAFTER CELL. LIMNOL. OCEANOGR. 16(1), 19-28. [POISSON DISTRIBUTION; TRANSFORMATION]

MCDONNELL, A. J. 1968. MULTIVARIATE ANALYSIS OF WATER-QUALITY DATA, PART 2. WATER SEWAGE WORKS 115(3), 119-123.

MCDONNELL, A. J. 1968. MULTIVARIATE ANALYSIS OF WATER-QUALITY DATA-PART 1. WATER SEWAGE WORKS 115(2), 72-74. [MULTIPLE REGRESSION; PRINCIPAL COMPONENT ANALYSIS; COMPUTER; CORRELATION MATRIX; EIGENVALUE]

MELHUISH, F. M. AND A. R. G. LANG. 1971. QUANTITATIVE STUDIES OF ROOTS IN SOIL: 2. ANALYSIS OF NON-RANDOM POPULATIONS. SOIL SCI. 112(3), 161-166. [COMPUTER]

MENG, C. H. 1972. A COMMENT ON THE PAPER BY CUNIA ON "SOME THEORY ON RELIABILITY OF VOLUME ESTIMATES IN A FOREST INVENTORY SAMPLE". FOREST SCI. 18(1), 95. [VARIANCE]

MICHELS, D. E. 1974. ANALYSIS OF PAIRED DATA SEQUENTIAL IN SPACE OR TIME AND THE RELATIONSHIPS TO SAMPLING CYCLIC DISTRIBUTIONS. AEC REPT. RFP-2165. DOW CHEMICAL U.S.A., ROCKY FLATS DIVISION, GOLDEN, COLORADO, 8 PP. [SOIL SAMPLES FOR CS-137]

MILLER, C. 1967. A STATISTICAL METHOD: SEQUENTIAL ANALYSIS. ANN. SCI. FOR., PARIS 24(4), 327-343. [FOREST RESEARCH]

MILNE, A. 1959. THE CENTRIC SYSTEMATIC AREA-SAMPLE TREATED AS A RANDOM SAMPLE. BIOMETRICS 15(2), 270-297.

MOLLER, F. AND A. ZATTERA. 1974. THE APPLICATION OF SEQUENTIAL ESTIMATION METHODS TO COUNTS OF PHYTOPLANKTON COMITATO NAZIONALE ENERGIA NUCLEARE, ROMA, ITALIA, REPORT RT/BIO(74)7. 75 PP. [COMPUTER; LIKELIHOOD RATIO; WALD'S PROBABILITY RATIO TEST; SEQUENTIAL SAMPLING; STEIN'S METHOD; SAMPLE SIZE]

MORISITA, M. 1964. APPLICATION OF I-INDEX TO SAMPLING TECHNIQUES. RES. POPUL. ECOL.-KYOTO 6(2), 43-53.

MORISITA, M. 1965. A REVISION OF THE METHODS FOR ESTIMATING POPULATION VALUES OF THE INDEX OF DISPERSION IN THE I-METHOD. RES. POPUL. ECOL.-KYOTO 7(2), 126-128. [STRATIFIED RANDOM SAMPLING]

MORISITA, M. 1966. THE RELATION BETWEEN VARIANCE-MEAN RATIO. RES. POPUL. ECOL.-KYOTO 8(1), 60-61. [INSECT]

MORRIS, R. F. 1960. SAMPLING INSECT POPULATIONS. ANNU. REV. ENTOMOL. 5, 243-264.

MOYLE, J. B. AND R. LOUND. 1960. CONFIDENCE LIMITS ASSOCIATED WITH MEANS AND MEDIANS OF SERIES OF NET CATCHES. TRANS. AMER. FISH. SOC. 89(1), 53-58. [NEGATIVE BINOMIAL DISTRIBUTION; NON-PARAMETRIC METHODS]

MUKERJI, M. K. 1973. THE DEVELOPMENT OF SAMPLING TECHNIQUES FOR POPULATIONS OF THE TARNISHED PLANT BUG, LYGUS LINEORALIS (HEMIPTERA: MIRIDAE). RES. POPUL. ECOL.-KYOTO 15(1), 50-63. [OPTIMUM SAMPLE SIZE; COST FUNCTION; MEAN CROWDING INDEX; REGRESSION; POISSON DISTRIBUTION; NEGATIVE BINOMIAL DISTRIBUTION; ANALYSIS OF VARIANCE]

MURPHY, G. I. AND R. I. CLUTTER. 1972. SAMPLING ANCHOVY LARVAE WITH A PLANKTON PURSE SEINE. U. S. NATL. MAR. FISH. SERV. FISH. BULL. 70(3), 789-798. [LARVAE DODGING MODEL]

MYERS, E. AND V. J. CHAPMAN. 1953. STATISTICAL ANALYSIS APPLIED TO A VEGETATION TYPE IN NEW ZEALAND. ECOLOGY 34(1), 175-185. [VARIANCE; LOGARITHMIC SERIES; FISHER'S INDEX OF DIVERSITY; ANALYSIS OF VARIANCE]

NELSON, T. C. AND F. A. BENNETT. 1965. A CRITICAL LOOK AT THE NORMALITY CONCEPT. J. FORESTRY 63(2), 107-109. [OPTIMAL SOLUTIONS]

NEUHOLD, J. M. AND K. H. LU. 1957. CREEL CENSUS METHOD. UTAH STATE DEPT. FISH GAME, SALT LAKE CITY, UTAH. PUBL. NO. 8. 36 PP. [STRATIFIED SAMPLING; RANDOM SAMPLING; NORMAL THEORY; CORRELATION COEFFICIENT; CONFIDENCE INTERVAL]

NEWMAN, E. I. 1966. A METHOD OF ESTIMATING THE TOTAL LENGTH OF ROOT IN A SAMPLE. J. APPL. ECOL. 3(1), 139-145. [EQUATIONS]

NEWTON, C. M., T. CUNIA AND C. A. BICKFORD. 1974. MULTIVARIATE ESTIMATORS FOR SAMPLING WITH PARTIAL REPLACEMENT ON TWO OCCASIONS. FOREST SCI. 20(2), 106-116. [STAND TABLES]

NORTON-GRIFFITHS, M. 1973. COUNTING THE SERENGETI MIGRATION WILDEBEEST USING TWO-STAGE SAMPLING. EAST AFR. WILDL. J. 11(2), 135-149. [JOLLY'S METHOD; SAMPLING ERROR]

O'REGAN, W. G. AND L. G. ARVANITIS. 1966. COST-EFFECTIVENESS IN FOREST SAMPLING. FOREST SCI. 12(4), 406-414. [ECONOMIC THEORY]

O'REGAN, W. G., D. W. SEEGRIST AND R. L. HUBBARD. 1973. COMPUTER SIMULATION AND VEGETATION SAMPLING. J. WILDL. MANAGE. 37(2), 217-222. [COST FUNCTION; CONFIDENCE INTERVAL; VARIANCE]

OAKLAND, G. B. 1950. AN APPLICATION OF SEQUENTIAL ANALYSIS TO WHITEFISH SAMPLING. BIOMETRICS 6(1), 59-67.

OLSON, F. C. W. 1959. QUANTITATIVE ESTIMATES OF FILAMENTOUS ALGAE. TRANS. AMER. MICROSC. SOC. 69(3), 272-279.

ORD, J. K. 1970. THE NEGATIVE BINOMIAL MODEL AND QUADRAT SAMPLING. PP. 151-163. IN, RANDOM COUNTS IN SCIENTIFIC WORK, VOLUME 2. RANDOM COUNTS IN BIOMEDICAL AND SOCIAL SCIENCES. (G. P. PATIL, ED.). PENNSYLVANIA STATE UNIVERSITY PRESS, UNIVERSITY PARK, PENNSYLVANIA. [BINOMIAL DISTRIBUTION; GENERALIZED POISSON DISTRIBUTION; INDEX OF DISPERSION; NEYMAN'S TYPE A DISTRIBUTION; POISSON DISTRIBUTION]

OSBORNE, J. G. 1942. SAMPLING ERRORS OF SYSTEMATIC AND RANDOM SURVEYS OF COVER-TYPE AREAS. J. AMER. STAT. ASSOC. 37(218), 256-264.

PALOHEIMO, J. E. AND L. M. DICKIE. 1963. SAMPLING THE CATCH OF A RESEARCH VESSEL. J. FISH. RES. BOARD CAN. 20(1), 13-25. [STRATIFICATION; CHI-SQUARE; SAMPLE SIZE]

PARCHEVSKAYA, D. S. 1970. SYSTEM OF STATISTICAL ANALYSIS AND PLANNING OF EXPERIMENT. PP. 26-28. IN, MARINE RADIOECOLOGY (G. G. POLIKARPOV, ED.). "NAUKOVA DUMKA", KIEV, U.S.S.R. ENGLISH TRANSLATION USAEC REPORT AEC-TR-7299. PP. 29-32.

PAYANDEH, B. 1970. RELATIVE EFFICIENCY OF TWO-DIMENSIONAL SYSTEMATIC SAMPLING. FOREST SCI. 16(3), 271-276.

PEDEN, D. G., R. M. HANSEN, R. W. RICE AND G. M. VAN DYNE. 1974. A DOUBLE SAMPLING TECHNIQUE FOR ESTIMATING DIETARY COMPOSITION. J. RANGE MANAGE. 27(4), 323-325. [COST FUNCTION; SAMPLE SIZE]

PERSSON, O. 1973. ON FOREST REGENERATION SURVEYS. INTERN. UNION OF FOREST RESEARCH ORGANIZATIONS, 4TH CONFERENCE OF THE ADVISORY GROUP OF FOREST STATISTICIANS, VANCOUVER, BRITISH COLUMBIA, CANADA, AUGUST 1973, 22 PP. [SAMPLING]

PETERSEN, R. G. AND L. D. CALVIN. 1965. SAMPLING. PP. 54-72. IN, METHODS OF SOIL ANALYSIS, PART 1. (C. A. BLACK, ET AL. EDS.). AMERICAN SOCIETY OF AGRONOMY, MADISON, WISCONSIN.

PHILLIPSON, J. 1971. OTHER ARTHROPODS. PP. 262-287. IN, METHODS OF STUDY IN QUANTITATIVE SOIL ECOLOGY: POPULATION, PRODUCTION AND ENERGY FLUX. (J. PHILLIPSON, ED.). BLACKWELL SCIENTIFIC PUBLICATIONS, OXFORD, ENGLAND. [SAMPLING; COST FUNCTION; PRODUCTION MODEL; ENERGY FLOW]

PINKAS, L., M. S. OLIPHANT AND C. W. HAUGEN. 1968. SOUTHERN CALIFORNIA MARINE SPORTFISHING SURVEY: PRIVATE BOATS, 1964; SHORELINE, 1965-66. CALIF. DEPT. FISH GAME FISH BULL. NO. 143. 42 PP. [PROBABILITY SAMPLING; VARIANCE; STRATIFIED SAMPLING]

POPE, J. A. 1956. AN OUTLINE OF SAMPLING TECHNIQUES. PP. 11-20. IN, PROBLEMS AND METHODS OF SAMPLING FISH POPULATIONS. RAPP. PROCES-VERB. REUNIONS J. CONS. PERM. INT. EXPLOR. MER 140(1), 11-20. [CLUSTER SAMPLING; BIAS; SIMPLE RANDOM SAMPLING; SAMPLE SIZE; TWO-STAGE SAMPLING; OPTIMUM ALLOCATION; STRATIFIED SAMPLING]

RAYNER, H. J. 1951. THE NUMBER OF SAMPLES NEEDED TO DETECT SIGNIFICANT FLUCTUATIONS IN THE ABUNDANCE OF INDICATOR PLANKTON ORGANISMS. TRANS. AMER. MICROSCOP. SOC. 70(1), 31-36. [ANALYSIS OF COVARIANCE; LOGARITHMIC TRANSFORMATION; ANALYSIS OF VARIANCE]

REEKS, W. A. 1956. SEQUENTIAL SAMPLING FOR LARVAE OF THE WINTER MOTH, OPEROPHTERA BRUMATA (LINN.). (LEPIDOPTERA: GEOMETRIDAE). CAN. ENTOMOL. 88(6), 241-246.

ROBSON, D. S. 1960. AN UNBIASED SAMPLING AND ESTIMATION PROCEDURE FOR CREEL CENSUSES OF FISHERMEN. BIOMETRICS 16(2), 261-277.

ROBSON, D. S. 1961. ON THE STATISTICAL THEORY OF A ROVING CREEL CENSUS OF FISHERMAN. BIOMETRICS 17(3), 415-437. [SAMPLING; ESTIMATION; PROBABILITY; BIAS; POISSON DISTRIBUTION; STOCHASTIC]

ROBSON, D. S. 1970. A SAMPLING DESIGN AND ESTIMATION FORMULA FOR A CREEL CENSUS OF STREAM FISHERMAN, A PRELIMINARY REPORT. BIOMETRICS UNIT, CORNELL UNIVERSITY, BU-275-M. MIMEO, 6 PP.

SAFRANYIK, L. AND K. GRAHAM. 1971. EDGE-EFFECT BIAS IN THE SAMPLING OF SUB-CORTICAL INSECTS. CAN. ENTOMOL. 103(2), 240-255.

SALTZMAN, B. E. 1972. SIMPLIFIED METHODS FOR STATISTICAL INTERPRETATION OF MONITORING DATA. J. AIR POLLUT. CONTROL ASSOC. 22(2), 90-95. [QUALITY CONTROL; SAMPLING; CONFIDENCE INTERVAL; NOMOGRAM]

SCHREUDER, H. T. 1966. UNEQUAL PROBABILITY AND DOUBLE SAMPLING IN FORESTRY. DISSERTATION. IOWA STATE UNIVERSITY, AMES, IOWA.

SCHREUDER, H. T. 1967. UNEQUAL PROBABILITY AND DOUBLE SAMPLING IN FORESTRY. IOWA STATE UNIVERSITY BULLETIN 66(9), 34-35.

SCHREUDER, H. T. 1970. A METHOD OF COUNT SAMPLING IN THE PLANE IN FORESTRY. (ABSTRACT). BIOMETRICS 26(3), 606.

SCHREUDER, H. T., G. L. TYRE AND G. A. JAMES. 1975. INSTANT- AND INTERVAL-COUNT SAMPLING: TWO NEW TECHNIQUES FOR ESTIMATING RECREATION USE. FOREST SCI. 21(1), 40-44. [PROBABILITY SAMPLING; VARIANCE]

SCHREUDER, H. T., J. SEDRANSK AND K. D. WARE. 1968. 3-P SAMPLING AND SOME ALTERNATIVES, I. FOREST SCI. 14(4), 429-454.

SCHREUDER, H. T., J. SEDRANSK AND K. D. WARE. 1971. FOREST SAMPLING IN THE ABSENCE OF A WELL-DEFINED FRAME. PP. 119-148. IN, STATISTICAL ECOLOGY, VOLUME 2, SAMPLING AND MODELING BIOLOGICAL POPULATIONS AND POPULATION DYNAMICS. (G. P. PATIL, E. C. PIELOU AND W. E. WATERS, EDS.). PENNSYLVANIA STATE UNIVERSITY PRESS, UNIVERSITY PARK, PENNSYLVANIA.

SCHREUDER, H. T., J. SEDRANSK, K. D. WARE AND D. A. HAMILTON 1971. 3-P SAMPLING AND SOME ALTERNATIVES, II. FOREST SCI. 17(1), 103-118. [MONTE CARLO]

SCHULTZ, V. 1952. A SURVEY DESIGN APPLICABLE TO STATEWIDE WILDLIFE SURVEYS. J. TENN. ACAD. SCI. 27(1), 60-66. [SAMPLING; CLUSTER SAMPLING; AREA SAMPLING; PROPORTIONAL ALLOCATION; STRATIFIED SAMPLING]

SCHULTZ, V. 1954. WILDLIFE SURVEYS - DISCUSSION OF A SAMPLING PROCEDURE AND A SURVEY DESIGN. TENN. GAME AND FISH COMM., NASHVILLE, TENNESSEE. 153 PP. (M. S. THESIS, VIRGINIA POLYTECHNIC INSTITUTE, BLACKSBURG, VIRGINIA, 1954.). [SAMPLING; CLUSTER SAMPLING; AREA SAMPLING; PROPORTIONAL ALLOCATION; STRATIFIED SAMPLING]

SCHUMACHER, F. X. 1945. STATISTICAL METHOD IN FORESTRY. BIOMETRICS 1(3), 29-32. [SAMPLING; DESIGN]

SEN, A. R. 1970. ON THE BIAS IN ESTIMATION DUE TO IMPERFECT FRAME IN THE CANADIAN WATERFOWL SURVEYS. J. WILDL. MANAGE. 34(4), 703-706. [SAMPLING]

SEN, A. R. 1970. RELATIVE EFFICIENCY OF SAMPLING SYSTEMS IN THE CANADIAN WATERFOWL HARVEST SURVEY, 1967-68. BIOMETRICS 26(2), 315-326. [CLUSTER SAMPLING; SINGLE-STAGE SAMPLING]

SEN, A. R. 1971. INCREASED PRECISION IN CANADIAN WATERFOWL HARVEST SURVEY THROUGH SUCCESSIVE SAMPLING. J. WILDL. MANAGE. 35(4), 664-668. [REGRESSION; OPTIMUM SAMPLE SIZE]

SEN, A. R. 1971. NON-SAMPLING ERRORS IN CANADIAN WATERFOWL SAMPLE SURVEYS. (ABSTRACT). BIOMETRICS 27(1), 260.

SEN, A. R. 1971. RESPONSE ERRORS IN CANADIAN WATERFOWL SURVEYS. (ABSTRACT). BIOMETRICS 27(2), 488.

SEN, A. R. 1971. SAMPLING TECHNIQUES FOR ESTIMATION OF MITE INCIDENCE IN THE FIELD WITH SPECIAL REFERENCE TO THE TEA CROP. PP. 253-265. IN, STATISTICAL ECOLOGY, VOLUME 2, SAMPLING AND MODELING BIOLOGICAL POPULATIONS AND POPULATION DYNAMICS. (G. P. PATIL, E. C. PIELOU AND W. E. WATERS, EDS.). PENNSYLVANIA STATE UNIVERSITY PRESS, UNIVERSITY PARK, PENNSYLVANIA.

SEN, A. R. 1971. SOME RECENT DEVELOPMENTS IN WATERFOWL SAMPLE SURVEY TECHNIQUES. APPL. STAT. 20(2), 139-147. [NON-SAMPLING ERRORS]

SEN, A. R. 1971. SOME RECENT DEVELOPMENTS IN WATERFOWL SAMPLE SURVEY TECHNIQUES. J. ROYAL STAT. SOC. SER. C 20(2), 139-147. [MAIL SURVEY]

SEN, A. R. 1973. ESTIMATION OF MEMORY BIAS IN WILDLIFE MAIL SURVEYS. BULL. INTERN. STAT. INST. 45(2), 401-411. [MODEL]

SEN, A. R. 1973. THEORY AND APPLICATION OF SAMPLING ON REPEATED OCCASIONS WITH SEVERAL AUXILIARY VARIABLES. BIOMETRICS 29(2), 381-385. [WATERFOWL MAIL SURVEY]

SEN, A. R. AND R. P. CHAKRABARTY. 1964. ESTIMATION OF LOSS OF CROP FROM PESTS AND DISEASES OF TEA FROM SAMPLE SURVEYS. BIOMETRICS 20(3), 492-504.

SEN, A. R., R. P. CHAKRABARTY AND A. R. SARKAR. 1966. SAMPLING TECHNIQUES FOR ESTIMATION OF INCIDENCE OF RED SPIDER MITE ON TEA CROP IN NORTHEAST INDIA. BIOMETRICS 22(2), 385-403.

SHAFER, E. L., JR. AND R. C. THOMPSON. 1968. MODELS THAT DESCRIBE USE OF ADIRONDACK CAMPGROUNDS. FOREST SCI. 14(4), 383-391. [FACTOR ANALYSIS; MULTIPLE REGRESSION MODEL]

SHANKS, R. E. 1953. BIASED FOREST STAND ESTIMATES DUE TO SAMPLE SIZE. SCIENCE 118(3077), 750-751. [SAMPLING; NORMAL DISTRIBUTION; TRANSFORMATION; QUADRAT; RANDOM PAIRS; PLOTLESS SAMPLES]

SHEPPERD, R. F. AND C. E. BROWN. 1971. SEQUENTIAL EGG-BAND SAMPLING AND PROBABILITY METHODS OF PREDICTING DEFOLIATION BY MALACOSOMA DISSTRITA (LASIOCAMPIDAE: LEPIDOPTERA). CAN. ENTOMOL. 103(10), 1371-1379. [SEQUENTIAL SAMPLING]

SHIUE, C.-J. 1960. SYSTEMATIC SAMPLING WITH MULTIPLE RANDOM STARTS. FOREST SCI. 6(1), 42-50.

SHIUE, C.-J. AND H. H. JOHN. 1962. A PROPOSED SAMPLING DESIGN FOR EXTENSIVE FOREST INVENTORY: DOUBLE SYSTEMATIC SAMPLING FOR REGRESSION WITH MULTIPLE RANDOM STARTS. J. FORESTRY 60(9), 607-610.

SHIUE, C.-J. AND R. BEAZLEY. 1957. CLASSIFICATION OF THE SPATIAL DISTRIBUTION OF TREES USING THE AREA SAMPLING METHOD. FOREST SCI. 3(1), 22-31.

SMITH, D. R. 1968. BIAS IN ESTIMATES OF HERBAGE UTILIZATION DERIVED FROM PLOT SAMPLING. J. RANGE MANAGE. 21(2), 109-110. [EQUATIONS]

SMITH, R. H., D. J. NEFF AND C. Y. MCCULLOCH. 1969. A MODEL FOR THE INSTALLATION AND USE OF A DEER PELLET GROUP SURVEY. ARIZONA GAME AND FISH DEPT., PHOENIX, ARIZONA. SPECIAL REPT. NO. 1, IV, 30 PP. [SAMPLING; TRANSECT; VARIANCE COMPONENTS; COST FUNCTION; CONFIDENCE INTERVAL]

STAGE, A. R. 1969. A GROWTH DEFINITION FOR STOCKING: UNITS, SAMPLING, AND INTERPRETATION. FOREST SCI. 15(3), 255-265. [FORESTRY; SUBSAMPLING]

STARK, R. W. 1952. SEQUENTIAL SAMPLING OF THE LODGEPOLE NEEDLE MINER. FORESTRY CHRON. 28(2), 57-60.

STROGONOV, A. A. 1970. ESTIMATION OF CONFIDENCE INTERVALS IN RADIOMETRY. PP. 42-45. IN, MARINE RADIOECOLOGY (G. G. POLIKARPOV, ED.)."NAUKOVA DUMKA," KIEV, U.S.S.R. (IN RUSSIAN). ENGLISH TRANSLATION USAEC REPORT AEC-TR-7299. PP. 49-52.

STROGONOV, A. A. AND A. V. KOVALEV. 1969. EVALUATION OF ACCURACY OF QUANTITATIVE SAMPLING OF A ZOOPLANKTON. HYDROBIOL. J. 5(2), 53-59. (ENGLISH TRANSL.). [CONFIDENCE INTERVAL; T-TEST]

STROGONOV, A. A. AND A. V. KOVALEV. 1970. STATISTICAL EVALUATION OF ACCURACY IN RESULTS FROM QUANTITATIVE ZOOPLANKTON HAULS. IN, MARINE RADIOECOLOGY. (G. G. POLIKARPOV, ED.). IZDATEL'STVO NAUKOVA DUMKA, KIEV. (ENGLISH TRANSL. U. S. AEC REPORT AEC-TR-7299; PP. 33-88. 1972). [SAMPLING; CONFIDENCE INTERVAL]

STRONG, F. E. 1960. SAMPLING ALFALFA SEED FOR CLOVER SEED CHALCID DAMAGE. J. ECON. ENTOMOL. 53(4), 611-615. [VARIANCE COMPONENTS; OPTIMUM SAMPLING; COST FUNCTION]

SUBCOMMITTEE ON RANGE RESEARCH METHODS OF THE AGRICULTURAL BOARD. 1962. EXPERIMENTAL DESIGNS. PP. 259-286. IN, BASIC PROBLEMS AND TECHNIQUES IN RANGE RESEARCH. NATIONAL ACADEMY OF SCIENCES RESEARCH COUNCIL, WASHINGTON, D. C. PUBL. 890.

SUBCOMMITTEE ON RANGE RESEARCH METHODS OF THE AGRICULTURAL BOARD. 1962. METHODS OF STUDYING VEGETATION. PP. 45-84. IN, BASIC PROBLEMS AND TECHNIQUES IN RANGE RESEARCH. NATIONAL ACADEMY OF SCIENCES RESEARCH COUNCIL, WASHINGTON, D. C. PUBL. 890. [QUANTIATIVE MEASUREMENTS; SAMPLING; PLOT; PLOTLESS METHODS; POINT ANALYSIS TECHNIQUES; TRANSECTS]

SUBCOMMITTEE ON RANGE RESEARCH METHODS OF THE AGRICULTURAL BOARD. 1962. SAMPLING METHODS WITH SPECIAL REFERENCE TO RANGE MANAGEMENT. PP. 223-257. IN, BASIC PROBLEMS AND TECHNIQUES IN RANGE RESEARCH. NATIONAL ACADEMY OF SCIENCES RESEARCH COUNCIL, WASHINGTON, D. C. PUBL. 890.

SUGIMOTO, T. 1969. PRELIMINARY STUDIES ON THE ESTIMATION OF THE POPULATION DENSITY OF AN AQUATIC MIDGE, TENDIPES DORSALIS MEIGEN (TENDIPEDIDAE: DIPTERA) IN BROOKS. JAPANESE J. ECOL. 19(1), 1-8. [NEGATIVE BINOMIAL DISTRIBUTION; SEQUENTIAL SAMPLING; VARIANCE; IWAO'S FORMULA]

SUKHATME, P. V., V. G. PANSE AND K. V. R. SASTRY. 1958. SAMPLING TECHNIQUE FOR ESTIMATING THE CATCH OF SEA FISH IN INDIA. BIOMETRICS 14(1), 78-96. [REGRESSION; CORRELATION COEFFICIENT; ANALYSIS OF VARIANCE; SUBSAMPLING; SYSTEMATIC SAMPLING; RANDOM SAMPLING]

SUKWONG, S., W. E. FRAYER AND E. W. MOGREN. 1971. GENERALIZED COMPARISONS OF THE PRECISION OF FIXED-RADIUS AND VARIABLE-RADIUS PLOTS FOR BASAL-AREA ESTIMATES. FOREST SCI. 17(2), 263-271. [NEYMAN'S SERIES; THOMAS'S SERIES; SAMPLING; POISSON POINT PROCESS; SPATIAL DISTRIBUTION]

SUMMERS, F. M. AND G. A. BAKER. 1952. A PROCEDURE FOR DETERMINING RELATIVE DENSITIES OF BROWN ALMOND MITE POPULATIONS ON ALMOND TREES. HILGARDIA 21(13), 369-382. [ANALYSIS OF VARIANCE; SAMPLING]

SUZUKI, O. 1973. A RATIONAL SAMPLING METHOD FOR ESTIMATION OF DEMERSAL FISH ABUNDANCE. BULL. JAPANESE SOC. SCI. FISH. 39(10), 1013-1019. [POLYA-EGGENBERGER DISTRIBUTION; INDEX OF PRECISION; SAMPLE SIZE]

TAFT, B. A. 1960. A STATISTICAL STUDY OF THE ESTIMATION OF ABUNDANCE OF SARDINE (SARDINOPS CAERULEA) EGGS. LIMNOL. OCEANOGR. 5(3), 245-264. [EQUATIONS; VARIANCE; LINEAR MODEL; NEGATIVE BINOMIAL DISTRIBUTION]

TAMAKI, G., J. U. MCGUIRE, JR. AND J. ONSAGER. 1973. SPATIAL DISTRIBUTION OF THE GREEN PEACH APHID USED IN ESTIMATING THE POPULATIONS OF GYNOPARAE. RES. POPUL. ECOL.-KYOTO 15(1), 64-75. [SAMPLING; POISSON DISTRIBUTION; NEGATIVE BINOMIAL DISTRIBUTION; LOGARITHMIC SERIES]

TANAKA, S. 1953. PRECISION OF AGE-COMPOSITION OF FISH ESTIMATED BY DOUBLE SAMPLING METHOD USING THE LENGTH FOR STRATIFICATION. BULL. JAPANESE SOC. SCI. FISH. 19(5), 657-670. [VARIANCE]

TAYLOR, C. C. 1953. NATURE OF VARIABILITY IN TRAWL CATCHES. U. S. FISH WILDL. SERV. FISH. BULL. 54, 145-166. [NEGATIVE BINOMIAL DISTRIBUTION; POISSON DISTRIBUTION; MODEL; PROBABILITY; TRANSFORMATION; EULERIAN DISTRIBUTION LOGARITHMIC SERIES; DIVERSITY INDEX]

TAYLOR, L. R. 1961. AGGREGATION, VARIANCE AND THE MEAN. NATURE 189(4766), 732-735.

TOMLINSON, P. K. 1971. SOME SAMPLING PROBLEMS IN FISHERY WORK. BIOMETRICS 27(3), 631-641. [STRATIFIED SAMPLING; TWO-STAGE SUBSAMPLING; RATIO ESTIMATOR]

TOSTOWARYK, W. AND J. M. MCLEOD. 1972. SEQUENTIAL SAMPLING FOR EGG CLUSTERS OF THE SWAINE JACK PINE SAWFLY, NEODIPRION SWAINEI (HYMENOPTERA: DIPRIONIDAE). CAN. ENTOMOL. 104(9), 1343-1347.

VAN DYNE, G. M. 1960. A METHOD FOR RANDOM LOCATION OF SAMPLE UNITS IN RANGE INVESTIGATIONS. J. RANGE MANAGE. 13(3), 152-153. [AREA EQUATION]

VAN DYNE, G. M. 1965. A FURTHER NOTE ON RANDOM LOCATIONS FOR SAMPLE UNITS IN CIRCULAR PLOTS. J. RANGE MANAGE. 18(3), 150-151. [SAMPLING]

VENRICK, E. L. 1971. THE STATISTICS OF SUBSAMPLING. LIMNOL. OCEANOGR. 16(5), 811-818. [VARIANCE; ANALYSIS OF VARIANCE; POISSON DISTRIBUTION; CHI-SQUARE; SAMPLING]

VOGL, R. J. 1969. QUANTITATIVE ECOLOGY: COMMENTS AND CRITICISM. THE BIOLOGIST 51(3), 85-90.

WADLEY, F. M. 1952. ELEMENTARY SAMPLING PRINCIPLES IN ENTOMOLOGY. U. S. DEPT. AGRICULTURE BUR. ENTOMOL. PLANT QUARANTINE REPT. ET-302, 18 PP. [BIAS; RANDOMNESS; SYSTEMATIC SAMPLING; SUBSAMPLING; STRATIFIED SAMPLING; WEIGHTING; COST FUNCTION]

WADLEY, F. M. 1967. CHP. 2. SAMPLING. PP. 31-46. IN, EXPERIMENTAL STATISTICS IN ENTOMOLOGY. (F. M. WADLEY). GRADUATE SCHOOL PRESS. U. S. DEPT. AGRICULTURE, WASHINGTON, D. C.

WALKER, F. T. 1950. SUBLITTORAL SEAWEED SURVEY OF THE ORKNEY ISLANDS. J. ECOL. 38(1), 139-165. [DENSITY-COVER EQUATION]

WARE, K. D. AND T. CUNIA. 1962. CONTINUOUS FOREST INVENTORY WITH PARTIAL REPLACEMENT OF SAMPLES. FOREST SCI. MONOGR. 3, 40 PP. [SAMPLING; VARIANCE; DOUBLE SAMPLING WITH REGRESSION; PLOT; OPTIMUM ALLOCATION]

WARREN, W. G. 1971. THE CENTER-SATELLITE CONCEPT AS A BASIS FOR ECOLOGICAL SAMPLING. PP. 87-116. IN, STATISTICAL ECOLOGY, VOLUME 2, SAMPLING AND MODELING BIOLOGICAL POPULATIONS AND POPULATION DYNAMICS. (G. P. PATIL, E. C. PIELOU AND W. E. WATERS, EDS.). PENNSYLVANIA STATE UNIVERSITY PRESS, UNIVERSITY PARK, PENNSYLVANNIA.

WATERS, W. E. 1955. SEQUENTIAL SAMPLING IN FOREST INSECT SURVEYS. FOREST SCI. 1(1), 68-79.

WENSEL, L. C. AND H. H. JOHN. 1969. A STATISTICAL PROCEDURE FOR COMBINING DIFFERENT TYPES OF SAMPLING UNITS IN A FOREST INVENTORY. FOREST SCI. 15(3), 307-317. [PROBABILITY; VARIANCE; MODEL; MATRIX; PLOT; WEIGHTING; COMPUTER]

WEST, O. 1938. THE SIGNIFICANCE OF PERCENTAGE AREA DETERMINATIONS YIELDED BY THE PERCENTAGE AREA OR DENSITY LIST METHOD OF PASTURE ANALYSIS. J. ECOL. 26(1), 210-217. [CORRELATION COEFFICIENT; PROBABLE ERROR]

WESTMAN, W. E. 1971. MATHEMATICAL MODELS OF CONTAGION AND THEIR RELATION TO DENSITY AND BASAL AREA SAMPLING TECHNIQUES. PP. 515-536. IN, STATISTICAL ECOLOGY, VOLUME 1. SPATIAL PATTERNS AND STATISTICAL DISTRIBUTIONS. (G. P. PATIL, E. C. PIELOU AND W. E. WATERS, EDS.). PENNSYLVANIA STATE UNIVERSITY PRESS, UNIVERSITY PARK, PENNSYLVANIA.

WESTRHEIM, S. J. 1967. SAMPLING RESEARCH TRAWL CATCHES AT SEA. J. FISH. RES. BOARD CAN. 24(6), 1187-1202. [CHI-SQUARE; REGRESSION; STRATIFIED SAMPLING; SERIAL SAMPLING]

WIEBE, P. H. AND W. R. HOLLAND. 1968. PLANKTON PATCHINESS, EFFECTS ON REPEATED NET TOWS. LIMNOL. OCEANOGR. 13(2), 315-321. [MODEL; SIMULATION; COMPUTER; SAMPLING; ARTIFICIAL POPULATION; POISSON DISTRIBUTION]

WILLIAMS, E. J. 1952. APPLICATIONS OF COMPONENT ANALYSIS TO THE STUDY OF PROPERTIES OF TIMBER. AUST. J. APPL. SCI. 3(2), 101-118. [COMPONENTS OF VARIANCE; ANALYSIS OF VARIANCE]

WILLIAMS, E. J. 1955. SAMPLING METHODS FOR TIMBER AND WOOD PRODUCTS. AUST. FORESTRY 19(1), 20-25. [RANDOM SAMPLING; STRATIFIED SAMPLING; SYSTEMATIC SAMPLING; COST FUNCTION]

WILLIAMS, R. M. 1956. THE VARIANCE OF THE MEAN OF SYSTEMATIC SAMPLES. BIOMETRIKA 43(1/2), 137-148. [TRANSECT; PLANTS]

WILLIAMS, W. T. 1971. STRATEGY AND TACTICS IN THE ACQUISITION OF ECOLOGICAL DATA. PROC. ECOL. SOC. AUSTRALIA 6, 57-62. [SAMPLING]

WINSOR, C. P. AND G. L. CLARKE. 1940. A STATISTICAL STUDY OF VARIATION IN THE CATCH OF PLANKTON NETS. J. MARINE RES. 3(1), 1-34. [COMPONENTS OF VARIANCE; ANALYSIS OF VARIANCE]

WINSOR, C. P. AND L. A. WALFORD. 1936. SAMPLING VARIATIONS IN THE USE OF PLANKTON NETS. J. CONS. PERM. INT. EXPLOR. MER 11(2), 190-204. [CHI-SQUARE; VARIANCE; PROBABILITY; POISSON DISTRIBUTION]

YAMAMOTO, T. 1955. SAMPLING METHODS USED IN JAPANESE FISHERIES CATCH STATISTICS. INDO-PACIFIC FISHERIES COUNCIL, BANGKOK. OCCASSIONAL PAPER, 55/2.

YAMANAKA, I. 1955. AN ECOLOGICAL INFERENCE OF STATISTICS CONCERNING THE SIZE COMPOSITION. ANN. REPT. JAPAN SEA REG. FISH. RES. LAB. 2, 43-54. [SAMPLING]

YATES, F. AND D. J. FINNEY. 1942. STATISTICAL PROBLEMS IN FIELD SAMPLING FOR WIREWORMS. ANN. APPL. BIOL. 29(2), 156-167.

YOKOTA, T., T. KITAGAWA AND T. ASAMI. 1953. BASIC STUDY OF FISH SCHOOL RESEARCH BY FISH FINDERS. BULL. JAPANESE SOC. SCI. FISH. 19(4), 341-371. [PROBABILITY; STRATIFIED SAMPLING]

YOUNGS, W. D. 1974. ESTIMATION OF THE FRACTION OF ANGLERS RETURNING TAGS. TRANS. AMER. FISH. SOC. 103(3), 616-618. [REGRESSION; MORTALITY MODEL]

ZINGER, A. 1964. SYSTEMATIC SAMPLING IN FORESTRY. BIOMETRICS 20(3), 553-565. [PROBABILITIES PROPORTIONAL TO SIZE; BIAS]

SURVIVAL AND MORTALITY

ABAKUKS, A. 1974. A NOTE ON SUPERCRITICAL CARRIER-BORNE EPIDEMICS. BIOMETRIKA 61(2), 271-275. [SURVIVORS; MODEL]

ABRAM, F. A. H. 1964. AN APPLICATION OF HARMONICS TO FISH TOXICOLOGY. AIR WATER POLLUT. 8(5), 325-338. [HARMONIC MEAN]

ALDERDICE, D. F. 1963. SOME EFFECTS OF SIMULTANEOUS VARIATION IN SALINITY, TEMPERATURE AND DISSOLVED OXYGEN ON THE RESISTANCE OF YOUNG COHO SALMON TO A TOXIC SUBSTANCE. J. FISH. RES. BOARD CAN. 20(2), 525-550. [BOX DESIGN]

ALDERDICE, D. F. AND J. R. BRETT. 1957. SOME EFFECTS OF DRAFT MILL EFFLUENT ON YOUNG PACIFIC SALMON. J. FISH. RES. BOARD CAN. 14(5), 783-785. [HARMONIC MEAN; REGRESSION]

ARMITAGE, P., G. G. MEYNELL AND T. WILLIAMS. 1965. BIRTH-DEATH AND OTHER MODELS FOR MICROBIAL INFECTION. NATURE 207(4997), 570-572.

BAILEY, N. T. J. 1973. THE ESTIMATION OF PARAMETERS FROM EPIDEMIC MODELS. PP. 253-267. IN, THE MATHEMATICAL THEORY OF THE DYNAMICS OF BIOLOGICAL POPULATIONS. (M. S. BARTLETT AND R. W. HIORNS, EDS.). ACADEMIC PRESS, NEW YORK, NEW YORK.

BAYLIFF, W. H. 1971. ESTIMATES OF THE RATES OF MORTALITY OF YELLOWFIN TUNA IN THE EASTERN PACIFIC OCEAN DERIVED FROM TAGGING EXPERIMENTS. BULL. INTER-AMER. TROP. TUNA COMM. 15(4), 381-418. [COEFFICIENT OF ANNUAL NATURAL MORTALITY; TAG-RECAPTURE; COEFFICIENT OF CATCHABILITY; MURPHY-TOMLINSON METHOD; BEVERTON-HOLT MODEL; ROBSON-CHAPMAN METHOD]

BEARD, R. E. 1959. NOTE ON SOME MATHEMATICAL MORTALITY MODELS. PP. 302-311. IN, THE LIFESPAN OF ANIMALS. CIBA FOUNDATION COLLOQUIA ON AGING, VOL. 5. (G. E. W. WOLSTENHOLME AND M. O'CONNOR, EDS.). LITTLE, BROWN AND CO., BOSTON, MASSACHUSETTS.

BEVERTON, R. J. H. AND J. A. GULLAND. 1958. MORTALITY ESTIMATION IN PARTIALLY FISHED STOCKS. PP. 51-66. IN, SOME PROBLEMS FOR BIOLOGICAL FISHERY SURVEY AND TECHNIQUES FOR THEIR SOLUTION. INTERN. COMM. NORTHWEST ATLANTIC FISH. SPEC. PUBL. NO. 1. [MODEL]

BEVERTON, R. J. H. AND S. J. HOLT. 1956. A REVIEW OF METHODS FOR ESTIMATING MORTALITY RATES IN EXPLOITED FISH POPULATIONS, WITH SPECIAL REFERENCE TO SOURCES OF BIAS IN CATCH SAMPLING. RAPP. PROCES-VERB. REUNIONS J. CONS. PERM. INT. EXPLOR. MER 140(1), 67-83. [MODEL]

BISHOP, Y. M. M. 1959. ERRORS IN ESTIMATES OF MORTALITY OBTAINED FROM VIRTUAL POPULATIONS. J. FISH. RES. BOARD CAN. 16(1), 73-90.

BLISS, C. I. 1940. THE RELATION BETWEEN EXPOSURE TIME, CONCENTRATION AND TOXICITY IN EXPERIMENTS ON INSECTICIDES. ANN. ENTOMOL. SOC. AMER. 33(4), 721-766. [ANALYSIS OF VARIANCE; CHI-SQUARE; REGRESSION; VARIANCE; LOG TRANSFORMATION]

BOTKIN, D. B. AND R. S. MILLER. 1974. MORTALITY RATES AND SURVIVAL OF BIRDS. AMER. NAT. 108(960), 181-192. [LIFE TABLE; MODEL]

BRETT, J. R. 1952. TEMPERATURE TOLERANCE IN YOUNG PACIFIC SALMON, GENUS ONCORHYNCHUS. J. FISH. RES. BOARD CAN. 9(6), 265-323. [ANALYSIS OF VARIANCE; ORTHOGONAL]

BROCK, V. E. AND R. H. RIFFENBURGH. 1960. FISH SCHOOLING: A POSSIBLE FACTOR IN REDUCING PREDATION. J. CONS. PERM. INT. EXPLOR. MER 25(3), 307-317. [MODEL]

BROWNIE, C. 1973. STOCHASTIC MODELS ALLOWING AGE-DEPENDENT SURVIVAL RATES FOR BANDING EXPERIMENTS ON EXPLOITED BIRD POPULATIONS. PH.D. THESIS, CORNELL UNIVERSITY, ITHACA, NEW YORK. 114 PP. [MAXIMUM LIKELIHOOD ESTIMATION; VARIANCE; MINIMAL SUFFICIENT STATISTICS]

BROWNIE, C. AND D. S. ROBSON. 1973. ESTIMATION OF ANNUAL MORTALITY RATES FROM WATERFOWL BANDING RECORDS. (ABSTRACT). TWENTY-FIFTH ANN. REPT. BIOMETRICS UNIT, CORNELL UNIVERSITY, ITHACA, NEW YORK. P. 81. (BU-444-M). [MAXIMUM LIKELIHOOD ESTIMATION; MODEL]

BROWNIE, C. AND D. S. ROBSON. 1974. MODELS ALLOWING FOR AGE-DEPENDENT SURVIVAL RATES FOR BAND-RETURN DATA. BIOMETRICS UNIT, CORNELL UNIVERSITY, BU-514-M. MIMEO. 34 PP. [TAG-RECAPTURE; MAXIMUM LIKELIHOOD ESTIMATION; SEBER-ROBSON-YOUNG MODEL MODIFICATION]

BROWNIE, C. AND D. S. ROBSON. 1974. TESTING FOR EQUALITY OF THE ANNUAL SURVIVAL RATES IN THE SRY MODEL FOR BAND RETURN DATA. BIOMETRICS UNIT, CORNELL UNIVERSITY, BU-521-M. MIMEO. 27 PP. [TAG-RECAPTURE; SEBER-ROBSON-YOUNG MODEL MODIFICATION]

CAUGHLEY, G. 1966. MORTALITY PATTERNS IN MAMMALS. ECOLOGY 47(6), 906-918. [LIFE TABLE]

CAUGHLEY, G. 1967. CALCULATIONS OF POPULATION MORTALITY RATE AND LIFE EXPECTANCY FOR THAR AND KANGAROOS FROM THE RATIO OF JUVENILES TO ADULTS. N. Z. J. SCI. 10(2), 578-584. [SHARPE AND LOTKA MATHEMATICS]

CHAPMAN, D. G. 1957. PROBLEMS OF ESTIMATION OF WILDLIFE MORTALITY RATES. (ABSTRACT). BIOMETRICS 13(4), 548-549. [TAG-RECAPTURE]

CHAPMAN, D. G. 1965. THE ESTIMATION OF MORTALITY AND RECRUITMENT FROM A SINGLE-TAGGING EXPERIMENT. BIOMETRICS 21(3), 529-542.

CHAPMAN, D. G. AND D. S. ROBSON. 1960. THE ANALYSIS OF A CATCH CURVE. BIOMETRICS 16(3), 354-368.

CHAPMAN, D. G. AND G. I. MURPHY. 1965. ESTIMATES OF MORTALITY AND POPULATION FROM SURVEY-REMOVAL RECORDS. BIOMETRICS 21(4), 921-935.

CHARLESWORTH, B. 1973. SELECTION IN POPULATIONS WITH OVERLAPPING GENERATIONS. V. NATURAL SELECTION AND LIFE HISTORIES. AMER. NAT. 107(954), 303-311. [GENETICS; INTRINSIC RATE OF INCREASE; SURVIVAL RATE; FECUNDITY RATE]

COHEN, J. E. 1973. SELECTIVE HOST MORTALITY IN A CATALYTIC MODEL APPLIED TO SCHISTOSOMIASIS. AMER. NAT. 107(954), 199-212.

COMFORT, A. 1957. SURVIVAL CURVES OF MAMMALS IN CAPTIVITY. PROC. ZOOL. SOC. (LONDON) 128(3), 349-364.

COOPER, A. C. 1965. THE EFFECT OF TRANSPORTED STREAM SEDIMENTS ON THE SURVIVAL OF SOCKEYE AND PINK SALMON EGGS AND ALEVIN. INTERN. PACIFIC SALMON FISH. COMM. BULL. 18. 69 PP. [TRANSPORT EQUATION]

CORMACK, R. M. 1964. ESTIMATES OF SURVIVAL FROM THE SIGHTING OF MARKED ANIMALS. BIOMETRIKA 51(3/4), 429-438. [MAXIMUM LIKELIHOOD; MODEL; TAG-RECAPTURE ESTIMATION]

COSTELLO, T. J. AND D. M. ALLEN. 1968. MORTALITY RATES IN POPULATIONS OF PINK SHRIMP, PENAEUS DUORARUM, ON THE SANIBEL AND TORTUGAS GROUNDS, FLORIDA. U. S. FISH WILDL. SERV. FISH. BULL. 66(3), 491-502. [TAG-RECAPTURE; BEVERTON-HOLT EQUATION]

DAVIS, D. E. 1952. DEFINITIONS FOR THE ANALYSIS OF SURVIVAL OF NESTLINGS. AUK 69(3), 316-320.

DAVIS, D. E. 1960. A CHART FOR ESTIMATION OF LIFE EXPECTANCY. J. WILDL. MANAGE. 24(3), 344-348.

DAVIS, D. E. 1970. EVALUATION OF TECHNIQUES FOR MEASURING MORTALITY. J. WILDL. DIS. 6(4), 365-375.

DAVISON, B. 1957. THE COMPATIBILITY OF THE SURVIVAL PLATEAUX HYPOTHESIS WITH GAUSSIAN DISTRIBUTION OF POPULATION. BULL. MATH. BIOPHYS. 19, 241-246.

DICKIE, L. M. 1963. ESTIMATION OF MORTALITY RATES OF GULF OF ST. LAWRENCE COD FROM RESULTS OF A TAGGING EXPERIMENT. PP. 71-80. IN, NORTH ATLANTIC FISH MARKING SYMPOSIUM. INTERN. COMM. NORTHWEST ATLANTIC FISH. SPEC. PUBL. NO. 4. [BEVERTON-HOLT EQUATION; RICKER'S EQUATION; TAG-RECAPTURE]

DODSON, S. I. 1972. MORTALITY IN A POPULATION OF DAPHNIA ROSEA. ECOLOGY 53(6), 1011-1023. [MODEL]

DOI, T. 1948. A STOCHASTIC METHOD FOR ESTIMATION OF THE SURVIVAL RATE. BULL. JAPANESE SOC. SCI. FISH. 14(2), 97-104. [CENTRAL LIMIT THEOREM]

DOI, T. 1973. A RAPID METHOD TO ESTIMATE THE OPTIMUM COEFFICIENT OF FISHING MORTALITY OF CHRYSOPHRYS MAJOR IN THE EAST CHINA SEA AND THE YELLOW SEA. BULL. JAPANESE SOC. SCI. FISH. 39(1), 1-5. [EQUATIONS]

EBERHARDT, L. L. 1972. SOME PROBLEMS IN ESTIMATING SURVIVAL FROM BANDING DATA. PP. 153-171. IN, POPULATION ECOLOGY OF MIGRATORY BIRDS: A SYMPOSIUM. U. S. FISH WILDL. SERV., BUR. SPORT FISH. WILDL., SPEC. SCI. WILDL. REPT. 2. [TAG-RECAPTURE]

EBERHARDT, L. L. 1974. REVIEW OF: SURVIVAL STUDIES OF BANDED BIRDS BY J. J. HICKEY. J. WILDL. MANAGE. 38(3), 579-580. [LIFE TABLE]

FARNER, D. S. 1949. AGE GROUPS AND LONGEVITY IN THE AMERICAN ROBIN: COMMENTS, FURTHER DISCUSSION, AND CERTAIN REVISIONS. WILSON BULL. 61(2), 68-81. [MORTALITY RATE]

FARRIS, J. S. 1968. SIGNIFICANCE TESTING AND CONFIDENCE INTERVALS FOR FIXED MORTALITY RATES. ECOLOGY 49(5), 994-996.

FINK, B. D. 1965. ESTIMATIONS, FROM TAGGING EXPERIMENTS, OF MORTALITY RATES AND OTHER PARAMETERS RESPECTING YELLOWFIN AND SKIPJACK TUNA. BULL. INTER-AMER. TROP. TUNA COMM. 10(1), 3-49. [GULLAND'S METHOD; TAG-RECAPTURE]

FORDHAM, R. A. AND R. M. CORMACK. 1970. MORTALITY AND POPULATION CHANGE OF DOMINICAN GULLS IN WELLINGTON, NEW ZEALAND WITH A STATISTICAL APPENDIX. J. ANIM. ECOL. 39(1), 13-27.

FRANCIS, R. C. 1974. RELATIONSHIP OF FISHING MORTALITY TO NATURAL MORTALITY AT THE LEVEL OF MAXIMUM SUSTAINABLE YIELD UNDER THE LOGISTIC STOCK PRODUCTION MODEL. J. FISH. RES. BOARD CAN. 31(9), 1539-1542.

FREDIN, R. A. 1965. SOME METHODS FOR ESTIMATING OCEAN MORTALITY OF PACIFIC SALMON AND APPLICATIONS. J. FISH. RES. BOARD CAN. 22(1), 33-51. [MODEL]

FREEMAN, D. C., L. G. KLIKOFF AND H. EYRING. 1974. APPLICATIONS OF THE SURVIVAL THEORY TO ECOLOGY. PROC. NATL. ACAD. SCI. U.S.A. 71(11), 4332-4335. [EYRING-STOVER THEORY]

GEIS, A. D. AND R. D. TABER. 1965. MEASURING HUNTING AND OTHER MORTALITY. PP. 284-298. IN, WILDLIFE INVESTIGATIONAL TECHNIQUES. (2ND REVISED EDITION). (H. S. MOSBY, ED.). THE WILDLIFE SOCIETY, WASHINGTON, D. C.

GOMPERTZ, B. 1825. ON THE NATURE OF THE FUNCTION EXPRESSIVE OF THE LAW OF HUMAN MORTALITY, AND ON A NEW MODE OF DETERMINING THE VALUE OF LIFE CONTINGENCIES. PHIL. TRANS. ROY. SOC. LONDON 115(1), 513-585.

GRAHAM, M. 1938. RATES OF FISHING AND NATURAL MORTALITY FROM THE DATA OF MARKING EXPERIMENTS. J. CONS. PERM. INT. EXPLOR. MER 13(1), 76-90. [MORTALITY RATE; MODEL; FISHING RATE]

GREEN, R. H. 1970. GRAPHICAL ESTIMATION OF RATES OF MORTALITY AND GROWTH. J. FISH. RES. BOARD CAN. 27(1), 204-208.

GRIB, I. V. 1972. ANALYSIS OF FISHKILLS IN SMALL RIVERS IN THE WESTERN POLES'YE ON THE UKRAINE. HYDROBIOL. J. 8(2), 29-34. (ENGLIS TRANSL.). [FISHKILL COEFFICIENT]

GULLAND, J. A. 1963. THE ESTIMATION OF FISHING MORTALITY FROM TAGGING EXPERIMENTS. PP. 218-227. IN, NORTH ATLANTIC FISH MARKING SYMPOSIUM. INTERN. COMM. NORTHWEST ATLANTIC FISH. SPEC. PUBL. NO. 4. [CATCHABILITY COEFFICIENT; TAG-RECAPTURE]

HALDANE, J. B. S. 1955. THE CALCULATION OF MORTALITY RATES FROM RINGING DATA. PROC. INTERN. ORNITHOL. CONGR. 11, 454-458.

HARCOURT, D. G. 1963. MAJOR MORTALITY FACTORS IN THE POPULATION DYNAMICS OF THE DIAMONDBACK MOTH, PLUTELLA MACULIPENNIS (CURT.) (LEPIDOPTERA: PLUTELLIDAE). MEM. ENTOMOL. SOC. CAN. 32, 55-66. [LIFE TABLE; MODEL]

HARRIS, E. K. 1958. ON THE PROBABILITY OF SURVIVAL OF BACTERIA IN SEA WATER. BIOMETRICS 14(2), 195-206. [THOMAS'S DISTRIBUTION; NORMAL DISTRIBUTION; GAMMA DISTRIBUTION; MODEL; BINOMIAL DISTRIBUTION; MAXIMUM LIKELIHOOD]

HAYASHI, S. 1968. PRELIMINARY ANALYSIS OF THE CATCH CURVE OF THE PACIFIC SARDINE, SARDINOPS CAERULEA GIRARD. U. S. FISH WILDL. SERV. FISH. BULL. 66(3), 587-598. [MODEL]

HENNY, C. J. 1967. ESTIMATING BAND-REPORTING RATES FROM BANDING AND CRIPPLING LOSS DATA. J. WILDL. MANAGE. 31(3), 533-538.

HICKEY, J. J. 1952. SURVIVAL STUDIES OF BANDED BIRDS. U. S. FISH WILDL. SERV. SPEC. SCI. REPT. WILDL. NO. 15. 177 PP.

HOLGATE, P. AND K. H. LAKHANI. 1967. EFFECT OF OFFSPRING DISTRIBUTION ON POPULATION SURVIVAL. BULL. MATH. BIOPHYS. 29, 831-839. [DISTRIBUTION]

HUGHES, R. D. 1962. A METHOD FOR ESTIMATING THE EFFECTS OF MORTALITY ON APHID POPULATIONS. J. ANIM. ECOL. 31(2), 389-396. [INTRINSIC RATE OF INCREASE; BIRTH RATE; GEOMETRIC PROGRESSION]

ITO, Y. 1959. A COMPARATIVE STUDY ON SURVIVORSHIP CURVES FOR NATURAL INSECT POPULATIONS. JAPANESE J. ECOL. 9, 107-115.

IVLEV, V. S. 1965. ON THE QUANTITATIVE RELATIONSHIP BETWEEN SURVIVAL RATE OF LARVAE AND THEIR FOOD SUPPLY. BULL. MATH. BIOPHYS. 27(SPECIAL ISSUE), 215-222.

IWAO, S. 1970. ANALYSIS OF CONTAGIOUSNESS IN THE ACTION OF MORTALITY FACTORS ON THE WESTERN TENT CATERPILLAR POPULATION BY USING M-M RELATIONSHIP. RES. POPUL. ECOL.-KYOTO 12(1), 100-110.

JENSEN, A. L. 1971. THE EFFECT OF INCREASED MORTALITY ON THE YOUNG IN A POPULATION OF BROOK TROUT, A THEORETICAL ANALYSIS. TRANS. AMER. FISH. SOC. 100(3), 456-459. [MODEL; MATRIX]

JENSEN, A. L. 1972. STANDARD ERROR OF LC 50 AND SAMPLE SIZE IN FISH BIOASSAYS. WATER RES. 6(1), 85-89. [PROBIT ANALYSIS]

JOHNSON, D. H. 1974. ESTIMATING SURVIVAL RATES FROM BANDING OF ADULT AND JUVENILE BIRDS. J. WILDL. MANAGE. 38(2), 290-297. [MODEL; MAXIMUM LIKELIHOOD ESTIMATION; COMPUTER]

JONES, R. 1956. THE ANALYSIS OF TRAWL HAUL STATISTICS WITH PARTICULAR REFERENCE TO THE ESTIMATION OF METHODS OF SAMPLING SURVIVAL RATES. PP. 30-39. IN, PROBLEMS AND METHODS OF SAMPLING FISH POPULATIONS. CONSEIL PERMANENT INTERN POUR L'EXPLORATION DE LA MER. RAPPORTS ET PROCES-VERBAUX DES REUNIONS 140(1). [TRANSFORMATION; CONFIDENCE INTERVAL; REGRESSION]

KEISTER, T. D. 1972. PREDICTING INDIVIDUAL TREE MORTALITY IN SIMULATED SOUTHERN PINE PLANTATIONS. FOREST SCI. 18(3), 213-217. [COMPUTER; COMPETITION INDEX]

KELKER, G. H. AND W. R. HANSON. 1964. SIMPLIFYING THE CALCULATION OF DIFFERENTIAL SURVIVAL OF AGE-CLASSES. J. WILDL. MANAGE. 28(2), 411.

KIRITANI, K. AND F. NAKASUJI. 1967. ESTIMATION OF THE STAGE-SPECIFIC SURVIVAL RATE IN THE INSECT POPULATION WITH OVERLAPPING STAGES. RES. POPUL. ECOL.-KYOTO 9(2), 143-152. [MODEL; LIFE TABLE; HOKYO AND KIRITANI'S METHOD]

KOLESNIK, YU. A. 1970. A METHOD FOR ESTIMATION OF THE INSTANTANEOUS NATURAL MORTALITY RATE OF THE WALLEYE POLLACK IN PETER THE GREAT BAY. J. ICHTHYOLOGY 10(6), 850-853. (ENGLISH TRANSL.). [MODEL]

KUPPER, L. L. 1971. SOME FURTHER REMARKS "ON TESTING HYPOTHESES CONCERNING STANDARDIZED MORTALITY RATIOS". THEOR. POPUL. BIOL. 2(4), 431-436.

KUPPER, L. L. AND D. G. KLEINBAUM. 1971. ON TESTING HYPOTHESES CONCERNING STANDARDIZED MORTALITY RATIOS. THEOR. POPUL. BIOL. 2(3), 290-298.

KURITA, S. 1948. A THEORETICAL CONSIDERATION ON THE METHOD FOR ESTIMATING THE YEARLY SURVIVAL RATE OF FISH STOCK BY USING THE AGE DIFFERENCE BETWEEN THE OLDEST AND THE AVERAGE. BULL. JAPANESE SOC. SCI. FISH. 14(1), 1-12. [TAUTI'S FORMULA; SURVIVAL RATE; SAMPLING]

LANDER, R. H. 1962. A METHOD OF ESTIMATING MORTALITY RATES FROM CHANGE IN COMPOSITION. J. FISH. RES. BOARD CAN. 19(1), 159-168.

LANDER, R. H. 1973. PROBLEM OF BIAS IN MODELS TO APPROXIMATE OCEAN MORTALITY, MATURITY, AND ABUNDANCE SCHEDULES OF SALMON FROM KNOWN SMOLTS AND RETURNS. U. S. NATL. MAR. FISH. SERV. FISH. BULL. 71(2), 513-525. [RICKER'S MODEL; FREDIN MODEL; CLEAVER MODEL; LIMIT-MEAN MODEL]

LESLIE, P. H. AND R. M. RANSON. 1940. THE MORTALITY, FERTILITY AND RATE OF NATURAL INCREASE OF THE VOLE (MICROTUS AGRESTIS) AS OBSERVED IN THE LABORATORY. J. ANIM. ECOL. 9(1), 27-52. [LIFE TABLE]

LOUGH, R. G. 1975. A REEVALUATION OF THE COMBINED EFFECTS OF TEMPERATURE AND SALINITY SURVIVAL AND GROWTH OF BIVALVE LARVAE USING RESPONSE SURFACE TECHNIQUES. U. S. NATL. MAR. FISH. SERV. FISH. BULL. 73(1), 88-94. [MULTIPLE REGRESSION MODEL]

MANLY, B. F. J. 1974. ESTIMATION OF STAGE-SPECIFIC SURVIVAL RATES AND OTHER PARAMETERS FOR INSECT POPULATIONS DEVELOPING THROUGH SEVERAL STAGES. OECOLOGIA 15(3), 277-285. [MODEL]

MATHEWS, C. P. AND D. F. WESTLAKE. 1969. ESTIMATION OF PRODUCTION BY POPULATIONS OF HIGHER PLANTS SUBJECT TO HIGH MORTALITY. OIKOS 20(1), 156-160.

MENSHUTKIN, V. V., L. A. ZHAKOV AND A. A. UMNOV. 1968. A MODEL METHOD EXAMINATION OF CAUSES OF DEATH AMONG YOUNG PERCH. PROBLEMS OF ICHTHYOLOGY 8(5), 704-712. [SIMULATION; COMPUTER]

MERTZ, D. B. AND R. B. DAVIES. 1968. CANNIBALISM OF THE PUPAL STAGE BY ADULT FLOUR BEETLES: AN EXPERIMENT AND A STOCHASTIC MODEL. BIOMETRICS 24(2), 247-275.

MERTZ, D. B., T. PARK AND W. J. YOUDEN. 1965. MORTALITY PATTERNS IN EIGHT STRAINS OF FLOUR BEETLES. BIOMETRICS 21(1), 99-114. [LIFE TABLE; GALTON RANK ORDER CURVES; SURVIVORSHIP CURVE]

MILLER, C. A. 1963. THE ANALYSIS OF PUPAL SURVIVAL IN THE UNSPRAYED AREA. PP. 63-70. IN. THE DYNAMICS OF EPIDEMIC SPRUCE BUDWORM POPULATIONS. (R. F. MORRIS, ED.). MEM. ENTOMOL. SOC. CAN. NO. 31. [MODEL; GRAPHIC ANALYSIS]

MILLER, D. R. 1970. THEORETICAL SURVIVAL CURVES FOR RADIATION DAMAGE IN BACTERIA. J. THEOR. BIOL. 26(3), 383-398.

MOORE, W. AND C. I. BLISS. 1942. A METHOD FOR DETERMINING INSECTICIDAL EFFECTIVENESS USING APHIS RUMICIS AND CERTAIN ORGANIC COMPOUNDS. J. ECON. ENTOMOL. 35(4), 544-553. [PROBIT; ANALYSIS OF VARIANCE; VARIANCE]

MORRIS, R. F. 1963. THE ANALYSIS OF GENERATION SURVIVAL IN RELATION TO AGE-INTERVAL SURVIVALS IN THE UNSPRAYED AREA. PP. 32-37. IN, THE DYNAMICS OF EPIDEMIC SPRUCE BUDWORM POPULATIONS. (R. F. MORRIS, ED.). MEM. ENTOMOL. SOC. CAN. NO. 31. [MODEL; REGRESSION]

MORRIS, R. F. 1963. THE DEVELOPMENT OF A POPULATION MODEL FOR THE SPRUCE BUDWORM THROUGH THE ANALYSIS OF SURVIVAL RATES. PP. 30-32. IN, THE DYNAMICS OF EPIDEMIC SPRUCE BUDWORM POPULATIONS. (R. F. MORRIS, ED.). MEM. ENTOMOL. SOC. CAN. NO. 31.

MORRIS, R. F. 1965. CONTEMPORANEOUS MORTALITY FACTORS IN POPULATION DYNAMICS. CAN. ENTOMOL. 97(11), 1173-1184.

MOTT, D. G. 1963. THE ANALYSIS OF THE SURVIVAL OF SMALL LARVAE IN THE UNSPRAYED AREA. PP. 42-52. IN, THE DYNAMICS OF EPIDEMIC SPRUCE BUDWORM POPULATIONS. (R. F. MORRIS, ED.). MEM. ENTOMOL. SOC. CAN. NO. 31. [SURVIVAL RATE; MULTIPLE REGRESSION MODEL]

OUNTFORD, M. D. 1971. OPULATION SURVIVAL IN A VARIABLE NVIRONMENT. J. THEOR. BIOL. 2(1), 75-79. [COHEN'S MODEL; OMPUTER]

UIR, B. S. AND H. WHITE. 1963. PPLICATION OF THE PALOHEIMO LINEAR QUATION FOR ESTIMATING MORTALITIES O A SEASONAL FISHERY. J. FISH. ES. BOARD CAN. 20(3), 839-840.

URAI, M. 1967. A MATHEMATICAL ODEL FOR THE SURVIVORSHIP CURVE. ES. POPUL. ECOL.-KYOTO 9(2), 5-82. [DEEVEY'S SURVIVORSHIP JRVE]

MURTON, R. K. 1966. A STATISTICAL EVALUATION OF THE EFFECT OF WOOD-PIGEON SHOOTING AS EVIDENCED BY THE RECOVERIES OF RINGED BIRDS. THE STATISTICIAN 16(2), 183-202. [MORTALITY RATE; MAXIMUM LIKELIHOOD]

MYHRE, R. J. 1967. MORTALITY ESTIMATES FROM TAGGING EXPERIMENTS ON PACIFIC HALIBUT. INTERN. PACIFIC HALIBUT COMM. REPT. NO. 42. 43 PP. [LINEAR REGRESSION MODEL; TAG-RECAPTURE; VARIANCE; CONFIDENCE INTERVAL]

NOSE, Y. 1963. ON THE TAGGED RATIO METHOD FOR ESTIMATING THE NATURAL MORTALITY COEFFICIENT BULL. JAPANESE SOC. SCI. FISH. 29(7), 663-666. [TAG-RECAPTURE; REGRESSION]

OLIFF, W. D. 1953. THE MORTALITY, FECUNDITY AND INTRINSIC RATE OF NATURAL INCREASE OF THE MULTIMAMMATE MOUSE, RATTUS (MASTOMYS) NATALENSIS (SMITH) IN THE LABORATORY. J. ANIM. ECOL. 22(2), 217-226. [LIFE TABLE; INTRINSIC RATE OF INCREASE]

ONDODERA, K. 1958. ON THE SURVIVAL OF TROUT FINGERLINGS STOCKED IN A MOUNTAIN BROOK - I. THE ESTIMATION OF SURVIVAL RATE BY OBSERVATION. BULL. JAPANESE SOC. SCI. FISH. 24(6/7), 428-434. [EQUATIONS]

PALOHEIMO, J. E. 1958. DETERMINATION OF NATURAL AND FISHING MORTALITIES OF COD AND HADDOCK FROM ANALYSIS OF TAG RECORDS OFF WESTERN NOVA SCOTIA. J. FISH. RES. BOARD CAN. 15(6), 1371-1381. [TAG-RECAPTURE; MORTALITY RATE]

PALOHEIMO, J. E. 1958. A METHOD OF ESTIMATING NATURAL AND FISHING MORTALITIES. J. FISH. RES. BOARD CAN. 15(4), 749-758.

PALOHEIMO, J. E. 1961. STUDIES ON ESTIMATION OF MORTALITIES. I. COMPARISON OF A METHOD DESCRIBED BY BEVERTON AND HOLT AND A NEW LINEAR FORMULA. J. FISH. RES. BOARD CAN. 18(5), 645-662.

PARKER, R. R. 1965. ESTIMATION OF SEA MORTALITY RATES FOR THE 1961 BROOD-YEAR PINK SALMON OF THE BELLA COOLA AREA, BRITISH COLUMBIA. J. FISH. RES. BOARD CAN. 22(6), 1523-1554. [TAG-RECAPTURE; EQUATIONS]

PARKER, R. R. 1968. MARINE MORTALITY SCHEDULES OF PINK SALMON OF THE BELLA COOLA RIVER, CENTRAL BRITISH COLUMBIA. J. FISH. RES. BOARD CAN. 25(4), 757-794. [SEBER-JOLLY; TAG-RECAPTURE; JOLLY; ESTIMATION; SAMPLING; VARIANCE]

PARKER, R. R. 1971. SIZE SELECTIVE PREDATION AMONG JUVENILE SALMONID FISHES IN A BRITISH COLUMBIA INLET. J. FISH. RES. BOARD CAN. 28(10), 1503-1510. [MODEL]

PAULIK, G. J. 1962. USE OF THE CHAPMAN-ROBSON SURVIVAL ESTIMATE FOR SINGLE- AND MULTI- RELEASE TAGGING EXPERIMENTS. TRANS. AMER. FISH. SOC. 91(1), 95-98.

PAULIK, G. J. 1963. ESTIMATES OF MORTALITY RATES FROM TAG RECOVERIES. BIOMETRICS 19(1), 28-57.

PAULIK, G. J. 1963. EXPONENTIAL RATES OF DECLINE AND TYPE (1) LOSSES FOR POPULATIONS OF TAGGED PINK SALMON. PP. 230-237. IN, NORTH ATLANTIC FISH MARKING SYMPOSIUM, INTERN. COMM. NORTHWEST ATLANTIC FISH. SPEC. PUBL. NO. 4. [TAG-RECAPTURE; MODEL]

PEARL, R. AND S. L. PARKER. 1921. EXPERIMENTAL STUDIES ON THE DURATION OF LIFE I. INTRODUCTORY DISCUSSION OF THE DURATION OF LIFE IN DROSOPHILA. AMER. NAT. 55(691), 481-509. [MODEL]

POLLOCK, K. H. AND D. L. SOLOMON. 1972. AN ASYMPTOTIC TEST FOR MORTALITY IN A K SAMPLE TAG-RECAPTURE EXPERIMENT. (ABSTRACT). BIOMETRICS 28(4), 1179. [ONE-TAILED BINOMIAL TEST]

POLLOCK, K. H., D. L. SOLOMON AND D. S. ROBSON. 1974. TESTS FOR MORTALITY AND RECRUITMENT IN A K-SAMPLE TAG-RECAPTURE EXPERIMENT. BIOMETRICS 30(1), 77-87. [ESTIMATION; MULTINOMIAL DISTRIBUTION; HYPERGEOMETRIC DISTRIBUTION; POISSON DISTRIBUTION; BINOMIAL DISTRIBUTION]

PRENTICE, R. L. AND A. EL SHAARAWI. 1973. A MODEL FOR MORTALITY RATES AND A TEST OF FIT FOR THE GOMPERTZ FORCE OF MORTALITY. APPL. STAT. 22(3), 301-314. [REGRESSION]

PRESTON, S. H. 1972. INTERRELATIONS BETWEEN DEATH RATES AND BIRTH RATES. THEOR. POPUL. BIOL. 3(2), 162-185.

QUICK, H. F. 1963. ANIMAL POPULATION ANALYSIS. PP. 190-228. IN, WILDLIFE INVESTIGATION TECHNIQUES. (2ND EDITION; REVISED). (H. S. MOSBY, ED.). THE WILDLIFE SOCIETY, WASHINGTON, D. C. [LIFE TABLE]

REGIER, H. A. 1962. ON ESTIMATING MORTALITY COEFFICIENTS IN EXPLOITED FISH POPULATIONS, GIVEN TWO CENSUSES. TRANS. AMER. FISH. SOC. 91(3), 283-294.

RICHARDS, O. W., N. WALOFF AND J. P. SPRADBERY. 1960. THE MEASUREMENT OF MORTALITY IN AN INSECT POPULATION IN WHICH RECRUITMENT AND MORTALITY WIDELY OVERLAP. OIKOS 11(2), 306-310.

RICKER, W. E. 1944. FURTHER NOTES ON FISHING MORTALITY AND EFFORT. COPEIA 1944(1), 23-44. [MODEL; RECRUITMENT; MORTALITY]

RICKER, W. E. 1952. NUMERICAL RELATIONS BETWEEN ABUNDANCE OF PREDATORS AND SURVIVAL OF PREY. CAN. FISH CULT. 13, 5-9.

ROBSON, D. S. 1961. A NOTE ON THE LIMITING DISTRIBUTION OF "JACKSON'S ESTIMATE" OF AVERAGE MORTALITY RATE. BIOMETRICS UNIT, CORNELL UNIVERSITY, BU-130-M. MIMEO, 3 PP.

ROBSON, D. S. AND D. G. CHAPMAN. 1961. CATCH CURVES AND MORTALITY RATES. TRANS. AMER. FISH. SOC. 90(2), 181-189.

RYSZKOWSKI, L. 1967. SHORT CUT METHODS FOR THE ESTIMATION OF MEAN LENGTH OF LIFE IN SMALL MAMMAL POPULATIONS. PP. 283-294. IN, SECONDARY PRODUCTIVITY OF TERRESTRIAL ECOSYSTEMS, VOLUME I. (K. PETRUSEWICZ, ED.). INSTITUTE OF ECOLOGY, POLISH ACAD. SCI., WARSAW, POLAND. [SURVIVORSHIP CURVE]

RYSZKOWSKI, L., W. WALKOWA AND T. WIERZBOWSKA 1967. ESTIMATION OF AVERAGE LENGTH OF LIFE OF MICE HAVING VARIABLE SURVIVAL RATES. EKOLOGIYA POLSKA-SERIA A 15(42), 791-801. [LIFE TABLE; COHORT; MORTALITY RATE]

SAILA, S. B. 1953. BIO-ASSAY PROCEDURES FOR THE EVALUATION OF FISH TOXICANTS WITH PARTICULAR REFERENCE TO ROTENONE. TRANS. AMER. FISH SOC. 83, 104-114. [GRAPHIC]

SANDERS, M. J. 1969. A METHOD OF DIRECTLY ESTIMATING NATURAL MORTALITY AND INITIAL TAGGING MORTALITY APPLICABLE TO CERTAIN EXPLOITED MOLLUSC POPULATIONS. J. CONS. PERM. INT. EXPLOR. MER 32(2), 416-418. [TAG-RECAPTURE]

SCHAEFER, M. B. 1943. THE THEORETICAL RELATIONSHIP BETWEEN FISHING EFFORT AND MORTALITY. COPEIA 1943(2), 79-82. [MODEL; RICKER'S MODEL; MORTALITY; RECRUITMENT]

EBER, G. A. F. 1970. STIMATING TIME-SPECIFIC SURVIVAL ND REPORTING RATES FOR ADULT BIRDS ROM BAND RETURNS. BIOMETRIKA 7(2), 313-318.

EBER, G. A. F. 1971. STIMATING AGE-SPECIFIC SURVIVAL ATES FROM BIRD-BAND RETURNS WHEN HE REPORTING RATE IS CONSTANT. IOMETRIKA 58(3), 491-497.

EBER, G. A. F. 1972. STIMATING SURVIVAL RATES FROM IRD-BAND RETURNS. J. WILDL. NAGE. 36(2), 405-413.

SILLIMAN, R. P. 1943. STUDIES ON THE PACIFIC PILCHARD OR SARDINE (SARDINOPS COERULA). 5. A METHOD OF COMPUTING MORTALITIES AND REPLACEMENTS. U. S. FISH WILDL. SERV. SPEC. SCI. REPT. 24. 10 PP.

SILLIMAN, R. P. 1945. DETERMINATION OF MORTALITY RATES FROM LENGTH FREQUENCIES OF THE PILCHARD OR SARDINE, SARDINOPS CAERULEA. COPEIA 1945(4), 191-196.

SSENTONGO, G. W. AND P. A. LARKIN. 1973. SOME SIMPLE METHODS OF ESTIMATING MORTALITY RATES OF EXPLOITED FISH POPULATIONS. J. FISH. RES. BOARD CAN. 30(5), 695-698.

STREHLER, B. L. AND A. S. MILDVAN. 1960. GENERAL THEORY OF MORTALITY AND AGING. SCIENCE 132(3418), 14-21. [GOMPERTZ; STOCHASTIC]

TAGUCHI, K. 1961. A TRIAL TO ESTIMATE THE INSTANTANEOUS RATE OF NATURAL MORTALITY OF ADULT SALMON (ONCORHYNCHUS SP.) AND THE CONSIDERATION OF RATIONALITY OF OFFSHORE FISHING - II. FOR RED SALMON (ONCORHYNCHUS NERKA) 1961. BULL. JAPANESE SOC. SCI. FISH. 27(11), 972-978. [POPULATION MODEL]

TAGUCHI, K. 1961. A TRIAL TO ESTIMATE THE INSTANTANEOUS RATE OF NATURAL MORTALITY OF ADULT SALMON (ONCORHYNCHUS SP.) AND THE CONSIDERATION OF RATIONALITY OF OFFSHORE FISHING -I. FOR CHUM SALMON (ONCORHYNCHUS KETA) 1961. BULL. JAPANESE SOC. SCI. FISH. 27(11), 963-971. [POPULATION MODEL]

TANAKA, S. 1953. THE METHOD TO CALCULATE PRECISION OF SURVIVAL RATE ESTIMATED FROM AGE-COMPOSITION, BY RATIO ESTIMATE - I. A METHOD APPLIED TO RANDOM SAMPLING. BULL. JAPANESE SOC. SCI. FISH. 18(8), 353-358.

TANAKA, S. 1953. THE METHOD TO CALCULATE PRECISION OF SURVIVAL RATE ESTIMATED FROM AGE-COMPOSITION, BY RATIO ESTIMATE - II. A METHOD APPLIED TO AGE-COMPOSITION OBTAINED USING LENGTH OF STRATIFICATION. BULL. JAPANESE SOC. SCI. FISH. 19(4), 279-282. [SAMPLING; VARIANCE; STRATIFICATION]

TANAKA, S. 1975. SOME CONSIDERATIONS ON THE METHODS FOR CALCULATING SURVIVAL RATE FROM CATCH PER UNIT EFFORT DATA. BULL. JAPANESE SOC. SCI. FISH. 41(2), 121-128.

TAYLOR, C. C. 1962. NATURAL MORTALITY RATE OF THE GEORGES BANK HADDOCK. U. S. FISH WILDL. SERV. FISH. BULL. 58, 1-7. [BEVERTON-HOLT EQUATION]

TESTER, A. L. 1955. ESTIMATION OF RECRUITMENT AND NATURAL MORTALITY RATE FROM AGE-COMPOSITION AND CATCH DATA IN BRITISH COLUMBIA HERRING POPULATIONS. J. FISH. RES. BOARD CAN. 12(5), 649-681. [EQUATIONS]

TINER, J. D. 1954. THE FRACTION OF PEROMYSCUS LEUCOPUS FATALITIES CAUSED BY RACCOON ASCARID LARVAE. J. MAMMAL. 35(4), 589-592. [EQUATIONS]

TYNDALE-BISCOE, C. H. AND R. M. WILLIAMS. 1955. A STUDY OF NATURAL MORTALITY IN A WILD POPULATION OF THE RABBIT, ORYCTOLAGUS CUNICULUS (L.). N. Z. J. SCI. TECHNOL. SECT. B 36(6), 561-580.

WATT, K. E. F. 1963. THE ANALYSIS OF THE SURVIVAL OF LARGE LARVAE IN THE UNSPRAYED AREA. PP. 52-63. IN, THE DYNAMICS OF EPIDEMIC SPRUCE BUDWORM POPULATIONS. (R. F. MORRIS, ED.). MEMOIRS OF THE ENTOMOLOGICAL SOCIETY OF CANADA NO. 31. [SURVIVAL MODEL]

WIDRIG, T. M. 1954. DEFINITIONS AND DERIVATIONS OF VARIOUS COMMON MEASURES OF MORTALITY RATES RELEVANT TO POPULATION DYNAMICS OF FISHES. COPEIA 1954(1), 29-32.

WIERZBOWSKA, T. AND K. PETRUSEWICZ. 1963. RESIDENCY AND RATE OF DISAPPEARANCE OF TWO FREE-LIVING POPULATIONS OF THE HOUSE MOUSE (MUS MUSCULUS L.). EKOLOGIYA POLSKA-SERIA A 11(24), 557-574. [MODEL; ANDERSON-DARLING GOODNESS OF FIT TEST]

WOHLSCHLAG, D. E. 1954. MORTALITY RATES OF WHITEFISH IN AN ARCTIC LAKE. ECOLOGY 35(3), 388-396. [MAXIMUM LIKELIHOOD]

YEAGER, J. F. AND S. C. MUNSON. 1945. THE RELATION BETWEEN POISON CONCENTRATION AND SURVIVAL TIME OF ROACHES INJECTED WITH SODIUM METARSENITE. ANN. ENTOMOL. SOC. AMER. 38(4), 559-600. [EQUATIONS; LAW OF MASS ACTION]

TAXONOMY

AMADON, D. 1949. THE SEVENTY-FIVE PER CENT RULE FOR SUBSPECIES. CONDOR 51(6), 250-258. [TAXONOMY]

ANDERSON, E. 1954. EFFICIENT AND INEFFICIENT METHODS OF MEASURING SPECIFIC DIFFERENCES. PP. 93-106. IN, STATISTICS AND MATHEMATICS IN BIOLOGY. (O. KEMPTHORNE, T. A. BANCROFT, J. W. GOWEN AND J. L. LUSH,EDS.). IOWA STATE COLLEGE PRESS, AMES, IOWA. [TAXONOMY; GRAPHIC]

ANDREYEV, V. L. AND V. I. DULEPOV. 1971. SALVELINUS LEUCOMAENIS FROM THE SOUTH KURILE ISLANDS. HYDROBIOL. J. 7(6), 59-66. (ENGLISH TRANSL.). [MAHALANOBIS' D2; INDEX OF SIMILARITY; MORTALITY RATE; PROBABILITY]

ANGEL, M. V. AND M. J. R. FASHAM. 1973. SOND CRUISE 1965: FACTOR AND CLUSTER ANALYSES OF THE PLANKTON RESULTS, A GENERAL SUMMARY. J. MAR. BIOL. ASSOC. U. K. 53(1), 185-231. [PRINCIPAL COMPONENTS ANALYSIS; DENDOGRAM; MATRIX]

ARAOZ, J., G. SARMIENTO AND M. MONASTERIO. 1971. AN ESSAY IN THE USE OF ASSOCIATION AND DISSOCIATION MEASURES IN PHYTOSOCIOLOGICAL CLASSIFICATION. J. ECOL. 59(1), 39-50. [CLUSTER ANALYSIS; COMPUTER; ALGORITHM; SORENSEN'S COEFFICIENT OF SIMILARITY]

BARLOW, C. A., J. E. GRAHAM AND S. ADISOEMARTO. 1969. A NUMERICAL FACTOR FOR THE TAXONOMIC SEPARATION OF PTEROSTICHUS PENSYLVANICUS AND P. ADSTRICTUS (COLEOPTERA: CARABIDAE). CAN. ENTOMOL. 101(12), 1315-1319. [DISCRIMINANT FUNCTION]

BEMIS, W. P., A. M. RHODES, T. W. WHITAKER AND S. G. CARMER. 1970. NUMERICAL TAXONOMY APPLIED TO CUCURBITA RELATIONSHIPS. AMER. J. BOT. 57(4), 404-412.

BEZDEK, J. C. 1974. NUMERICAL TAXONOMY WITH FUZZY SETS. J. MATH. BIOL. 1(1), 57-71. [CLUSTER ANALYSIS]

BLACKITH, R. E. 1957. POLYMORPHISM IN SOME AUSTRALIAN LOCUSTS AND GRASSHOPPERS. BIOMETRICS 13(2), 183-196.

BLACKITH, R. E. 1965. MORPHOMETRICS. PP. 225-249. IN, THEORETICAL AND MATHEMATICAL BIOLOGY. (H. J. MOROWITZ AND T. H. WATERMAN, EDS.). BLAISDELL PUBLISHING CO., NEW YORK, NEW YORK. [EUCLIDEAN SPACE; CANONICAL ANALYSIS; MULTIVARIATE, MATRIX; DISCRIMINANT FUNCTION; TAXONOMY]

BOTTOMLY, J. 1971. SOME STATISTICAL PROBLEMS ARISING FROM THE USE OF THE INFORMATION STATISTIC IN NUMERICAL CLASSIFICATION. J. ECOL. 59(2), 339-342. [TAXONOMY]

BUCHANAN-WOLLASTON, H. J. 1933. SOME MODERN STATISTICAL METHODS: THEIR APPLICATION TO THE SOLUTION OF HERRING RACE PROBLEMS. J. CONS. PERM. INT. EXPLOR. MER 8(1), 7-47. [MOMENTS; FOURIER SERIES; CURVE OF ERROR; MAXIMUM LIKELIHOOD; VARIANCE; ANALYSIS OF VARIANCE; CHI-SQUARE; MULTIPLE REGRESSION MODEL; METHOD OF LEAST SQUARES]

BUCK, R. C. AND D. L. HULL. 1969. REPLY TO GREGG. SYST. ZOOL. 18(3), 354-357. [LOGIC; SETS; TAXONOMY]

BURNABY, T. P. 1966. GROWTH-INVARIANT DISCRIMINANT FUNCTIONS AND GENERALIZED DISTANCES. BIOMETRICS 22(1), 96-110. [TAXONOMY; CANONICAL VECTOR; MATRIX]

CAIRNS, J., JR. AND R. L. KAESLER. 1971. CLUSTER ANALYSIS OF FISH IN A PORTION OF THE UPPER POTOMAC RIVER. TRANS. AMER. FISH. SOC. 100(4), 750-756. [JACCARD'S COEFFICIENT; COPHENETIC CORRELATION COEFFICIENT]

CARMICHAEL, J. W. AND P. H. A. SNEATH. 1969. TAXOMETRIC MAPS. SYST. ZOOL. 18(4), 402-415. [MODEL; SIMILARITY; MATCHING COEFFICIENT; COLE'S COEFFICIENT OF ASSOCIATION; COMPUTER]

CARMICHAEL, J. W., J. A. GEORGE AND R. S. JULIUS. 1968. FINDING NATURAL CLUSTERS. SYST. ZOOL. 17(2), 144-150. [ALGORITHM; COMPUTER]

CAZIER, M. A. AND A. L. BACON. 1949. INTRODUCTION TO QUANTITATIVE SYSTEMATICS. BULL. AMER. MUS. NAT. HIST. 93(ART. 5), 343-388. [COEFFICIENT OF VARIANCE; STANDARD ERROR; MODE; MEDIAN; CORRELATION]

CIESIELSKA, B. AND W. KUPSC. 1968. CRANIOMETRIC VARIATIONS IN A POPULATION OF THE SPOTTED SOUSLIK. ACTA THERIOL. 13(11), 151-176. [DISCRIMINANT FUNCTION]

COLLESS, D. H. 1967. AN EXAMINATION OF CERTAIN CONCEPTS IN PHENETIC TAXONOMY. SYST. ZOOL. 16(1), 6-27. [COEFFICIENT OF ASSOCIATION; DISTANCE COEFFICIENT]

COLLESS, D. H. 1970. THE PHENOGRAM AS AN ESTIMATE OF PHYLOGENY. SYST. ZOOL. 19(4), 352-362. [TAXONOMY; EVOLUTIONARY MODEL; CLUSTER ANALYSIS]

CORRUCCINI, R. S. 1972. ALLOMETRY CORRECTION IN TAXIMETRICS. SYST. ZOOL. 21(4), 375-383. [Q-MODE CORRELATION COEFFICIENTS; COMPUTER]

CRAWFORD, R. M. M. AND D. WISHART. 1967. A RAPID MULTIVARIATE METHOD FOR THE DETECTION AND CLASSIFICATION OF GROUPS OF ECOLOGICALLY RELATED SPECIES. J. ECOL. 55(2), 505-524. [COMPUTER; QUADRAT; MODEL; CZEKANOWSKI'S COEFFICIENT; MULTIVARIATE ANALYSIS]

CROVELLO, T. J. 1970. ANALYSIS OF CHARACTER VARIATION IN ECOLOGY AND SYSTEMATICS. PP. 55-98. IN, ANNUAL REVIEW OF ECOLOGY AND SYSTEMATICS, VOLUME 1. (R. F. JOHNSTON, P. W. FRANK AND C. D. MICHENER, EDS.). ANNUAL REVIEWS, INC., PALO ALTO, CALIFORNIA.

CURTIS, J. T. AND R. P. MCINTOSH. 1950. THE INTERRELATIONS OF CERTAIN ANALYTIC AND SYNTHETIC PHYTOSOCIOLOGICAL CHARACTERS. ECOLOGY 31(3), 434-455. [LOG-NORMAL DISTRIBUTION; OCTAVE; DISTRIBUTION]

DAGNELIE, P. 1965. L'ETUDE DES COMMUNAUTES VEGETALES PAR L'ANALYSE STATISTIQUE DES LIAISONS ENTRE LES ESPECES ET LES VARIABLES ECOLOGIQUES: PRINCIPES FONDAMENTAUX. BIOMETRICS 21(2), 345-361.

DAGNELIE, P. 1971. SOME IDEAS ON THE USE OF MULTIVARIATE STATISTICAL METHODS IN ECOLOGY. PP. 167-174. IN, STATISTICAL ECOLOGY, VOLUME 3, MANY SPECIES POPULATIONS, ECOSYSTEMS, AND SYSTEMS ANALYSIS. (G. P. PATIL, E. C. PIELOU AND W. E. WATERS, EDS.). PENNSYLVANIA STATE UNIVERSITY PRESS, UNIVERSITY PARK, PENNSYLVANIA.

DALE, M. B. 1968. ON PROPERTY STRUCTURE, NUMERICAL TAXONOMY AND DATA HANDLING. PP. 185-197. IN, MODERN METHODS IN PLANT TAXONOMY. (V. H. HEYWOOD, ED.). ACADEMIC PRESS, NEW YORK, NEW YORK.

DALE, M. B. AND D. J. ANDERSON. 1972. QUALITATIVE AND QUANTITATIVE INFORMATION ANALYSIS. J. ECOL. 60(3), 639-653. [SIMILARITY MEASURE; NUMERICAL CLASSIFICATION; SHANNON MEASURE]

DALE, M. B. AND D. WALKER. 1970. INFORMATION ANALYSIS OF POLLEN DIAGRAMS I. POLLEN SPORES 12(1), 21-37. [INFORMATION ANALYSIS; FORTRAN; COMPUTER]

DALE, M. B., G. N. LANCE AND L. ALBRECHT. 1971. EXTENSIONS OF INFORMATION ANALYSIS. AUST. COMPUT. J. 3(1), 29-34.

DAVIDSON, R. A. AND R. A. DUNN. 1966. A NEW BIOMETRIC APPROACH TO SYSTEMATIC PROBLEMS. BIOSCIENCE 16(8), 528-536.

DAY, B. B. AND M. M. SANDOMIRE. 1942. USE OF THE DISCRIMINANT FUNCTION FOR MORE THAN TWO GROUPS. J. AMER. STAT. ASSOC. 37(220), 461-472. [DEER; MULTIVARIATE ANALYSIS]

DICE, L. R. 1952. QUANTITATIVE AND EXPERIMENTAL METHODS IN SYSTEMATIC ZOOLOGY. SYST. ZOOL. 1(3), 97-104. [TAXONOMY; SAMPLE SIZE; SAMPLING]

DICE, L. R. AND H. J. LERAAS. 1936. A GRAPHIC METHOD FOR COMPARING SEVERAL SETS OF MEASUREMENT. CONTRIB. LAB. VERTEBR. GENETICS, UNIVERSITY OF MICHIGAN, ANN ARBOR, MICHIGAN, NO. 3. 3 PP.

DOYEN, J. T. AND C. N. SLOBODCHIKOFF. 1974. AN OPERATIONAL APPROACH TO SPECIES CLASSIFICATION. SYST. ZOOL. 23(2), 239-247. [TAXONOMY; ECOLOGICAL DISTANCE; GRAPHIC MODEL]

EADES, D. C. 1970. THEORETICAL AND PROCEDURAL ASPECTS OF NUMERICAL PHYLETICS. SYST. ZOOL. 19(2), 142-171. [TAXONOMY; CLUSTER ANALYSIS; SIMILARITY COEFFICIENTS; COMPUTER]

EBERHARDT, L. L. 1968. AN APPROXIMATION TO A MULTIPLE-COMPARISON TEST. COPEIA 1968 (2), 314-319.

ESTABROOK, G. F. 1966. A MATHEMATICAL MODEL IN GRAPH THEORY FOR BIOLOGICAL CLASSIFICATION. J. THEOR. BIOL. 12(3), 297-310. [TAXONOMY]

ESTABROOK, G. F. AND D. J. ROGERS. 1966. A GENERAL METHOD OF TAXONOMIC DESCRIPTION FOR A COMPUTED SIMILARITY MEASURE. BIOSCIENCE 16(11), 789-793. [COMPUTER]

FAEGRI, K. AND P. OTTESTAD. 1949. STATISTICAL PROBLEMS IN POLLEN ANALYSIS. UNIV. BERGEN ARBOK, NATURVITENSK. 1948, NR. 3, 1-29.

FARRIS, J. S. 1968. CATEGORICAL RANKS AND EVOLUTIONARY TAXA IN NUMERICAL TAXONOMY. SYST. ZOOL. 17(2), 151-159. [MEASURE OF DIVERGENCE; MATRIX; RANK OF TAXON]

FARRIS, J. S. 1969. ON THE COPHENETIC CORRELATION COEFFICIENT. SYST. ZOOL. 18(3), 279-285. [PHENOGRAM; MATRIX; MAXIMIZATION; OPTIMALITY CRITERION; TAXONOMY]

FARRIS, J. S. 1969. A SUCCESSIVE APPROXIMATIONS APPROACH TO CHARACTER WEIGHTING. SYST. ZOOL. 18(4), 374-385. [TAXONOMY; CLADISTICS]

FARRIS, J. S. 1970. METHODS FOR COMPUTING WAGNER TREES. SYST. ZOOL. 19(1), 83-92. [TAXONOMY; COMPUTER]

FARRIS, J. S. 1972. ESTIMATING PHYLOGENETIC TREES FROM DISTANCE MATRICES. AMER. NAT. 106(951), 645-668. [WAGNER PROCEDURE]

FARRIS, J. S. 1973. ON COMPARING THE SHAPES OF TAXONOMIC TREES. SYST. ZOOL. 22(1), 50-54. [CLUSTER DISTORTION METHOD]

FARRIS, J. S. 1973. A PROBABILITY MODEL FOR INFERRING EVOLUTIONARY TREES. SYST. ZOOL. 22(3), 250-256. [MAXIMUM LIKELIHOOD]

FARRIS, J. S., A. G. KLUGE AND M. J. ECKARDT. 1970. A NUMERICAL APPROACH TO PHYLOGENETIC SYSTEMATICS. SYST. ZOOL. 19(2), 1972-189. [CLUSTER ANALYSIS; WEIGHTED INVARIANT STEP STRATEGY]

FARRIS, J. S., A. G. KLUGE AND M. J. ECKARDT. 1970. ON PREDICTIVITY AND EFFICIENCY. SYST. ZOOL. 19(4), 363-372. [THROCKMORTON'S INDICES; TAXONOMY; CONCORDANCE]

FELSENSTEIN, J. 1973. MAXIMUM LIKELIHOOD AND MINIMUM-STEPS METHODS FOR ESTIMATING EVOLUTIONARY TREES FROM DATA ON DISCRETE CHARACTERS. SYST. ZOOL. 22(3), 240-249 [ALGORITHM]

FIELD, J. G. 1969. THE USE OF THE INFORMATION STATISTIC IN THE NUMERICAL CLASSIFICATION OF HETEROGENEOUS SYSTEMS. J. ECOL. 57(2), 565-569. [HYPERGEOMETRIC DISTRIBUTION; PROBABILITY; MAXIMUM LIKELIHOOD]

FIELD, J. G. AND G. MCFARLANE. 1968. NUMERICAL METHODS IN MARINE ECOLOGY 1. A QUANTITATIVE "SIMILARITY" ANALYSIS OF ROCKY SHORE SAMPLES IN FALSE BAY, SOUTH AFRICA. ZOOLOGICA AFRICANA 3(2), 119-137 [CLUSTER ANALYSIS; JACCARD'S COEFFICIENT; CZEKANOWSKI'S COEFFICIENT; COMPUTER]

FISHER, D. R. AND F. J. ROHLF. 1969. ROBUSTNESS OF NUMERICAL TAXONOMIC METHODS AND ERRORS IN HOMOLOGY. SYST. ZOOL. 18(1), 33-36. [CORRELATION COEFFICIENT; RANDOM ERRORS; COMPUTER; CLUSTER ANALYSIS; DISTANCE COEFFICIENT]

FISHER, R. A. 1936. THE USE OF MULTIPLE MEASUREMENTS IN TAXONOMIC PROBLEMS. ANN. EUGEN. 7(2), 179-188.

FLAKE, R. H. AND B. L. TURNER. 1968. NUMERICAL CLASSIFICATION FOR TAXONOMIC PROBLEMS. J. THEOR. BIOL. 20(2), 260-270. [SET]

FUJII, K. 1969. NUMERICAL TAXONOMY OF ECOLOGICAL CHARACTERISTICS AND THE NICHE CONCEPT. SYST. ZOOL. 18(2), 151-153. [COMPUTER; SIMILARITY; CORRELATION MATRIX]

GHENT, A. W. 1963. KENDALL'S "TAU" COEFFICIENT AS AN INDEX OF SIMILARITY IN COMPARISONS OF PLANT OR ANIMAL COMMUNITIES. CAN. ENTOMOL. 95(6), 568-575.

GINSBURG, I. 1938. ARITHMETICAL DEFINITION OF THE SPECIES, SUBSPECIES AND RACE CONCEPT, WITH A PROPOSAL FOR A MODIFIED NOMENCLATURE. ZOOLOGICA 23(3), 253-286. [TAXONOMY]

GINSBURG, I. 1939. THE MEASURE OF POPULATION DIVERGENCE AND MULTIPLICITY OF CHARACTERS. J. WASH. ACAD. SCI. 29(8), 317-330.

GINSBURG, I. 1940. DIVERGENCE AND PROBABILITY IN TAXONOMY. ZOOLOGICA 25(1), 15-31.

GOODALL, D. W. 1968. AFFINITY BETWEEN AN INDIVIDUAL AND A CLUSTER IN NUMERICAL TAXONOMY. BIOMETRIE-PRAXIMETRIE 9(1), 52-55.

GOODALL, D. W. 1968. IDENTIFICATION BY COMPUTER. BIOSCIENCE 18(6), 485-488. [TAXONOMY]

GOODALL, D. W. 1974. NUMERICAL CLASSIFICATION. PP. 575-615. IN. ORDINATION AND CLASSIFICATION OF COMMUNITIES. (R. H. WHITTAKER, ED.). DR. JUNK PUBLISHERS, THE HAGUE, NETHERLANDS. [SIMILARITY; PROBABILITY; MODEL; SAMPLE SIZE; CLUSTERING; MULTIVARIATE ANALYSIS; INFORMATION ANALYSIS; ORDINATION; ASSOCIATION ANALYSIS]

GOODMAN, M. AND G. W. MOORE. 1971. IMMUNODIFFUSION SYSTEMATICS OF THE PRIMATES I. THE CATARRHINI. SYST. ZOOL. 20(1), 19-62. [COMPUTER; TAXONOMY; SET THEORY; TAXONOMIC DISTANCE]

GOODMAN, M. W. 1972. DISTANCE ANALYSIS IN BIOLOGY. SYST. ZOOL. 21(2), 174-186. [PRINCIPAL COMPONENTS; SOKAL DISTANCE; GENERALIZED DISTANCE; TAXONOMY]

GOWER, J. C. 1969. A SURVEY OF NUMERICAL METHODS USEFUL IN TAXONOMY. ACAROLOGIA(PARIS) 11(3), 357-375. [CLUSTER ANALYSIS]

GOWER, J. C. 1974. MAXIMAL PREDICTIVE CLASSIFICATION. BIOMETRICS 30(4), 643-654. [NUMERICAL TAXONOMY PROCEDURE]

GREGG, J. R. 1968. BUCK AND HULL: A CRITICAL REJOINDER. SYST. ZOOL. 17(3), 342-344. [REFERS TO SYST. ZOOL. 15, 97-111; SET THEORY]

HALL, A. V. 1969. AVOIDING INFORMATIONAL DISTORTION IN AUTOMATIC GROUPING PROGRAMS. SYST. ZOOL. 18(3), 318-329. [SIMILARITY COEFFICIENTS; JACCARD'S COEFFICIENT; CZEKANOWSKI'S COEFFICIENT; EUCLIDEAN DISTANCE; HETEROGENEITY; CORRELATION COEFFICIENT; COMPUTER]

HALL, A. V. 1970. A COMPUTER-BASED SYSTEM FOR FORMING IDENTIFICATION KEYS. TAXON 19(1), 12-18. [TAXONOMY]

HANSELL, R. I. C. AND D. A. CHANT. 1973. A METHOD FOR ESTIMATING RELATIVE WEIGHTS APPLIED TO CHARACTERS BY CLASSICAL TAXONOMISTS. SYST. ZOOL. 22(1), 46-49. [PROBABILITY]

HAZEL, J. E. 1972. ON THE USE OF CLUSTER ANALYSIS IN BIOGEOGRAPHY. SYST. ZOOL. 21(2), 240-242. [DICE COEFFICENTS; Q-MODE ANALYSIS; COMPUTER]

HENDRICKSON, J. A., JR. 1968. CLUSTERING IN NUMERICAL CLADISTICS: A MINIMUM-LENGTH DIRECTED TREE PROBLEM. MATH. BIOSCI. 3, 371-381. [EVOLUTIONARY LATTICE]

HILL, D. R. 1959. SOME USES OF STATISTICAL ANALYSIS IN CLASSIFYING RACES OF AMERICAN SHAD (ALOSA SAPIDISSIMA). U. S. FISH WILDL. SERV. FISH. BULL. 59, 269-286. [ANALYSIS OF VARIANCE; DISCRIMINANT FUNCTION]

HUBBS, C. L. AND A. PERLMUTTER. 1942. BIOMETRIC COMPARISON OF SEVERAL SAMPLES, WITH PARTICULAR REFERENCE TO RACIAL INVESTIGATIONS. AMER. NAT. 76(767), 582-592. [TAXONOMY; VARIANCE]

HUBBS, C. L. AND C. HUBBS. 1953. AN IMPROVED GRAPHICAL ANALYSIS AND COMPARISON OF SERIES OF SAMPLES. SYST. ZOOL. 2(2), 49-57. [TAXONOMY]

HUGHES, R. E. AND D. V. LINDLEY. 1955. APPLICATION OF BIOMETRIC METHODS TO PROBLEMS OF CLASSIFICATION IN ECOLOGY. NATURE 175(4462), 806-807. [MULTIVARIATE ANALYSIS; MAHALANOBIS' D2; DISCRIMINANT ANALYSIS]

HUHEEY, J. E. 1965. A MATHEMATICAL METHOD OF ANALYZING BIOGEOGRAPHICAL DATA. I. HERPETOFAUNA OF ILLINOIS. AMER. MIDL. NAT. 73(2), 490-500. [DIVERGENCE FACTOR]

HULL, D. L. AND D. P. SNYDER. 1969. CONTEMPORARY LOGIC AND EVOLUTIONARY TAXONOMY: A REPLY TO GREGG. SYST. ZOOL. 18(3), 347-354. [MODEL]

HUTCHINSON, M., K. I. JOHNSTONE AND D. WHITE. 1969. TAXONOMY OF THE GENUS THIOBACCILLUS: THE OUTCOME OF NUMERICAL TAXONOMY APPLIED TO THE GROUP AS A WHOLE. J. GEN. MICROBIOL. 57(3), 397-410.

JACKSON, D. M. AND L. J. WHITE. 1971. THE WEAKENING OF TAXONOMIC INFERENCES BY HOMOLOGICAL ERROR. MATH. BIOSCI. 10, 63-116. [JACCARD'S SIMILARITY FUNCTION; RUSSEL AND RAO SIMILARITY FUNCTION; DICE SIMILARITY FUNCTION; KULCZYNSKI'S SIMILARITY FUNCTION]

JACKSON, J. F. 1970. LOGNORMAL SCALE COUNTS. SYST. ZOOL. 19(2), 194-196. [TAXONOMY]

JARDINE, C. J., N. JARDINE AND R. SIBSON. 1967. THE STRUCTURE AND CONSTRUCTION OF TAXONOMIC HIERARCHIES. MATH. BIOSCI. 1, 173-179. [TAXONOMY]

JARDINE, N. AND R. SIBSON. 1968. A MODEL FOR TAXONOMY. MATH. BIOSCI. 2, 465-482.

JAYASINGHE, S. D. D. AND T. KAWAKAMI. 1974. RACE SEPARATION OF DEEP SEA SMELT OF JAPAN SEA. BULL. JAPANESE SOC. SCI. FISH. 40(3), 255-260. [COMPUTER; DISCRIMINANT FUNCTION]

JOHNSON, L. A. S. 1970. RAINBOW'S END: THE QUEST FOR AN OPTIMAL TAXONOMY. SYST. ZOOL. 19(3), 203-239. [ORDINATION; NUMERICAL TAXONOMY]

JOLICOEUR, P. AND P. BRUNEL 1966. APPLICATION DU DIAGRAMME HEXAGONAL A L'ETUDE DE LA SELECTION DE SES PROIES PAR LA MORUE. VIE ET MILIEU 17(1-B), 419-433.

KASHYAP, R. L. AND S. SUBAS. 1974. STATISTICAL ESTIMATION OF PARAMETERS IN A PHYLOGENETIC TREE USING A DYNAMIC MODEL OF THE SUBSTITUTIONAL PROCESS. J. THEOR. BIOL. 47(1), 75-101. [EVOLUTION; NEYMAN'S MODEL]

KENDRICK, W. B. AND L. K. WERESUB. 1966. ATTEMPTING NEO-ADANSONIAN COMPUTER TAXONOMY AT THE ORDINAL LEVEL IN THE BASIDIOMYCETES. SYST. ZOOL. 15(4), 307-329.

KERFOOT, W. C. AND A. G. KLUGE. 1971. IMPACT OF THE LOGNORMAL DISTRIBUTION ON STUDIES OF PHENOTYPIC VARIATION AND EVOLUTIONARY RATES. SYST. ZOOL. 20(4), 459-464.

KLAUBER, L. M. 1943. A GRAPHIC METHOD OF SHOWING RELATIONSHIPS. BULL. ZOOL. SOC. SAN DIEGO NO. 18, 61-76. [LOG TRANSFORMATION; PEARSON'S COEFFICIENT OF RACIAL LIKENESS; COEFFICIENT OF DIVERGENCE; TAXONOMY]

KLUGE, A. G. AND J. S. FARRIS. 1969. QUANTITATIVE PHYLETICS AND THE EVOLUTION OF ANURANS. SYST. ZOOL. 18(1), 1-32. [WAGNER TREE; INDEX OF CONSISTENCY; WEIGHTING; COMPUTER]

KOSTITZIN, V. A. 1940. SUR LA SEGREGATION PHYSIOLOGIQUE ET LA VARIATION DES ESPECES. ACTA BIOTHEOR. 5(3), 160-168. [COEFFICIENT OF MORTALITY; MODEL]

LANCE, G. N. AND W. T. WILLIAMS. 1965. COMPUTER PROGRAMS FOR MONOTHETIC CLASSIFICATION ("ASSOCIATION ANALYSIS"). COMPUTER J. 8(3), 246-249.

ANCE, G. N. AND W. T. WILLIAMS. 966. COMPUTER PROGRAMS FOR IERARCHICAL POLYTHETIC LASSIFICATION ("SIMILARITY NALYSES"). COMPUTER J. 9(1), 0-64. [INFORMATION STATISTIC; R; AXONOMY]

LANCE, G. N. AND W. T. WILLIAMS. 1967. MIXED-DATA CLASSIFICATORY PROGRAMS. I. AGGLOMERATIVE SYSTEMS. AUST. COMPUT. J. 1(1), 15-20. [COMPUTER; METRIC MEASURES; SIMILARITY MEASURE; INFORMATION STATISTIC; TAXONOMY]

LE QUESNE, W. J. 1969. A METHOD OF SELECTION OF CHARACTERS IN NUMERICAL TAXONOMY. SYST. ZOOL. 18(2), 201-205. [DATA MATRIX; COEFFICIENT OF CHARACTER STATE RANDOMNESS]

LE QUESNE, W. J. 1972. FURTHER STUDIES BASED ON THE UNIQUELY DERIVED CHARACTER CONCEPT. SYST. ZOOL. 21(3), 281-288. [COEFFICIENT OF CHARACTER STATE RANDOMNESS; TAXONOMY; NORMAL DEVIATE; METHOD OF ELIMINATION; COMPUTER]

LE QUESNE, W. J. 1974. THE UNIQUELY EVOLVED CHARACTER CONCEPT AND ITS CLADISTIC APPLICATION. SYST. ZOOL. 23(4), 513-517. [SETS; EQUATIONS]

LEUSCHNER, D. AND R. HEINE. 1973. CRITERIA OF INFORMATION THEORY IN NUMERICAL TAXONOMY. BIOMETRISCHE ZEITSCHRIFT 15(6), 393-401.

LEWONTIN, R. C. 1966. ON THE MEASUREMENT OF RELATIVE VARIABILITY. SYST. ZOOL. 15(2), 141-142. [COEFFICIENT OF VARIABILITY]

LIETH, H. AND G. W. MOORE. 1971. COMPUTERIZED CLUSTERING OF SPECIES IN PHYTOSOCIOLOGICAL TABLES AND ITS UTILIZATION FOR FIELD WORK. PP. 403-422. IN, STATISTICAL ECOLOGY. VOLUME 1. SPATIAL PATTERNS AND STATISTICAL DISTRIBUTIONS. (G. P. PATIL, E. C. PIELOU AND W. E. WATERS, EDS.). PENNSYLVANIA STATE UNIVERSITY PRESS, UNIVERSITY PARK, PENNSYLVANIA.

LUBISCHEW, A. A. 1962. ON THE USE OF DISCRIMINANT FUNCTIONS IN TAXONOMY. BIOMETRICS 18(4), 455-477.

LUNDBERG, J. G. 1972. WAGNER NETWORKS AND ANCESTORS. SYST. ZOOL. 21(4), 398-413. [DISTANCE MATRIX]

LYSENKO, O. AND P. H. A. SNEATH. 1959. THE USE OF MODELS IN BACTERIAL CLASSIFICATION. J. GEN. MICROBIOL. 20(2), 284-290. [TAXONOMY]

LYUBISHCHEV, A. A. 1959. THE APPLICATION OF BIOMETRICS IN TAXONOMY. VESTNIK LENINGRADSKOGO UNIVERSITETA 14(9), 128-136. (ENGLISH TRANSL. U. S. DEPT. COMMERCE REPORT OTS-60-51147).

MAILLEFER, A. 1929. LE COEFFICIENT GENERIQUE DE P. JACCARD ET SA SIGNIFICATION. MEM. SOC. VAUDOISE SCI. NAT. 3, 113-183.

MANISCHEWITZ, J. R. 1973. PREDICTION AND ALTERNATIVE PROCEDURES IN NUMERICAL TAXONOMY. SYST. ZOOL. 22(2), 176-184. [TRANSFORMATION; SIMILARITY COEFFICIENTS; COMPUTER]

MARCHI, E. AND R. I. C. HANSELL. 1973. A FRAMEWORK FOR SYSTEMATIC ZOOLOGICAL STUDIES WITH GAME THEORY. MATH. BIOSCI. 16, 31-58. [MATRIX; ALGORITHM]

MARCHI, E. AND R. I. C. HANSELL. 1973. GENERALIZATIONS ON THE PARSIMONY QUESTION IN EVOLUTION. MATH. BIOSCI. 17, 11-34. [CLADISTIC TREE]

MARCUS, L. F. AND J. H. VANDERMEER. 1966. REGIONAL TRENDS IN GEOGRAPHIC VARIATION. SYST. ZOOL. 15(1), 1-13. [METHOD OF LEAST SQUARES; POLYNOMIAL; COEFFICIENT OF DETERMINATION; POLYNOMIAL REGRESSION SURFACE; COMPUTER GRAPHICS; TREND-SURFACE ANALYSIS]

MARR, J. C. 1955. THE USE OF MORPHOMETRIC DATE IN SYSTEMATIC, RACIAL AND RELATIVE GROWTH STUDIES IN FISHES. COPEIA 1955(1), 23-31. [TAXONOMY]

MCINTOSH, W. B. 1955. THE APPLICABILITY OF COVARIANCE ANALYSIS FOR COMPARISON OF BODY AND SKELETAL MEASUREMENTS BETWEEN TWO RACES OF THE DEERMOUSE, PEROMYSCUS MANICULATUS. CONTRIB. LAB. VERTEBR. BIOL., UNIVERSITY OF MICHIGAN, ANN ARBOR, MICHIGAN, NO. 72, 54 PP. [REGRESSION; COMPONENTS OF VARIANCE; MODEL; BARTLETT'S TEST OF HOMOGENEITY OF VARIANCES; TYPE I AND II ERRORS; CHI-SQUARE]

MINKOFF, E. C. 1965. THE EFFECTS ON CLASSIFICATION OF SLIGHT ALTERATIONS IN NUMERICAL TECHNIQUE. SYST. ZOOL. 14(3), 196-213. [CORRELATION COEFFICIENT; DISTANCE COEFFICIENT; COMPUTER; TAXONOMY]

MISRA, R. K. 1972. ANALYSIS OF A MERISTIC TRAIT FOR HYBRIDIZATION IN SUNFISHES (GENUS LEPOMIS) FRESHWATER BIOL. 2(4), 321-324. [VARIANCE; ALLOMETRY; HYBRID INDEX]

MISRA, R. K. AND C. HOLDSWORTH. 1972. A RELATIVE GROWTH INDEX OF HYBRIDIZATION IN SUNFISHES (GENUS LEPOMIS, CENTRARCHIDAE). FRESHWATER BIOL. 2(4), 325-335. [HYBRID INDEX; VARIANCE]

MISRA, R. K. AND L. L. SHORT. 1974. A BIOMETRIC ANALYSIS OF ORIOLE HYBRIDIZATION. CONDOR 76(2), 137-146. [COEFFICIENT OF ALLOMETRY; VARIANCE; HYBRID INDEX]

MISRA, R. K., D. W. MATSON AND M. H. A. KEENLEYSIDE. 1970. STATISTICAL METHODS FOR ANALYSIS OF HYBRID INDICES, WITH AN EXAMPLE FROM FISH POPULATIONS. J. STAT. RES. 4(1), 50-70.

MOORE, G. W., W. S. BENNINGHOFF AND P. S. DWYER. 1967. A COMPUTER METHOD FOR THE ARRANGEMENT OF PHYTOSOCIOLOGICAL TABLES. PROC. ASSN. COMP. MACHIN. 22, 297-299.

MOORE, J. J., S. J. P. FITZSIMMONS, E. LAMBE AND J. WHITE. 1970. A COMPARISON AND EVALUATION OF SOME PHYTOSOCIOLOGICAL TECHNIQUES. VEGETATIO 20(1-4), 1-20. [RANDOM SAMPLING; COMPUTER; INDEX OF SIMILARITY; CLUSTER ANALYSIS; PRINCIPAL COMPONENT ANALYSIS]

MOSIMANN, J. E. AND R. L. GREENSTREET. 1971. REPRESENTATION-INSENSITIVE METHODS FOR PALEOECOLOGICAL POLLEN STUDIES. PP. 23-54. IN, STATISTICAL ECOLOGY, VOLUME 1. SPATIAL PATTERNS AND STATISTICAL DISTRIBUTIONS. (G. P. PATIL, E. C. PIELOU AND W. E. WATERS, EDS.). PENNSYLVANIA STATE UNIVERSITY PRESS, UNIVERSITY PARK, PENNSYLVANNIA.

MOSS, W. W. AND J. A. HENDRICKSON, JR. 1973. NUMERICAL TAXONOMY. ANNU. REV. ENTOMOL. 18, 227-258.

MOTTLEY, C. M. 1941. THE COVARIANCE METHOD OF COMPARING THE HEAD-LENGTHS OF TROUT FROM DIFFERENT ENVIRONMENTS. COPEIA 1941(3), 154-159. [COVARIANCE OF ANALYSIS; LOG TRANSFORMATION]

NASTANSKY, L., S. M. SELKOW AND N. F. STEWART. 1974. AN IMPROVED SOLUTION TO THE GENERALIZED CAMIN-SOKAL MODEL FOR NUMERICAL CLADISTICS. J. THEOR. BIOL. 48(2), 413-424. [EVOLUTIONARY TREE; ALGORITHM]

ORLOCI, L. 1967. DATA CENTERING: A REVIEW AND EVALUATION WITH REFERENCE TO COMPONENT ANALYSIS. SYST. ZOOL. 16(3), 208-212. [R-TECHNIQUE; Q-TECHNIQUE; MATRIX]

PAGE, L. M. 1974. THE SUBGENERA OF PERCINA (PERCIDAE: ETHEOSTOMATINI). COPEIA 1974(1), 66-86. [COMPUTER; NUMERICAL TAXONOMY]

PARR, A. E. 1956. ON THE ORIGINAL VARIATES OF TAXONOMY AND THEIR REGRESSIONS UPON SIZE IN FISHES. BULL. AMER. MUS. NAT. HIST. 110(5), 369-398.

PETERS, J. A. 1971. A NEW APPROACH IN THE ANALYSIS OF BIOGEOGRAPHIC DATA. SMITHSON. CONTRIB. ZOOL. NO. 107. 28 PP. [COMPUTER; TAXONOMY]

PETERS, J. A. 1972. ON THE USE OF CLUSTER ANALYSIS OF BIOGEOGRAPHY: A REPLY. SYST. ZOOL. 21(2), 242-244. [COMPUTER; SYST. ZOOL. 21(2), 240-242.]

PHIPPS, J. B. 1971. DENDROGRAM TOPOLOGY. SYST. ZOOL. 20(3), 306-308. [TAXONOMY; COPHENETIC DISTANCE]

PROCTOR, J. R. 1966. SOME PROCESSES OF NUMERICAL TAXONOMY IN TERMS OF DISTANCE. SYST. ZOOL. 15(2), 131-140. [MATCHING COEFFICIENT; DISTANCE COEFFICIENT; CLUSTER ANALYSIS; CENTROID COEFFICIENT]

ROBINS, J. D. AND G. D. SCHNELL. 1971. SKELETAL ANALYSIS OF THE AMMODRAMUS - AMMOSPIZA GRASSLAND SPARROW COMPLEX: A NUMERICAL TAXONOMIC STUDY. AUK 88(3), 567-590. [DENDOGRAM; COMPUTER; R-TYPE ANALYSIS; Q-TYPE ANALYSIS; CLUSTER ANALYSIS; MATRIX; SIMILARITY COEFFICIENTS; CORRELATION COEFFICIENT; PRINCIPAL COMPONENTS; PHENOGRAM; MULTIVARIATE]

ROGERS, D. J. AND H. FLEMING. 1964. A COMPUTER PROGRAM FOR CLASSIFYING PLANTS. II. A NUMERICAL HANDLING OF NON-NUMERICAL DATA. BIOSCIENCE 14(9), 15-28. [CLUSTER ANALYSIS; TAXONOMY; DISTANCE FUNCTION; MODEL; ENTROPY; INDEX OF SIMILARITY]

ROGERS, D. J. AND T. T. TANIMOTO. 1960. A COMPUTER PROGRAM FOR CLASSIFYING PLANTS. SCIENCE 132(3434), 1115-1118. [TAXONOMY]

ROGERS, J. S. 1972. DISCRIMINANT FUNCTION ANALYSIS OF MORPHOLOGICAL RELATIONSHIPS WITHIN THE BUFO COGNATUS SPECIES GROUP. COPEIA 1972(2), 381-383. [COMPUTER; CANONICAL VARIABLE]

ROHLF, F. J. 1965. MULTIVARIATE METHODS IN TAXONOMY. PP. 3-13. IN, PROC. IBM. SCI. COMP. SYMP. STATISTICS, IBM, WHITE PLAINS, NEW YORK.

ROHLF, F. J. 1968. STEREOGRAMS IN NUMERICAL TAXONOMY. SYST. ZOOL. 17(3), 246-255. [PRINCIPAL COMPONENT ANALYSIS; MATRIX; COMPUTER]

ROHLF, F. J. 1970. ADAPTIVE HIERARCHICAL CLUSTERING SCHEMES. SYST. ZOOL. 19(1), 58-82. [TAXONOMY; PRINCIPAL COMPONENTS ANALYSIS; NON-METRIC SCALING; ALGORITHM; CLUSTER ANALYSIS; MATRIX; DISTANCE FUNCTION]

ROHLF, F. J. 1974. METHODS OF COMPARING CLASSIFICATIONS. PP. 101-113. IN, ANNUAL REVIEW OF ECOLOGY AND SYSTEMATICS, VOLUME 5. (R. F. JOHNSTON, P. W. FRANK, AND C. D. MICHNER, EDS.). ANNUAL REVIEWS, INC., PALO ALTO, CALIFORNIA. [TAXONOMY; MATRIX; COPHENETIC CORRELATION COEFFICIENT; ORDINATION]

ROHLF, F. J. AND D. R. FISHER. 1968. TESTS FOR HIERARCHAICAL STRUCTURE IN RANDOM DATA SETS. SYST. ZOOL. 17(4), 407-416. [CLUSTER ANALYSIS; COPHENETIC CORRELATION COEFFICIENT; DISTANCE COEFFICIENT; MONTE CARLO; RANDOM NUMBER GENERATOR; COMPUTER]

ROHLF, F. J. AND R. R. SOKAL. 1962. THE DESCRIPTION OF TAXONOMIC RELATIONSHIPS BY FACTOR ANALYSIS. SYST. ZOOL. 11(1), 1-16.

ROHLF, F. J. AND R. R. SOKAL. 1967. TAXONOMIC STRUCTURE FROM RANDOMLY AND SYSTEMATICALLY SCANNED BIOLOGICAL IMAGES. SYST. ZOOL. 16(3), 246-260. [COMPUTER; DISTANCE COEFFICIENT; MATRIX]

ROHWER, S. A. AND D. L. KILGORE, JR. 1973. INTERBREEDING IN THE ARID-LAND FOXES, VULPES VELOX AND V. MACROTIS. SYST. ZOOL. 22(2), 157-165. [PRINCIPAL COMPONENTS ANALYSIS; DISCRIMINANT FUNCTION]

ROTH, H. D. 1970. CLUSTER ANALYSIS FOR THE BIOLOGICAL AND SOCIAL SCIENCES. SMITHSONIAN INSTITUTION INFORMATION SYSTEMS INNOVATIONS 2(2). I, 35 PP. [TAXONOMY; DENDOGRAM; COMPUTER]

ROTHSCHILD, B. J. 1963. GRAPHIC COMPARISONS OF MERISTIC DATA. COPEIA 1963(4), 601-603. [TAXONOMY]

ROYCE, W. F. 1964. A MORPHOMETRIC STUDY OF YELLOWFIN TUNA THUNNUS ALBACARES (BONNATERRE). U. S. FISH WILDL. SERV. FISH. BULL. 63(2), 395-443. [REGRESSION]

RUBIN, J. 1966. AN APPROACH TO ORGANIZING DATA INTO HOMOGENEOUS GROUPS. SYST. ZOOL. 15(3), 169-182. [TAXONOMY; HILL-CLIMBING ALGORITHM; SIMILARITY COEFFICIENTS; MEASURE OF OBJECT STABILITY; COMPUTER]

RUBIN, J. 1967. OPTIMAL CLASSIFICATION INTO GROUPS: AN APPROACH FOR SOLVING THE TAXONOMY PROBLEM. J. THEOR. BIOL. 15(1), 103-144. [ALGORITHM; SET; STABILITY; COEFFICIENT OF SIMILARITY; MAHALANOBIS' D2]

RUBIN, J., AND H. P. FRIEDMAN. 1967. A CLUSTER ANALYSIS AND TAXONOMY SYSTEM FOR GROUPING AND CLASSIFYING DATA. REPORT AND PROGRAM, IBM CORPORATION, NEW YORK SCIENTIFIC CENTER, 590 MADISON AVE., NEW YORK, NEW YORK. PP. 1-219.

SAILA, S. B. AND J. M. FLOWERS. 1969. GEOGRAPHIC MORPHOMETRIC VARIATION IN THE AMERICAN LOBSTER. SYST. ZOOL. 18(3), 330-338. [MULTIVARIATE ANALYSIS; CENTOUR CONCEPT; DISCRIMINANT FUNCTION; BAYES' THEOREM; DISPERSION MATRIX]

SHEALS, J. G. 1969. COMPUTERS IN ACARINE TAXONOMY. ACAROLOGIA(PARIS) 11(3), 376-394. [CLUSTER ANALYSIS; PRINCIPAL COMPONENT ANALYSIS]

SHIPP, W. S. AND G. M. BARNWELL. 1959. THE THEORY AND METHODS OF A NEW APPROACH TO THE MATHEMATICAL TAXONOMY OF FISHES WITH A HYPOTHETICAL DESCRIPTION OF THEIR GROWTH PHENOMENA. PP. 31-41. IN, RESEARCH PAPERS, REPORTS OF RESEARCH IN PROGRESS IN THE PHYSICAL, BIOLOGICAL, AND ENGINEERING SCIENCES. (W. L. EVANS, ED.). UNIVERSITY OF ARKANSAS, FAYETTEVILLE, ARKANSAS.

SHMIDT, V. M. 1964. THE BIOMETRICAL METHOD IN BOTANICAL TAXONOMY. BOTANICHESKIY ZHURNAL, LENINGRAD 49(1), 85-93. (ENGLISH TRANSL. U. S. DEPT. OF COMMERCE JPRS 25, 353.).

SIBSON, R. 1970. A MODEL FOR TAXONOMY. II. MATH. BIOSCI. 6, 405-430.

SIMS, R. W. 1969. OUTLINE OF AN APPLICATION OF COMPUTER TECHNIQUES TO THE PROBLEM OF THE CLASSIFICATION OF THE MEGASCOLECOID EARTHWORMS. PEDOBIOLOGIA 9(1/2), 35-41. [MATRIX; COEFFICIENT OF SIMILARITY; PRINCIPAL COMPONENT ANALYSIS]

SMARTT, P. F. M., S. E. MEACOCK AND J. M. LAMBERT. 1974. INVESTIGATIONS INTO THE PROPERTIES OF QUANTITATIVE VEGETATIONAL DATA. J. ECOL. 62(3), 735-759. [TRANSFORMATION; NUMERICAL MODELS; COPHENETIC CORRELATION COEFFICIENT; WILLIAMS AND CLIFFORD'S D; ORLOCI'S INFORMATION COEFFICIENT; MEACOCK'S COEFFICIENTS; COMPUTER]

SMIRNOV, E. S. 1968. ON EXACT METHODS IN SYSTEMATICS. SYST. ZOOL. 17(1), 1-13. [INDICES; E-MODEL DISTRIBUTION; MULTIMODAL DISTRIBUTION; BIMODAL DISTRIBUTION; T COEFFICIENT]

SMITH, D. W. 1969. A TAXIMETRIC STUDY OF VACCINIUM IN NORTHEASTERN ONTARIO. CAN. J. BOT. 47(11), 1747-1759. [TAXONOMY]

SNEATH, P. H. A. 1957. THE APPLICATION OF COMPUTERS TO TAXONOMY. J. GEN. MICROBIOL. 17(1), 201-226. [SIMILARITY; BACTERIA]

SNEATH, P. H. A. 1966. ESTIMATING CONCORDANCE BETWEEN GEOGRAPHICAL TRENDS. SYST. ZOOL. 15(3), 250-252. [TREND-SURFACE ANALYSIS]

SNEATH, P. H. A. AND R. R. SOKAL. 1962. NUMERICAL TAXONOMY. NATURE 193(4818), 855-860.

SOKAL, R. R. 1961. DISTANCE AS A MEASURE OF TAXONOMIC SIMILARITY. SYST. ZOOL. 10(2), 70-79. [COEFFICIENT OF ASSOCIATION; MAHALANOBIS' D2; COEFFICIENT OF RACIAL LIKENESS; CORRELATION COEFFICIENT]

SOKAL, R. R. 1965. STATISTICAL METHODS IN SYSTEMATICS. BIOL. REV. 40(3), 337-391. [TAXONOMY; MULTIVARIATE ANALYSIS; CANONICAL ANALYSIS; R; DISCRIMINANT FUNCTION]

SOKAL, R. R. 1966. NUMERICAL TAXONOMY. SCI. AMER. 215(6), 106-116. [COMPUTER]

SOKAL, R. R. 1970. ANOTHER NEW BIOLOGY. BIOSCIENCE 20(3), 152-159.

STOWER, W. J., D. E. DAVIES AND I. B. JONES. 1960. MORPHOMETRIC STUDIES OF THE DESERT LOCUST, SCHISTOCERCA GREGARIA (FORSK.). J. ANIM. ECOL. 29(2), 309-339. [MULTIVARIATE ANALYSIS; MATRIX; GENERALIZED DISTANCE]

STRAUSS, D. J. 1971. INDEPENDENCE OF VARIABLES IN NUMERICAL TAXONOMY. SYST. ZOOL. 20(4), 470-471. [PRINCIPAL COMPONENTS ANALYSIS]

STROUD, C. P. 1953. AN APPLICATION OF FACTOR ANALYSIS TO THE SYSTEMATICS OF KALOTERMES. SYST. ZOOL. 2(2), 76-92. [MATRIX; TAXONOMY]

TALKINGTON, L. 1967. A METHOD OF SCALING FOR A MIXED SET OF DISCRETE AND CONTINUOUS VARIABLES. SYST. ZOOL. 16(2), 149-152. [TAXONOMY; EUCLIDEAN DISTANCE; SIMILARITY COEFFICIENTS]

THROCKMORTON, L. H. 1968. CONCORDANCE AND DISCORDANCE OF TAXONOMIC CHARACTERS IN DROSOPHILA CLASSIFICATION. SYST. ZOOL. 17(4), 355-387. [COEFFICIENT OF SIMILARITY; RUNS]

TOBLER, W. R., H. W. MIELKE AND T. R. DETWYLER. 1970. GEOBOTANICAL DISTANCE BETWEEN NEW ZEALAND AND NEIGHBORING ISLANDS. BIOSCIENCE 20(9), 537-542.

UNDERWOOD, R. 1969. THE CLASSIFICATION OF CONSTRAINED DATA. SYST. ZOOL. 18(3), 312-317. [TAXONOMY; TRANSFORMATION; COMPUTER; PROBIT; SIGMOID FORMS; RATIO CORRELATION]

VASICEK, Z. AND R. JICIN. 1972. THE PROBLEM OF SIMILARITY OF SHAPE. SYST. ZOOL. 21(1), 91-96. [SHAPE CURVES; TAXONOMY]

VOGT, W. G. AND D. G. MCPHERSON. 1972. THE WEIGHTED SEPARATION INDEX: A MULTIVARIATE TECHNIQUE FOR SEPARATING MEMBERS OF CLOSELY-RELATED SPECIES USING QUALITATIVE DIFFERENCES. SYST. ZOOL. 21(2), 187-198. [TAXONOMY]

VORIS, H. K. 1971. NEW APPROACHES TO CHARACTER ANALYSIS IS APPLIED TO THE SEA SNAKES (HYDROPHIIDAE). SYST. ZOOL. 20(4), 442-458. [INFORMATION THEORY; CHI-SQUARE; TAXONOMY]

WILKINSON, C. 1970. ADDING A POINT TO A PRINCIPAL COORDINATES ANALYSIS. SYST. ZOOL. 19(3), 258-263. [TAXONOMY]

WILKINSON, C. 1974. NUMERICAL CLASSIFICATION: SOME QUESTIONS ANSWERED. CAN. ENTOMOL. 106(5), 449-464. [DENDOGRAM; CLUSTER ANALYSIS; ADD-A-POINT PROGRAM; PRINCIPAL CO-ORDINATE ANALYSIS]

WILLIAMS, W. T. 1963. COMPUTERS AS BOTANISTS. NATURE 197(4872), 1047-1049.

WILLIAMS, W. T. 1967. THE COMPUTER BOTANIST. AUST. J. SCI. 29(8), 266-271.

WILLIAMS, W. T. 1967. NUMBERS, TAXONOMY, AND JUDGEMENT. BOT. REV. 33(4), 379-386.

WILLIAMS, W. T. 1971. PRINCIPLES OF CLUSTERING. PP. 303-326. IN, ANNUAL REVIEW OF ECOLOGY AND SYSTEMATICS, VOLUME 3. (R. F.JOHNSTON, P. W. FRANK AND C. D. MITCHENER, EDS.). ANNUAL REVIEWS, INC., PALO ALTO, CALIFORNIA. [MODEL; WEIGHTING; MEASURING]

WILLIAMS, W. T. AND G. N. LANCE. 1968. CHOICE OF STRATEGY IN THE ANALYSIS OF COMPLEX DATA. THE STATISTICIAN 18, 31-43. FLAG.

WILLIAMS, W. T. AND H. T. CLIFFORD. 1971. ON THE COMPARISON OF TWO CLASSIFICATIONS OF THE SAME SET OF ELEMENTS. TAXON 20(4), 519-522. [TAXONOMY; COPHENETIC CORRELATION COEFFICIENT]

WILLIAMS, W. T. AND J. M. LAMBERT. 1961. MULTIVARIATE METHODS IN TAXONOMY. TAXON 10, 205-211.

WILLIAMS, W. T. AND M. B. DALE. 1965. FUNDAMENTAL PROBLEMS IN NUMERICAL TAXONOMY. ADV. BOT. RES. 2, 35-68. [ASSOCIATION ANALYSIS; INDEX OF SIMILARITY; CORRELATION COEFFICIENT; MAHALANOBIS' D2; MODEL; INFORMATION; PROBABILITY; EUCLIDEAN MEASURE OF DISTANCE]

WILLIAMS, W. T., M. B. DALE AND L. G. MOCKETT. 1964. DISSIMILARITY ANALYSIS: A NEW TECHNIQUE OF HIERARCHAL SUB-DIVISION. NATURE 202(4936), 1034-1035.

WILLIAMS, W. T., M. B. DALE AND P. MACNAUGHTON-SMITH. 1964. AN OBJECTIVE METHOD OF WEIGHTING IN SIMILARITY ANALYSIS. NATURE 201(4917), 426.

WILSON, E. O. 1965. A CONSISTENCY TEST FOR PHYLOGENIES BASED ON CONTEMPORANEOUS SPECIES. SYST. ZOOL. 14(3), 214-220. [SET THEORY; PROBABILITY]

WOOL, D. 1973. QUANTITATIVE EVALUATION OF TEMPORAL SIMILARITY IN ABUNDANCE OF ANIMAL SPECIES. RES. POPUL. ECOL.-KYOTO 15(1), 90-98. [PRINCIPAL COMPONENTS ANALYSIS; CORRELATION MATRIX; INDEX OF SIMILARITY; CORRELATION COEFFICIENT; NUMERICAL TAXONOMY PROCEDURE; MATRIX; PHENOGRAM]

BOOKS, BIBLIOGRAPHIES, AND EDUCATIONAL REFERENCES

.. BOOKS

ANDREWARTHA, H. G. 1971. INTRODUCTION TO THE STUDY OF ANIMAL POPULATIONS. 2ND. EDITION. UNIVERSITY OF CHICAGO PRESS, CHICAGO, ILLINOIS. XIV, 283 PP.

ANDREWARTHA, H. G. AND L. C. BIRCH. 1954. THE DISTRIBUTION AND ABUNDANCE OF ANIMALS. UNIVERSITY OF CHICAGO PRESS, CHICAGO, ILLINOIS. XV, 782 PP. [MODEL; PREDATOR-PREY MODEL; COMPETITION; LOGISTIC; LIFE TABLE]

ANONYMOUS. 1972. 3RD CONFERENCE ADVISORY GROUP OF FOREST STATISTICIANS. INSTITUT NATIONAL DE LA RECHERCHE AGRONOMIQUE, PARIS, FRANCE. I.N.R.A. PUBL. 72-3. 332 PP.

BAILEY, N. T. J. 1957. THE MATHEMATICAL THEORY OF EPIDEMICS. CHARLES GRIFFIN AND CO., LONDON, ENGLAND. VIII, 194 PP.

BAILEY, N. T. J. 1964. THE ELEMENTS OF STOCHASTIC PROCESSES WITH APPLICATIONS TO THE NATURAL SCIENCES. JOHN WILEY AND SONS, NEW YORK, NEW YORK. XI, 249 PP.

BAILEY, N. T. J. 1967. THE MATHEMATICAL APPROACH TO BIOLOGY AND MEDICINE. JOHN WILEY AND SONS, NEW YORK, NEW YORK. XIII, 296 PP.

BARTLETT, M. S. 1960. STOCHASTIC POPULATION MODELS IN ECOLOGY AND EPIDEMIOLOGY. METHUEN AND CO., LONDON, ENGLAND. X, 90 PP.

BARTLETT, M. S. 1966. AN INTRODUCTION TO STOCHASTIC PROCESSES. 2ND EDITION. CAMBRIDGE UNIVERSITY PRESS, LONDON, ENGLAND. XVI, 362 PP. [POPULATION DYNAMICS]

BARTLETT, M. S. AND R. W. HIORNS. (EDS.). 1973. THE MATHEMATICAL THEORY OF THE DYNAMICS OF BIOLOGICAL POPULATIONS. ACADEMIC PRESS, NEW YORK, NEW YORK. XII, 347 PP.

BEVERTON, R. J. H. AND S. J. HOLT. 1957. ON THE DYNAMICS OF EXPLOITED FISH POPULATIONS. FISH. INVEST. MINIST. AGRIC. FISH., FOOD (GREAT BRITAIN) SER. 2, VOL. 19. 533 PP.

BHARUCHA-REID, A. T. 1962. REVIEW OF: STOCHASTIC POPULATION MODELS IN ECOLOGY AND EPIDEMIOLOGY BY M. S. BARTLETT. BIOMETRICS 18(2), 253-254.

BLACKITH, R. E. AND R. A. REYMENT. 1971. MULTIVARIATE MORPHOMETRICS. ACADEMIC PRESS, NEW YORK, NEW YORK. IX, 412 PP.

BROOKS, C. E. P. AND N. CARRUTHERS. 1953. HANDBOOK OF STATISTICAL METHODS IN METEOROLOGY. HER MAJESTY'S STATIONARY OFFICE, LONDON, ENGLAND. VIII, 412 PP.

CHASTON, I. 1971. MATHEMATICS FOR ECOLOGISTS. BUTTERWORTHS, LONDON, ENGLAND. VIII, 132 PP.

COHEN, J. E. 1966. A MODEL OF SIMPLE COMPETITION. HARVARD UNIVERSITY PRESS, CAMBRIDGE, MASSACHUSETTS. X, 138 PP.

COHEN, J. E. 1971. CASUAL GROUPS OF MONKEYS AND MEN; STOCHASTIC MODELS OF ELEMENTAL SOCIAL SYSTEMS. HARVARD UNIVERSITY PRESS, CAMBRIDGE, MASSACHUSETTS. XIII, 175 PP.

COLE, A. J. (ED.). 1969. NUMERICAL TAXONOMY. ACADEMIC PRESS, NEW YORK, NEW YORK. XV, 324 PP.

D'ANCONA, U. 1954. THE STRUGGLE FOR EXISTENCE. E. J. BRILL, LEIDEN, NETHERLANDS. XII, 274 PP.

DAVIES, R. G. 1971. COMPUTER PROGRAMMING IN QUANTITATIVE BIOLOGY. ACADEMIC PRESS, NEW YORK, NEW YORK. XI, 492 PP.

DE WIT, C. T. AND J. GOUDRIAAN. 1974. SIMULATION OF ECOLOGICAL PROCESSES. CENTRE FOR AGRICULTURAL PUBLISHING AND DOCUMENTATION, WAGENINGEN, NETHERLANDS. 159 PP. [MODEL; COMPUTER]

ELLIOTT, J. M. 1971. SOME METHODS FOR THE STATISTICAL ANALYSIS OF SAMPLES OF BENTHIC INVERTEBRATES. FRESHWATER BIOL. ASSOC., AMBLESIDE, WEST MORELAND, UNITED KINGDOM, SCI. PUBL. NO. 25. 144 PP.

EMLEN, J. M. 1973. ECOLOGY: AN EVOLUTIONARY APPROACH. ADDISON-WESLEY PUBLISHING CO., READING, MASSACHUSETTS. XIV, 493 PP.

EVANS, G. C. 1972. THE QUANTITATIVE ANALYSIS OF PLANT GROWTH. UNIVERSITY OF CALIFORNIA PRESS, BERKELEY, CALIFORNIA. XXVI, 734 PP.

FRANSZ, H. G. 1974. THE FUNCTIONAL RESPONSE TO PREY DENSITY IN A ACARINE SYSTEM. P.U.D.O.C., WAGENINGEN CENTRE FOR AGRICULTURAL PUBLISHING AND DOCUMENTATION, NETHERLANDS. VIII, 144 PP. [COMPUTER SIMULATION; DETERMINISTIC MODEL; STOCHASTIC MODEL; HOLLING MODEL]

FREESE, F. 1967. ELEMENTARY STATISTICAL METHODS FOR FORESTERS. U. S. FOREST SERVICE, WASHINGTON, D. C. AGRICULTURE HANDBOOK 317. IV, 87 PP.

FRETWELL, S. D. 1972. POPULATIONS IN A SEASONAL ENVIRONMENT. PRINCETON UNIVERSITY PRESS, PRINCETON, NEW JERSEY. XXIII, 217 PP. [MODEL]

GAUSE, G. F. 1934. THE STRUGGLE FOR EXISTENCE. WILLIAMS AND WILKINS CO., BALTIMORE, MARYLAND. IX, 163 PP.

GILPIN, M. E. 1975. GROUP SELECTION IN PREDATOR-PREY COMMUNITIES. PRINCETON UNIVERSITY PRESS, PRINCETON, NEW JERSEY. XIII, 108 PP.

GOEL, N. S. AND N. RICHTER-DYN. 1974. STOCHASTIC MODELS IN BIOLOGY. ACADEMIC PRESS, NEW YORK, NEW YORK. X, 269 PP.

GOUNOT, M. 1969. METHODES D'ETUDE QUANTITATIVE DE LA VEGETATION. MASSON ET CIE EDITEURS, PARIS, FRANCE. 314 PP.

GREEN, R. F. 1974. REVIEW OF: NUMERICAL TAXONOMY BY P. H. A. SNEATH AND R. R. SOKAL. BIOMETRICS 30(2), 372-373.

GREGG, J. R. 1954. THE LANGUAGE OF TAXONOMY. COLUMBIA UNIVERSITY PRESS, NEW YORK, NEW YORK. IX, 70 PP. [SET THEORY]

GREIG-SMITH, P. 1964. QUANTITATIVE PLANT ECOLOGY. 2ND EDITION BUTTERWORTHS, LONDON, ENGLAND. XII, 256 PP.

GRIFFITH, A. L. AND B. S. RAM. 1947. THE SILVICULTURE RESEARCH CODE. VOL. 2. THE STATISTICAL MANUAL. THE MANAGER OF PUBLICATIONS, DELHI, INDIA. VII, 214 PP.

GRIFFITH, A. L. AND J. PRASAD. 1949. THE SILVICULTURE RESEARCH CODE. VOL. 3. THE TREE AND CROP MEASUREMENT MANUAL. THE MANAGER OF PUBLICATIONS, DELHI, INDIA. VIII, 229 PP.

H.T.H.P. 1931. REVIEW: "LECONS SUR LA THEORIE MATHEMATIQUE DE LA LUTTE POUR LA VIE" BY V. VOLTERRA. NATURE 128(3240), 963.

HAMMACK, J. AND G. M. BROWN, JR. 1974. WATERFOWL AND WETLANDS: TOWARD BIOECONOMIC ANALYSIS. THE JOHNS HOPKINS UNIVERSITY PRESS, BALTIMORE, MARYLAND. 95 PP. [ECONOMIC AND POPULATION DYNAMIC MODELS; MULTIPLE REGRESSION; BEVERTON-HOLT EQUATION; MAXIMIZATION; STOCHASTIC SIMULATION; OPTIMAL MANAGEMENT MODELS; COMPUTER]

HASSELL, M. P. 1973. REVIEW OF: THE ANALYSIS OF BIOLOGICAL POPULATIONS BY M. WILLIAMSON. J. ANIM. ECOL. 42(2), 470-471.

HASSELL, M. P. 1975. REVIEW OF: MODELS IN ECOLOGY BY J. MAYNARD SMITH. J. ANIM. ECOL. 44(1), 344-345.

HASSELL, M. P. 1975. REVIEW OF: STABIBLITY AND COMPLEXITY IN MODEL ECOSYSTEMS BY R. M. MAY. J. ANIM. ECOL. 44(3), 931-932.

HORN, H. S. 1971. THE ADAPTIVE GEOMETRY OF TREES. PRINCETON UNIVERISTY PRESS, PRINCETON, NEW JERSEY. XI, 144 PP. [MODEL]

HUSCH, B. 1963. FOREST MENSURATION AND STATISTICS. ROLAND PRESS, NEW YORK, NEW YORK. VIII, 474 PP.

INTERNATIONAL ATOMIC ENERGY AGENCY. 1974. COMPUTER MODELS AND APPLICATION OF THE STERILE-MALE TECHNIQUE. IAEA, VIENNA, AUSTRIA. PUBLICATION STI/PUB/340. III, 195 PP. [SIMULATION]

IOSIFESCU, M. AND P. TAUTU. 1973. STOCHASTIC PROCESSES AND APPLICATIONS IN BIOLOGY AND MEDICINE II: MODELS. SPRINGER-VERLAG NEW YORK INC., NEW YORK, NEW YORK. 337 PP. [EVOLUTION; POPULATION DYNAMICS; POPULATION GROWTH]

IVLEV, V. S. 1961. EXPERIMENTAL ECOLOGY OF THE FEEDING OF FISHES. YALE UNIVERSITY PRESS. NEW HAVEN, CONNECTICUT. 302 PP.

JARDINE, N. AND R. SIBSON. 1971. MATHEMATICAL TAXONOMY. JOHN WILEY AND SONS, NEW YORK, NEW YORK. XVIII, 286 PP.

JEFFERS, J. N. R. 1959. EXPERIMENTAL DESIGN AND ANALYSIS IN FOREST RESEARCH. ALMQVIST AND WIKSELL, STOCKHOLM, SWEDEN. 172 PP.

JEFFERS, J. N. R. (ED.). 1972. MATHEMATICAL MODELS IN ECOLOGY. BLACKWELL SCIENTIFIC PUBLICATIONS, OXFORD, ENGLAND. VI, 398 PP.

JOLLY, G. M. 1974. REVIEW OF: THE ESTIMATION OF ANIMAL ABUNDANCE AND RELATED PARAMETERS BY G. A. F. SEBER. J. ANIM. ECOL. 43(2), 603-605.

KERNER, E. H. 1972. GIBBS ENSEMBLE: BIOLOGICAL ENSEMBLE. THE APPLICATION OF STATISTICAL MECHANICS TO ECOLOGICAL, NEURAL, AND BIOLOGICAL NETWORKS. GORDON AND BREACH SCIENCE PUBLISHERS, NEW YORK, NEW YORK. XI, 167 PP.

KERSHAW, K. A. 1974. QUANTITATIVE AND DYNAMIC PLANT ECOLOGY. 2ND EDITION. AMERICAN ELSEVIER PUBLISHING CO., NEW YORK, NEW YORK. X, 308 PP.

KEYFITZ, N. 1968. INTRODUCTION TO THE MATHEMATICS OF POPULATION. ADDISON-WESLEY PUBLISHING CO., READING, MASSACHUSETTS. XIV, 450 PP.

KEYFITZ, N. AND W. FLIEGER. 1971. POPULATION: FACTS AND METHODS OF DEMOGRAPHY. W. H. FREEMAN AND CO., SAN FRANCISCO, CALIFORNIA. XI, 613 PP.

KHIL'MI, G. F. 1957. THEORETICAL FOREST BIOGEOPHYSICS. IZDATEL'STVO AKADEMII NAUK, SSSR, MOSKVA (ENGLISH TRANSL., 1962. U. S. DEPT. OF COMMERCE, OTS 61-31089. 155 PP.).

KOSTITZIN, V. A. 1939. MATHEMATICAL BIOLOGY. (TRANSL. BY T. H. SAVORY). GEORGE G. HARRAP AND CO., LONDON, ENGLAND. 238 PP.

KRANZ, J. (ED.). 1974. EPIDEMICS OF PLANT DISEASES: MATHEMATICAL ANALYSIS AND MODELING. SPRINGER-VERLAG NEW YORK INC., NEW YORK, NEW YORK. X, 170 PP.

KUCZYNSKI, R. R. 1969. THE MEASUREMENT OF POPULATION GROWTH: METHODS AND RESULTS. DEMOGRAPHIC MONOGRAPHS, VOL. 6. GORDON AND BREACH SCIENCE PUBLISHERS, INC., NEW YORK, NEW YORK. 255 PP.

LEIGH, E. G., JR. 1968. MAKING ECOLOGY AN APPLIED SCIENCE. (REVIEW OF: ECOLOGY AND RESOURCE MANAGEMENT, A QUANTITATIVE APPROACH BY K. E. F. WATT). SCIENCE 160(3834), 1326-1327.

LEVIN, S. A. (ED.). 1975. ECOSYSTEM ANALYSIS AND PREDICTION. PROCEEDINGS OF A SIAM-SIMS CONFERENCE HELD AT ALTA, UTAH, JULY 1-5, 1974. XIV, 337 PP. [MODELING]

LEVINS, R. 1968. ECOLOGICAL ENGINEERING: THEORY AND TECHNOLOGY. (REVIEW OF: ECOLOGY AND RESOURCE MANAGEMENT, A QUANTITATIVE APPROACH BY K. E. F. WATT). QUART. REV. BIOL. 43(3), 301-305.

LEVINS, R. 1968. EVOLUTION IN CHANGING ENVIRONMENTS: SOME THEORETICAL EXPLORATIONS. PRINCETON UNIVERSITY PRESS, PRINCETON, NEW JERSEY. IX, 120 PP.

LEVINS, S. A. AND C. A. S. HALL. 1973. REVIEW OF: SYSTEMS ANALYSIS AND SIMULATION IN ECOLOGY, VOL. I. BY B. C. PATTEN. BIOMETRICS 29(4), 832-833.

LEWIS, T. AND L. R. TAYLOR. 1967. INTRODUCTION TO EXPERIMENTAL ECOLOGY. ACADEMIC PRESS, NEW YORK, NEW YORK. XI, 401 PP.

LOTKA, A. J. 1925. ELEMENTS OF PHYSICAL BIOLOGY. WILLIAMS AND WILKINS CO., BALTIMORE, MARYLAND. XXX, 460 PP. (REISSUED AS ELEMENTS OF MATHEMATICAL BIOLOGY. DOVER, 1956.).

LUDWIG, D. 1974. STOCHASTIC POPULATION THEORIES. LECTURE NOTES IN BIOMATHEMATICS 3. SPRINGER-VERLAG NEW YORK INC., NEW YORK, NEW YORK. VI, 108 PP.

MACARTHUR, R. H. 1972. GEOGRAPHICAL ECOLOGY: PATTERNS IN THE DISTRIBUTION OF SPECIES. HARPER AND ROW, PUBLISHERS, NEW YORK, NEW YORK. XVIII, 269 PP. [PREDATOR-PREY MODEL; COMPETITION; DIVERSITY MEASURE; SPECIES PACKING]

MACARTHUR, R. H. AND E. O. WILSON. 1967. THE THEORY OF ISLAND BIOGEOGRAPHY. PRINCETON UNIVERSITY PRESS, PRINCETON, NEW JERSEY. XI, 203 PP.

MAY, R. M. 1973. STABILITY AND COMPLEXITY IN MODEL ECOSYSTEMS. PRINCETON UNIVERSITY PRESS, PRINCETON, NEW JERSEY. IX, 235 PP.

MAYNARD SMITH, J. 1968. MATHEMATICAL IDEAS IN BIOLOGY. CAMBRIDGE UNIVERSITY PRESS, LONDON, ENGLAND. VII, 152 PP.

MAYNARD SMITH, J. 1974. MODELS IN ECOLOGY. CAMBRIDGE UNIVERSITY PRESS, NEW YORK, NEW YORK. XII, 146 PP.

MEAD, R. 1973. REVIEW OF: MATHEMATICAL MODELS IN ECOLOGY. EDITED BY J. N. R. JEFFERS. J. APPL. ECOL. 10(2), 667-668.

METZ, J. A. J. 1973. REVIEW OF: POPULATIONS IN A SEASONAL ENVIRONMENT BY D. FRETWELL. ACTA BIOTHEOR. 22(4), 207-210. [MODEL]

MILNER, C. AND R. E. HUGHES. 1968. METHODS FOR THE MEASUREMENT OF THE PRIMARY PRODUCTION OF GRASSLAND. IBP HANDBOOK NO. 6. BLACKWELL SCIENTIFIC PUBLICATIONS, OXFORD, ENGLAND. XII, 70 PP. [SAMPLE SIZE; LEAF AREA; STANDING CROP; EQUATIONS]

MONTEITH, J. L. 1973. PRINCIPLES OF ENVIRONMENTAL PHYSICS. AMERICAN ELSEVIER PUBLISHING CO., NEW YORK, NEW YORK. XIII, 241 PP.

MUELLER-DOMBOIS, D. AND H. ELLENBERG. 1974. AIMS AND METHODS OF VEGETATION ECOLOGY. JOHN WILEY AND SONS, NEW YORK, NEW YORK. XX, 547 PP. [BOOK; SAMPLING; ORDINATION; INDICES]

MUNN, R. E. 1970. BIOMETEOROLOGICAL METHODS. ACADEMIC PRESS, NEW YORK, NEW YORK. XI, 336 PP.

NEELEY, P. M. 1972. REVIEW OF: COMPUTER PROGRAMMING IN QUANTITATIVE BIOLOGY BY R. G. DAVIES. SYST. ZOOL. 21(4), 451-460.

NEWBOULD, P. J. 1967. METHODS FOR ESTIMATING THE PRIMARY PRODUCTION OF FORESTS. IBP HANDBOOK NO. 2. BLACKWELL SCIENTIFIC PUBLICATIONS, OXFORD, ENGLAND. IX, 62 PP. [EQUATIONS]

NISHIDA, T. AND T. TORII. 1970. A HANDBOOK OF FIELD METHODS FOR RESEARCH ON RICE STEM-BORERS AND THEIR NATURAL ENEMIES. IBP HANDBOOK, NO. 14, BLACKWELL SCIENTIFIC PUBLICATIONS, OXFORD, ENGLAND, X, 132 PP. [SEQUENTIAL SAMPLING; LIFE TABLE]

NIX, H. A. (ED.). 1971. QUANTIFYING ECOLOGY. PROC. ECOL. SOC. AUSTRALIA, VOL. 6, I, 243 PP.

O'NEILL, R. V. 1973. REVIEW OF: MATHEMATICAL MODELS IN ECOLOGY EDITED BY J. N. R. JEFFERS. ECOLOGY 54(2), 459-460.

ORLOCI, L. 1975. MULTIVARIATE ANALYSIS IN VEGETATION RESEARCH. DR. JUNK PUBLISHERS, THE HAGUE, NETHERLANDS. VIII, 276 PP.

PANKHURST, R. J. (ED.) 1975. BIOLOGICAL IDENTIFICATION WITH COMPUTERS. ACADEMIC PRESS, NEW YORK, NEW YORK. X, 333 PP.

PARKINSON, D., T. R. G. GRAY AND S. T. WILLIAMS, (EDS.). 1971. METHODS FOR STUDYING THE ECOLOGY OF SOIL MICROORGANISMS. IBP HANDBOOK NO. 19. BLACKWELL SCIENTIFIC PUBLICATIONS, OXFORD, ENGLAND, XI, 116 PP. [SAMPLING; BIOMASS; ACTIVITY]

PARSONS, T. R. AND M. TAKAHASHI. 1973. BIOLOGICAL OCEANOGRAPHIC PROCESSES. PERGAMON PRESS, NEW YORK, NEW YORK. X, 186 PP. [PHOTOSYNTHESIS MODEL; FEEDING MODEL; PRODUCTION MODEL; INDEX OF STABILITY]

PATIL, G. P., E. C. PIELOU AND W. E. WATERS, (EDS.). 1971. STATISTICAL ECOLOGY, VOLUME 2. SAMPLING AND MODELING BIOLOGICAL POPULATIONS AND POPULATION DYNAMICS. PENNSYLVANIA STATE UNIVERSITY PRESS, UNIVERSITY PARK, PENNSYLVANIA. VI, 420 PP.

PATIL, G. P., E. C. PIELOU AND W. E. WATERS, (EDS). 1971. STATISTICAL ECOLOGY, VOLUME 1. SPATIAL PATTERNS AND STATISTICAL DISTRIBUTIONS. PENNSYLVANIA STATE UNIVERSITY PRESS, UNIVERSITY PARK, PENNSYLVANIA. XXVIII, 582 PP.

PATIL, G. P., E. C. PIELOU AND W. E. WATERS, (EDS). 1971. STATISTICAL ECOLOGY, VOLUME 3. MANY SPECIES POPULATIONS, ECOSYSTEMS, AND SYSTEMS ANALYSES. PENNSYLVANIA STATE UNIVERSITY PRESS, UNIVERSITY PARK, PENNSYLVANIA. IV, 462 PP.

PATTEN, B. C. (ED.). 1971. SYSTEMS ANALYSIS AND SIMULATION IN ECOLOGY, VOLUME I. ACADEMIC PRESS, NEW YORK, NEW YORK. XV, 607 PP.

PATTEN, B. C. (ED.). 1972. SYSTEMS ANALYSIS AND SIMULATION IN ECOLOGY, VOLUME II. ACADEMIC PRESS, NEW YORK, NEW YORK. XVI, 592 PP.

PAULIK, G. J. 1969. REVIEW OF: ECOLOGY AND RESOURCE MANAGEMENT BY K. E. F. WATT. TRANS. AMER. FISH. SOC. 98(4), 729-730.

PAVLIDIS, T. 1973. BIOLOGICAL OSCILLATORS: THEIR MATHEMATICAL ANALYSIS. ACADEMIC PRESS, NEW YORK, NEW YORK. XIII, 207 PP. [CIRCADIAN RHYTHMS; POPULATION]

PETRUSEWICZ, K. AND A. MACFADYEN. 1970. PRODUCTIVITY OF TERRESTRIAL ANIMALS: PRINCIPLES AND METHODS. IBP HANDBOOK NO. 13. BLACKWELL SCIENTIFIC PUBLICATIONS, OXFORD, ENGLAND. XII, 190 PP. [MODEL; TAG-RECAPTURE; DEMOGRAPHY; FEEDING MODEL; PRODUCTION MODEL; ENERGY]

PIELOU, E. C. 1969. AN INTRODUCTION TO MATHEMATICAL ECOLOGY. JOHN WILEY AND SONS, NEW YORK, NEW YORK. VIII, 286 PP.

PIELOU, E. C. 1974. POPULATION AND COMMUNITY ECOLOGY: PRINCIPLES AND METHODS. GORDON AND BREACH SCIENCE PUBLISHERS, NEW YORK, NEW YORK. VIII, 424 PP.

PIELOU, E. C. 1975. ECOLOGICAL DIVERSITY. JOHN WILEY AND SONS, NEW YORK, NEW YORK. VIII, 165 PP.

POOLE, R. W. 1974. AN INTRODUCTION TO QUANTITATIVE ECOLOGY. MCGRAW-HILL BOOK CO., NEW YORK, NEW YORK. XII, 532 PP.

PRATT, J. W. (ED.). 1974. STATISTICAL AND MATHEMATICAL ASPECTS OF POLLUTION PROBLEMS. MARCEL DEKKER, INC., NEW YORK, NEW YORK. XXVII, 392 PP.

PRODAN, M. 1968. FOREST BIOMETRICS. PERGAMMON PRESS, NEW YORK, NEW YORK. XI, 447 PP.

REDDINGIUS, J. 1971. GAMBLING FOR EXISTENCE. E. J. BRILL, LEIDEN, NETHERLANDS. VI, 208 PP.

REYMENT, R. A. 1971. INTRODUCTION TO QUANTITATIVE PALEOECOLOGY. ELSEVIER PUBLISHING CO., NEW YORK, NEW YORK. XIII, 226 PP.

RICKER, W. E. 1948. METHODS OF ESTIMATING VITAL STATISTICS OF FISH POPULATIONS. INDIANA UNIVERSITY PUBLICATIONS, SCIENCE SERIES NO. 15. V, 101 PP.

RICKER, W. E. 1958. HANDBOOK OF COMPUTATIONS FOR BIOLOGICAL STATISTICS OF FISH POPULATIONS. FISH. RES. BOARD CAN. BULL. 119. OTTAWA, CANADA. 300 PP.

RICKER, W. E. (ED.) 1968. METHODS FOR ASSESSMENT OF FISH PRODUCTION IN FRESH WATERS. IBP HANDBOOK NO. 3, BLACKWELL SCIENTIFIC PUBLICATIONS, OXFORD, ENGLAND. XIII, 313 PP.

ROHLF, F. J. 1972. REVIEW OF: MULTIVARIATE MORPHOMETRICS BY R. E. BLACKITH AND R. A. REYMENT. SYST. ZOOL. 21(3), 348-349.

ROSE, C. W. 1966. AGRICULTURAL PHYSICS. PERGAMON PRESS, NEW YORK, NEW YORK. XVI, 226 PP.

SCHUMACHER, F. X. AND R. A. CHAPMAN. 1948. SAMPLING METHODS IN FORESTRY AND RANGE MANAGEMENT. SCHOOL OF FORESTRY, DUKE UNIVERSITY. DURHAM, NORTH CAROLINA. BULL. 7 (REVISED). 222 PP.

SEBER, G. A. F. 1973. THE ESTIMATION OF ANIMAL ABUNDANCE AND RELATED PARAMETERS. CHARLES GRIFFIN AND CO., LONDON, ENGLAND. XII, 506 PP.

SHIMWELL, D. W. 1972. THE DESCRIPTION AND CLASSIFICATION OF VEGETATION. UNIVERSITY OF WASHINGTON PRESS, SEATTLE, WASHINGTON. XIV, 322 PP. [PLOTLESS SAMPLING; POISSON SERIES; PRESTON'S OCTAVE; ORDINATION ANALYSIS; PRINCIPAL COMPONENTS ANALYSIS; JACCARD'S COEFFICIENT; SORENSEN'S COEFFICIENT; ORLOCI'S COEFFICIENT; BRAY-CURTIS ORDINATION]

SHUGART, H. H. 1974. REVIEW OF: STABILITY AND COMPLEXITY IN MODEL ECOSYSTEMS BY R. M. MAY. ECOLOGY 55(6), 1429-1430.

SIMPSON, G. G., A. ROE AND R. C. LEWONTIN. 1960. QUANTITATIVE ZOOLOGY. REVISED EDITION. HARCOURT, BRACE AND CO., NEW YORK, NEW YORK. VII, 440 PP.

SKELLAM, J. G. 1975. REVIEW OF: ECOLOGICAL STABILITY, EDITED BY M. B. USHER AND M. H. WILLIAMSON. J. ANIM. ECOL. 44(3), 935-937.

SLOBODKIN, L. B. 1961. GROWTH AND REGULATION OF ANIMAL POPULATIONS. HOLT, RINEHART AND WINSTON, NEW YORK, NEW YORK. VIII, 184 PP.

SNEATH, P. H. A. AND R. R. SOKAL. 1973. NUMERICAL TAXONOMY: THE PRINCIPLES AND PRACTICE OF NUMERICAL CLASSIFICATION W. H. FREEMAN AND CO., SAN FRANCISCO, CALIFORNIA. XV, 573 PP.

SOKAL, R. R. AND P. H. A. SNEATH. 1963. PRINCIPLES OF NUMERICAL TAXONOMY. W. H. FREEMAN AND CO., SAN FRANCISCO, CALIFORNIA. XVI, 359 PP.

SOUTHWOOD, T. R. E. 1966. ECOLOGICAL METHODS. METHUEN AND CO., LONDON, ENGLAND. XVIII, 391 PP.

STEELE, J. H. 1974. THE STRUCTURE OF MARINE ECOSYSTEMS. BLACKWELL SCIENTIFIC PUBLICATIONS, LONDON, ENGLAND X, 128 PP. [MODEL; DIVERSITY; STABILITY; SIMULATION; PREDATOR-PREY MODEL]

TORII, T. 1956. THE STOCHASTIC APPROACH IN FIELD POPULATIONS ECOLOGY: WITH SPECIAL REFERENCE TO FIELD INSECT POPULATIONS. JAPAN SOCIETY FOR PROMOTION OF SCIENCE, TOKYO, JAPAN. 277 PP., PLUS CHART.

ULLYETT, G. C. 1953. BIOMATHEMATICS AND INSECT POPULATION PROBLEMS - A CRITICAL REVIEW. MEMOIR ENTOMOL. SOC. OF S. AFRICA, PRETORIA, SOUTH AFRICA. NO. 2. 89 PP.

USHER, M. B. AND M. H. WILLIAMSON. (EDS.). 1974. ECOLOGICAL STABILITY. HALSTED PRESS, NEW YORK, NEW YORK. XII, 196 PP.

VAN WIJK, W. R. (ED.). 1966. PHYSICS OF PLANT ENVIRONMENT 2ND EDITION. NORTH-HOLLAND PUBLISHING CO., AMSTERDAM, NETHERLANDS. XVI, 382 PP.

VARLEY, G. C., G. R. GRADWELL AND M. P. HASSELL. 1973. INSECT POPULATION ECOLOGY. AN ANALYTICAL APPROACH. BLACKWELL SCIENTIFIC PUBLICATIONS, LONDON, ENGLAND. X, 212 PP. [MODEL; PREDATOR-PREY MODEL; LIFE TABLE; COMPETITION]

VOLLENWEIDER, R. A. 1969. A MANUAL ON METHODS FOR MEASURING PRIMARY PRODUCTION IN AQUATIC ENVIRONMENTS. IBP HANDBOOK NO. 12, BLACKWELL SCIENTIFIC PUBLICATIONS, OXFORD, ENGLAND. XVI, 213 PP. [SAMPLING; EQUATIONS; STATISTICS]

VOLTERRA, V. 1931. LECONS SUR LA THEORIE MATHEMATIQUE DE LA LUTTE POUR LA VIE. GAUTHIERS-VILLARS, PARIS, FRANCE. VI, 214 PP. (COMPILED BY M. BRELOT).

WADLEY, F. M. 1967. EXPERIMENTAL STATISTICS IN ENTOMOLOGY. GRADUATE SCHOOL PRESS, U. S. DEPT. AGR., WASHINGTON, D. C. VIII, 133 PP.

WAGNER, F. H. 1974. REVIEW OF: STABILITY AND COMPLEXITY IN MODEL ECOSYSTEMS BY R. M. MAY. ECOLOGY 55(6), 1428-1429.

WALTMAN, P. 1974. DETERMINISTIC THRESHOLD MODELS IN THE THEORY OF EPIDEMICS. LECTURE NOTES IN BIOMATHEMATICS VOL. 1. SPRINGER-VERLAG NEW YORK INC., NEW YORK, NEW YORK. V, 101 PP.

WARREN, J. A. 1969. REVIEW OF: FOREST BIOMETRICS BY M. PRODAN. BIOMETRICS 25(1), 182.

WATT, K. E. F. 1958. MATHEMATICAL FISH POPULATION DYNAMICS. (REVIEW OF: ON THE DYNAMICS OF EXPLOITED FISH POPULATIONS BY J. H. BEVERTON AND S. J. HOLT.). ECOLOGY 39(4), 777.

WATT, K. E. F. 1968. ECOLOGY AND RESOURCE MANAGEMENT. A QUANTITATIVE APPROACH. MCGRAW-HILL BOOK CO., NEW YORK, NEW YORK. XII, 450 PP.

WATT, K. E. F. (ED.). 1966. SYSTEMS ANALYSIS IN ECOLOGY. ACADEMIC PRESS, NEW YORK, NEW YORK. XIII, 276 PP.

WEATHERLEY, A. H. 1972. GROWTH AND ECOLOGY OF FISH POPULATIONS. ACADEMIC PRESS, NEW YORK, NEW YORK. X, 293 PP.

WHITTAKER, R. H. (ED.). 1973. ORDINATION AND CLASSIFICATION OF COMMUNITIES. DR. JUNK PUBLISHERS, THE HAGUE, NETHERLANDS. X, 737 PP.

WILLIAMS, C. B. 1964. PATTERNS IN THE BALANCE OF NATURE AND RELATED PROBLEMS IN QUANTITATIVE ECOLOGY. ACADEMIC PRESS, NEW YORK, NEW YORK. VII, 324 PP.

WILLIAMSON, M. 1972. THE ANALYSIS OF BIOLOGICAL POPULATIONS. EDWARD ARNOLD LTD., LONDON, ENGLAND. XI, 180 PP.

WILLIAMSON, M. 1974. REVIEW OF: GEOGRAPHICAL ECOLOGY. PATTERNS IN THE DISTRIBUTION OF SPECIES BY R H MACARTHUR. J. ANIM. ECOL. 43(2), 601-602.

WILSON, E. O. AND W. H. BOSSERT. 1971. A PRIMER OF POPULATION BIOLOGY. SINAUER ASSOCIATES, STAMFORD, CONNECTICUT. 192 PP.

WINBERG, G. G. (ED.). (TRANSL. BY A. DUNCAN). 1971. METHODS FOR THE ESTIMATION OF PRODUCTION OF AQUATIC ANIMALS. ACADEMIC PRESS, NEW YORK, NEW YORK. XII, 175 PP.

ZAIKA, V. E. 1972. UDEL'NAYA PRODUKTSIYA VODNYKH BESPOZVONOCHNYKYH. IZDATEL'STVO "NAUKOVA DUMKA", KIEV. (ENGLISH TRANSL. BY A. MERCADO AND EDITED BY B. GOLLEK. 1973. SPECIFIC PRODUCTION OF AQUATIC INVERTEBRATES. HALSTED PRESS, NEW YORK, NEW YORK. VI, 154 PP.). [MODEL; INTRINSIC RATE OF INCREASE; EQUATIONS]

.. BIBLIOGRAPHIES

ANDERSON, D. R. 1972. BIBLIOGRAPHY ON METHODS OF ANALYSING BIRD BANDING DATA WITH SPECIAL REFERENCE TO THE ESTIMATION OF POPULATION SIZE AND SURVIVAL. U. S. FISH AND WILDL. SERV., SPEC. SCI. REPT. WILDL. NO. 156. II, 13 PP.

ANONYMOUS. 1956. SCIENTIFIC AND TECHNICAL PAPERS BY ALFRED J. LOTKA. PP. 442-447. IN, ELEMENTS OF MATHEMATICAL BIOLOGY.(A. J. LOTKA). DOVER PULICATIONS, NEW YORK.

ARGENTESI, F., G. DI COLA AND N. VERHEYDEN. 1973. BIOSYSTEMS MODELING: A PRELIMINARY BIBLIOGRAPHIC SURVEY. EUR-4966 COMMISSION OF THE EUROPEAN COMMUNITIES, BRUSSELS, BELGIUM. 44 PP. [BIBLIOGRAPHY]

EDITORS. 1973. OBITUARY. ROBERT HELMER MACARTHUR (1930-1972). THEOR. POPUL. BIOL. 4(3), 249-250.

GOODALL, D. W. 1962. BIBLIOGRAPHY OF STATISTICAL PLANT SOCIOLOGY. EXCERPTA BOTANICA SEC. 4, 253-322.

KADLEC, J. A. 1971. A PARTIAL ANNOTATED BIBLIOGRAPHY OF MATHEMATICAL MODELS IN ECOLOGY. ANALYSIS OF ECOSYSTEMS, IBP, SCHOOL OF NATURAL RESOURCES, UNIVERSITY OF MICHIGAN, ANN ARBOR, MICHIGAN. 10 PP. PLUS BIBLIOGRAPHY.

MCMULLIN, B. B. 1973. A SURVEY OF PAPERS ON ECOSYSTEMS ANALYSIS FROM 1947-1971 IN THE JOURNAL ECOLOGY. AEC REPT. ORNL-EIS-72-19. OAK RIDGE NATL. LAB., OAK RIDGE, TENNESSEE. I, 50 PP.

MORRIS, M. J. 1967. AN ABSTRACT BIBLIOGRAPHY OF STATISTICAL METHODS IN GRASSLAND RESEARCH. U. S. FOREST SERVICE, WASHINGTON, D. C. MISC. PUBLICATION NO. 1030. 222 PP.

NEWELL, W. T. AND J. NEWTON. 1968. ANNOTATED BIBLIOGRAPHY ON SIMULATION IN ECOLOGY AND NATURAL RESOURCES MANAGEMENT. QUANTITATIVE SCIENCE PAPER NO. 1, CENTER FOR QUANTITATIVE SCIENCE IN FORESTRY, FISHERIES AND WILDLIFE, UNIVERSITY OF WASHINGTON, SEATTLE, WASHINGTON. MIMEO, 32 PP.

O'NEILL, R. V., J. M. HETT AND N. F. SOLLINS. 1970. A PRELIMINARY BIBLIOGRAPHY OF MATHEMATICAL MODELING IN ECOLOGY. AEC REPT. ORNL-IBP-70-3. OAK RIDGE NATL. LAB., OAK RIDGE, TENNESSEE. VI, 87 PP.

PARKER, R. A. AND D. ROOP. 1975. SURVEY OF AQUATIC ECOSYSTEM MODELS. REPORTS RECEIVED THROUGH 10/31/74. THE INSTITUTE OF ECOLOGY, WASHINGTON, D. C. XIII, 131 PP.

POSPAHALA, R. S. 1969. A LITERATURE REVIEW ON COMPUTERS IN WILDLIFE BIOLOGY. COLORADO GAME, FISH AND PARKS DIV. SPEC. REPT. 23. IV, 15 PP.

SCHULTZ, V. 1958. STATISTICAL METHODOLOGY. PP. 67. IN, TEN-YEAR INDEX TO THE JOURNAL OF WILDLIFE MANAGEMENT, VOLUMES 11-20, INCLUSIVE 1947-1956.

SCHULTZ, V. 1960. STATISTICAL METHODOLOGY IN THE JOURNAL OF WILDLIFE MANAGEMENT, VOLUMES 1-10. J. WILDL. MANAGE. 24(2), 230.

SCHULTZ, V. 1961. AN ANNOTATED BIBLIOGRAPHY ON THE USES OF STATISTICS IN ECOLOGY-A SEARCH OF 31 PERIODICALS. AEC REPT. TID-3908. VII, 315 PP.

SCHULTZ, V. 1971. A BIBLIOGRAPHY OF SELECTED PUBLICATIONS ON POPULATION DYNAMICS, MATHEMATICS, AND STATISTICS IN ECOLOGY. PP. 417-425. IN, STATISTICAL ECOLOGY, VOLUME 3, MANY SPECIES POPULATIONS, ECOSYSTEMS, AND SYSTEMS ANALYSIS. (G. P. PATIL, E. C. PIELOU AND W. E. WATERS, EDS.). PENNSYLVANIA STATE UNIVERSITY PRESS, UNIVERSITY PARK, PENNSYLVANIA.

SCOTT, T. G., M. B. NORGREN AND W. S. OVERTON. 1967. STATISTICAL METHODOLOGY. PP. 196-198. IN, TEN-YEAR INDEX TO THE JOURNAL OF WILDLIFE MANAGEMENT, VOLUMES 21-30, 1957-1966.

SCUDO, F. M. 1971. REFERENCES TO VOLTERRA'S WORKS. PP. 21-22. IN, VITO VOLTERRA AND THEORETICAL ECOLOGY. (F. M. SCUDO, ED.). THEOR. POPUL. BIOL. 2(1), 1-23.

SWAN, G. W. 1974. A BIBLIOGRAPHY OF MATHEMATICAL BIOLOGY. PP. 273-280. IN, MATHEMATICAL PROBLEMS IN BIOLOGY, VICTORIA CONFERENCE. LECTURE NOTES IN BIOMATHEMATICS, VOL. 2. (R. VAN DEN DRIESSCHE, ED.). SPRINGER-VERLAG NEW YORK INC., NEW YORK, NEW YORK.

WATT, K. E. F. 1957. A BIBLIOGRAPHY ON COMPUTERS, OPERATIONS RESEARCH, MATHEMATICS AND STATISTICS FOR THE BIOLOGIST. ONTARIO DEPARTMENT OF LANDS AND FORESTS, DIVISION OF RESEARCH, CANADA. MIMEO, 7 PP.

.. EDUCATIONAL REFERENCES

CHAPMAN, D. G., G. J. PAULIK, G. M. VAN DYNE, W. E. WATERS AND J. WILSON. 1971. PANEL DISCUSSION: TRAINING AND RESEARCH PROBLEMS IN QUANTITATIVE ECOLOGY. PP. 427-462, IN, STATISTICAL ECOLOGY, VOLUME 3, MANY SPECIES POPULATIONS, ECOSYSTEMS, AND SYSTEMS ANALYSIS. (G. P. PATIL, E. C. PIELOU AND W. E. WATERS, EDS.). PENNSYLVANIA STATE UNIVERSITY PRESS, UNIVERSITY PARK, PENNSYLVANIA.

FRAYER, W. E. (ED.). 1970. PROCEEDINGS OF THE SYMPOSIUM ON THE DEVELOPMENT AND IMPLEMENTATION OF COURSES AND CURRICULA IN NATURAL-RESOURCES BIOMETRY. COLORADO STATE UNIVERSITY, FT. COLLINS, COLORADO. 210 PP.

LACKEY, R. T. 1975. TEACHING WATER RESOURCE MANAGEMENT WITH THE AID OF A COMPUTER-IMPLEMENTED SIMULATOR. VPI-WRRC-BULL 78, VIRGINIA WATER RESOURCES RESEARCH CENTER, BLACKSBURG, VIRGINIA. V, 98 PP. [MODEL]

MILLER, G. H. 1968. ECOLOGY AND MATHEMATICS. BULL. ECOL. SOC. AMER. 49(4), 138-142.

MILLER, G. H. 1968. WHAT MATHEMATICS COURSES SHOULD A BIOLOGIST TAKE? BIOSCIENCE 18(6), 489-491.

OSTER, G. F. 1974. A SIMPLE ANALOG FOR TEACHING DEMOGRAPHIC CONCEPTS. BIOSCIENCE 24(4), 212-213. [MODEL; COHORT; PROBABILITY; LOTKA'S EQUATION]

PATTEN, B. C. 1966. SYSTEMS ECOLOGY: A COURSE SEQUENCE IN MATHEMATICAL ECOLOGY. BIOSCIENCE 16(9), 593-598.

PAULIK, G. J. 1969. COMPUTER SIMULATION MODELS FOR FISHERIES RESEARCH, MANAGEMENT, AND TEACHING TRANS. AMER. FISH. SOC. 98(3), 551-559.

PAULIK, G. J. 1969. DIGITAL SIMULATION MODELLING IN RESOURCE MANAGEMENT AND THE TRAINING OF APPLIED ECOLOGISTS. QUANTITATIVE SCIENCE PAPER NO. 6. CENTER FOR QUANTITATIVE SCIENCE IN FORESTRY, FISHERIES AND WILDLIFE, UNIVERSITY OF WASHINGTON, SEATTLE, WASHINGTON. MIMEO, 56 PP.

PAULIK, G. J. 1972. DIGITAL SIMULATION MODELING IN RESOURCE MANAGEMENT AND THE TRAINING OF APPLIED ECOLOGISTS. PP. 373-418. IN, SYSTEMS ANALYSIS AND SIMULATION IN ECOLOGY, VOLUME II. (B. C. PATTEN, ED.). ACADEMIC PRESS, NEW YORK, NEW YORK. [MODEL; COMPUTER; MATRIX]

SCHRECK, C. B. AND W. H. EVERHART. 1973. COMPUTER TEACHING EXERCISES IN FISHERIES SCIENCE. TRANS. AMER. FISH. SOC. 102(3), 656-657.

SCHULTZ, A. M., R. P. GIBBENS AND L. DEBANO. 1961. ARTIFICIAL POPULATIONS FOR TEACHING AND TESTING RANGE TECHNIQUES. J. RANGE MANAGE. 14(5), 236-242. [SAMPLING]

SCHULTZ, V. 1970. MATHEMATICS AND APPLIED STATISTICS TRAINING FOR ECOLOGISTS WITH COMMENTS ON THE PROGRAM AT WASHINGTON STATE UNIVERSITY. PP. 157-168. IN, PROCEEDINGS OF THE SYMPOSIUM ON THE DEVELOPMENT AND IMPLEMENTATION OF COURSES AND CURRICULA IN NATURAL-RESOURCES BIOMETRY. (W. E. RAYER, ED.). COLORADO STATE UNIVERSITY, FT. COLLINS, COLORADO.

COTT, G. A. M. 1969. THE EACHING OF QUANTITATIVE ECOLOGY. ROC. N. Z. ECOL. SOC. 16, 73-75.

ITLOW, F. B. AND R. T. LACKEY. 972. COMPUTER ASSISTED NSTRUCTION IN NATURAL RESOURCE ANAGEMENT. PROC. ANN. CONF. S. . ASSOC. GAME AND FISH OMMISSIONERS 26, 500-505.

ITLOW, F. B. AND R. T. LACKEY. 974. DAM: A COMPUTER-IMPLEMENTED TER RESOURCE TEACHING GAME. ANS. AMER. FISH. SOC. 103(3), 1-609. [MODEL]

VAN DYNE, G. M. 1969. IMPLEMENTING THE ECOSYSTEM CONCEPT IN TRAINING IN THE NATURAL RESOURCE SCIENCES. PP. 327-367. IN, THE ECOSYSTEM CONCEPT IN NATURAL RESOURCE MANAGEMENT. (G. M. VAN DYNE, ED.). ACADEMIC PRESS, NEW YORK, NEW YORK. [SYSTEMS ANALYSIS; MODEL]

WATT, K. E. F. 1965. AN EXPERIMENTAL GRADUATE TRAINING PROGRAM IN BIOMATHEMATICS. BIOSCIENCE 15(12), 777-780.

WERNER, P. A. AND E. D. GOODMAN. 1974. NEW ACADEMIC TRAINING FOR ECOLOGISTS AND ENGINEERS. AIBS EDUCATION REVIEW 3(4), 1-5. [COMPUTER; MODELING]

WILLIAMSON, M. H. 1967. INTRODUCING STUDENTS TO THE CONCEPTS OF POPULATION DYNAMICS. PP. 169-176. IN, THE TEACHING OF ECOLOGY. (J. M. LAMBERT, ED.). BLACKWELL SCIENTIFIC PUBLICATIONS, OXFORD, ENGLAND.

WOODRUFF, N. H. 1940. MATHEMATICS USED IN BIOLOGY. J. TENN. ACAD. SCI. 15, 175-266.

YOUNG, H. E. 1967. MATHEMATICS, STATISTICS AND COMPUTER PROGRAMMING IN THE FORESTRY CURRICULUM. J. FORESTRY 65(1), 36-37.

YOUNG, H. E. 1972. BASIC QUANTITATIVE INSTRUCTION IN THE FORESTRY CURRICULUM. J. FORESTRY 70(3), 161-162.